Construction Project Management

The role of the project manager continues to evolve, presenting new challenges to established practitioners and those entering the field for the first time. This second edition of Peter Fewings' groundbreaking textbook has been thoroughly revised to recognise the increasing importance of sustainability and lean construction in the construction industry. It also tackles the significance of design management, changing health and safety regulation, leadership and quality for continuous improvement of the service and the product.

Using an integrated project management approach, emphasis is placed on the importance of effectively handling external factors in order to best achieve an on-schedule, on-budget result, as well as good negotiation with clients and skilled team leadership. Its holistic approach provides readers with a thorough guide in how to increase efficiency and communication at all stages while reducing costs, time and risk. Short case studies are used throughout the book to illustrate different tools and techniques.

Combining the theories underpinning best practice in construction project management, with a wealth of practical examples, this book is uniquely valuable for practitioners and clients as well as undergraduate and graduate students for construction project management.

Peter Fewings has wide experience of project management and is a learning and teaching fellow at the University of the West of England leading courses on construction and project management, researching and writing in the area of private public partnerships, health and safety and ethics. He is also a practitioner in Myers Briggs.

Martyn Jones is a leading authority on supply chain management and innovation in the construction industry. Under the auspices of Constructing Excellence he works with clients and supply-side organisations to unlock the value in their supply chains.

Construction Project Management

An Integrated Approach

Peter Fewings

Routledge
Taylor & Francis Group

LONDON AND NEW YORK

First edition published 2005
by Spon Press

This edition published 2013
by Routledge
2 Park Square, Milton Park, Abingdon, Oxon OX14 4RN

Simultaneously published in the USA and Canada
by Routledge
711 Third Avenue, New York, NY 10017

Routledge is an imprint of the Taylor & Francis Group, an informa business

© 2013 Peter Fewings

The right of Peter Fewings to be identified as author of this work has
been asserted by him in accordance with sections 77 and 78 of the
Copyright, Designs and Patents Act 1988.

British Library Cataloguing in Publication Data
A catalogue record for this book is available from the British Library

Library of Congress Cataloging-in-Publication Data
Fewings, Peter.
   Construction project management: an integrated approach/
   Peter Fewings. – 2nd ed.
   p. cm.
   Includes bibliographical references and index.
   1. Construction industry–Management. 2. Project management.
   3. Building–Superintendence. I. Title.
   TH438.F48 2012
   690.068'4–dc23
   2012003781

ISBN13: 978-0-415-61344-6 (hbk)
ISBN13: 978-0-415-61345-3 (pbk)
ISBN13: 978-0-203-83119-9 (ebk)

Typeset in Sabon and Gill Sans by
Florence Production Ltd, Stoodleigh, Devon

Dedicated to my dear wife and main supporter Lin who has given me inspiration.

Contents

Figures and tables

Figures

Tables

Acknowledgements

This book would not have been possible without the help of a lot of people. I have been very lucky to have so many who have helped me including those who have offered me case study material and opportunities to interview them. I would like to particularly thank Martyn Jones who was able to substantially rework his specialist Chapter 12 and input into Chapters 6 and 14, Tony Westcott who kindly let me use his material on Andover North Site Redevelopment again, Buchan Concrete who gave me pictures to use, Carol Graham who patiently read through new parts of the script, Christian Henjewele who made helpful comments on Chapter 15, Simon Timms who did the same for Chapter 11 and Simon Lee who kept me on a balanced path for the work on lean construction in Chapter 5. I am very grateful for their generosity and hard work. If there are any mistakes they are mine and not theirs. I am grateful for all the others who gave me inspiration in the writing and who I have acknowledged in endnotes.

Peter Fewings

Preface for second edition

In expanding the book we have now become even more convinced that the way forward is to integrate the construction project management process so that the managers and the industry in which they work are prepared for the information revolution which has begun to make significant changes since the first edition. I will cover how these changes have and will impact project management in construction. I have essentially added two chapters to the book. Chapter 11 has been added to cover the growing influence of sustainability on the delivery of projects so concentrates on making sure sustainable designs are effectively delivered and how they can meet global and national targets for better environmental and social issues in an economic way. Chapter 5 which previously mainly covered the measurement of performance has been expanded to cover the more radical productivity issues which need to be in place to make sure that the client demands for more (sustainability and better quality) for less (more productive) are better understood and set out as an opportunity rather than a threat. Case studies have been expanded substantially to illustrate the concepts and older ones have been retained where they are still relevant. Other developments are perceived in supply chain management (SCM), the use of building information modelling (BIM) as a requirement for public projects, the development of project leadership and the wider use of public private partnership (PPP) procurement in variant forms to the private finance initiative which was the only focus before. Martyn has developed the area of SCM and leadership and given more examples of its use in construction. Quality and sustainability have a bigger emphasis throughout the book in the recognition that these are critical to the life cycle business case of clients and the need to reduce resources usage in the future and maximise the performance of buildings where most of the business cost is. Ethics have been introduced in Chapter 6 as a clear guidance in times when professional behaviour needs better definition, competitive tactics need to be fair and corruption has been identified as restricting development opportunities in procurement of international projects with multinational teams and is seriously monitored by international organisations and the new legislation in some countries.

Health and safety issues continue to challenge and, although accidents have been reduced significantly where they are managed, there is a constant pressure to work towards zero accidents and Chapter 10 has added some more about a health and safety culture which is able to be bottom up and will also be effective in joining all parties and examining incident causes. Value management has partly been moved into design management to indicate its importance at this stage, but has retained its relationship with risk in Chapter 8.

Some of the book expansion has also come from the desire to give some more explanatory material so that this edition will be even more useful in university courses as well as for practitioners. For example, some more explanation of the use of total quality management systems which underpin lean construction is given. More case studies have been provided throughout illustrating best practice and should also help understanding. I have added material to recognise the PRINCE2 approach to project management in the introduction.

Peter Fewings

Foreword for second edition

For a book to stand out in a sea of construction project management titles you need a really distinguishing feature. This second edition promotes the importance of process integration so that there is a unity of purpose in delivering the client objectives from all the key players.

The integral participation of the client is clearly an important part of the decision making process supported by a broad, but specialist project team. Integrated construction is the way forward for better value for all the stakeholders thriving for excellence.

Buildings have developed rapidly in the last seven years since this title was first launched and sustainable smart building now provides a central focus for both design and construction which has been included as an integrated thread throughout the book.

Two new chapters on lean and sustainable construction have been included to show how project management has the potential to take the imperatives of a sustainable built environment and deliver efficient and effective infrastructure for the 21st century.

The emphasis on better planning at an early stage with the introduction of Building Information Management (BIM) has also greatly expanded the chapter on design management which recognises that construction and design must be harmonised in the delivery of successful projects and that best value in delivering projects needs collaboration at an early stage to combine the building database and to get the construction and design team working together earlier than the traditional models.

Quality, risk and construction time, cost, project organisation and supply chain, health and safety management and systems improvement are all included and play their part in the controlled implementation of the project, but central to this is the development and motivation of people to create synergy and commitment in project team.

Uniquely the project manager works like the skipper of a racing yacht challenging, trimming, co-ordinating and inspiring human effort and creativity so that all players feel part of the same team and work for each other crossing the finish line proud of their achievements and craftsmanship.

The key ingredients of this book complement the Chartered Institute of Building's own Code of Practice for Project Management for Construction and Development to the extent that the key underlying theme of integration and harmonisation within the Code of Practice is also ingrained within and expanded upon in this book. Together with the Code of Practice, this book is a useful resource for practitioners to develop their knowledge of integration, and expand their skill set to continue to achieve excellence in performance within the fast moving construction industry which has become much more knowledge and value oriented, in keeping with other aspects of 21st Century.

Alan Crane CBE, FCIOB
President of the Chartered Institute of Building
April 2012

Introduction

Many books have been written on project management and there are two approaches. One deals mainly with the tools and techniques of project management and provides instruction on what they are and how to use them. The other approach takes a managerial viewpoint and is concerned more with the context and the way in which decisions are made and the tools which are most appropriate in that situation. The revised edition of this book remains allied to the integrated managerial approach, analysing how techniques have been applied in traditional and best practice and synthesising additional guidance on evaluating contextual factors that make projects unique and meet construction client requirements. These are summarised as PROMOTE in Chapter 16.

Project management process and product

In construction there is a long history of project management and standard systems have been set up which have become comfortable, but have not always produced best value for the client. Every project is different and has a unique location and due to the varying time and budget constraints the final product is an untested prototype, which has been subject to continuing design variations. Right first time is therefore a particular challenge to an industry that has not standardised its products. The industry is also quite fragmented with many inexperienced clients and separate design and construction organisations. The supply chain can often be quite long with some detailed design provided at a second and third tier contract level and little direct labour provided by the main contractor. This presents additional challenges to the construction project manager who needs to co-ordinate the design and construction sides and make decisions based on the promises of others.

It is a particular challenge to deliver to tight time, cost and quality targets that are set by most clients. In response, construction has had to adopt a much more client-orientated view. This view allows alternative procurement strategies where design and construction are much more integrated and opportunities for the development of the brief to take account of project constraints and ongoing business opportunities are recognised. Risk management is about managing uncertainty and opportunity in the context of increasing value. Recession is also likely to affect supply chain relationships and may strengthen or weaken collaboration between the client and the contractor depending on their foundational base. In boom times the boot is on the other foot and the supply chain can be more selective in what it chooses.

It is important to look at both the function of the product (effectiveness) and the efficiency of the process and this means a closer relationship between the client and the whole project team including definition (briefing), strategy, design and construction. Project management has an important role to play here by understanding the client's business as well as the project objectives and incorporating these in the design, method and sequence of the finished construction. Chapters on business development, risk, design and supply chain management will cover this issue of driving value for money into the business case. Value management in its simplest sense is a way of reducing waste in order to maximise productive output. This waste comes in all forms, and reports such as Latham (1994), Atkins (1994, Europe wide), Egan (1998 and 2002) have particularly recognised inefficiencies in the construction process and set targets for improvement.[1] Toyota, who are well known for pioneering lean manufacturing techniques, have recognised that there is as much as 60 per cent unproductive activity in their system so there is always room to improve. More recently Wolstenholme (2009) suggests there are blockers in both the business and economic model used by clients to drive new facilities development and in supplier capability for innovation and bringing in new blood, driving out waste in design and construction and not integrating standard delivery models (procurement).[2] These issues are not new, but project managers should address them as not being inevitable.

People

Project management is at least 50 per cent people management. This book emphasises the importance of people skills, organisation, leadership and trust. Interpersonal skills are gained from experience so the purpose of discussion is to raise awareness of the importance of these skills and the ways in which they might be developed in different situations. It will cover the issues which arise in managing a supply network (Chapter 12) and communicating across a broad inter-professional team. It will support the concept of leadership and innovation and look at ways of developing competence in negotiation, delegation and motivating teamwork in the specific context of construction and engineering projects. It evaluates the formal initiatives such as Respect for People and the EFQM excellence model, which recognises the increasingly key role of developing people for the project. With leadership and better communication these have the potential to improve teamwork and interpersonal skills in construction projects where there is complex diverse specialism. The development of leadership in the context of project structures and culture is covered in Chapter 6 and synergistic teamwork is of particular interest in Chapter 7 on engineering the psycho-productive environment.

The structure of the book is based on the life cycle of a construction project as portrayed in the Code of Practice for Project Management for Construction and Development (CPCPM).[3] This is so that it will be easy to cross reference the material in this book with the Code of Practice. The CIOB have welcomed this, but this book is designed to stand alone in offering guidance to best practice for construction professionals and illustrative case study material, and our views are not necessarily those held by the Working Group for the Code of Practice.

The managerial approach focuses on decision making. In order to support this, the authors have drawn on the Gateway Review® process published originally by the Office of Government Commerce (OGC) and reference to this can be made through the Treasury website.[4] This provides a generic framework of five decision 'gateways' allied to the life cycle of a capital expenditure project and identifies the necessary activity to achieve these from a client perspective. Detailed notes specify the relevant government procedures, but the model is useful to the project manager in identifying the information and its timing to elicit important client and project decisions for any type of project. The approach will be discussed in Chapter 1.

The chapters are roughly in life cycle order, but some chapters have also been added to develop specific themes such as sustainable delivery, lean construction, health and safety, risk and value management, supply chain management, and leading and managing teams. These thematic chapters will particularly take the structure discussed above.

The key features which together will differentiate the book from other texts are:

- an integrated approach to hard and soft management issues at all stages of the construction project cycle
- an open systems approach emphasising the role of managing external factors and sustainability in gaining project success
- the key role of new forms of procurement, such as public private partnerships, as systems to encourage the culture of partnering and which can be broadly adopted in integrating the life cycle of capital provision and maintenance
- recognition of the need to engineer the psycho-productive environment to improve the efficiency and delivery of a project.

Definitions, standards and codes

A *project* is defined in BS 6079-1 Guide to Project Management as:

> A unique set of co-ordinated activities with definite starting and finishing points, undertaken by an individual or organisation to meet *specific objectives* within defined schedule, cost and performance parameters.[5]

It is important to understand how the definition of a project can apply to lots of types of activities and not just construction. If you understand this then you will see the differences between project management and general management and why project management is used in construction.

Features of a project include change requirements, unique prototype, non-repetitive, goal orientated, holding a particular set of constraints, measurable output and handover.[6]

Project management may be defined as

> The overall planning, co-ordination and control of a project from inception to completion, aimed at meeting a client's requirements in order to produce a functionally and financially viable project that will be completed safely, on time, within authorised cost and to the required quality standards.[7]

This definition underlines the importance to recognise the independent role of mediating the client's interests with the specifics of managing and integrating the project team. The *project manager*'s objectives are:

- providing a cost effective and independent service
- selecting, correlating, integrating and managing different disciplines and expertise
- satisfying the object and provisions of the project brief from inception to completion . . . to the client's satisfaction
- safeguarding client's interests at all times and, where possible, giving consideration to the needs of the eventual user of the facility.[8]

Programme management is the management of several related parallel or sequential projects. A programme is often identified to provide co-ordinated management and may be many related small or large projects. It helps strategically for the client to manage resources, standardise outputs and allow supply chain management to look at issues which are common between projects to develop value and to capitalise on past knowledge gained. Alternatively the programme will relate construction project/s to IT projects, moving people projects and managing space projects that they need, setting up new systems to make everything work. To manage these separately would court disaster.

Portfolio management is a senior management process to make sure that the right programmes and projects are selected to realise the optimum benefits to the organisation. It is a co-ordinated collection of strategic processes and decisions that together enable the most effective balance of organisational change and business as usual.[9] It is about choosing the right change and timing of subsequent projects and not the change management process itself. This is assumed in Chapter 2.

Project definition as in Chapter 3 is the development of the project brief and scope up to the planning application stage, so that the risks have been identified and the value for money has been optimised to suit business needs.

Body of Knowledge (BoK) and methodologies

The two well-known institutions supporting and certifying the area of project management have produced supportive documentation called BoK defining generic competencies of a project manager. They also offer project management training in competently acting as a member of a team and in assessing their level of competence.

The APM BoK (5th edition)[10] breaks down 52 competencies into seven sections – project management in context, planning the strategy, executing, techniques, business and commercial, organisation and governance, people and professions – and recognises three competencies areas – technical, behavioural and contextual. It is internationally recognised through the International Project Management Association (IPMA). The APMP exam is a test on the knowledge of these areas internationally accepted at level D. There are other assessments for levels C and B. Level A recognises mature advanced practice.

The American Project Management Institute BoK (PMI BoK 4th edition)[11] talks about five basic processes which are the initiating, planning, executing, monitoring, and controlling and closing. These are clearly linked to the life cycle of a project mentioned earlier. Processes overlap and interact in each phase and include inputs, tools and techniques applied and outputs. There are nine project management knowledge areas – integration, scope, time, cost, quality, human resource, communications, risk and procurement. These PM knowledge areas refer to the processes relevant to that area. There are equivalent tests of competence in knowledge and practice in the PMI.

PRINCE2™ methodology[12]

PRINCE2:2009 is termed a 'method' and stands for PRojects IN Controlled Environments and was authored by the OGC.[13] It describes procedures, roles and responsibilities to co-ordinate people and activities in a project including designing, supervising and change management. It recognises six variables of time, cost, quality, scope, risks and benefits. The common

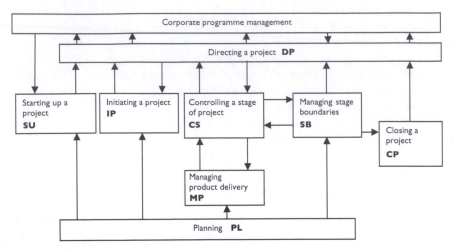

Figure 1 Structure of PRINCE2 showing different elements
Adapted, Crown Copyright 2007

language is useful as it is widely used, although it is not so well used in construction projects. It does not deal with competencies and as such is not a body of knowledge. It gives a flexible framework to work within for all sizes of project, but does not tell you how to do it. It identifies life cycle stages of starting up, initiating, controlling, managing stage boundaries and closing a project. The format of PRINCE2™ is as adapted in Figure 1.

Directing a project is delegated to the project manager for the whole length of the project and they formally start up the project by producing the risk log and the project brief, and initiating a project team and an outline plan for the initiation stage. The initiation stage develops the business case and the detailed project plan making sure of a management structure. They work with and report to the project board.

Managing stage boundaries is a scoping, viability and progress exercise to ensure the brief and other parameters are up to date, whilst *controlling a stage of a project* is a much more detailed control of the life cycle stage in procuring packages to do a work stage and reviewing time, cost, quality, risk and value in an integrative way. The *managing stage boundaries* report may cause the project board to terminate a project. The *managing product delivery* is a subset of control with ongoing managing of the detail of the package programme, cost and quality to deliver product. *Closing a project* means bringing it to an end and handing over to the user in an ordered way.

The *planning* process is product based and determines the products needed, their content and their sequencing. However it does consider things such as risk and change control needed to manage the product sequence, cost and specification. This leads to a consideration of PRINCE2™ components now known as themes. In PRINCE2™ these are a mixture of management and technical stages.

- *Management components* are planning, organisation and controls. Management stages are a project stage which spans between two important management decision points identified on the overall programme and they identify roles, detailed product sequences and regular controls and reporting on progress to facilitate decisions such as change and work package authorisation.
- *Technical components* are required to develop the products through the technical life cycle stages of design, build and implementation and deal with the quality reviews and costing of the various work packages. This requires the application of quality reviews and configuration and risk management for each product outcome.

There is an extensive glossary of terms which has been concisely defined for use in PRINCE2™ projects. For example an exception report is one that explains an issue which falls outside the band of tolerance for its performance. This common language is a useful tool where so many terms have different applications in different project environments. However this benefit also leads to it being described as a bureaucratic system. This can be tackled by simplifying it to suit the size and uniqueness of a particular project. PRINCE2™ is used internationally.

The project management codes mentioned previously as BS6079 and the CIOB's CPCPM 2010 represent best practice in the UK and as such have a very similar role to PRINCE2™ above. The BOKs have been derived though to define the competency of its members and their use is therefore different even though it covers a lot of the same material. The profusion of different guides can cause confusion due to the different use of language so the language of the CPCPM 2010 has been primarily used as it is sector specific. Some comparison with the other sources has been made in order to illuminate best practice. An international project management code is in preparation.

Uses

The text should be of great use to those who are completing their studies in construction and project management. To support this aim, key concepts have been introduced and developed at the beginning of the chapter. This will also give a proper foundation for the development of innovative practice which will be discussed at the end of relevant chapters, drawing on current research and appraising the way forward for practitioners. Although much reference is made to documentation and case studies in the UK and European context, the research carried out suggests that the practical applications and challenges for an integrated project management approach apply to most countries, so it is hoped that a wider readership will be able to use the book. Any comments will be welcome.

It is assumed that the reader has some knowledge of construction terms, but a wider glossary has been compiled to make these clearer. Key definitions and concepts have been covered in the text and a particular feature of the text is to relate construction project management to management theory where relevant for greater understanding of the context.

Peter Fewings
December 2011

References

1　Latham M. (1994) *Constructing the Team: Final Report of The Government/ Industry Review of Procurement and Contractual Arrangements in the UK Construction Industry.* UK, Department of the Environment; Atkins W.S. (1994) *SECTEUR, Strategic Study on the Construction Sector: Strategies: Final Report* 111/4173/93 EC Directorate General: Industry; Egan J. (1998) *Rethinking Construction.* UK, Department of Environment, Transport and the Regions; Egan J. (2002) *Accelerating Construction.* UK, Strategic Forum for Construction.

2　Wolstenholme A. (2009) *Never Waste a Good Crisis: A Review of Progress since Rethinking Construction and Thoughts for Our Future.* London, Constructing Excellence (with forwards by Latham, Egan and Raynsford).

3　CIOB (2010) *Code of Practice for Project Management for Construction and Development.* 4th edn. Oxford, Blackwell Publishing.

4　HM Treasury website: http://www.hm-treasury.gov.uk.

5　British Standards Institution (2010) *BS6079-2 Guide to Project Management.* UK, BSI.

6　Maylor H. (2003) *Project Management.* Harlow, Pearson Education.

7　CIOB (2010) *Code of Practice for Project Management for Construction and Development.* 4th edn. Oxford, Blackwell Publishing, p.xvii

8　CIOB (2010) *Code of Practice for Project Management for Construction and Development.* 4th edn. Oxford, Blackwell Publishing, p.5.

9　OGC (2011) *OGC Management of Portfolios.* London, The Stationery Office.

10　Dixon M. (Ed.) (2008) *Project Management Body of Knowledge.* High Wycombe, The Association for Project Management.

11　Project Management Institute (2008) *A Guide to the Project Management Body of Knowledge.* Upper Darby, PA, PMI.

12　ILX Group (2009) *PRINCE2® 2009 Foundation User Guide.* Nantwich, ILX.

13　OGC (2009) *PRINCE2:2009 Refresh.* UK, OGC.

Chapter 1

Project life cycle and success

Project management is not a new concept, but it has emerged since the Second World War as a methodology that can be applied to intensive periods of work with a specific objective, which can be isolated from general management so that expenditure can be ring fenced and the synergy of a team is engaged. However not all managers are able to cope with the dynamic nature of projects, where decisions have to be made fast and planning and control have to be very tight. Large projects such as the NASA space programme, the Polaris submarine programme and the Channel Tunnel have developed techniques for project management that have set a pattern for subsequent ones. These projects have also had to develop specific roles and create management structures to suit and satisfy various interests, both within the project and contract and outside. Many tools and techniques are specific to PM, but some have been borrowed from general management. Construction work particularly lends itself to project management because of the temporary and unique nature of the work. PM though is an effective management process used in many contexts.

This chapter will look at the project as a whole from inception to completion. The objectives are:

- to define the contruction project life cycle
- to distinguish specific project management activities and allocating roles in the life cycle
- to investigate factors which affect the way that projects are managed
- to understand complexity and maturity models for project management
- to determine the critical factors for project success.

Project life cycle

The life cycle of a construction project from a client's point of view really starts when there is a formal recognition of project objectives, generally termed *inception*, when a project team is assembled, through to the delivery of these objectives called *completion* or project delivery. Activities relating

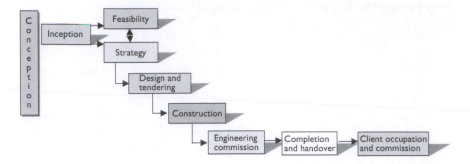

Figure 1.1 Life cycle of construction projects

Source: Based on the CIOB (2010) *Code of Practice for Project Management* (Figure 3.1). Used with permission[1]

to the conception of a project take place over an extended period before the project starts. A diagram indicating the main elements of its life cycle is shown in Figure 1.1. At the end of the life cycle there is a commissioning process in order to ensure everything is working and to hand over the completed project with its associated documentation so that it can be used efficiently.

The life cycle of a project varies depending upon the viewpoint of the participant. Different parts of a life cycle are often managed by different people and not all organisations are involved in the project all the way through from inception to completion of a building project. For example, a main contractor gets involved from tendering through to handover of the building before the client's fitting out – just two parts of the client's project life cycle. For them it is a complete project with an inception and a completion with handover of their work. Traditional procurement puts an emphasis on this fragmented view of project life cycle where most are involved for only a part of the life cycle.

There is a need to be flexible in the view of the construction life cycle, as there are now many different forms of procurement, which put a different emphasis on different stages of the cycle such as Design, Build, Finance and Operate (DBFO), which has a strong contractor provider involvement in the inception, feasibility and operation phases. This broader, more integrated viewpoint helps to recognise the client's position as a developer or user of an asset, which is essential to the success of any project. The project life cycle model in Figure 1.1 by indicating a possible overlap of the phases is robust enough to cover a wide variety of procurement approaches and allows for the development of innovative approaches to meet client requirements best. The development cycle is wider and tracks the building or structure to a change of use or to demolition and recycling for the next development opportunity (this is better shown in Figure 11.1, Chapter 11). The end of a construction project is the handover to a facilities management team, who maintain the structure and services.

The life cycle of construction projects starts at *inception* at the stage where a client's business case for a building or refurbishment is communicated to a professional team to develop the constraints. Outline planning consent may have been achieved, but only if a site has already been chosen. The inception process may be an extended period to outline a business case.

In order to proceed a client has to test the *feasibility* of the business case they have. The fundamental go ahead would have been reviewed by the beginning of the inception stage and the main focus for the project manager at this stage is to define options for project feasibility and their financial viability or benefits. The feasibility stage can include investigation of alternative site locations, funding options, design option appraisal, value enhancement, comparative estimates and life cycle costing. It considers the associated project constraints and marketing implications and is closely tied up with the strategy for the project. At the end of this period a feasibility study needs to test affordability (fit within an outline budget and cost plan) and meet constraints (an option to achieve planning permission and give best value for the client). It may reduce scope to meet budget constraints and the client needs to make critical decisions which suit the needs of their business. A key part of this is to identify and allocate risks and to carry out a functional analysis to optimise value for money. The user or facilities management groups may be involved at this stage. Typical outputs at this stage are a funding source, a basic risk assessment of external and internal factors, a design concept statement and drawings which are discussed with the planning authority and a discounted net cash flow or cost benefit analysis within a budget.

Strategy deals with how a project is carried out and controlled, such as choosing the correct procurement route, value management of the budget and a cost break down, developing the control and quality management systems, and a methodology for construction. Strategy is a partly parallel activity to feasibility as the viability is often dependent on the strategy. For example, the funding of the project is tied up with the programme time and the cash flow availability. Strategy also needs to identify the right procurement method and determine the organisation structure of the project. A key output at this stage is the project execution plan (PEP) that fully analyses and allocates the risk issues. It also specifies how the project is going to be planned and organised through the subsequent stages of the project life cycle. The brief needs to be developed to ensure a full understanding of the client's requirements and the design and construction strategies need to be co-ordinated within the project constraints. The PEP is also useful as a baseline document that gives a master plan for the work to be done. It is free to be changed as more information is gained and the design is developed, but with those changes, budget, programme, risks and cost benefit need to be re-evaluated so that the PEP plays a useful role as a control document in predicting accurately the value to the client of making the change. This integration of design with other strategic impacts is a core PM role. A master programme shows key dates for approvals. If a construction manager can be brought in at this stage, more reliable information is available for construction planning and methodology.

Pre-construction (design and tendering) appoints the full design and construction team and includes the full development of the design scheme, detailed drawings, tendering and mobilisation of resources for construction. There is a clear responsibility to manage design and early procurement lead times and to identify a start date for construction which is related to the handover and occupation of the building. Risk and value factors continue to be managed so that the client gets best value. Outputs include further statutory permissions such as building regulations, integrated design drawings, tender documents and tender information for a later health and safety plan, contractor appointment, an agreed contract price and contract programme and a risk register. It is notoriously difficult to control diverse design activity to meet deadlines, to anticipate the timing and nature of statutory consents and to predict a market price which will comply with budget constraints.

The *construction* phase is self-explanatory, but it has a particular emphasis on the detailed control of time, quality and cost and the management of many other issues such as supply chain, health and safety planning, the environment and change. Outputs here will include construction stage programmes, construction health and safety plans, method statements, cash flow forecasts, quality assurance schemes and change orders. In taking on a contractor, there is a risk of conflict if information is not available, things get changed a lot, or the project is delayed. Conflict management, leadership and team building skills are used a lot in this stage.

Engineering commissioning comes at the latter end of construction. It is distinctive as its outputs should include the efficient functioning of the building. The management of the process includes the signing off of various regulatory requirements such as building regulations, fire and water certificates, gas and electrical tests and meeting the conditions for product warranties. Output is working systems.

Practical completion is certified by the project manager for the formal *handover* of the fabric and systems to the facilities team. Liability is not limited by occupation and there is a responsibility to put defects right if and when they occur after handover. Documentation and a health and safety file are handed over for the safe and efficient use of the building's systems and maintenance schedules. Handover is sometimes called *close out*, because it suggests a focused period of preparation to ensure the project and the documentation are in order and the facilities team and users are properly briefed and inducted. A defects liability period protects the client from unexpected malfunctions or defects. The main output, apart from documentation, is to pass on the knowledge for running the building safely and efficiently. The CDM co-ordinator and principle contractor work hard to get the documents together in a useful format and to train the facilities staff.

The *client's occupational fit out* follows full or sectional completion of the contractor work and may well involve a new project team. This period often has intensive collaboration with user groups and facilities management teams. During this period there is a need to commission equipment, move personnel and induct occupiers in the use of the building and in its emergency procedures. Outputs include fitting out, space management and the production of health and safety policies, user manuals and training programmes.

The final stage is *post project appraisal* and review. The objective is to evaluate success in meeting the objectives as set out in the business case/ project brief and to look at lessons learnt and to carry forward improvements, where relevant, to the next project or phase. Outputs from this stage are client satisfaction surveys, benefit evaluation, production incentives and project process reports, which may inform projects for the future.

Project management

It is important to distinguish project management responsibilities and the following definitions help:

> The planning, monitoring and control of all aspects of a project and the motivation of all those involved to achieve the project objectives on time and to cost, quality and performance.[2]

> The art of directing and co-ordinating human and material resources through the life of a project by using modern management techniques to achieve predetermined goals of scope, cost, time, quality and participant satisfaction.[3]

The common elements of project management in these definitions and the CIOB definition mentioned in the introduction are time, cost and quality management and these can be viewed as a triangle as shown in Figure 1.2. Scope and satisfaction could be added as the PMI definition.

These three dimensions of control – time, cost and quality – represent the specific project efficiency factors. They are managed for the satisfaction of the client's requirements, but in themselves are secondary to the client's business needs, which are likely to be determined by the market. For example, programme control is a subset of finishing in time for the Christmas sales period when dealing with a retail client. Quality is not absolute, but related to the need for a building's cladding to efficiently keep out the weather and still look good for 10 years before the next refurbishment. It is likely that the client will prioritise one or two dimensions more than another. Thus an Olympic stadium will prioritise time over cost and it will also be a showcase

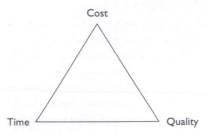

Figure 1.2 The three dimensions of project management control

so will need to look good. Capital cost is important, but if income suffers then it dissolves into relative insignificance.

Another important aspect of the PMI definition is the management and motivation of the 'human resource' and achieving 'participant satisfaction'. Understanding how participants from different organisations work together and having the skill to direct them towards common project goals, without upsetting them, is a very important part of project management.

Project management is about balancing the needs of all the participants in the project. Forcing the pace of design for the sake of construction or vice versa may mean inefficiencies later. There is a need to integrate the business needs of the contractors and consultants with the requirements of the client. However the project manager's prime concern is to try and get value for the client and on occasions conflict may arise within the project team. On these occasions conflict needs to be fairly managed so as not to de-motivate the project team. This in itself is counter-productive.

Programme and portfolio management

Today project management also has to be viewed in the light of the client's programme of projects. This is because there is a strategic planning aspect (normally called portfolio management) that requires resources in the commissioning organisation to be balanced between the different projects. This applies to the way in which personnel are allocated in the client organisation as well as the sequencing to suit the cash flow and overall organisation budget.

Programme management is a series of related projects which are controlled by a programme manager to help this process and adjudicate where there are conflicting demands on personnel and on the budget. A larger client will have a project office to facilitate this. The project office may also standardise processes and make sure that quality standards are equitable and appropriate for the work to be done. Clients will also exercise training.[4]

Project team roles

The project manager is the leader of the team and acts on behalf of the client as well as trying to maintain an efficient project team. It is acknowledged above that the leadership of the construction project may change during the project life cycle under some types of procurement. For example in UK traditional procurement it is most likely that the architect or the engineer will take the lead in the inception and design stages and will act on behalf of the client. During construction the main contractor will have a leading role. For large or complex projects the client appoints an executive project manager with direct leadership of the project team through whom the client communicates.

In this case the project manager will be a single point of contact for the client co-ordinating the design, construction and other professional roles shown in Figure 1.3. This defines the communication channels, but the actual contract is signed by the client with individual organisations and so there are differing contractual links and communication links.

The supply chain is managed day to day through the communication links and things can go badly wrong if the client is allowed to pursue direct contact through their contractual links, thus undermining the co-ordination role of their project manager and the design and construction managers. The construction and design sides need to co-ordinate their operations right through the life cycle and this model of an executive project manager feeding through to respective design and construction co-ordinators ensures that this happens.

In the traditional system the design brief taker tends to take this role by default and hands on the baton to the main contractor during the post tender stage. This creates problems when the design is changed and the impacts on

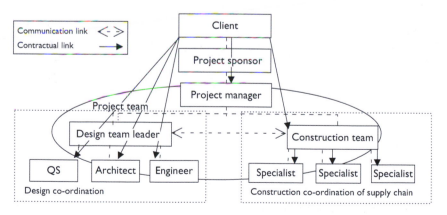

Figure 1.3 Project structure diagram – executive project management model

construction time, quality and cost and health and safety are not reassessed. On a complex project the extra cost of a project manager is well worth it.

The design manager's role is to co-ordinate the various design functions and if necessary specialist design expertise as and when needed. This role is also traditional. The main contractor or construction manager's role is to tender specialist packages, set up site procedures and integrate the construction programme and the interfaces between specialist packages. They have accountability for meeting the contract programme and budget and working with the design team to implement a quality product.

Where there is a lack of trust and respect between the designer and the contractor, or the contractor and the client, the culture can become one of 'them and us' and costs and time constraints are often at risk to the disadvantage of the client. Conciliation is a key role for the project manager to manage disputes, discrepancies, omissions and changes so that the team works smoothly.

It is normal in construction for roles to be assigned to different members of an inter-organisational team e.g. quantity surveyor, engineer/architect, construction manager and client. The project manager is the leader of the project team with a single point of contact with the client.

It is important to try standardising some definitions in order to understand project management as people use terms to mean a lot of different things. We need to consider the roles in the context of the *client's* total view of a project i.e. a finished building or facility, not *'our view'*, which covers just the length of a participant's involvement. For example a contractor and a subcontractor both call the person they appoint to run the contract for them the project manager, the former in a traditional contract only spans the construction phase and the latter only covers the length of time the specialist installation is on the site. Some of the most important of these definitions are given in the glossary at the back of the book.

Table 1.1 illustrates how different members of the team play leading roles at activities which are predominant in each of the life cycle stages assuming a project structure using an executive project manager. This is not meant to be an exhaustive coverage of the activities which the team performs. However it does give you an idea of the main connections between the different players. It is based on the RIBA Plan of Work stages which are similar to life cycle elements.

The client has a continuing interaction with the project team and may even take a lead role in project management. Their essential role is to sign off each stage so that work may proceed to the next stage. These are called client approval points or gateway reviews. For a more specific reference to the public procurement gateway review system there is more information in Chapter 2 on project definition. The final gateway review is the post project appraisal.

Table 1.1 Roles of project team at each stage of implementation

Stage	Role	Client	Approvals Leader
Inception	Client objectives interpreted to strategic brief.		Client advisor
	Professional interpretation & development of brief to determine value and performance.		PM
	Outline planning.	CA1	Architect
Feasibility	Test for viability and/or option appraisal. Project risks assessed.		PM/QS
	Outline design and cost plan.		Architect/QS
	Funding and location.		PM
	Client go ahead on scheme.	CA2	
	Appoint professional design team.		PM/Architect
Strategy	Decide on procurement route, risk management, cost control and quality management.	CA2a	PM/(CM)
Scheme Design	A scheme design and planning application.	CA3	PM/Design team leader
	Cost plan and cost checks – iterative with client.		
	Buildability testing. Building regulations approval.		QS
	Health and safety co-ordination.		PM/CM CDM Co-ordinator
Tender	Prepare detail design and bill of quantities.		Architect/QS
	Tender documents.	CA4	Client
	Pre tender health and safety plan.	CA4a	CDM Co-ordinator
Construction	Appoint contractor(s).	CA5	CM/QS/ Architect
	Mobilisation of construction process, tender subcontractors, health and safety plan.		CM
	Time, quality and cost control.		CM/QS
Commissioning	Test and snag all systems.		CM/Architect
	Ensure equipment compliance and efficiency to meet client objectives.		
Post project review	– feed back into future projects.		
	– lessons for client.	CA6	Client
	– lessons for PM.		PM
Occupation	Recheck 'under use' conditions.		Client/User
	Manuals and training.		

KEY: PM = project manager; CM = construction manager; QS = quantity surveyor; PS = planning supervisor; CA = client approval

Project manager: skills and functions

Choosing the project manager is important to the culture of the project and will be the responsibility of the client with guidance from advisors. It is quite common for legal or other professions to provide early advice in these areas. So what is required of the project manager?

The Association of Project Management (APM) lists eight characteristics of a certified project manager.[5] These are an open, positive, 'can do' attitude, common sense, open mindedness, adaptability, inventiveness, prudent risk taker (weighing up the risk), fairness and commitment. Many of these skills are associated with managing people and themes such as partnership, negotiated tenders and building the team have been a feature of periodic industry wide reports.

From a client's point of view these qualities will make the project manager easier to work with, reliable and realistic. However, they might want to add 'a willingness to see those things which are important to their business' and to adapt the system and parameters of the project to suit. They would also need to have assurance of general competence and experience especially in dealing with projects that have similar characteristics to their own. From a project team point of view being open minded and taking the lead in fairness and commitment are expected of someone who has high expectations of their performance and has to deserve their loyalty.

The project manager's role will be related to the procurement route chosen by the client. In the case of the executive structure the project manager manages all aspects of the project from inception to completion. In a traditional procurement the project manager changes in different stages of the project life cycle. Either way there are four key tasks for the project manager acting on behalf of the client:

1   Guide and advise the client.
2   Manage the resources to carry out project activities.
3   Build the project team.
4   Ensure customer requirements are met.

In traditional procurement systems where there is a changing of project leadership at different stages of the project life cycle, there is potential for (1), (3) and (4) to be dealt with in a piecemeal way. The appointment of a dedicated project manager should help to provide customer focus and to draw the client into the team. This gives more certainty to the client approval process.

Meeting customer requirements means gaining knowledge of the customer's business, sharing problems at an early stage, so that trust is built up, and reviewing project goals at regular intervals to make sure that the developing brief meets expectations. Guiding and advising the client is of particular importance, referring back to the need to reconcile the client's brief with construction constraints. Building the project team is about developing relationships between professionals and ensuring efficient communications between diverse and numerous participating professionals and contractors.

The resources managed will have a different focus at different stages of the life cycle. For example at the inception stage managing funding is critical; at the design stage it is a matter of managing information; at the construction stage physical resources such as materials, plant and labour become important.

So how important is it that a project manager has a technical background? Can a construction project be managed by a generic project manager? Are there certain types of people or personalities which are better at managing construction projects?

Certainly (3) and (4) could be fulfilled by any experienced project manager and production management experience will help with (2). (1) is likely to require industry knowledge of how procedures work and what the economic, technological and legal constraints are. Because of the technical knowledge required the project manager cannot be completely impartial when making decisions and an ethical code needs to be established between professional conduct as a construction profession and decision making as a manager based on co-ordinating expert response. Here it is probable that a competent technical training is not sufficient for reading economic situations or providing broad enough advice to a client who wants to push technological frontiers. In this sense technical knowledge is important, but very much part of a range of skills and experience which will apply to construction projects as much as they will to any project. It is useful at this point to remember the APM BoK which tries to deconstruct 52 different areas of competence.[6]

Ethical project leadership

Ethics in construction is often connected with the professional status of a manager to give client confidentiality and updating knowledge under a code of professional conduct. In respect to leadership there is a strong point to be made for a code of ethics. Ireland *et al.* suggest accountability in maintaining high standards, in conduct and leadership of the team, in relations with employers and clients, and accountability to the wider community and the reputation of the profession.[7] Meredith and Mantel see that the ethical position for a project manager is in the middle, representing the interests of all parties fairly to one another.[8] RIBA helpfully identifies the three categories of honesty and integrity, competence and professional relationships.[9] The first deals with areas such as confidentiality, impartiality and fair judgements, the second considers adequate research and knowledge and experience with an awareness of wider stakeholders and ability to balance requirements. The last includes dispute management, fair competition, respecting social diversity and communicating transparently in terms of abilities and weaknesses of a solution.

Sustainable delivery is also an important ethical consideration which will be dealt with in a later chapter. The trend is for greater standards of sustainability to meet regulatory and political targets. Respect for equality and diversity is also a pressing ethical issue which is being examined more closely by legislation, but is also an ethical area for leadership to go beyond compliance in being fair and establishing equal opportunities.

Professional institutional codes of conduct are not always effective. They are often hard to police because of very subjective statements and commercial forces are given undue weighting so that client demands may be used to justify action which is much less acceptable to the community. This may particularly occur in the area of sustainable development where every building newly constructed or refurbished should be able to demonstrate an ability to be appropriately sustainable, not only in compliance to current codes, but to aid future tougher targets by exceeding current ones. Consequently some institutions such as the CIOB, the Royal Academy of Engineering and the Society for Construction Law have introduced the idea of an additional ethical code which suggests action which is more than compliant and is the result of more significant involvement in the design and briefing process for contractors and the involvement in the construction and maintenance process for designers.

Ethical leadership is discussed by Fewings and covers the two key areas of openness to learn and courage to stand up against traditional and currently compliant practice which is not ethical.[10] Differentiating a company or a project in this way on the basis of ethical principles may not be financially easy in a highly competitive environment, but awareness and action to deal early with future ethical issues can also give market leadership and in the case of a client may provide a reputation for future work. It may also be protective against unexpected retrospective action such as the Office of Fair Trading (OFT) investigation into bidding practices in the UK and previously the EU requirements for more transparent tendering. It may also make it easier to get onto local authority tender lists if foresight in introducing environment management systems is exercised. For example Hamzah *et al.*'s (2007)[11] survey indicated that construction quality in Malaysia suffered due to unethical acts to do with procurement and lackadaisical supervision of work completed. They describe the construction industry as having a propensity for unethical behaviour because of the low price mentality and fierce competition and low margins leading contractors with evidence that contractors unethically undercut their bids and collude. In addition Transparency International (2005) has identified the construction industry as being one of the worst perpetrators of bribery in their report,[12] and the OECD has created strong measures to curb multinational companies offering bribes in the procurement of international projects, by signing up developing countries to sanctions such as withdrawing export credit cover.[13] The CIOB's own survey also identified an unexpected large awareness of corrupt practice in the UK industry.[14] The new Bribery Act 2010 was subsequently introduced in the UK to require appropriate preventative governance to reasonably raise awareness and to dis-incentivise bribery.

Case study 1.1 Considerate Constructors Scheme

The use of a Considerate Constructors Scheme in the UK works to engage community stakeholders, and worker and public health and safety is enhanced and site work is delivered more sustainably. This has been a popular voluntary scheme which has been rewarded by prizes for best points achieved in one of six categories. Over a number of years the scheme has been adopted by a majority of contractors in urban areas on the basis that it builds relationships in the community which pay back dividends in reducing conflict with them whilst reducing nuisance value of construction work. It also raises reputation with public and private clients who also retain their own reputation and helps value driven tenders. Sustainably it helps to encourage recycling and reduce waste which meets other objectives.

Project complexity

The degree of complexity is not directly dependent on value. The overall rating has been shown to be useful for assessing the experience of staff required, the workload generated and the degree of systemisation and formality required. Maylor speaks of organisational resource and technical complexity.[15] He makes an overall rating by multiplying them together for comparison of project complexity.

Organisational complexity increases with the number of organisations and stakeholders that are involved and the degree of integration of their work – high for a construction project even of a small size. According to Gracunias's theoretical formula, six organisations, quite basic for construction, gives 222 inter-relationships between the different supervisors and roles that are played.[16] Non-standard building contracts take people out of their comfort zone and create further complexity.

Resource complexity increases with the value of the project and significance and the range of the resources. In terms of construction size ranges from very small to extremely large. A site manager will be assigned full time to a project with a value which exceeds £1–2m.

Technical complexity increases with the use of non-standard technology, the building constraints which exist and the complexity of the technology. Compare, for example, the repair of a historic building's roof compared with the use of prefabricated trusses. The combination of high ratings for at least two areas makes a lot of difference. Innovation requires additional research and testing.

Political complexity is another factor defined as an external project environment which is sensitive and leads to unclear goals, interference or a complicated communication process. The project is not stand alone and requires lots of external inputs. Such a project would include the construction of a bridge close to a bird sanctuary which would require careful negotiation with pressure groups, innovative methods to reduce traffic noise and compensation to pay for disruption and protection of the reserve. Political impacts also occur because of cost reduction in the choice and impact of cuts to the projects.

Difficult technical solutions make a lot of difference to construction projects as resources – craftsmen and materials in particular – are geared up for standard building construction.

Case study 1.2 Construction complexity

If we compare two smaller projects we can illustrate the complexity index which emerges using Maylor's overall complexity measure in Table 1.2.[17]

The first project has demanding requirements, but is not unusual for most contractors. The rating will go up with the combination of greater speed and innovation and quality. The complexity factors are increased for the second case study because of the organisational and technical factors and these are despite the size of the project.

Organisational complexity is very dependent on the unique features of the site and often quite small projects offer almost insuperable problems as illustrated. The fragmentation between design and construction often means that insufficient buildability has been achieved at design stage, exacerbating the execution stages with later design adjustments that can create conflict. On a conservation project this is usual, but targets are harder to achieve and need more experience and supervision for the size of the job.

Table 1.2 Comparison of building complexity

Complexity factor case study description	Organisational	Resource	Technical	Combined index
Construction of a primary care doctor's surgery/ pharmacy. Features atrium, low energy usage and novel design. Fast programme on new site.	There is a full range of trade organisations working closely together on a range of different spaces. There is storage and access on the car park, where offices can be accommodated.	Value £1.5m Fast programme Good quality Wide range of resources	The innovative design has taken a long design development and went through planning with some qualifications for redesign. The building has modern materials and prefabricated components. Redesign to improve BREEAM rating. Rating 5/10	4 x 6 x 5 = 120
	Rating 4/10	Rating 6/10		
Refurbishment of historic church hall. Features new roof, repair of stonework, demolition and creation of new basement area under the floor.	There are a limited number of subcontractors, say 12, involved as there is a limited range of finishes. Space is very limited and the site is cramped with no outside storage in the city centre.	Value £600,000 Range of resources is quite low Programme affects use of the building so time is tight High quality	The design of the roof is complex, uses large non standard timbers and made to measure stonework with advanced craft techniques. Innovative technology for forming the basement. Long gestation period, but planning acceptable. Rating 9/10	6 x 4 x 9 = 216
	Rating 6/10	Rating 4/10		

Source: Based on Maylor's (2003) methodology[18]

Project management maturity model

Both clients and contractors have to have maturity for successful projects. For an immature client they are likely to be fire fighting reactively rather than being proactive to problems and the quality of the outcomes (e.g. over-runs of programme and cost and poor quality or inappropriate specification) is more likely to be compromised through a lack of preparedness. The UK Office of Government Commerce (OGC) portfolio, programme and project maturity model measures the six processes of management control: benefits management, financial management, stakeholder management, risk management, organisational governance and resource management. A maturity model is used to assess what level of capability an organisation is at with regard to defining and managing its projects and also defining the level of capability needed to manage a particular project.

The common causes of failure for programmes and projects are design and definition failure including scope, decision making failure connected with shortage of senior management commitment, discipline failure connected with poor risk management, supplier management failure connected with poor contract terms and understanding of commercial imperatives, and people failure where there is poor stakeholder management, cultural issues and lack of ownership.[19]

A mature organisation requires management processes, clear roles and a proven control system to evaluate and regulate the status and health of the project. It will also have learned and be learning from its mistakes and continuously improving. It will have a sound briefing and value/risk assessment process to assure best value in what it asks for.

There are five levels of maturity in the OGC model in Figure 1.4 and these measure an organisation's capability so a low level indicates that help is needed, a medium level can deal with simple to standard projects and repeat successes and an advanced level can deal with complex and fast paced projects with changing requirements and can optimise its projects and continuously improve.

Determining the critical factors for success

If we take success as the delivery of a product that meets the expectations of the client at the same time as giving profitable business to the provider, the facts are clear that the construction industry has a flawed record in the delivery of projects. Major projects have been subject to cost and time overruns and the performance of the product has often fallen short of the criteria of the client let alone the expectations. To add to this the profitability of the industry has been in question and major investors are reconsidering their exposure to certain sectors of it.

Level	Process documents	Repeatability of success	Training in place	Standard terminology	Planning and control	Risk management	Top management commitment
Level 1 Awareness	Subjective, few documents	Poor	Minimal	None	Based on individuals	None	
Level 2 Repeatable process	Basic resource and budget tracking	Hit and miss	Generic	Yes	Major milestones	Significant	Some leads Not consistent
Level 3 Defined process	Good	Some understanding of the process	Established and specific	Able to tailor to suit specifics	Quality management better	More co-ordinated	Consistent engagement
Level 4 Managed process	Full set of measurable requirements	Quantitative management	Established and more specific	Yes	Fuller metrics for improvement and adapt	Good	Proactive and innovative
Level 5 Optimised	Full set	Full understanding of cause and external factors	Anticipate future needs	Innovative	Quantitatively managed	Knowledge management	Seen as exemplars

Figure 1.4 Project maturity levels (Based on the P3M3 OGC model)

How do you define a project? What is it that makes a project a success for the various parties involved? Are there inherent factors common to all projects or is it in the nature of the project? Are the factors within the control of the project team or are they, as in the case of political factors, to be responded to? Do you select projects which have built in success and by definition avoid others?

Industry-wide studies

The factors of success have been reviewed in the context of many different reports in recent years. Many of them are industry wide. The results of these reports have often depended on the viewpoint of the party who has commissioned them. Many wider consensus reports have been heralded in principle, but have failed when there has been an attempt to implement the details and in the interpretation of the reports, which has been harder to harmonise between different parties to the contract. They are summarised in Table 1.3. The reports connect success to productivity, reduction or successful management of conflict, greater efforts to be client focused, reduced defects/waste and strategic management to respond to external factors.

Critical success factors (CSF) are often categorised in different ways. Pinto and Slevin's 10 factors of success, as determined from a survey of project managers, are shown in Figure 1.5 and were ranked.[20] The 10 factors were grouped as:

- strategic factors critical in the early stages of the life cycle
- tactical factors (shaded) which became most important in the latter stages.

Top management support, detailed programmes and budgets for control were the strategic factors and these remained critical and most important throughout the project. Interestingly the next most important issue was the involvement of the client in the project team. Troubleshooting was ranked lowest.

However others have said that factors are different for different industries and also qualified that failure may be defined in different ways.[21] Pinto and Mantel say that the fundamental causes across industries for failure are poor planning, insufficient senior management support and getting the wrong project manager.[22]

Applying this to construction it is clear that the brief and its interpretation, effective scheduling and control, together with continued client involvement at the highest levels, are critical. The idea has been broached that success can be seen from different stakeholders' points of view, the main ones being

Table 1.3 Some reports indicating the need to improve construction success in the UK

Report and title	Broad findings
Morris and Hough (1987) Anatomy of Major Projects	A review of five major project case studies and the issues which caused success or failure indicate planning for external factors.
Slevin and Pinto (1986–1988) Several reports on Determination of critical success factors	A 50 item instrument has been developed to measure a project's score on each of 10 factors on over 400 projects. Report concludes that strategic issues are important throughout the whole life cycle with tactical issues only being equal in the later stages.
Reading Building 2000 Report (1990) Reading Centre for Strategic Studies in Construction	Identifies the major issues to address by the year 2000 to keep UK industry competitive world wide. This was eclipsed by the Latham and Egan Reports.
The Linden Bovis Report (1992)	Unfavourably comparing productivity between the industries of the UK and the USA.
The Latham Report (1994) Constructing the Team	Suggested that a productivity improvement of 30% was needed and possible by re-engineering the construction process to eliminate confrontational contracts and relationships and introduce partnering and contract change in the UK Housing, Construction and Regeneration Act.
The Egan Report (1998) Rethinking Construction	Supports value and process improvements including introducing selection on value to reduce defects, increase safety and introduce benchmarking to make it possible to track continuous improvement. Introduces five drivers for success and the enhancement of client value for money. Targets set similar to Latham.
The Construction Clients Forum Report (1998) Constructing Improvement	Calls for a pact to be made between the client and supply side designed to improve value for money and the profitability of the supply side.
Strategic Forum for Construction (2002) Accelerating Construction (Egan update report)	Measures improvements over the intervening four years since 1998, indicating that demonstration projects where Egan principles have been practised have reached the targets set and far exceeded industry averages.
Strategic Forum for Construction (2003)	Sets up a toolkit for the integration of the project team in the better delivery of a project in partnership. The parameters that it sets are very wide, but it represents a methodology to integrate the client and the team in a continuous improvement culture in pursuance of Egan and Clients Forum Reports.
Modernising Construction (1999) NAO	National Audit office identified major inefficiencies in public procurement with 3/4 of buildings exceeding their budgets by 50% and 2/3 exceeding their programme by 63%.
Construction Matters (2008)[23] for the Parliamentary Select Committee	Looks at the impact of the built environment as a whole (20% of GDP) and considers the need for sustainability in this sector as key to meeting the targets for a sustainable Britain and more efficient construction.
Wolstenholme Report (2010)[24] for Constructing Excellence	Determines the level of industry progress rated at 4/10 since Egan and to define the improvement agenda for the next 10 years.

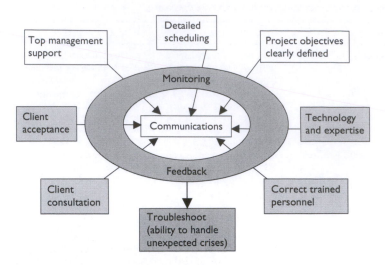

Figure 1.5 Factors of success

Source: Adapted from Slevin and Pinto (1986)[25]

the client/user and the project management team. Morris and Hough identify responding to external factors as a critical issue in the success of projects and cite the Thames Barrier project as a successful project even though it took twice the planned time and cost four times the budget, because it provided profit for most contractors.[26] Toor and Ogunlana believed that success was dependent on underlying relationships and came up with the critical COMs: Comprehension, Competence, Commitment and Communication.[27] This forms a very simple analysis which can easily be done and worked on by the project manager.

Wateridge in his research on measuring success compares the major criteria as seen from the viewpoint of the users and the project manager.[28] Table 1.4 indicates a subtle difference between the two points of view for the top five criteria.

Table 1.4 indicates that although meeting user requirements is not in contention between users and project team, the idea of making them happy takes a different status in the two points of view and is replaced in the project team's mind by commercial success (making the project team happy!). It is interesting that budget and time are more important to users. Perhaps quality has been defined in a different way by the users as it is not otherwise mentioned in the top five.

Table 1.4 Project success factors from different viewpoints

Users' Criteria		Project Manager's Criteria	
Successful projects should	%	Successful projects should	%
Meet user requirements	96	Meet user requirements	86
Make users happy	71	Ensure commercial success	71
Meet budget	71	Meet quality	67
Meet time	67	Meet budget	62
Achieve purpose	57	Achieve purpose	62

Adapted from J. Wateridge (1998)[29]

Conclusion

Project management has been around for a long time in construction, but has not always delivered the value that clients have been promised. The unique nature of the product needs to be properly planned for success, understanding the client business and not ignoring the key role of external events. The move is for large clients to seek to enhance value and not to tolerate under-performance. Integrating the events in the project life cycle more fully (especially design and construction) moves away from traditional procurement methods so it is important to look at more innovative ways of delivering projects. Clients need to be clear about the impact of the approvals and decisions they are making. Later chapters connect risk management (RM) and value management (VM) and also look more closely at design management and the supply chain and how procurement can be managed in a better way.

The professional roles in projects are changing, but there is a need to manage a fragmented process seamlessly and to this end there is a special chapter on the development of PPP schemes. The development of a strategic plan is important for all areas of project management indicating the organisational structure, culture and leadership, the strategy for more sustainable building and the promotion of effective teams. The credibility of a project manager depends not just on their technical ability, but on their ability to provide an ethically sensitive service to all the parties to the contract and to deliver client satisfaction and advice.

Project success is the subject of the recent industry reports, but it is clear that success depends upon the degree of strategic planning, the ability to integrate and getting the right person to match the job, in order to achieve the process-focused improvements which are being recommended by the reports.

References

1 CIOB (2010) *Code of Practice for Project Management for Construction and Development*. 4th edn. Oxford, Blackwell Publishing.
2 BSI (2010) *Guide to Project Management*. BS6079-1, UK, British Standards Institute.
3 PMI (2008) *A Guide to the PMBOK*. 5th edn. Darby, California PMI Publishing.
4 Maylor H. (2003) *Project Management*. 3rd edn. Harlow, Pearson Education, pp.56–59.
5 Dixon M. (Ed.) (2006) *Body of Knowledge*. High Wycombe, Association for Project Management.
6 PMI (2008) *A Guide to the PMBOK*. 5th edn. Darby, California PMI Publishing.
7 Ireland L.R., Pike W.J. and Shrock J.L. (1982) *Ethics for Project Managers*. Proceedings of the 1982 PMI Seminar/Symposium on Project Management. Toronto, Canada.
8 Meredith J. and Mantel J. (2006) *Project Management: A Managerial Approach*. 6th edn. Oxford, John Wiley.
9 RIBA (2005) *Code of Professional Conduct: For Members of the Royal Institute of Architects*. UK, Architects Registration Board.
10 Fewings P. (2008) *Ethics for the Built Environment*. Abingdon, Taylor and Francis, p.109.
11 Hamzah A.R., Saipol B.A.K, Mohd S.M.D., Mohammed A.B. and Yap X.W. (2007) *Does Professional Ethic Affect Construction Quality?* Proceedings of Quantity Surveying International Conference 4–5 September. Kuala Lumpur, Malaysia.
12 Transparency International (2005) *Global Report on Construction*. March.
13 OECD (2009) Convention for Combating Bribery Against Foreign Public Officials in *International Business Transactions*, 17 December, and signed by the OECD countries and Brazil, Argentina, Bulgaria, Chile and South Africa. Revised edition.
14 CIOB (2006) *Corruption in the Construction Industry Survey*. Ascot, CIOB.
15 Maylor H. (2003) *Project Management*. 3rd edn. Harlow, Pearson Education, p.37.
16 Gracunias V.A. (1937) 'Relationship in organisation'. In *Papers on the Science of Administration*. University of Columbia. As quoted in Mullins L.J. (2003) *Management and Organisational Behaviour*. 3rd edn. London, Pitman, p.336.
17 Maylor H. (2003) *Project Management*. 3rd edn. Harlow, Pearson Educational.
18 Maylor H. (2003) *Project Management*. 3rd edn. Harlow, Pearson Educational, Table 2.5.
19 OGC (2008) *Portfolio, Programme and Project Management Maturity Model*, Public Consultation Draft v2.0. Available online: http://www.p3m3-officialsite.com/P3M3Model/P3M3Model.asp (accessed 4 March 2012).
20 Pinto J.K. and Slevin D.P. (1987) 'Critical factors in successful project implementation', *IEEE Transactions on Engineering Management*, EM-34, February, 22–27.
21 Baker N.R., Bean A.S., Green S.G., Blank W. and Tadisina S.K. (1983) *Sources of First Suggestion and Project Failure/Success in Industrial Research*. Conference on the Management of Technological Innovation. Washington DC.
22 Pinto J.K. and Mantel S.J. (1990) 'The causes of project failure', *IEEE Transactions on Engineering Management*. February.
23 House of Commons Business and Enterprise Committee (2008) *Construction Matters*, 9th Report 2007–8. Available online: http://www.publications.parliament.

uk/pa/cm200708/cmselect/cmberr/127/127i.pdf (accessed 12 December 2011). House of Commons Committee 16 July, Chaired Peter Luff.

24 Wolstenholme A. (2010) *Never Waste a Good Crisis*. London, Constructing Excellence.

25 Pinto J.K. and Slevin D.P. (1987) 'Critical factors in successful project implementation', *IEEE Transactions on Engineering Management*, EM-34, February, 22–27.

26 Morris P. and Hough G. (1986). *Anatomy of New York, Major Projects*. New York, Wiley.

27 Toor S. and Ogunlana S.O. (2008) 'Critical COMs of success in large-scale construction projects in Thailand construction industry', *International Journal of Project Management*, 26(4): 420–430

28 Wateridge J. (1998) 'How can IS/IT projects be measured for success?', *International Journal of Project Management*, 16(1): 62.

29 Wateridge J. (1998) 'How can IS/IT projects be measured for success?', *International Journal of Project Management*, 16(1): 62.

Chapter 2

Building the client business case

This chapter looks at the concept of using project management to fulfil a client's business objectives. Clients may be classified into public and private, profit and non-profit making. The term business case implies the need to justify a need. In this sense all categories of client will have a business case and this will define the client objectives in the context of the project. Construction projects are complex and expensive and need to justify the expenditure and need.

The objectives of this chapter are:

- to understand client objectives as an outcome of client values and type
- to balance project constraints and objectives
- to know how to present a business case
- to present the context of decision making using stage gate reviews
- to benchmark and build value in the business case
- to identify and manage project stakeholders.

These issues are supplemented with some case studies as a way of illustrating the theory and presenting current practice in commissioning building work. Business planning advice has been well documented by many agencies and the Gateway Review programme and the Government procurement guides available through the Office for Government Commerce website are a good example of best practice.[1] For private inexperienced clients publications such as the Construction Clients Confederation 'starter' charter will help focus the advice.[2] A balanced approach is needed in the context of client type, building uses, project size, public versus private clients and the unique stakeholders of each. There is also a need to view business planning as an open system that is heavily affected by project and environmental constraints. The business case is a starting point or a benchmark for the level of performance required. It is very easy to erode the value of the business case, but the integrated approach allows for working together with the project team to preserve and improve that value.

Table 2.1 The difference between client and project objectives

Project objectives	Client objectives
efficiency, to given time, cost and quality levels	statement of need
teamwork	functional facility
technical task	financially viable

According to Egan (2002) value improvements require that performance measurement is carried out.[3] Maylor reminds us that it is easy to measure the wrong things and that real improvements are made by long term measurements across projects so that supplier behaviour changes are permanent and not just reactive.[4] The process of feasibility and funding appraisal is dealt with in the next chapter.

Project constraints and client objectives

The job of the project manager is to understand the client objectives and to ascertain the priorities. It is also to provide a professional service which not only develops the business case by applying the right tests to the assumptions made, but can also advise on the specific project constraints and promote technical project objectives. Table 2.1 indicates a balance of project efficiency with client objectives.

These project constraints come in the form of external circumstances and site characteristics which constrain the design and construction process. Project constraints can be classified as:

- Economic factors, which affect funding and market prices. Market prices for the tendering of construction work vary significantly according to supply and demand. Positively, local authorities may recognise employment opportunities and provide tax breaks for certain locations.
- Ethical and environmental choice to suit sustainable project.
- Physical site constraints. Site access might cause expense, restrictions or limitations on the positioning of the building. Ground conditions determine foundations, topography affects the design and boundaries the shape and orientation. Alternative locations could be considered.
- Resource availability such as labour skills and choice of materials.
- Time constraints for achievable goals to completion and phasing.
- Technical/design issues, cost versus quality balance, life cycle costing.
- Planning constraints which exist to the type of developments noted, making some locations easier than others to gain permission or apply conditions, for example, building height restrictions on certain sites. All developments must show sustainable strategies to reduce carbon use.

- Local councils, through the planning system, may seek to impose agreements (planning gain) on developers. They are designed to contribute towards the community created by new development, in return for the benefit gained to the client business. Highway authorities may require new layouts to improve traffic flow such as additional road widening, improvement of junctions or the provision of traffic signalling.
- Neighbour concerns.
- Health and safety issues.
- Legal requirements such as durability, contamination and sustainability covered by the Building Regulations and various environmental Acts.

Some of these arise out of a technical knowledge of the construction process so will appear strange to the client. The main role of the project manager is to make known these constraints to the client and guide them to best value, whilst maintaining the real essence of the client's objectives.

Figure 2.1 gives the context for developing client and project objectives which are different, but complement each other.

| A clear statement of objectives | Project manager | A clear statement of constraints |

Figure 2.1 Balancing constraints and objectives

Client objectives

Company goals are a framework and provide some guidance for the direction of the business, for example the type and location of markets to be in, the investment needs and the growth rates of the business. Strategic planning does not define the building project, but should give some justification for a project. The client needs to be sure of testing key assumptions, giving clear requirements, having a good management commitment, sufficient skilled resources and flexibility of contractual arrangements to cope with change.

Figure 2.2 shows the steps in the process in moving from client objectives to the project brief. The latter part from design brief where the brief needs to be defined is covered in Chapter 3.

Client objectives for a new building project define the business case for it and lead to the development of a project brief. SMART is a well-known mnemonic that can be applied overhauling a client's objectives to ensure effective implementation.

Specific – they need to identify the outcomes clearly (e.g. business volumes, returns and markets)

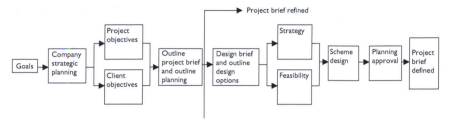

Figure 2.2 Process of developing brief from objectives

Measurable – so it is feasible to monitor attainment, for example a budget or a functional space (e.g. so many beds spaces or car parking spaces)

Achievable – check that different objectives such as space and budget or complexity and time scale do not clash. Ownership of the objective also helps commitment to it

Realistic – to the available resources of funding, time scale, materials, labour etc.

Time bound – with a programme for delivery of the objectives. This will be a high level programme indicating key dates for attainment of the objectives, such as decanting or occupation. They should be in the context of the business need

Client objectives are the starting point for defining the project, but may need to be clarified by the project team. Value can be enhanced by optimising solutions to meet essential requirements and to separate out desirables as a 'wish list'. Client objectives recognise external constraints and will inform the commissioning of a specific project. For example, Dyson have an objective to expand their business and to do so they have moved production from Wiltshire to Malaysia, because they recognise that production is cheaper there and because the conditions for planning permission in expanding on their current site are considered more onerous. Building in Malaysia is cheaper and they have calculated that even with export costs, they can undercut their competitors in European markets.

Different types of clients have different types of objectives. The main client types are private, public and developer. So it is important to 'get into the client's shoes' in order to understand their objectives and the basis of the client brief. Below are a few simplistic examples of generic objectives for specific client types:

- A *manufacturer* needs functional efficiency, value for money, to meet performance criteria and to start production as soon as possible. They will be a secondary client as the building is a means to an end.
- A *developer* needs a cheap, quick and attractive building to sell or to rent economically. This will be a primary use as the building is being used as a commodity in itself.

- A *public body* needs a building that lasts a long time, is efficient use of tax payers' money, is within yearly budget and is low cost to run. This is again a secondary use of the building.

Masterman and Gameson have classified clients on a two dimensional grid as experienced or inexperienced and primary (wanting buildings because they trade in buildings) and secondary (those who use buildings to house their business).[5]

The experience of the client will determine the degree of involvement that can be expected from the client in developing objectives. It will also determine their understanding of the project constraints. A small manufacturer is unlikely to have built very often and so is inexperienced. BAA on the other hand is experienced and has strong influence on the procurement of its built assets. This is quite simplistic and Green believes that the consumer led market has led to a more organic iterative approach to arriving at the brief.[6]

Project objectives

These objectives are associated with the efficiency and effectiveness of the project process. The project manager has particular responsibility to meet these as well as to help the client to meet their own business objectives. They are traditionally to do with project budget, quality and programme, but there are also other aspects of project objectives which are important to the success of the project and these will be considered later.

The time–cost–quality triangle in Figures 2.3 to 2.5 indicates the need to understand the balance between each of the parameters in agreement with the client requirements. It is assumed that all three priorities are equal, but some may be more equal than others to make it easier to make decisions. A single priority is shown by 'pushing the ball into one corner'. A double priority (quite normal) is shown by 'pushing the ball' to the middle of one side. It becomes much more difficult to manage if a triple priority is given. The first two do not imply that the third factor is unimportant.

Time, cost and quality are all important to a client, but one may be more important than another. For example a local authority is almost always tied to the lowest price. So once the budget is set for a school it is embarrassing for the budget to be exceeded, because it means going back to central government, providing fewer school places, or hijacking money from another scheme. Other things such as durability costs may become important.

Alternatively an Olympic stadium must be ready in time for the event and the quality is important to the ambassadorial role that is so important to a government electing to hold the Olympics. This means that a budget will be secondary to the quality and the programme as shown in Figure 2.5. Other things such as security are also important.

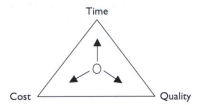

Figure 2.3 Project objectives time, cost and quality

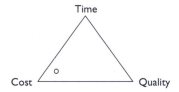

Figure 2.4 A single priority for the local authority

Figure 2.5 A double priority for the National Olympic Organising Committee

Project objectives also include *managing people* and creating synergy from the team so that there is greater productivity. An open and blame free culture, a conducive working environment, good communication with regular short meetings and agreed document information distribution all help to create this type of win–win environment.

Poor management of people and lack of trust produces the 'them and us' approach. This approach may occur between the contractor and the client, between the design and construction teams, or between the contractor and their supply chain. In all cases the effect is unproductive use of time. For example, a failure to look at ways of getting over problems to mutual advantage and time spent trying to apportion blame and making claims leads to one or both parties out of pocket.

At tender stage the 'them and us' approach produces a minimum bid and claims for extras. This produces bad feelings and a 'win–lose' or 'lose–lose' environment between the client and the contractor. This can be avoided by an open attitude on both sides and a commitment to collaboration.

The project manager may also agree some *task-orientated objectives* such as zero defects, planning to substantially improve health and safety risks and shared bonuses for reducing costs below budget. This particular agenda would relate to the improvements which are being recommended by the Egan Reports (1998 and 2002) and others.[7]

Presenting a business case

A business case is presented at the inception stage of a construction project and will confirm the expected need for new construction works to take place. In addition it will contain constraints such as budget limits, date when required and some performance requirements.

Business budget constraints are different from project constraints and refer to at least making a profit margin which covers the risks of the investment and a return to the shareholders if it is a profit making venture. Public ventures need to objectively value benefits against the costs if they are to make the justification. Non-profit ventures do not have to cover shareholder expectations, but still need to cover risk of not breaking even and costs of new investment.

Time constraints are often indirectly connected with returns. A new super-market can afford greater capital costs if it is able to start trading earlier to a ready market. Every day saved will equal profit bonus even if shared with the contractor. If the money is reinvested into capital cost that gives life cycle saving then continuous cost benefit is achieved.

Quality constraints are based on balancing the durability of higher cost materials and best quality workmanship with the reduced maintenance costs. Cheaper materials may be used where less durability is required or there is a shorter term interest in the asset or where capital funds are scarce. The additional benefits of quality are that it projects an attractive image to the customer.

According to the CIOB,[8] a sound business case prepared for presentation at project inception will:

- be driven by needs
- be based on sound information and reasonable estimation
- contain rational processes
- be aware of associated risks
- have flexibility
- maximise the scope obtaining best value from resources
- utilise previous experiences
- incorporate sustainability cost effectively.

In doing this the key questions are: What is a reasonable and affordable budget cost? What are the investment and funding opportunities? What time scale is required? We shall now look at Case study 2.1 to test the strength of the business case.

Case study 2.1 Student accommodation feasibility

The business case for the provision of new university student accommodation is to replace outdated facilities and possibly enhance the location of the residences by reducing distance from the university and making them useful for alternative use out of term time. One problem the university has is the unavailability of borrowed capital credits and so it has to generate capital by selling assets or it has to commission provision on the basis of a revenue charge, for example, leasing or PFI. It also has to consider that because students now contribute fully to their accommodation and to some of their fees that accommodation has to be attractive to draw students to the university. The university has investigated the alternative routes which are open to them without capital finance and compared them with the cost of doing nothing. The option to proceed with a leaseback arrangement with a housing association was eventually adopted.

The objectives of the university are to provide 400 student places in a safe, secure and reliable environment under the overall control of the university. It proposes single or grouped en suite rooms with the provision for catering services to widen use out of term time for conferences and the provision of laundry and common rooms for communal purposes. The issues which are covered in the business case are:

- The opportunities for sites – some existing sites have restrictive covenants on redevelopment so newer sites may be looked at.
- The performance of the service will be in accordance with the university design standards and outside provision of FM services will be judged against the university's own FM standards.
- Initial costs and rents must be affordable to students and competitive in the private market.
- Management of the flats by others, but not on matters of discipline, pastoral care and leadership which will involve the university.
- Risks judged to be significant were construction cost, timing overruns and subsequent defects in construction, the funding and ownership of assets, the standard of management of the service, the long-term maintenance and equipment replacement, levels of

student occupation and out of term use and bad debts from unpaid rents.

- Transfer and retention of risk – to transfer construction problems, the level of availability and standards of service and out of term use fully to the provider by delegating FM services as well as property ownership in return for a fixed service charge. In addition, to transfer the risk for below 80 per cent term time occupancy to the provider. To retain risk for student discipline, pastoral care, rent collection and the provision of wardens. To provide affordable rents for the students.

In order to keep the service charge down they also considered the permanent transfer of obsolete land to be used for independent development by the provider.

Analysis

If we compare Case study 2.1 with the criteria for a sound business case indicated by the CIOB Project Management Code of Practice then we can make some favourable comparisons with the points made:

- Driven by needs – the university was commissioning new residences because of the changing demands of more market orientated students and the rising costs of maintaining a catered service and shared rooms. The desire for renewing stock alone is not a sufficient driver to attract students, who also have changing needs.
- Flexibility – the danger is that the fashion for single flatted rooms may change again e.g. students might prefer to pay less and share or have catering facilities. Conference use does give some flexibility.
- Sound information – reasonable estimation has been used to assess the market, but general trends have been quoted and not specific market research for the project. In this case a number of options have been presented in the outline business case document (OBC). Affordable rents are considered.
- Rational processes – a written business case and the use of option analysis using discounted cash flow for comparison with each other and doing nothing is rationale as long as the disadvantages of doing nothing are also quantified and risk transfer is quantified objectively. A tendering process which is competitive or negotiated from a strong position is also important. In this case there was some concern that only one housing association was prepared to bid.

- Awareness of associated risk – a risk register associated with probability index means that the comparative impact of risk can be assessed. In this case a range of business, delivery, property management, occupational and user risks were considered and transfer of ownership risks.
- Scope and best value of resources – as the university did not favour rate capital credits or borrowing rights it was considering a number of options which provided provision other than immediate ownership. The release of land to a developer kept rents down, but was it a short-term view and did they get value for money for it?
- Previous experience – this was based on some favourable reviews of a development in similar circumstances.
- Sustainability – the only real references are to social aspects of sustainability in this case and these refer to the enhancing the communal facilities in the accommodation for student quality of life and could be extended to choice of location to suit neighbourhood concerns. Planning requirements today would need evidence of carbon reductions and the use of renewable energy where possible. This is paid back long term. The earlier this is done the lower the cost is.

It would be fair to say that there is, on balance, a sound business case. The university has a strong sense of its requirements and responsibilities.

Business improvement

The Construction Clients' Group describe their role 'as promoting best practice as a construction client that result in best business improvement through a construction outcome' and have set up the 2012 Client Commitments to define the best practice role of clients against the benefits it brings to their business.[9] They recommend:

- A transparent procurement decision making with selection on best value, early contractor involvement, collaborative working, fair payment, risk management and non-adversarial.
- A commitment to people including community involvement, training and development, equal opportunities and considerate sites.
- Client leadership championing best practice with a clear vision and detailed brief, clear financial objectives, client integration in the project team and adequate client resource.
- Sustainable buildings which address environmental, social and economic aspects of projects with waste minimisation and low carbon performance, enhancing and protecting the natural environment and that of the community.
- Design quality so that designs suit the practical, functional and operational requirements of the building to meet client and user needs and to

ensure that whole life value and cost effective solutions are delivered. Design will be tested by third party reviews and other tools for testing.

• Health and safety to ensure a risk register of key risks and that projects aspire to be injury and accident free. They have a set up a working group for improvements.

These commitments reflect the Construction Commitments 2012 for all parties in the construction industry which have integrated improvement criteria, for example a 10 per cent increase of projects using the design quality indicator year on year.[10]

Managing change in the business case

There is a need in projects to build flexibility into the system, as in Case study 2.2, so there is not an 'all or nothing' approach to projects at the business case stage. Many business cases suffer from optimism bias even after they are incepted to a construction stage. The role of a continuous review process during delivery by senior sponsors allowing withdrawal of funds may seem draconian, but these reviews can be used as an early warning to adjust funding and design to meet objective. A change management process can be built into the project procurement and contract that predicts cost of change ahead of the need to make a decision. Reporting which continuously monitors benefit realisation at the business development stage and responds strategically to benefit erosion or enhancement is needed.

Case study 2.2 Olympic business legacy

In the 2012 London Olympic Games, project legacy is a key issue to justify the cost for a short use of facilities. The business case for the Olympics is to reuse facilities by the reconfiguration of the stadium, student village and media centre for the improvement of national facilities and of the East London residents in the long term. A clear design was commissioned which allowed the 90,000 seat central stadium to be partially dismantled for a reduced seating national athletic stadium which was subsequently not paying back sufficiently and needed to be sold to a football club who would allow use for athletics out of season. As a football club they agreed a football ground with a track and the public legacy is preserved. To ensure legacy the competitive price is impaired that could be gained for the site and thus for funding other public projects for the good of East Londoners. The final solution is to maintain a reduced size national stadium.

Transformational change management

The process of using a project to bring about fundamental change is well established. Construction projects often partly represent the means by which change is achieved in business transformation in enabling different spatial configurations (as in Case study 2.3 and Table 2.2), recognising more flexible working patterns and allowing a more low carbon building, but also sustainable working in social and environmental terms. Transformational change manifests itself in the external and internal product integrity as defined above and makes the building itself and the way that it is delivered an integral part of the equation for business success.

Case study 2.3 Birmingham City Council 2006[11]

An example of transformational change is the project delivered in Birmingham City Council in 2006 to transform customer services across their public buildings and housing provision. It has a number of phases, 1–7, to deliver specific named benefits starting from an identified need for change. Phases 2–4 deal with the implementation stages and 5–7 make sure that it works and is stabilised in use. The final phase is a formal feedback on how fully benefits have actually been realised in use. The system CHAMPS2 has been generalised to be used in similar business change projects. Its effectiveness is based on claims to reduce costs, to be a flexible model and to ensure leadership control.

The outcomes achieved in Birmingham City Council were spread across three areas, as in Table 2.2.

This project would have invested in the rationalisation of property to develop spatial configurations which were more efficient as well as investment in people and information systems.

Table 2.2 Birmingham City Council business transformation

Project name and investment (£m)	Value of benefits achieved (£m)	Value of cashable benefits (£m)	Notes
Customer first project 175	321	197	Single point of contact and real time progress information
Excellence in people management 81	289	148	Create a flexible, agile and competent workforce to drive efficiency
Corporate services transformation 144	860	518	Cross council procurement, better supplier contract, automation and training

The following system is more widely used in public projects in the UK and abroad and acts on a similar phased system to focus client decision making.

Developing value in construction – the Gateway framework for decision making

The Gateway Review™ is a project life cycle procurement guide for public contracts developed initially by the UK Treasury, in order to ensure consistency and value for money in government and other public contracts.[12] It is used here as a generic model to underline the key client decision points. There are six decision points called gateways shown below and in Figure 2.6.

- Gateway 0 establishes whether there is a business need.
- Gateway 1 assesses the high level business case and budget and makes way for expenditure on outside consultants.
- Gateway 2 assesses the feasibility study and proposed procurement strategy and gives the critical go ahead to proceed with contract documentation for the implementation of the project. A proper project execution plan should be in place to proceed.
- Gateway 3 assesses the contractor tender bids report and provides the go ahead for detailed design and construction. (In this integrated model it is assumed the scheme design is done by the contractor, but a separate design contract could be procured prior to tender competitions by the contractor.) There are two additional decision points at completion of planning application and detail design stages.
- Gateway 4 is the acceptance of the finished project either for a separate client fit out contract or for occupation.
- Gateway 5 is a benefits evaluation which takes place after occupation.

These gateways are the basis for getting approval to proceed to the next phase of the project cycle and making sure that there is adequate information available for client decision making for the viability of the project. This system has several points to assess value, but note the early and later value management (VM) opportunity for gateways 1 and 3.

This system has been set up for the client, but the project manager is appointed after gateway 1. At this stage a feasibility study is commissioned with an outline design and a risk assessment and then a more general value management opportunity with the project team appointed. A *project execution plan* is prepared which covers the strategy for the job and considers the programme, cash flow, procurement strategy, project organisation, health and safety plan, environmental impact and design management.

The plan establishes how the project is going to be carried out whilst the feasibility establishes how the project can be delivered viably and within

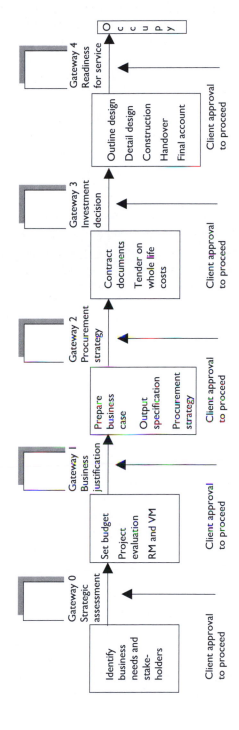

Figure 2.6 Diagrammatic of the Gateway Procurement Lifecycle: integrated process

Based on OGC (2004) Achieving Excellence in Construction Procurement Guide[13]

budget time and quality constraints. A project manager is the key person in providing good advice on the most suitable procurement strategy and in developing a master plan for the delivery of the project to the customer's needs and for the best value.

The system in Figure 2.6 which has been used by the UK government in promoting better procurement efficiency would suit some types of non-traditional procurement.

Value for money is increasingly involving an appraisal of whole life cycle costing (WLC), which analyses the cost between capital cost (CAPEX) and operating and maintenance costs (OPEX). This puts into question the acceptance of the lowest capital price tender and forces clients to think through the impact of a building's design, energy costs, location and financial benefits as a single entity with the capital value of the building created. The true consideration depends on reliable figures being available and the value of WLC to the client will vary according to their post-contract interest in the building. Some procurement options such as prime contracting and DBFO put the CAPEX and OPEX risk with the provider and this is another driver for WLC appraisal. OPEX costs are also recognised as needing to be sustainable in a climate where scarce resources are being increasingly recognised. The process of value management and life cycle costing are further discussed in Chapter 9.

Project stakeholders

Stakeholders are those who have an interest in the project process or outcome. The obvious parties to the contract have an interest in the *outcome* of the project. The client wants to get a building which meets expectations, therefore generating expected returns, the member organisations of the project team want to make good returns, gain experience and build a reputation to get further work and users/employees want a resource which is comfortable and convenient. Those interested in the *process* might be community based and the employees of the organisation. The community want a minimal disruption for neighbours, courteousness by project workers and an interesting project, which employs people or gives something back to the community.

Stakeholders are often classified as internal or external. Internal stakeholders are defined as members of the project coalition and include the project delivery team, the client, suppliers, users and those who provide finance and insurance. External stakeholders are those outside this inside circle that have a stake in the outcome of the project or may be affected by it in a significant way. This can be shown in Figure 2.7. They are unlikely to be contracted into the organisation, but bring pressure to bear in indirect ways such as planning appeals or raised insurance premiums.

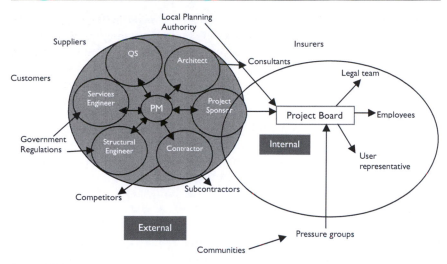

Figure 2.7 Internal and external stakeholders

As a project manager working on behalf of the client and responsible for the project dynamics, both the client's stakeholders and the project's stakeholders become important to manage. The client's stakeholders, which are established at verification of need stage, need to be made known to the project manager and will influence the project management approach indirectly. We will mainly deal here with management of the client's stakeholders, but some of the stakeholders are related to the execution of the project and not to the business need.

In addition there are many external stakeholders who have a direct or indirect impact on decision making and consultation and it is wise to get to know their views. In the community these amount to neighbours, users of services affected, local authorities including planning authorities and regulatory enforcers such as building control, highways and social services. Developers can expect a lot of discussion prior to the planning phase to determine planning gains and highway adjustments to reduce traffic impacts of major development.

Johnson and Scholes have developed a mapping matrix shown in Figure 2.8 which helps to identify what information should be available to different stakeholders based on their interest and also the influence and therefore the priorities which should be put on managing the stakeholders.[14] They suggest that there are different requirements for management of stakeholders depending on their degree of influence (power) and significance, their predictability and interest.

Newcombe suggests that there are different requirements for the management of stakeholders depending on their degree of predictability as well as their interest.[15] There is a need often to deal with the conflicting requirements of stakeholders – so who do you satisfy and who do you disappoint? A project manager needs to satisfy the client's ultimate interests, which means making decisions to ensure the project's efficiency objectives don't obliterate the business effectiveness objectives.

Power may issue from a stakeholder's ability to take action which would be helpful or detrimental to the project outcome. A good example is the Planning Authority who can hold up or stop the project and also any legally enforceable action such as compliance to fire regulations. These people may not have excessive interaction, but may have significant adverse impacts.

Interest issues from the effect on working relationships and the amount of interaction that there is between the stakeholder and the project team. These can have a slowing down and souring effect which affects the productivity of the project where information is unavailable or unreliable. An example of this type of stakeholder would be a key specialist contractor who provides less than full resource requirements.

Unpredictability is an important aspect to manage as 'sleeping interest' may emerge where no concerns existed previously. What they ask for may be at a late stage and therefore become disruptive and powerful stakeholders who are unpredictable, for example planning committee members who disagree with planning officer views, can delay programme and escalate cost or even stop a project.

The community is also a stakeholder and may be able to have considerable power where a group is able to obstruct progress. A client may feel this pressure and make instructions for change as a powerful stakeholder. A project team will be affected by delays, for example archaeological finds. Case study 2.4 covers a real case.

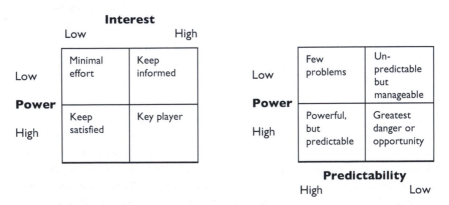

Figure 2.8 Matrix for mapping stakeholders

Source: Based on Johnson and Scholes, and Newcombe[16]

Stakeholder conflict management

A project needs to be further defined by reference to those who will later have an interest in it. The briefing process is already complex and developmental in order to take on board the constraints, available resources and opportunities that arise during the developmental period. Hence it is important at this stage to get past mapping and simple consultation to practiced engagement with multiple parties.

Stakeholders and users in particular need to be handled carefully if they are involved in the briefing process. Focus groups are good, but if the focus groups are not handled carefully then different users will get into conflict with each other and compromise may not be easy. If the wrong questions are asked – 'What would you like?' – then expectations can be raised above the ability of the budget or other constraints to meet. It is better to err on the negative side.[17] Conflict management is another aspect of stakeholder management as any change made due to one stakeholder's preferences may be against the interests of another as Case study 2.4 shows. Opportunities arise at a subsequent stage which impact on funding as Case study 3.1 shows in the next chapter.

Case study 2.4 Stakeholder conflict

The proposal for sports facilities in the extension plan for a university revolved around the conditional provision of a community swimming pool based on shared cost between the university and the local authority and would become available for community use also. With the opportunity of a new football stadium built on university land for no extra cost and the facility for university football and rugby matches free, the provision of a swimming pool was seen to be surplus to requirements when there was no local funding forthcoming for the pool.

How is this managed? Which stakeholders will gain at the expense of others? Will both facilities be expected? These questions could be brought back to stakeholder consultation.

User stakeholders present a more social set of requirements from internal stakeholders who will have clearly defined their business need.[18] This requires a different type of consulting and management which is based more on listening, innovation and mutual acceptable compromise. The proximate siting of wind farms or telephone transmitters near communities is a typical example of this due to strong feelings involved. Specific problems and objectives must be identified and divorced from subjective statements by appointing representatives to resolve conflict.

Case study 2.5 Community stakeholder pressure

When a water company put in its new infrastructure for sewage treatment in Weston-super-Mare, there was an assumed improvement in the handling of waste products with a new cleaner discharge from the treated sewage. However because this discharge was now deemed suitable to discharge into the sea a pressure group acting on behalf of surfers became concerned with the quality of the seawater that they were now operating in. Information was developed to reassure these stakeholders and to limit some of the discharge. Subsequently new work is to be done to improve discharge quality and also to limit raw storm water discharge by storm water settlement tanks in more serious storm weather and to meet the previous rather than the minimum bathing water standards. There is also a new water habitat for voles to be built on the site as it was built on important ecological grassland.

Minor stakeholder views may be seen as important in the long term as illustrated in Case study 2.5.

Stakeholder analysis will help project managers:

- understand the differing impact of stakeholders and the conflicts between them, though some will always be dark horses especially where they are little known
- harness support from powerful stakeholders
- meet project mission and objectives
- understand community and society impacts.

Through this project managers begin to develop a strategy which will be most beneficial to the client and make sure they satisfy the key players and manage risk and uncertainty better so that unpredictable actions are reduced.

Managing stakeholders

Managing this phenomenon is important and consists of a mixture of selling the project objectives, satisfying the customer and compensating the stakeholders who have become dissatisfied. Making the logic of the project clear (selling) is a longer term solution as it attempts to draw stakeholders behind the project and saves abortive resources by being preventative rather than reactive. Compensating stakeholders may be expensive, but may bring a better project solution if other benefits are given for lost amenity.

Chinyio and Olomolaiye point out that there are multiple ways of categorising and mapping project stakeholders and suggest that a multi-dimensional plot is needed to catch the full complexity of the interactions of stakeholders before they can be understood and managed.[19] One particular management requirement to bring the negatively and neutrally positioned stakeholders onside is Maister's first law of service.[20] It is stated as:

satisfaction = perception − expectation

It is a good benchmark for managing stakeholders, as it has an effect on the initial planning. This law distinguishes between what the stakeholder sees as a product outcome or a level of service with what they were expecting from the project. In achieving stakeholder satisfaction it is not enough to

Case study 2.6 Internal stakeholder, relocate HQ

A major insurance company makes a strategic decision to move its office from the city centre to a peri-urban position to gain more efficient running expenses, pull all personnel together into the same building and make efficiency savings. It also took tax breaks being offered by the new local authority to gain employment opportunities. The customers are not affected as most business is done remotely, the business becomes more efficient and so produces better returns to shareholders, but the existing employees are mainly disappointed because many of them have to travel further and do not have access to the benefits of city centre shopping and amenities in their lunch breaks and after hours. The local community is residential and is extremely worried that a large new facility in the vicinity will cause even worse traffic congestion and parking problems and loss of visual amenity and green fields.

What does the company do?

In their new building plans they offer a gymnasium and sports amenities and ample car parking space and an interesting 'street' environment to compensate the employees. They also develop their grounds to provide a lake, ample planting with walks and picnic sites that will provide alternative lunch time activities and encourage wildlife to stay in the area. This will both partly satisfy the local community and satisfy the Local Planning Authority, who have qualified the planning permission to meet local objections in the planning appeal. The shareholders have suffered no long-term loss to their returns and the solution is integrated to partly alleviate loss to the other stakeholders.

Case study 2.7 External stakeholders for a new swimming pool

A well-publicised objective for a local authority (LA) to supply a 25m swimming pool during 2003 to supplement sports facilities in the area may have helped them to be elected, but the programme slipped to 2004 because of budget availability. In this case the stakeholders are the council tax paying public, neighbours and the LA.

When they open the swimming pool in the summer of 2004 they decide that they will compensate by fitting out a new gym suite for fitness training to offset the disappointment of being late. This came out of earlier local consultation during the planning stages which was not promised in the final scheme. Stakeholder satisfaction has been achieved by exceeding outcomes and offering people more ownership by consulting them. A basic 25m swimming pool in autumn 2003 might have got them re-elected. A basic pool in 2004 would have caused frustration and possibly dissatisfaction because people would have made plans and have had to alter them.

The LA budget is a major constraint to the project. They wanted to include the fitness gym suite, preferably within the same budget. Budget savings may be made by considering one or more of the following:

- a value management exercise to question other requirements in the brief, such as the amount of car parking and the orientation of the building, which may reduce the length of the access road
- a life cycle approach would take the opposite view to achieve reduced running costs and increased component life, and this could be used as an argument for an increased capital budget
- a later cost cutting exercise would, in contrast, reduce the roof specification and the thickness of the tarmac and generally reduce the quality and increase the running costs of the building.

The first two now become a specific project design issue and must be levered into the brief development stage. In terms of managing stakeholders the client needs to be directed to make these decisions early.

perform satisfactorily as this will simply neutralise expectation and provide zero satisfaction. Stakeholder management is about making the sum positive by exceeding expectations in time, cost and quality. In practice this means advertising the core objectives that will be the main outcomes and what must be achieved *and* setting about achieving some extras.

From a contractor's point of view competitive tendering leaves no room for optional extras for the client, which leaves the area of service as the most powerful way of exceeding expectations. It is also clear from the second Egan Report (2002)[21] that the exceeding expectations principle is clearly enshrined in their vision statement so that customer value is optimised. Negotiated tenders (also espoused by Egan) leave more room for the client to arrange a 'win–win' situation.

Certainly research (Office of Fair Trading 2008)[22] is clear that the way in which a service is provided and the importance of being able to do what you promise is most important in keeping stakeholders on your side. Continuing service improvement is also important in keeping the stakeholders satisfied. Case study 2.6 refers to the efforts of the client to satisfy internal and external stakeholders in moving their headquarters (HQ).

Case study 2.7 looks at the controlling factors for a public client and the need to manage the various aspects so the satisfaction and project constraints are kept in balance.

Conclusion

The business planning process begins before the inception stage of the project and informs the client about the feasibility of the project in outline terms. The briefing and ongoing development of the brief take place in the next phase of feasibility testing when a solution is engineered by clearly communicating the project objectives and reconciling them with project constraints. This stage optimises the value and reviews the effect on the stakeholders to mitigate the conflicts which may arise.

Egan and the CCG have issued reports which indicated that the integration of the client into the construction process is a necessary and not an optional development that must be made if value is to be built into the process. This has brought about the wider use of negotiation during the procurement process in order to make the client's objectives more accessible to the whole supply side and build in value. The integrated project team is seen as the maintenance of supply chains from project to project so that the team are not always learning on the job. This depends a lot on the greater involvement of the client in choosing limited partners for repeat work and naming suppliers and a lot of work needs to be done to convince one-off clients to be more involved. It also means making a strategic move away from the single stage competitive tendering with selection of the lowest price. This system ignores other aspects which bring value to the business case such as earlier finish, guaranteed fixed prices, flexibility, sustainability and lesser life cycle costs. More strategic long term relationships are possible by adopting forms of contract that allow more direct contact between the contractors and the client such as prime contracting, construction management, design and build and PFI where appropriate.

Stakeholder mapping and management is seen as increasingly important as external stakeholders and building users can impact unpredictably on the project outcomes if they are ignored. One way of managing is to make some concessions or compensations as a clear negotiation for their support to the main project objectives. Sustainable outcomes may sometimes clash in the eyes of different stakeholders and the social outcomes may have to be varied to provide community gains which are not partial to one party at the expense of another.

References

1   The Office for Government Commerce (OGC) (2004) *Online Gateway Review*™ *and Achieving Excellence Guides*. UK, OFT.
2   Construction Clients Confederation (2002) *Charter*. Available online at: http://www.clientsuccess.org/WhatsInvolved.htm (accessed 4 March 2012).
3   Egan J. (2002) *Accelerating Change*. UK, Strategic Forum for Construction.
4   Maylor H. (2003) *Project Management*. 3rd edn. Harlow, Pearson Education.
5   Masterman J. and Gameson P. (2002) *Client Characteristics and Needs in Relation to their Selection of Building Procurement Systems*. Proceedings of CIB 92 Symposium East Meets West: Procurement Systems.
6   Green S. (1999) 'Partnering: the propaganda of corporatism', *Journal of Construction Procurement*, 5(2): 177–186.
7   Egan J. (1998) *Rethinking Construction*. UK, Department of Environment Transport and the Regions; and Egan J. (2002) *Accelerating Change*. UK, Department of Trade and Industry. (This is a progress update on the Egan targets set in 1998.)
8   CIOB (2010) *Code of Practice for Project Management Construction and Development*. 4th edn. Oxford, Blackwell Publishing, pp.3–4.
9   Construction Clients' Group (2008) *Client Commitments Best Practice Guide*. UK, Constructing Excellence.
10 Strategic Forum for Construction (2008) *The Construction Commitments 2012*. London, Strategic Forum for Construction.
11 Lane K. (2011) 'Giving transformational change a better name', *Project Manager Today*. March.
12 OGC (2004) *Achieving Excellence Procurement Guide, Project Procurement Lifecycle: The Integrated Process*. UK, OGC.
13 OGC (2004) *Achieving Excellence Procurement Guide, Project Procurement Lifecycle: The Integrated Process*. UK, OGC.
14 Johnson G. and Scholes K. (2002) *Exploring Corporate Strategy*. Harlow, Pearson Education.
15 Newcombe R. (1999) *Stakeholder Management*. Keynote Address to the CIB65 Conference on Customer Satisfaction. Cape Town University.
16 Johnson G. and Scholes K. (2002) *Exploring Corporate Strategy*. Harlow, Pearson Education; Newcombe R. (1999) *Stakeholder Management*. Keynote Address to the CIB65 Conference on Customer Satisfaction. Cape Town University
17 Barrett P.S. and Stanley C.A. (1999) *Better Construction Briefing*. Oxford, Blackwell Science.
18 Rittel H.W.J. and Webber M.M. (1972) *Dilemmas in a General Theory of Planning*. Working Paper No. 72-194, University of California, Berkeley, CA.

19 Chinyio E. and Olomolaiye P. (2010) 'Introducing stakeholder management'. In E. Chinyio and P. Olomolaiye (Eds) *Construction Stakeholder Management*. Oxford, Blackwell-Wiley.

20 Maister D.H. (1993) *Managing the Professional Service Firm*. New York, Free Press.

21 Strategic Forum for Construction (2002) *Accelerating Change*. DTI. A document which reviews and updates the progress since the Construction Taskforce (Egan) Report (1998) *Rethinking Construction*. UK, DTI.

22 Office of Fair Trading/Ipsos Mori (2008) *Consumer Detriment: Assessment of the Impact of Consumer Problems with Goods and Services*. April. London, HMSO.

Chapter 3

Project definition

Every construction project needs to have a clear briefing which is realistic and derives from the client's objectives and satisfies the business case. The initial brief at inception is developed in the feasibility and strategy stages. The concept of *project definition* is a managed stage trying to reach an agreed design scheme and methodology. The main objectives of the chapter are:

- determining the elements of project definition
- mapping the construction process
- assessing project feasibility and affordability
- managing the project scope
- dealing with external factors
- balancing risk and value and allocating risk
- project evaluation techniques.

The design brief

Gray *et al.* talk about three distinct types of knowledge controlling design.[1] These originate from:

- the client in the early stages of inception
- the individual designers in the concept and scheme design stage
- the design/construction manager in the detail design, specialist and construction stages.

The first two types of knowledge are particularly relevant to project definition, but by no means exclude the third type.

Hellard sees that there are four possibly conflicting elements to the brief:[2]

- Function: technical and physical requirements to meet the business case
- Aesthetic: satisfaction of human subjective aspects
- Cost: both capital and running costs
- Time: the logistic requirements for commercial completion and occupation.

The client may wish to determine some or all of these elements in the outline brief, depending on the degree of innovation and flexibility the client wishes to give to the team.

Product scope definition

The CIOB Code of Practice recognises project scope as a key requirement of the outline project brief, but this is only the starting point for scope management. During project definition it is useful to state what the brief does not cover, as an aid to setting the boundaries for the project.

The PMI BoK identifies five stages of scope management which are client brief, scope definition, work breakdown and control, change control and scope verification. The key issue here in project definition is determining the scope. Scope refers to two aspects:

- Product scope – the definition of the product features and functions. This is provided in initial form by the client and is developed by the project manager and the design team. It is closely connected with the design brief.
- Project scope management – refers to the work which must be done in order to deliver the project with the features and functions specified and is the prime concern of the project management team, guided by client and project constraints (covered in Chapter 4).

Product scope may be in prescriptive or performance terms. The more the brief is described in performance terms, the more there is room for developing it and adding value. This gives a sliding scale in construction from a client who describes their business requirements such as 'produce 100 cars per week', to a client who hands over the drawings defining the location, the building type and preferred material specifications. In practice a client will use their experience to specify key production components and put building design in the hands of a team of specialised designers.

Determining the elements of project definition

Project definition is carried out in the period from receiving the performance specification during the inception stages up until the receipt of full planning permission. This is the right stage for a value management workshop. Only then can a working brief be established.

The RIBA Plan of Work lists work stages to reach the planning application stage when the client's approval allows a scheme to be submitted. These stages are strongly related to the traditional forms of contract and procurement and assume a two-way relationship between the client and their designers. In Figure 3.1 the RIBA plan has been linked to the relevant project life cycle stages.

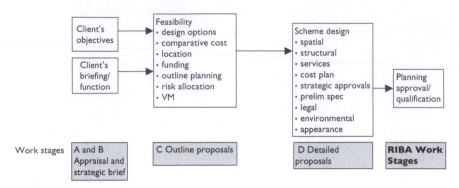

Key

A Appraisal is the identification of the client's requirements and possible constraints on development and feasibility. It considers a probable procurement route.

B Strategic briefing preparation by confirming key requirements and constraints. It also identifies organisational structure and the range of consultants which might be appointed.

C Outline proposals, taking account of any feasibility studies which have been produced. Estimate of cost. Review of the procurement route.

D Detailed proposals: complete development and completion of the detail brief. Application for full development control approval.

Figure 3.1 Development of brief showing early RIBA Plan of Work stages

Source: RIBA (2001)[3]

The RIBA work stages provide a managed process with a greater emphasis on the design process. Traditionally the architect has led the management of the project definition stage and this may vary with different procurement systems. It is important also to consider the cross over of the strategy stage with feasibility and to understand that the programme, the client funding cash flow, risk and value assessments and organisation of the project may have important impacts on the design and feasibility. The normal way to proceed with the development is to include option appraisal or an iterative process to develop the design based on the client and site constraints considered.

The outcome should be tested to most effectively deliver client business case whilst keeping capital, maintenance and operating costs efficient. The RIBA considers that the client should freeze information from the brief at Stage D, Detailed proposals, freezing the design at the end of the stage.[4]

A fuller picture

The CIOB Code of Practice therefore talks about an outline brief moving towards a detailed brief and adds a strategy stage as in Figure 3.2.[5]

The outline brief should cover the client's objectives, functional requirements and project scope business constraints. The development of the

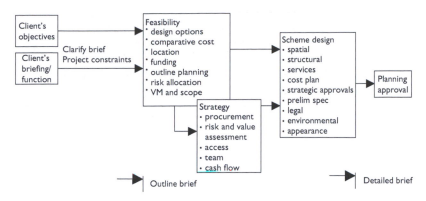

Figure 3.2 Project definition process

brief is a critical stage at which the scope of the project is properly established and verified by the project constraints (Chapter 2), which ultimately leads to the planning permission to proceed. This stage is iterative and requires the involvement of the client with the designer and ideally, if procurement method allows, the involvement of the construction manager. The detailed brief provides resource requirements, risk and value studies, costing, site constraints, feasibility and a full planning application preferably discussed with the planning officer. At the end of the project definition stage the detailed project brief should have determined scope as fully as possible, because beyond this stage changes to scope become much more expensive. Sustainable design and an indication of its impact on carbon reduction must be included and are crucial to planning consent.

The crossover between strategy and feasibility provides an axis for checking between the process and the product. It means that:

- Design development to be cost checked informs the designer of the impact of design change on budget.
- Project organisation and communications are properly considered.
- Procurement method can be evaluated to suit the client's unique requirements.
- The impact of design on construction method and construction time can be assessed.
- A master plan for the key time constraint of programme and cash flow is in place.
- Risk is identified and allocated.
- Best value is managed.

The project manager has a key responsibility to co-ordinate the whole process and to ensure that the strategic/feasibility issues are properly integrated with the development of the design proposals. This is also recognised as a separate skill by the RIBA. The following are managed in relation to the external environment:

- clarification of the brief
- feasibility and affordability
- scope management, change and contingency
- risk assessment of external factors
- funding and location
- design management
- stakeholder management
- organisation and culture.

In the Gateway Review™ system project definition comes between gateways 1 and 2 and identifies the production of a feasibility study, a procurement strategy and an output based (performance) specification.

The APM BoK focus on the project definition stage comes under requirements management.[6] They define this as the process of capturing, analysing and testing the documented statement of stakeholder and user wants and needs. Requirements are a statement of the need that a project has to satisfy, and should be comprehensive, clear, well structured, traceable and testable.

Managing the client

Failings are often connected with the perennial problems of people relationships and good practice often fails because of this. Barrett and Stanley studied a number of problems with briefing for construction projects and suggested five key solutions which went beyond normal good practice.[7] These were empowering the client supported by appropriate user involvement, managing the project dynamics supported by appropriate team building and *appropriate* visualisation techniques that supports both. Visualisation was an example of 'good practice' failing; for example, 3D walk-through without customised context to the client was often more confusing to the client than photographs of previous buildings. Another statement they made in the context of not knowing was the need 'for the briefing process to manage the client through this journey from uncertainty to certainty'. This requires project managers to envisage the client's business needs in assessing alternatives and values that are critical.

In getting to know the client, Boyd and Chinyio talk about the uniqueness of different types of clients and in particular their reaction to change which is represented by the commissioning of a construction project.[8] They recommend a process consultation with the client which helps to see how they react to the industry, what their main business concerns are, how they manage conflict and uncertainty, what drives their objectives, what frustrated them and what they consider is a successful outcome. Their research indicated that different types of clients have different levels of uncertainty and different means for dealing with it.

This knowledge of the client may be useful in the physical, emotional and psychological process of good project management. Process is an important area of concern and leads to understanding client values and the ability to optimise the value of the project. More is written on this in Chapter 8 on value and risk management.

Process mapping

Process mapping is used to model and test the system that is intended. In projects it is used to make sure that all operations carried out in the project life cycle are planned and covered. Formally this can be drawn as a flow chart indicating the activities (in rectangles) and key decision points (in diamonds). It is often connected with a series of different pro forma documents. Electronically these are accessed as links across the intranet or the project extranet, making it a good communication tool and knowledge base across all members of the project team. Problems which arise are related to the compatibility of systems and the shortfalls which arise in the degree of information accessible. Sophisticated systems may access comprehensive external sources through knowledge portals. More integrated internal systems may be linked to a knowledge management intranet, which identifies a wide range of experience and resources within the organisation.

The advantage of the formal system is that it works well in quality assurance. The disadvantage is that the system may become inflexible and discourage innovation and improvement, though it is possible to build in review and improvement stages. Case study 3.1 discusses an FM organisation carrying out maintenance, and smaller projects not exceeding £300,000 for a client who has several complex facilities at different locations.

Process mapping is a starting point for quality improvement systems because it allows effective communication systems and also can easily incorporate the measurement of KPIs to assess productivity so that different project teams may be compared and lessons learnt. Suggestions for adjustments to the system should always be encouraged from the workforce so that they feel involved and do not simply tolerate flaws, inconsistencies or blocks in the system.

Analysis

The case study process mapping indicates the integration of the approval process with the design process. This indicates a need to have good communications with prompt responses for the project to proceed efficiently.

Case study 3.1 Managing the estate

An FM company operates a three-year rolling contract extendable on the basis of performance. The value of work on one of their sites is £5.5m per year on which they are paid a fee. The amount of work may vary each year and the fee adjusted on a pro rata basis. Their work is broken down into maintenance and projects. The proportion by value is approximately 50 per cent for projects and 25 per cent each for responsive and planned maintenance. There are approximately eight managers and technical officers are employed to supervise projects. There are 33 documents, which are used to progress the project from inception to completion, and also statutory forms to cover building regulations and the CDM Regulations procedures. Contract information has been grouped together in a process map which covers the areas of quality assurance, health and safety, programmed work, planned maintenance, emergency callout, non-programmed work, financial management, works management stores and resource management.

The system is linked directly then to each of the relevant forms in order to ensure complete use of the system. Looking specifically at the projects the process map could be constructed as follows. There are four key players and two audit agencies. The process is mapped in Figure 3.3.

The process map in Figure 3.3 indicates four specific approval stages from initial client request up to the pre-start meeting and indicates also the responsibilities. This map could also be linked to the specific contract pro formas that are relevant at each of these stages so that relevant checklists are processed and approvals signed up. There should also be an opportunity for review of this process. For example when the job is logged, there is a checklist of 11 items for the extent of the design services covering such items as:

- Does the brief need amplification with the client?
- Does programme need amplification with the client?
- Is contact with the utilities required?
- Is a policy on hazardous material required?
- Has a planning supervisor been appointed?

There is also a sign off for the procurement manager to consider before the subcontractor starts work. This is called the initial quality check, which covers, for example, safety checks, statutory requirements and use and protection of existing services.

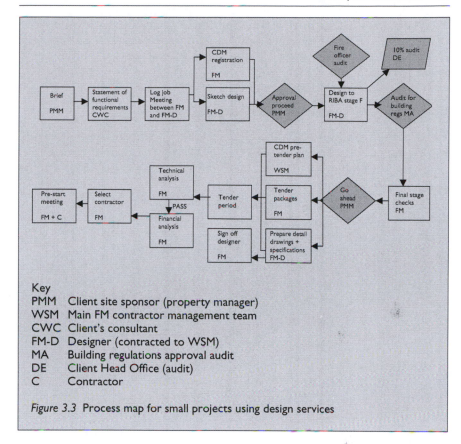

Key

PMM   Client site sponsor (property manager)
WSM   Main FM contractor management team
CWC   Client's consultant
FM-D   Designer (contracted to WSM)
MA   Building regulations approval audit
DE   Client Head Office (audit)
C   Contractor

Figure 3.3 Process map for small projects using design services

It is a useful system to have where there are repeat projects and there is a single designer and some regular subcontractors, as in the case study. Where there is a broader range then a more flexible system may become necessary to cater for different reporting and approvals in delivering the final project brief.

The case study mapping has benefits for ensuring that a system is adhered to, but decisions should be made when adopting a quality assurance system, so that bottom up improvements may be built into the system otherwise frustration will set in. This system has two feedbacks. One is from the customer in the form of a customer satisfaction annual survey, inviting the client to provide a scoring of 1–10 against different characteristics of the service given. These were measured against tangibles, reliability, responsiveness and other questions about the assurance engendered and empathy with the customer such as making the effort to know the customers' needs.

Another form was called 'An opportunity for improvement' referring to the ISO 9001 quality system in place and to identifying technical improvements or cost savings relevant to other contracts. Members of staff at the 'coal face' take the initiative and responsibility for filling the form in. The attractiveness of the form is that it invites staff to propose a solution as well as identify a problem or query. To promote use, responses should be made to the initiator.

Building information modelling (BIM)

The Wolstenholme Report (2009) identified that on average between 28 and 58 per cent of construction projects came in on time and in the 10-year period (improving) since the Egan Report 1998 between 48 and 50 per cent came in on cost or better.[9]

Building information modelling is a sharing of information through the design, construction and whole life cycle of buildings, managing it in an integrated way by object modelling. An object corresponds directly with a component or element of a building. It integrates layers of information starting with a 3D design model which is broken down into objects such as a wall or floor or a window or roof. Other information relevant to the objects such as information on cost, time taken for construction, options for specification and maintenance information can be allocated in a single object. These objects provide common 'building blocks' for passing information between different parties such as the architect, civil engineer, contractor and other consultants. Conventionally complete designs have to be considered to do complete cost checks and handed to the contractor to do complete programmes. BIM allows all sources of information to be standardised for multiple solutions or option generation. Its key attribute is a much more rounded delivery of information answering feasibility and value related questions. Its major problem is that information is not currently available across the board in the form required and has to be generated, meaning it will take time to build a wide data of standard information.

The power of BIM is that when compiled with other modules it can provide standardised building or design options with costing and programming and maintenance, but may be used to comprehensively assess the impact of a change in one or more of the information source areas to optimise design, method and value. Options may be tested in terms of resource alternatives and standardisation may be developed and incrementally improved in prototype. BIM software and preplanning needs to be cost effective, but according to Wolstenholme, and Barrett and Stanley,[10] traditional solutions fall far short and meet unpredictable outcomes in areas such as time and cost which can generate huge wastage. With regards to briefing there is a less tangible, but significant, benefit in terms of the communication vehicle which could be a powerful tool for communication, empowering the client and managing project dynamics more effectively in its essence as an appropriate visualisation tool and in its dependence on team building.

Claimed benefits for BIM vary from stage to stage in the project life cycle.[11] In the design stage they are likely to produce 8–10 per cent savings because of understanding and spatial co-ordination. During the technical specialist stage of design 8–10 per cent savings can be made on the co-ordination of the trade contractor in tendering and the workshop design post tender where this is complex. In the mobilisation and construction stage this level of saving should be maintained because of the transparency of the design with construction method and the 4D aspect of adding time issues and getting productivity co-ordination between different trades on site. Health and safety should be improved in the planning stage. In the operational stages there is little data, but strong expectations of early co-ordination with FM software to manage the assets knowledgeably and reduce energy and other resource usage and prevent work stoppage. In all cases duplicate information can be avoided by not re-entering information. BIM has the power to gain value in excess of its cost, certainly in the long term. In the short term it requires a culture change with associated training and software costs.

Case study 3.2 is a contractor generated hospital BIM and 3.3 is the remodelling of Manchester Central Library and involved the advanced modelling of an existing building.

In the UK it is proposed that improvements could eliminate mistakes and save time by a better flowing integrated information system and suggested that larger public projects should all use BIM by 2015. This was adopted by the UK government for all significant public projects, not just large ones.

BIM is not without its problems. It is susceptible to 'rubbish in, rubbish out' and this rests in the evolutionary development of reliable information sources for the relationship between time and methodology, cost and design in very wide component ranges and choice. It also is expensive to install compatible systems and to train people to use them consistently with good inter-operability across the whole supply chain (see Chapter 12). Willingness and ability of diverse parties to make their inputs compatible with a different

Case study 3.2[12] **BIM hospital**

Work by a PFI contractor on a major hospital in London claimed savings on design co-ordination reviews, automatic material takeoffs, access time to information and site records administrators. This was offset against the cost of two BIM specialists, IT hardware and software and database development. They used BIM for virtual briefing, clash prevention, safety reviews, room data sheets, 4D construction planning, room progress monitoring, medical equipment and plant verification and material takeoffs. Tablet computers were used to bring the system to the workface.

Case study 3.3 Manchester public library

This project has invested in a digitised measurement of an existing historic building which was loaded into the BIM software. This template was then circulated to the architect to use as a basis to remodel the space layouts to suit a more modern use of the library. These remodelled layouts were then overlaid with the additional structural solution and services and interactively tested with the architectural layouts to give an integrated 3D solution. This model has automatically eliminated any clashes and has proved an accurate source for the addition of detailed specialist design. As all the rooms are modelled together CAD FM models are also to be generated to identify the inventories and dimensions for room data sheets for fitting out and maintenance. 'What if' scenarios can be carried out to optimise the integrated design.

system for each project depends on more standardisation especially for small jobs where the proportionate set up costs are otherwise very high for each project and may not get obvious payback. BIM may depend upon the resolve of knowledgeable clients to insist on compliance. A transition process is obvious with client champions and incentives developed to gain momentum.

Feasibility and affordability

Looking at it proactively, feasibility seeks a solution that is possible within the applicable constraints, whilst meeting the client objectives. Investigations are carried out that will give an overall picture of the costs and constraints of the project and whether it is viable.

In the CIOB Code of Practice (2010) there are several issues that are mentioned as part of the feasibility stage.[13] These are risk assessments, public consultation, geo-technical study, environmental impact assessment (if applicable), health and safety study, legal/statutory/planning constraints or requirements, estimates of capital and operating costs, assessments of potential funding and potential site assessments. These studies will give an overall picture of the costs and constraints of the project and whether it is viable.

Two different types of feasibility

At the outline business case stage the client, before inception, considers the viability of the project – whether they need the project and whether there is a suitable return to make an investment in bricks and mortar. This will often take the form of a development appraisal where the estimated costs of the development are weighed against the income and benefit of the investment and over what period they will break even and make a return to the business. It will assess whether the investors' minimum profit margin has been generated. Profits must justify the risk and the additional effort of investment. There is little difference with a non-profit or public organisation. We shall call this affordability.

At project feasibility stage the information originates from the outline project brief and the business case. Essentially it is a second stage of feasibility and looks to either optimise value within the parameters which have been set or provide an option appraisal of different designs, locations, funding and methods. Viable alternatives can now be assessed against client values. For our purposes we shall call this feasibility.

Client value is an important concept in feasibility and refers to the underlying beliefs of the organisation and may also be emphasised in the individual beliefs of senior managers responsible for the project. Client values determine the priorities and underlying rationale for the decisions taken. Typically the balance between image and aesthetics and function/utility will determine design acceptance. Other issues are the degree of weight that is afforded to environmental issues and sustainability. Political and stakeholder values will also have an influence on the design and choice.

Public view – cost benefit analysis

Cost benefit analysis is a way of taking into account factors other than income that are included in the wider term of benefits used to justify choice. For example, a bridge may cut down the journey times of thousands of commuters and save money and time to the economy. Financial costs are recognised, such as the cost of the bridge and maintenance, and non-financial costs should also be recognised such as loss of environmental facility and homes blighted by additional noise. Benefits are recognised as cutting down journey times for thousands of commuters and the return on tolls. A cost benefit analysis is more relevant in the investment appraisal of a public project that offers social intangible facilities and would otherwise not produce an accountable return. At inception stage the main consideration is that there is an acceptable business case taking into account direct and indirect benefits.

Cost benefit analysis is typically viewed as a soft management approach to the issue of feasibility, but in public projects, which are not driven by commercial concerns, the benefits have to be justified. If comparable projects are to be assessed then it is important that factors included are formally validated as the same for all projects and intangible benefits have a standard valuation in comparable projects or options. Different benefits may cross over and care is needed not to double count. The system is categorised and valued as in Figure 3.4.

Direct costs £120 000	Direct benefits £90 000
Indirect costs £10 000	Indirect benefits £60 000

Figure 3.4 Cost benefit matrix

If the costs and benefits are as the matrix (Figure 3.4), then:

Costs = 120,000 + 10,000 = £130,000

Benefits = 90,000 + 60,000 = £150,000

Then the net benefit = £20,000

The example in Table 3.1 indicates the way in which costs are balanced against benefits for a flood defence scheme. Direct costs and benefits are those which are incurred from the building and direct revenue for the service. Indirect benefits are those accrued to third parties such as the benefits for owners now being able to insure or sell their properties. In practice a lot of weight is put upon well populated areas and less urgency. Therefore benefit is put on scattered populations.

For flood defence work there is a government guide for flood defence, identifying value for benefits such as the number of properties which would be saved from flooding by major proposed flood defence works, with a weighting put against factories and offices. Currently agricultural land gets a very low weighting. There are also some measurable social benefits, which may be relevant to certain schemes in order to make sure that urban schemes are not the only schemes that get the go ahead on the basis of a priority list and scarce resources. Third party costs are those that the Environmental Agency incurs in dealing with floods.

Table 3.1 Comparative examples of the elements of cost benefit analysis

	Benefit	Cost
Tangible (measurable in financial accounts)	e.g. Homes protected from flooding Increased value of houses	Construction costs, land, fees compulsory purchase etc., maintenance
Intangible (measurement of financial worth requires discretion and may vary according to client values)	Better wildlife habitats Gain on prime development land not flooded	Loss of developable land and visual amenity

Cost benefit analysis and value

At feasibility a public body may carry out a cost benefit analysis where there is a shortfall between the cost and income. This should be a rigorous and realistic assessment of the financial worth of the benefit and the cost of intangibles. The Millennium Dome is an example where a grossly over-optimistic assessment of the visitor figures was made and there was justified sharp criticism of the decision made. On the other hand the Sydney Opera House went badly over budget to a factor of 10 and has paid for itself over and over again as a perennial visitor attraction and income earner.

The difference between the public and private developer is the worth of the benefit which is used to weigh the balance. In the case of a sports centre for a private developer the income from fees will offset the building costs, consultant fees and land. However a private developer will pay substantial premiums for residential land near water – here there is a direct benefit from the saleable price. An indirect benefit is the regeneration of contaminated land around old docks and recreational access without public cost. A public body can price the benefit of providing the community with a social amenity which meets council objectives and reduces crime amongst young people. In both cases the project is designed at minimum cost which will allow the appropriate level of performance and levels of safety and meet the objectives. The objectives are different.

In the case of social cost this is often transferred to the private developer by the legal system. Under the planning system, a new housing estate may have to provide a proportion of affordable housing, contributions towards a new school, a better road junction or open grassed areas. Under environmental law a new owner is responsible for clearing up site contamination and may have to provide acoustic barriers, holding ponds and flood defences, and actually improve visual amenity.

In many cases it makes sense to set up a public private partnership for developments which are more socially beneficial. Here a developer is offered an incentive such as tax breaks for development in unemployment black spots. Or public money is provided for infrastructure improvements that make the area newly accessible and clear up major contamination. The public contribution is based on the final value to the private party – a host of regeneration monies are available for former industrial and contaminated areas.

Funding and investment appraisal

A balance between value and development cost should be achieved to make a project viable. This may be carried out either by measuring cost and assessing the value, or alternatively by calculating the end value and working out a budget to suit. In either case the value must exceed the cost to justify the project. The method and source of funding is important in order to assess cost. There are several tools that may be used to assess this on construction projects.

Payback method

This is a measure for assessing project incomes against project capital cost and possibly running cost. The payback period is the point at which future incomes equalise the capital costs expended. So in a production profit of £300,000 per year, providing a payback to cover the capital cost at the end of the fourth year of income. In tabular form the cumulative income looks like Table 3.2, which is a simple example of a new building whose cost including fees and fitting out is £1.2m and there is a rental income each year.

Table 3.2 Payback method

Capital cost	Income	Cumulative income	Net cumulative	
year 0			−1,200,000	
year 1	300,000	300,000	−900,000	
year 2	300,000	600,000	−600,000	
year 3	300,000	900,000	−300,000	
year 4	300,000	1,200,000	0	Payback
year 5	300,000	1,500,000	300,000	Total income = £1,500,000

This method shows how long it takes to payback (year 4), but fails to indicate, unless incomes are progressively predicted, what final profit is made. It also does not indicate the declining value of a sum of money which is paid later. This means that unless values are discounted then there is an unrealistic evaluation of the profit received as money in the hand can be invested. This applies just as much to the use of company reserves as it does to borrowed capital.

Accountancy rate of return

This method calculates the sum of all the income flows (+ve) and the capital cost outflow (-ve) and calculates any net return which comes to the business. This is usually expressed on a yearly basis as a percentage. Thus in the above example:

The net profit would be

£1,500,000 − 1,200,000 = 300,000 over 5 years which is:

£300,000/5 = £60,000/year

The return is therefore

(300,000/1,200,000) × 100 per cent

= 25 per cent return on outlay

Again this method takes no account of the declining value of a sum of money received in the future and does not take account of how many years it takes to get a profit. The figure of 25 per cent return again is misleading, but is directly comparable with other projects calculated the same way.

Discount cash flow method

Discount cash flow (DCF) method is similar to accountancy rate of return, but it operates a discount value equal to the cost of capital. So if money can be borrowed on average by the company at 10 per cent then the present value (i.e. the value today of money received in n years' time) is the reciprocal of compound interest.

Compound interest can be expressed as $P(1 - r)^n$

Where n = number of years invested; r = rate of interest expressed as a fraction; P = principal sum invested.

Therefore present value can be expressed as the inverse of interest rate i.e.

$$\frac{1}{P(1 - r)^n}$$

This will be a discounted value taking into account the number of years in the future it is received. The rate of discount of 10 per cent shown in column five of Table 3.3 reflects the return available from investing the cash and also includes a premium to allow for the level of risk of the investment. This rate is set by companies with reference to the cost of borrowing for them, even if the interest rates change. Column four shows the undiscounted net cash flow.

When the discounted income and the costs for all years are added up cumulatively we have what is called net present value (NPV). This is summarised in column five. Because the income is received in the future each year's income has been adjusted downwards using the formula above or

Table 3.3 Net present value method for capital cost and five year income

Capital cost	Income	Cumulative income	Cumulative net	NPV@10%
year 0			−1,200,000	
year 1	300,000	300,000	−900,000	
year 2	300,000	600,000	−600,000	
year 3	300,000	900,000	−300,000	
year 4	300,000	1,200,000	0	
year 5	300,000	1,500,000	300,000	−£57,058.15

tables which are available. This rate reflects the rates of interest and when inflation rises proportionately with interest rates it is a reliable indicator for comparison purposes.

Now we can see that the payback over five years at a 10 per cent discount rate is negative compared with the undiscounted column four and reflects therefore the cost of not receiving the income until future periods. The NPV method is much more robust than the two previous methods for incomes that are received over an extended period of time, though there are problems in being sure that the predicted discount rate is correctly reflecting the cost of capital, over the period of five years projected. If inflation moves with interest rates in the same direction (normal), then there is no problem. If as in some economic scenarios it moves counter to interest rates i.e. getting larger as interest rates stay the same then this cannot be assumed. It is also possible that cost of capital may rapidly change, moving away from the predicted amount, and if a range of conditions is likely then a sensitivity analysis must be carried out to see how sensitive the return is to an upward change in capital.

Internal rate of return

Internal rate of return is the discount rate in the method above which will return a nil NPV i.e. breakeven. This is often used by investment analysts to compare with the cost of capital which the company has. The company will take a view on what return over and above the cost of capital they require, that added to the cost of capital is called the *hurdle rate*. So a cost of capital of 10 per cent and an expected profit level of 10 per cent means a 20 per cent hurdle rate. This will pay a contribution to company overheads of say 2 per cent and provide a net contribution to profits of 8 per cent. However a project with a greater risk of not receiving the returns will have a higher premium rate. Another risk which might be reflected at the time of making the decision is how volatile the interest rates are. Both these may cause adjustments to the hurdle rate.

Taking the example above and allowing another year of income so that we have a positive result for the NPV, it can be seen in Table 3.4 that the effect of adding another 300,000 income at the 10 per cent income rate has added only approximately £154,000 (57,058 + 96,889) to the NPV, because of the substantial discount at the sixth year.

In order to bring the positive NPV to break even in the sixth year, i.e. 0, we raise the discount rate until that happens. In our example that will be nearly 13 per cent and this is the internal rate of return (IRR) for the project. This suggests that the project will only make an additional 3 per cent towards overheads and profit, which may not be enough for the company to go ahead until the cost of capital for them reduces or the project income can be enhanced or capital cost reduced.

Table 3.4 Example of IRR on the example above, but with 6 year income

Capital cost (year)	Income	Cumulative income	Cumulative net income	NPV@10%	IRR (rate at which NPV = 0)
0			−1,200,000		
1	300,000	300,000	−900,000		
2	300,000	600,000	−600,000	96,889.28	12.98%
3	300,000	900,000	−300,000		
4	300,000	1,200,000	0		
5	300,000	1,500,000	300,000		
6	300,000	1,800,000	600,000		

The IRR can be misleading for the same reasons as the discounted cash flow method. When comparing different projects or options it is important to test how sensitive each option is to changes in any of the factors. If a small change say in interest rate indicates a large change in the NPV then this project becomes more risky, especially if the returns are achieved over the long term.

Case study 3.4 is an example of a comparative appraisal which has been carried out using discounted cash flow and it focuses mainly on the direct financial costs and benefits. It considers several options for a university to see which has the best discounted cash flow.

Case study 3.4 Option appraisal on university accommodation

Looking again at the university example in Chapter 2 (Case study 2.1) for student accommodation it initially applies the DCF method for option appraisal, but also looks at other issues which are relevant to selection.

The requirement here is for providing good quality up-to-date accommodation for 400 student beds. All options have been reviewed against the 'do nothing' option as a base for comparing their suitability. Five options for providing accommodation on the existing site and one on a new site have been considered and the costs of these options are set over 27 years. The existing site has 72 hall spaces and is big enough to redevelop for all 400 residences, but this would mean knocking down existing academic services. Options are reviewed including the cost of reinstating the academic services elsewhere if required. The options are shown in Table 3.5.

All the results in Table 3.5 are given as net present value compared with 'do nothing' and they are compared for three different types of

Table 3.5 Alternative choices for funding

Option 1	'Do nothing.' *This is the standard to compare against.*
Option 2	Get rid of all existing halls of residence and then pay leasing costs to use privately provided residential units. This would emerge as a lease charge for the university plus the management costs.
Option 3A	A split site service where 72 hall spaces are refurbished on an existing site and 328 are new built elsewhere.
Option 3B	A split site service where existing halls are demolished and 150 spaces are rebuilt on existing site and 250 new built elsewhere.
Option 4A	A single site service where 72 hall spaces are refurbished and 328 new spaces are built by demolishing rest of site.
Option 4B	A single site service where existing buildings are demolished and 400 spaces are rebuilt completely on the existing site.
Option 4C	A new residential site for all bed spaces with existing bed space accommodation on the existing site converted for academic space.

Table 3.6 Comparison of options using NPV

		NPV over 27 years compared with the 'do nothing' costs £'000			Procurement options
		UBR	LB/FM	DBFO	
Option 1	Do nothing	0*			This option actually is the most expensive, except for conventional funding on the two new build options
Option 2	Head lease	2,962	–	–	
Option 3A	Split site refurbish and new	827	3,537	3,337	
Option 3B	Split site new	692	3,647	3,072	
Option 4A	Single site refurbish and new	1,568	4,061	4,080	The 'non-own' options here create the biggest savings
Option 4B	Single site new	–622	3,322	2,686	The all new DBFO options are more expensive than the head lease
Option 4C	Single alternative site	–794	3,080	2,185	

*Positive figures mean costs are less than 'do nothing'

KEY: UBR = conventional loan finance procurement (actually unavailable to university). LB/FM = private build with facilities management and lease back to the university with a service charge. DBFO = PFI agreement with private build and facilities management with building reverting back to the university after capital/service charge has been levied for 25 years.

procurement. The 'do nothing' option assumes that there will be high ongoing maintenance costs and these must be carried out by the university. In Table 3.6 they are reduced to nothing to make other costs more easily comparable. The NPV values on all 3 and 4 options are carried out over 27 years including a two year build period before occupation, a 7.28 per cent discount rate, which represents real cost to the university, and a positive allowance for transferred risk in the case of LB/FM and DBFO options.

Analysis of option appraisal

Although conventional loan finance is not available it is interesting to note that when all the costs are taken into account over a 20 year period this is not actually a cheap option and the university might as well use a DBFO on option 4A to allow outside private provision of accommodation for 25 years.

Head lease will give the university a newer facility than the 'do nothing' option and is cheap, but it will not give the university any residual value for an asset, the costs will continue after 25 years at a high level and the university will also retain quite a lot more risks and the responsibility for maintaining the properties. The head lease is more expensive than the other 'non-own' options.

The private build option 4A with leaseback compares favourably with the DBFO option. However with DBFO some opportunity savings may be made if the university is able to sell off land, which can be developed profitably by the DBFO provider as an alternative project e.g. luxury housing. Also the provider might be able to use the student accommodation for holiday or conference lets, giving dual use on catering, communal or leisure facilities made available on the site, during the student holidays, and reducing the DBFO service charge. These financial analyses should also be subject to a sensitivity analysis on the main variables as in Table 3.7. These may affect the rankings and Table 3.7 indicates the extent of the changes tested.

In the sensitivity analysis the construction costs increasing favoured the head lease option more as PFI pays back over a long period, but essentially the DBFO and LB/FM options still remained the cheapest and are the least affected by uncertainty. The DBFO contractor takes the interest rate risk.

This type of appraisal gives a strong financial base but other benefits or circumstances may also influence the final decision to go ahead. The sensitivity analysis is of particular importance to test the effect of variations in the assumptions made and in public projects estimates may be put on non-tangible benefits as suggested in the first example. The operational and through life costs are discussed in more detail in Chapter 8, but these will be important to the development decision especially if the client is an owner

Table 3.7 Sensitivity analysis for financial factors on option appraisal

Variable	Change tested
Construction costs	+30%
Operational costs	+20%
Student rental levels	+15%
Occupancy levels	+ 3%
Land values	+20%
Interest rates	+50%

and user of the building. Discounting already incorporates assumptions for inflation and therefore maintenance and running costs well into the future feature less strongly in the equation where there is a large discount rate. In private development projects the net returns are compared with the company hurdle rates (breakeven plus a suitable margin) to ensure that the risks of development are adequately covered by the margins expected.

A project with more than normal risk shown by the volatility of a sensitivity analysis is likely only to be approved in the case of enhanced margins in the appraisal. Lending institutions will also apply risk assessment when considering borrowing which is purely secured against the project outcomes. Private development is often more short term in its view of returns, looking for payback after a short period.

Developer's budget

In a developer's investment decision for a project it is made in the feasibility stage. The basic costs of a construction project are the land, the design fees, the building costs and the fitting out costs and there are often costs for additional advice. Other things however may influence project costs such as inflation, the interest rates available for borrowing money and the risk associated with getting a return (this will inflate the rate of interest which is offered for loans). Cash flow is also important and if payments are up front they will mean an extended borrowing period. For example land often has to be bought early, but if a deposit can secure an option on the land with payment later at an agreed rate this will be preferable.

Residual method. A developer's budget sets out the costs and expected income for a project and includes the percentage return which a developer would want from the sale or rent of the building. The value of the land is the residual sum which is the difference between the costs and the income. A typical developer's budget is laid out in Table 3.8 for an office block on city centre land at £1m/acre. The figures assume £10m on average is borrowed during the build programme for two years assuming a straight

Table 3.8 Developer's budget for office block

Amount of land required 0.9 acres	£'000
Building costs	20,000
Design fees @15%	3,000
Other consultant fees @ 3%	600
Statutory fees @ 2%	400
Infrastructure costs 1% of building cost	200
Building total	24,200
Expected return on costs 20% to cover risks	4,840
Borrowing costs £10,000 @ 5% over 2 years	1,025
Building costs and returns	30,065
INCOME	−31,000
Residual value of land	935

line of time versus spend. It also assumes that the building is sold immediately and that the loan is paid off.

If the building is kept for rent then borrowing costs continue and rent income builds up more slowly. Because of the time value of money, both income and borrowings are then discounted to give net present value. Borrowing costs are often judged to be neutral with present value gains. The building becomes an asset on the balance sheet. The value of the building may increase over and above the discount value, but this could only be recognised in the accounts for profit if it is prudently revalued, or the building is sold.

The cost of a project will double in 7.25 years at an inflation rate of 10 per cent (this can easily be ascertained from the same tables that are used to assess present value).

External factors in feasibility assessment

Construction projects do not exist in a vacuum. They are influenced by external political, economic, technological and social factors which either directly affect the project conditions, such as the price of materials or the going rate for labour, or affect the client's business and impact upon the scope and specifications of the project. For example the need to get to market earlier may reduce the timescales for the project to take place. Clients are constantly reviewing their investment to increase or reduce their capacity to suit market demand. If interest rates go up it may well affect the volume and speed of house building that takes place. If there is a new regulation which makes building more expensive then this may reduce a developer's capacity to invest. The government may also increase real estate investment by giving tax incentives for building in certain areas.

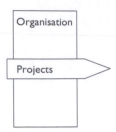

Figure 3.5 Projects done for external clients

The APM *Body of Knowledge* refers to external factors as the project environment and defines it as the context within which the project is formulated, assessed and realised.[14] It is important for the project manager to understand the influence these have on the feasibility, strategy, design implementation and outcomes of the project and they should be evaluated when recommending certain courses of action and when assessing options.

External factors in construction projects arise because they do not normally exist inside a single business. They are likely to have an external client and they are likely to work with other external organisations in order to deliver the outputs which the client needs. This model is shown in Figure 3.5.

It is also very likely that a building organisation works on more than one project; for example an architect is likely to get more than one commission at any one time, and a larger contractor will succeed in winning more than one contract. Thus we can see that the model should be developed to show this and to show that more than one organisation is involved.

In the model in Figure 3.6 the architect, engineer and main contractor are the same on three of the projects, but the quantity surveyor shares only two.

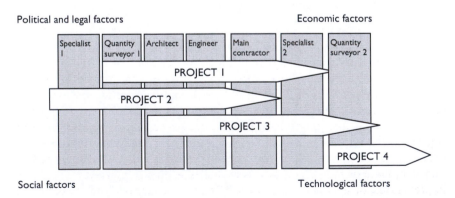

Figure 3.6 Complex relationships

The specialists will be assigned to projects in accordance with the need. The same external environment influences all four of these projects. The four external influences are often known as the PEST or STEP factors.

Political and legal factors

Political factors are connected with government policies which might have an influence on the project. These policies cover all sorts of areas, but a few of them are given below:

- Fiscal policy covers government tax and spending plans affect the viability of building work, and the incentives there are to build.
- The level of skills in the industry can, in the long term, be affected by the policies for training, although many companies are not facing up to their own responsibilities to sponsor training.
- Regeneration policy means that there are plenty of incentives to invigorate and develop city centres. Many companies have been drawn into dealing with brown field sites and decontamination of land.
- Energy policies and subsidies provide shorter payback on renewable energies.
- Building regulation changes have forced more concern with energy conservation.
- Landfill tax has encouraged recycling and less waste and aggregates taken to landfill sites.
- The government desire to renew school stock and to provide priority funds makes this sector desirable, but vulnerable to government change of mind. This affects all public sectors.

These policies and regulations have a major effect on building design, types of work available and the methods and resources which are used by contractors.

Economic issues

The economy covers inflation and interest rates and these in turn have an influence on the growth of the economy (GDP), the ability of clients to invest and spend money, the level of house prices and tenders for business and commercial contracts, the value of stocks and shares, the rate of employment, exchange rates and the funding which is most economically available. Other things which might affect the way that companies invest, or do business, are the borrowing limits that they have and these are directly related to the profits they can make or the bank credit available. The government either controls interest rates, or allows the central bank to set these according to economic need. Credit facilities affect decisions to invest.

Sociological factors

These are related to the fashions that people have and can therefore affect the market demand and the proportion of money that is put into housing or other spending. Whether to buy or to rent redistributes the market. Communities may also put pressure on developers, contractors and designers to meet societal norms which they feel are acceptable. Environmental concerns show themselves in more energy saving designs, using environmentally sensitive materials and living more simply. Governments might try and influence this by the use of fiscal policy and incentives.

Technological factors

These are issues which relate to the advances in technology which can affect the methods and materials which are used in construction. They may be the prerogative of the client, the designer or individual specialists in response to new opportunities. These factors allow for innovative factors and they may or may not be important to the client's future business and are unlikely to be so influential in the post design stages of the project. Prefabrication and sustainability are bringing innovative design and intelligent buildings.

IT systems affect the communication capacity on the project and integration of individual systems can be important for efficient information flow. BIM has already been discussed and extranets, social networking, mobile technology, GIS and intelligent buildings are others which are having an influence on how clients use buildings and what project managers need to know.

Sustainability factors

Sustainability is defined in Chapter 11, but briefly covers a holistic agenda to make our use of resources appropriate so as not to harm future generations. Sustainable construction covers environmental factors in a balanced way and therefore is integrated with the social, economic and technological factors above. It is recognised that sustainable buildings may have greater capital cost, though this factor will reduce as products and manufacturing costs reduce with wider use and fundamental building design changes become the norm rather than 'tweaking' an existing one. Client and professional attitudes are important and there is now a professional ethic to provide sustainable options and to make legal and moral responsibilities clear to the client. Many of these relate to the social aspect which has a deeper meaning in referring to better quality of life at home and work. Leading companies recognise that fair dealing with employees includes updating their work environment as well as other conditions of employment to meet this end and may expect payback through productivity improvements. They recognise life cycle cost reductions in sustainable buildings.

A long sighted client will still need to balance their business case and briefing is connected with the existing sustainable values of the client and the legal requirements in legislation such as the Climate Change Act in the UK which requires a demonstration of environmental improvements and carbon reduction in use. The ultimate goal in this Act is in reaching neutral carbon requirements for 2016 for residential and 2019 for other forms of building. Some clients wish to be industry leaders in meeting targets earlier. BREEAM compliance is discussed in Chapter 11, but will impact on the briefing process as well as later on.

Building value and risk allocation

Eliminating, reducing or protecting from risk at this early stage is closely allied with the value management process, which seeks to adjust the brief in order to optimise value for money. There is scope here to produce radical solutions that reduce or eliminate risk by more holistic solutions that are properly tied in to client objectives. Later on when the design is firmed up there is less scope to provide alternatives. A problem solving rather than a solution generating culture is now in place. This is done by using value management workshops (see Chapter 9).

Figure 3.7 shows the relationship between scope to change, cost of change and the stage of the project life cycle. However there is no guarantee that initial assumptions for the scope, use of the building, will not change with the demand for the product. Part of risk management is to build in flexibility and con-tingency so that change is possible. Some building projects are more prone to this than others. More is covered on risk management in Chapter 8.

Changes need to take place at the feasibility and strategy stage of the project otherwise substantial abortive costs are incurred due to the effort which is wasted in redesigning or worse still in abortive orders or construc-tion work. Value management can be used to test the original assumptions made in the brief and early design.

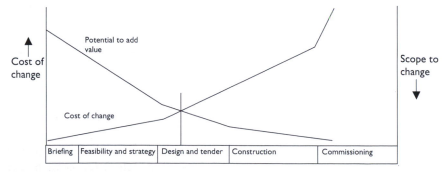

Figure 3.7 The cost of change

Risks are assessed in the feasibility stage and may be mitigated by applying strategies which reduce them or by choosing different options in order to make feasibility and viability more robust. Risk allocation is particularly important for making sure that residual risks are the responsibility of those who are best able to manage them. This increases the probability of keeping the risk low and also those who have the most experience in managing the risks will price them competitively without a cover contingency. Chapter 4 covers the identification and allocation of risk and Chapters 8 and 9 deal with the process of risk and value management in more detail.

Conclusion

Project definition is all about communicating a project brief well, being flexible, carefully assessing scope and delivering a product which suits purpose, but which will also maximise value and be effective. Many projects have failed due to inappropriate design of buildings which have failed to accommodate client expectations. It is clear that there are many strategic issues that should be considered in the project definition stage to inform the design so the early involvement of advice on the construction process and specialist design is good practice.

Project appraisal is used to test ongoing financial feasibility, and is useful for comparing different options, but have the right options been chosen? Quantitative financial methods are often presented as infallible, but clients are not warned of the factors which are particularly sensitive to changes in the market such as interest rate changes, technological and statutory updates and obsolescence over quite short periods which then impact heavily on the feasibility. Taking external factors into account is vital to the success of the project definition stage. Political, economic, social and technological factors can easily change so it is wise to build in flexibility and contingency factors.

There is often a lack of communication at the briefing stage in particular because of client uncertainty and this needs to be managed in the project definition period. Modern technology has the capability to produce good quality 3D visualisation so that clients are empowered and spatial planning and other client concerns are facilitated and signed off knowledgeably. Building information technology (BIM) is beginning to find some client support now and has further capability in helping the team to package information in an integrated understandable way, in an object orientated format that more closely simulates the way a building is assembled. 4D allows an instant appraisal of the time and cost impact of design options and updates in the briefing process to serve the client's concern to optimise value and to efficiently manage the life cycle costs. BIM allows a complete handover of building information to the facilities management team. User involvement in the briefing process remains important and a balance of continuing stakeholder management needs to be struck to avoid raising expectations too far.

Clients are sometimes unaware of the balance in increasing capital costs so that running costs can be substantially reduced, but many would connect the latter with the need to meet sustainability criteria in their built assets to reduce energy use. Responsible clients only accept capital price rises for sustainable buildings if they can prove the business case and the payback on such changes. They expect buildings to be innovative and economic which are also more sustainable. It is important then to integrate the economic criteria with environmental and social credentials to improve productivity and retention of staff. Clients need to be empowered to plan their capital and ongoing budgets carefully if they are to avoid expensive late changes.

References

1 Gray C., Hughes W. and Bennett J. (1994) *The Successful Management of Design: A Handbook of Building Design Management*. Centre for Strategic Studies in Construction, Reading University.
2 Baden Hellard (1997) *Total Quality Management in Construction*. London, Thomas Telford.
3 Royal Institution of British Architects (2001) *The RIBA Plan of Work* (as amended 24 May 2001). London, RIBA Services.
4 RIBA (2000) *The Architects' Plan of Work*. London, RIBA Enterprises, p.15.
5 CIOB (2010) *Code of Practice for Project Management for Construction and Development*. 4th edn. Oxford, Blackwell.
6 APM (2006) *APM Body of Knowledge Definitions*. 4th edn. High Wycombe, APM.
7 Barrett P. and Stanley C. (1999) *Better Construction Briefing*. Oxford, Blackwell Publishing.
8 Boyd D. and Chinyio E. (2006) *Understanding the Construction Client*. Oxford, Blackwell.
9 Wolstenholme, A. (2009) *Never Waste a Good Crisis: A Review of Progress since Rethinking Construction and Thoughts for Our Future*. London, Constructing Excellence.
10 Wolstenholme A. (2009) *Never Waste a Good Crisis: A Review of Progress since Rethinking Construction and Thoughts for Our Future*. London, Constructing Excellence; Barrett P. and Stanley C. (1999) *Better Construction Briefing*. Oxford, Blackwell Publishing.
11 Penn C. (2010) *BIM A Process for Clients and the Whole Construction Industry*. Presentation at BSIRIA Conference at Ecobuild March 2010.
12 Throssell D. (2010) *IM Case Study Presentation* at BSIRIA Conference at Ecobuild March 2010.
13 CIOB (2010) *Code of Practice for Project Management in Construction and Development*. 4th edn. Oxford, Blackwell.
14 Association for Project Managers (2006) *Body of Knowledge*. 4th edn. High Wycombe, APM.

Chapter 4

Strategic issues

This chapter moves on from looking at the product and its definition and introduces the major decisions which decide *how* the project is carried out. It gives an overall view of strategic planning and control *systems* of the project. The overall strategy is sometimes pulled together in a project execution plan (PEP) or master plan, which guides the team on the organisation, time, cost, quality, information and risk allocation and identifies the planning and control systems. An approach that recognises the relationship between the strategic subsystems will help to provide a more integrated approach and smoothes the path of the project manager who needs to co-ordinate these systems.

The objectives of the chapter are to:

- define strategic project control system and its subsystems
- develop scope and change management
- develop the master plan and establish the role of control systems
- identify strategic risks and their allocation
- understand programme, cost and quality control
- choose a procurement strategy
- evaluate and improve information flow.

Project strategy is defined as 'the route to get to where we have determined to go'. Strategy also concerns the organisation of people and tasks. It follows feasibility and runs alongside design.

- For people the questions are 'How do we achieve effective communications?', 'Who is responsible for what?' and 'How can we achieve the quality standards consistently?'
- For tasks the questions are 'How shall we procure the work and materials?', 'What are the sequencing priorities?', 'How shall we manage cost constraints?' and 'What control systems will suit best?'

The project manager has to ensure the objectives the client has prioritised and to do so in a safe, sustainable and orderly way. Progress should be reported in a way that gives consistency and confidence. Strategy has to consider the planning and control of procurement, cost, time, resource, health and safety, risk and quality. We shall refer to the final developed strategy as the master plan. The planning and control needs to respond to external pressures, which are often less predictable. Strategy should take an open systems view and heed:

- the impact of the external environment forces
- the effect of the project on overall organisational objectives
- the effect on the community.

Not taking external issues into account has often led to the failure of projects. Although the external environment is not in the control of the project manager, it can be predicted so that a response can be made. For example, the shortage of skilled labour locally could lead to the incorporation of a wider subcontractor tender list to ensure a proper service.

Figure 4.1 shows the main elements to be established in the strategy stage, together with the employment of the project team to suit. Strategy depends upon organisation and control. Key control processes are managing project scope, the method for procuring the contract, setting up control systems, managing time and cost, and safety. The control is interlinked by the constraints and the scope definition. The tools and choices in row three must not operate in a vacuum.

Now we need to consider how a project breaks down into systems.

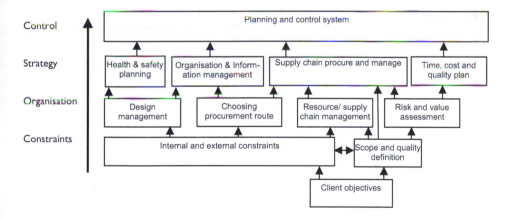

Figure 4.1 Elements of the strategy stage

Project systems

The term 'systems' is used as an analytical tool to work out what inputs are needed to achieve the objectives. Each system has an input, a managed transforming process and an output. An open systems approach recognises that systems are influenced by what is going on in the external environment, for example, the availability of skilled labour or sufficient plant and the impact of the economy.

Figure 4.2 shows what is called an open systems approach recognising the environmental influence. There is an overall system and there are subsystems. In the case of construction projects, for example, the construction process has an *input* of raw materials, a *transformation* – a process of assembly – and an *output*, which is a finished building as in Figure 4.3(a). Figure 4.3(b) shows how the overall project might look for achieving client objectives.

The construction assembly system 4.3(a) inputs consist of resources to carry out the construction. In system 4.3(b) the inputs are the client brief, the project constraints and the development of that brief to get satisfaction and further commissions. Both systems are managed by the project manager.

Project subsystems

The project overall is indicated with its subsystems in Figure 4.4 showing project life cycle phases of concept, design, assembly and handover. Each of the subsystems has its own inputs dependent on another subsystem and

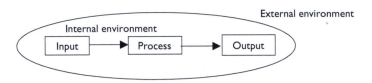

Figure 4.2 The basic system

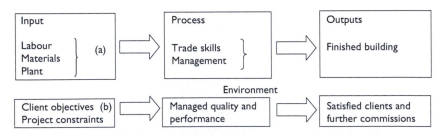

Figure 4.3 Overall construction system – a) assembly b) satisfaction

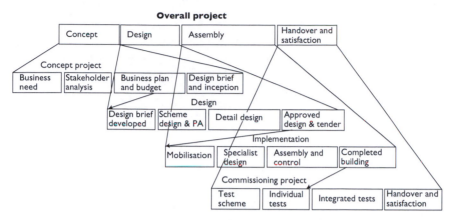

Overall project

Figure 4.4 The project subsystems. Adapted for construction project management from Burke (2003)[1]

develops these through two processes to have a relevant output. The concept has been developed through to the business plan by the clients and their advisors. The design brief is handed to the design team who develop a functional design to comply with the clients' brief and with the planning and statutory controls. This is cost checked to ensure that the market tender complies with budget.

The construction stage continues with the mobilisation of the contractor to work with the design to assemble the building and control cost and time and quality. The completed building or phase is handed over to the commissioning project and the individual parts are tested. The structure, finishes and services are integrated to be airtight and fit for purpose with working services to the satisfaction of the client. The breakdown of each life cycle has been simplified. It shows the interconnection of product subsystems in a simplistic way, as it is quite possible that the different phases of the whole project overlap. These systems help give a good foundation for understanding the project scope.

Project scope – planning and controlling

The PMI describes five process groups.[2] These are initiating, planning, executing, controlling and closing that *integrate* the design and construction phases and help to control scope. These are related across the life cycle phases in Figure 4.5. Scope is given in the inception stages as a client brief, defined in the feasibility stage, broken down into its elements and responsibilities in the strategy stage and executed in the design and construction stage. It is essential that scope is known at all stages and the impact of any changes is recalculated and agreed as these essentially change the project charter or contract.

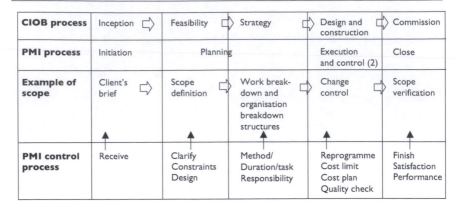

CIOB process	Inception ⇨	Feasibility ⇨	Strategy ⇨	Design and construction ⇨	Commission
PMI process	Initiation	Planning		Execution and control (2)	Close
Example of scope	Client's brief ⇨	Scope definition ⇨	Work break-down and organisation breakdown structures ⇨	Change control ⇨	Scope verification
PMI control process	Receive	Clarify Constraints Design	Method/ Duration/task Responsibility	Reprogramme Cost limit Cost plan Quality check	Finish Satisfaction Performance

Figure 4.5 Integration of project process subsystems with life cycle

Figure 4.5 clearly shows how scope management is applied across the project life cycle and also indicates the various controls that would need to be applied to keep to contract and to ensure that scope does not 'creep' and lose value. *Product* scope in construction is often referred to as the client's design brief and was considered under project definition (Chapter 3).

Project scope management covers a much wider remit and includes at least the time, cost and quality management as well as the implementation of the design to the right level of quality and the resource planning for product delivery. The feasibility stage will also be concerned with establishing project scope. There are five stages in scope management mentioned which are hierarchical:

1   Establishing the project contract or charter.
2   Scope definition.
3   Scope planning – work breakdown structure (WBS) and organisation breakdown structure (OBS) and task responsibility.
4   Scope change control on, for example, business capacity, functional efficiency, appearance, accessibility for customers and staff, user comfort, environmental performance and low maintenance and running costs for success too.
5   Scope verification.

The *project contract or charter* formally recognises the existence of a project and provides formal reference to the business need that the project addresses and by reference to other documents, the product description or project brief. Constraints are recognised in the contract conditions and will refer for instance to a contract sum, a defined time period, recognised behaviour and protocol, and assumptions for managing the contract. Key documents in construction are the contract which defines the time period and standard conditions of contract, drawings, specification, a contract sum received from the contractor and a contract programme.

In construction it is easy to see the building as only a product, but construction projects are also a service to the client and success depends upon delivering the building on programme, fit for purpose, on budget, with no significant defects, easily maintainable and without a poor safety record.

Scope planning and definition are the next levels of detail in order to facilitate control of the project and break down the project into different functions or sections. Scope planning identifies the relationship between groups of activities at definition stage to co-ordinate the numerous activities and facilitate decision making, giving a basis for control. In particular it should identify the project deliverables and project objectives which can be monitored. For example, spatial design and services design are two elements of a breakdown related to user comfort. As such they are separately delivered, but also need to be co-ordinated. WBS also creates a hierarchy of subsequent detail as indicated in the clean room project in Figure 4.6.

The primary aim of the lower levels of the WBS is to provide a structure for allocating tasks, though individual responsibility needs to be connected. At the higher levels it helps to identify stakeholder requirements. The organisation breakdown structure (OBS) will be used for allocating organisational or senior management responsibility. Breakdowns may not be functional as indicated in Figure 4.6 but may be related to phasing or to subcontract packages.

A task responsibility matrix (TRM) matches the task with the responsible person. This is useful to do at each level of the WBS. So for Figure 4.6 at scope planning level we will have a design manager responsible for design plan and at scope definition level we will have separate designers such as project architect and structural designer responsible for respective designs. However the TRM can pick up different responsibilities such as drawing register, risk register, and health and safety file which work across the tasks.

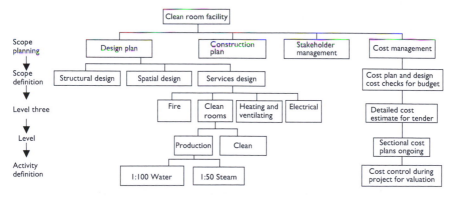

Figure 4.6 Work breakdown structure for a clean room facility design

Scope verification is checking the achievement of the breakdown structure and should be connected to a checking and monitoring procedure including progress review, audit, testing and inspection. Formal inspection takes place at the end of a project or a phase.

Scope change control manages changes in the scope of work that was initially established in the budget. This becomes most critical after the scheme design, but significant changes in scope will have a knock on effect in the scheme design and detail brief. They may occur due to:

- external events such as a new government regulation
- error or omission on the part of the client in defining the scope of the product required, such as the need for additional welfare facilities to meet staff agreements
- value adding change, such as reducing running costs by taking on a technology which was not available when the scope was first defined
- planning permission condition or qualification, such as the requirement for more underground parking or a change in brick type.

These may in turn disrupt general progress and impact budget, sequence, design fees or finish time. A scope change management system needs to track these changes and their budget implications. Such changes need to be fed back through the planning process such as health and safety implications, co-ordination of different aspects of design, drawing adjustments, and cash flow. It does *not* include those changes which are of a developmental nature during the design period. These changes can be defined as a natural part of a continuous improvement culture to optimise a solution or to make choices between options. Change control is covered in detail in Chapter 9.

The development of the master plan

The APM BoK links the project master plan with the key success criteria and suggests that key performance indicators should be established in order to monitor the plan as a basis for control and evaluate its success. It is a plan for action and establishes the control criteria. In the PMI BoK and the OGC Gateway Review™ the master plan is called the project execution plan (PEP). The plan answers three main questions:

- How? The plan investigates the procurement process, establishes a quality, sustainability and health and safety plan, and tracks the flow of inform-ation, and the use of IT to meet client objectives.
- How much? Develops a cost plan and a cash flow to suit the client's budget and fund availability.
- When? Establishes a programme of key activities and events (milestones) with calendar dates to establish key decision points (gateways) to ensure an agreed finish date is achievable and to determine resource usage.

ISO 10006 and the OGC gateway process give some guidance as to what should appear in the project execution plan, but this is an individual preference.[3]

The master plan is a combination of these individual time schedules and budgets adjusted to suit the overall budget and programme constraints. A 'hardening' of the plan takes place after consultation and approval by the client or project sponsor. The main stages of approval are project go ahead (after feasibility), planning submission and detail design.

The iterative process will take the form of re-establishing targets and basic areas of performance responsibility which are *accepted* by each party with budgets determined for each area. Freeze dates for any design changes need to be agreed on a procurement programme in order to avoid abortive costs and to establish client commitment dates. The freeze date should be directly proportional to the procurement lead in times and how this relates to ordering and execution date of the process. The iterative process depends upon good teamwork between many participants. At this stage further significant changes are formalised by a change order and should be at the prerogative of the client.

Reiss regards a plan as having three main uses:[4]

- a thinking mechanism (50 per cent)
- a communication tool (25 per cent)
- a yardstick (25 per cent).

He puts the emphasis of planning on thinking through the work ahead in a systematic way and enhancing the manager's communication ability.

Risk planning – identification and allocation

It is important to recognise that uncertainty will exist and should be assessed, should be reduced to acceptable levels and each risk should be managed by the party who is best able to deal with it. Risk transfer can be expensive if it means that risks are managed by those who are inexperienced, or who do not have the influence to reduce its impact or probability or it is difficult to get anyone to ensure it.

Typical risk areas are client changes due to business predictions or user requirements, subcontractor default, technical failure, strikes or other labour shortages, bad weather, critical task sequences, tight deadlines, resource limitations, complex co-ordination requirements, and new complex or unfamiliar tasks. External PEST factors can also affect a current contract and they are not controllable from within, but may be predictable. Many of these are not covered under insurance.

It is important to assess and manage risk in the context of the feasibility of the construction project. Risks which appear in the business case can be put into four categories according to Smith;[5] these are financial, revenue, implementation risks, and risks connected with the operation of the finished facility. These equally cover external and internal risks which directly impact as in design delays or indirectly impact as in client revenue reduction which might on the viability of the project.

It is very difficult to get a universal categorisation and experienced clients such as the NHS have developed standard risk registers that categorise risk to allow comparison, but allow flexibility for the project team to add and subtract as appropriate. Every project is different and needs a unique register of risks generated from a knowledge of the contract and past experience. Edwards has categorised these risk factors and these are used below.[6]

Political and legal risks are shared between the client and their professional teams. Legal liability could be allocated to any of the parties and insurance is used to cover events such as professional indemnity, third party, fire and damage which are compulsory. Political events often affect the project financially in its operating or financial structure. An example of this type of risk is illustrated by Case study 4.1

Financial risks are allocated to the financial sponsor of the project, who is usually the client in the traditional construction procurement. Other risks include interest level predictions changing significantly. Equity funding may not be taken up as expected and stock market conditions can destabilise company borrowing levels and cause lenders to be more wary and more expensive.

Case study 4.1 Building Schools for the Future (BSF)

> BSF was a multibillion pound policy to renew all the schools in the UK and they were let in groups by locally based educational partnerships. This was an attractive proposition to the contractors and consultants that worked hard to get on the framework partnerships as it gave work into the future and a learning curve that could maximise profits and share them with the client. Contractors invested in specialist systems to suit the rules. However with the coming of a new political regime it was considered that there were inefficiencies in the system and all partnerships were stopped quite suddenly, leaving contractors and consultants with investments for the future that were surplus to requirements and in some cases too much dependency on this source of income.
>
> It is not easy to allow for this type of risk except by keeping a balanced portfolio of projects.

Revenue risks are allocated to the client or the facilities manager. Traditionally revenue is a risk for the project client, but this is important for DBFO projects which have toll collecting agreements; although there are often let out clauses, the main issue is cash flow. For example, the Second Severn Crossing is run by a private consortium that charge tolls, but are restricted to toll escalation within inflation index linked limits. However in order to strike the deal with the Skye Bridge Crossing, the time span for collection of tolls before handing back to the Scottish Office was extended to cover the extra costs of certain environmental design changes. Market changes or poor market conditions can severely affect revenues.

Implementation risks are those most closely associated with the construction process of design, tender, construction and commission. They are traditionally shared between the client and the project team according to standard conditions of contract. Broader forms of procurement move all but scope changes back into the realm of the designers and contractors. Design and build and design and manage lock the design and construction responsibility together and provide much better protection for the client from design development changes, which often create abortive work with no increase in value to the client. This area will be followed up in Chapter 8 on risk and value management.

Technology risks are to do with teething problems for new products or designs, or with the use of IT systems which go wrong. These are usually at the risk of the client, but are related to the design role also, where the project team has a performance brief and is dealing with the development of new products, or using them in innovative ways.

Operational risks are shared between the designer and the client and relate to weak and inappropriate design, poor briefing, lack of co-ordination between design and fitting out requirements and poor use or misuse of the building. Plant performance has a major effect and again is related to design. However the fitting out and quality standards specified by the client may have led to operational unreliability, expensive breakdowns and downtime, excessive running costs or even a de-motivated workforce. Case study 4.2 illustrates this type of risk.

Allocation of risks in procurement

Allocation of risks is very dependent on the procurement system and the chosen contract conditions.

In Table 4.1 the risks have been allocated between three parties in a contract to provide student accommodation on either traditional contract or builder build, lease and operate. It shows how risk is allocated quite differently for different procurement strategies.

The allocation of risks to the party that can best manage them is an important principle. It has the potential to move from a risk aversion culture,

Case study 4.2 Design risks

An operational problem occurred in a brand new office block for a government agency where an open plan, sealed and air-conditioned building had been designed and built. The commissioned building was handed over to the client without fitting out. When the computers were installed there were far more than designed for and the extra heat generated meant that, in hot weather, the air conditioning was unable to adequately cool the building, so staff were faced with difficult working conditions during the summer.

A major electrical breakdown was also caused by flooding into the basement electrical switch gear, putting systems out over a frenetic weekend. Insufficient back up, lack of co-ordination and operational changes gave the whole project a bad name and wrecked initial productivity in the new office. Managing risk for the operational conditions is one of the hardest jobs for the project manager as they often do not have access to the maintenance team at the design stage and dependence is placed on standard building loads.

This type of risk is again difficult to assess as it would seem a one-off. The key issue was the leaving out of the maintenance team from the design process.

Table 4.1 Table to show breakdown of allocated risk

	Client	Design	Contractor	Shared
Planning delay	X			0
Design change	X	0		
Client scope change	X 0			
Planning qualification	X		0	
Higher cost of construction	X		0	
Industrial action			X 0	
Latent defect			X 0	
Heating failure		X 0		
Level of use of rooms	X 0			

Key: Traditional contract (X); build, lease and operate (0)

Case study 4.3 Risk control at Heathrow Airport Terminal 5

This was a huge project worth about £4 billion and work was let to a construction management contractor with direct payment of the specialist contractors by the client. The client was fully integrated into the management of the project and was heavily experienced. The business case for the contract meant that the terminal was going to earn a lot of money as soon as it was opened and had to be finished on time or early. Liquidated damages transferred to lower levels in the supply chain to the lowest level of incompetence as is common would not be affordable by those smaller companies and so money would be lost in dispute. There was room for a new approach to risk.

The client agreed with the managing contractor and the first tier level of the supply chain that risk would be completely shared, with the client taking full responsibility on condition that collaborative approaches were binding in the contracts. This took the emphasis off the contractors to push blame around for mistakes which would have unaffordable consequences to working to help each other avoid risk and get the airport terminal open on or before time. In itself a high risk strategy as it only needed one non-co-operation to create havoc, the terminal was finished on time and money was spent up front to avoid risk of dispute later. For example, very good remuneration was offered to ensure productive work.

where risk is passed on to the first level of incompetence, to a value inducing culture where the risk allocation has been properly agreed and negotiated. A radical example of risk management is illustrated in Case study 4.3

It is reasonable for an inexperienced client to pass on all the construction related risk, but it is expensive for a client to pass on the risk of, say, the level of occupancy of the student accommodation to the contractor, unless those rooms are leased from them. The contractor would charge a high premium for a risk in room letting shortfalls for which they have no knowledge or prior experience. The project manager is in a strong position to guide and advise the client on allocation of risks for improving the value for money.

As the project progresses towards the scheme design, so the scope for change reduces and after the brief is frozen the cost of any change rises rapidly (see Figure 3.8 in Chapter 3) including those connected with allocating risk. At the inception stages there is the greatest level of uncertainty and risk evaluation becomes firmer as more is known about the project. Risk evaluation and management is covered in Chapter 8.

Project control

The control system is critical to the health of the project and its choice should influence the planning processes rather than the other way round. Initially the client will specify project progress reports which are critical to maintaining their confidence in meeting project objectives for design, cost, performance and programme. Frequently, experienced clients also see the project team as critical to effective control, wishing to vet the personnel at interview stage. It is also likely that they will insist on approving key suppliers for proper control of the project

There are two approaches to control discussed here – feedback and feed forward systems. In construction the former method is used much more often and is shown in Figure 4.7. Unless instantaneous feedback is obtained the system may not produce information quickly enough to take action in a unique project situation. This is because a project's activities take place over a relatively short period of time and if control is based on finding out when things have gone wrong then corrective action might arrive too late.

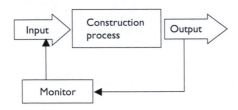

Figure 4.7 Feedback control system

The feedback control system briefly consists of setting up targets, monitoring the performance, measuring the gap between planned and achieved, identifying the problem and taking corrective action.

There are a number of issues which inhibit the efficiency of the feedback control system:

- It can take time before an output is reliably measurable. Frequent checks are expensive.
- Monitoring is not normally continuous with immediate correction. For example, a thermostat would give immediate feedback of a change of temperature and simultaneously switch the heating back on.
- Control is reactive and time is spent on investigation. For example, labour productivity, plant suitability and availability and materials delivery are all very different problems.

- Projects have limited repetition and delayed correction means that action will be applicable for a limited period only.

Feed forward control indicated in Figure 4.8 shows how monitoring mainly anticipates the problems prior to taking avoiding action.

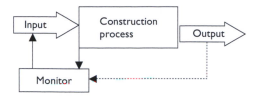

Figure 4.8 Feed forward system

In the feed forward system the key components are to:

- set up targets
- anticipate problems with meeting targets by discussing with the team, who will know their concerns, at regular intervals before the event
- take prior management action in order to avoid or mitigate the problem
- monitor performance to make sure action has worked (feedback)
- make further adjustments if necessary.

A typical feed forward system is operated through the establishment of advance meetings to co-ordinate the activities of different specialists and to consider resources and information availability and the resolution of current problems. It tests the robustness of individual plans and concentrates on interfaces where the performance of one contractor will have an effect on another. It operates on the latest information available and allows for contingency planning where there are any doubts over performance or issues outside the control of the project staff.

In feed forward the emphasis is on thinking through the implications with others so that communications are implicit. The feed forward method (prediction) is more efficient as long as there is an openness to discuss problems and to deal with them realistically with a commitment to flexible planning. There are other methods of control that are helpful in considering effective management.

Other methods of control

Management by exception is a simple principle for releasing management time monitoring the formal control system. It only requires inputs outside

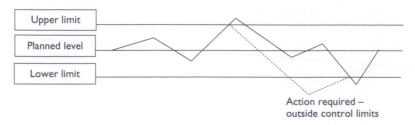

Figure 4.9 Control bands

a + or − band of conformity to target as in Figure 4.9. Detailed control is delegated to the first line manager.

This system allows for a band of control within which the project manager does not interfere. It recognises that 'actual' will vary from 'planned' at times, but that it is the responsibility of others to let the project manager know where performance is outside the certain limits. This is still a feedback system, but its strength is that the project manager can have more time for strategic forward thinking. Delegation of this nature has a motivating effect in the right climate, allowing responsibility and development of more junior staff. The weakness especially for construction projects is the variable quality of supervision in the context of specialist contractors and the critical effect on lots of parallel activities if one goes out of control. It is also difficult with one-off projects to accurately monitor and give *timely* feedback on time, cost and quality conformity. Therefore it may not be realised that an activity has gone outside its control band.

Management by objectives (MBO) has a similar goal in transferring responsibility to a lower level and cutting down the level of supervision required. Here a junior manager will be required to identify performance targets for their section of work for which they are accountable and to report back at longer intervals on their achievement. This can be motivating for the section manager, as it empowers them to make important decisions. It also requires perception and trust on the part of the senior manager in agreeing achievable targets, which stretch and do not strain the manager. It works best where there are clear objectives.[7]

Programme control

The time schedule or programme is often presented in the form of a Gantt chart with a list of grouped activities shown as a bar to represent their duration and on a time scale in order to indicate their sequence. The schedule shows the best estimate prediction of events and durations but not their logical relationships which means they can be misused by overlapping or squashing activities more than the logic would allow and showing a foreshortening of the programme. Linked Gantt charts are superior in that they indicate the critical path sequence (see Glossary) with a vertical line. Modern even simple programming software allows logic constraints and resource loading to be attached to activities and resource limits to be imposed. The better ones start from the work breakdown structures so that the responsibility and grouping structures are thought through.

There is no one method to do a programme and an essential part of using it as a baseline for progress is to agree it with the management team as the method determines the sequence. A prefabricated panel of bricks instead of brickwork may eliminate scaffold, as bricks can be treated like cladding panels and fixed from scissor lifts. Progress is recorded as in the Appendix to this chapter.

The time schedule, like design, is an iterative process. Iterative means that better schedules supersede not so good ones. Programmes are reprocessed as better information becomes available or details are changed. The concept of the 'last planner' is an iterative one developed by Ballard and Howell to look realistically at the real completion of activities and not those that were on the wish list of an unrealistic plan.[8] Their research showed that work actually completed on a weekly basis achieved between 27 and 68 per cent of the master programme allocation. In practice last planner means the development of realistic weekly plans of work that work towards a flexible re-planning to suit achievable progress each week, but provide a planned ability to pull back time lost on the master programme.

This process accords with the feed forward control process above which may also be used for cost and design control. Successful operation depends on openness in discussing problems, a 'no blame' team culture to overcome problems and a commitment to the project as a whole. Case study 4.4 shows an example of the control of slippage in the design stage of a programme.

Contingency planning

Contingency means putting in activity buffers on the critical path, or in some very volatile projects 'a plan B', when things go differently from originally planned. Contingency planning is a way of allowing for adverse changes and generally means making time allowance for items beyond control or introducing a formal risk management system. However, flexible programming, risk transfer or mitigation by insurance may also be used individually or together. The approach should be specified and the potential problems should be identified together with the reasonable precautions taken.

Cost planning and control

Cost management aims to control the project budget which is set after the value management process and to predict the timing of the payments (client

Case study 4.4 Slippage in design programme

Figure 4.10 illustrates a typical situation for a project falling behind programme and not meeting the critical construction start date on site. In a feedback system the programme will be measured at time now and the programme will be discovered to have dropped behind. Reprogramming the project to finish the stage on time depends upon feeding in extra resources (plant or labour), re-sequencing the logical progression or overlapping activities to a greater extent to catch up. This is a traditional response.

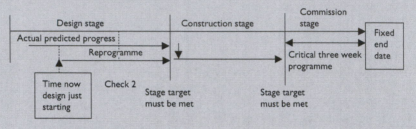

Figure 4.10 Reprogramming within the critical stage dates

In feed forward, communications will be better and design problems will be picked up before irretrievable lateness becomes a problem and although extra resources are fed in they are not trying to make up time already lost. It is wise to investigate the reasons for under resourcing in the first place so the original problem does not persist.

Traditional contracts mean design must be largely complete to go out to tender, which means that additional designers must be drafted in to solve this problem under both systems. Both systems can fall down if there is a lack of communication, then the targets will not be monitored tightly or be over optimistic, or misleading information will be given.

cash flow). Budgets set for the design and construction stages should be monitored and kept within the cost and cash flow limits set. The key areas of change are design changes and client scope changes. Again there are two systems of control – feedback and feed forward.

Many systems are based on creating a project cost plan with elements or cost centres, collecting cost data by monitoring actual costs and providing a monthly feedback report to the client and then looking for ways to make savings at a late stage where there is an overrun on a cost centre. This conforms to a feedback system and creates problems in projects where there are relatively short activities with limited opportunities to make savings within the latter part of the element. Cost control is often applied to a specialist package as a whole. Problems are:

- Contingency sums are built in as a sort of cushion with resultant uncontrolled extras.
- It provides clear information about problems, but often too late.
- Clients are made to pay for lack of planning by the project team.

Once the budget breakdown has been allocated to specialist elements, early expectations of cost breakdowns are used to limit expenditure in these proportions. If radical value management takes place, contortion can occur and a second stage of cost planning is needed, which allows new information about actual prices to adjust elemental budget allocations, at the expense of other budget heads as a dynamic control system applied to the design development.

In a feed forward system there is forward planning and warnings can be made. It predicts areas of critical pressure by assessing the risk and focusing cost management efforts on the 20/80 Pareto principle. A process of control stages is indicated in Table 4.2, which encourages the early prediction of cost critical areas.

Table 4.2 Cost management as a feed forward system

Phase	Process	Life cycle stage
Outline cost plan	Elemental cost breakdown based on outline specification. Cost checks on design as developed.	Feasibility
Scheme cost plan	Elemental cost plan based on approved design, following VM process. Cost checks on detailed design. Value engineering.	Strategy and design
Cash flow forecast	Monthly commitment based on procurement programme.	Detailed design
	Update client on contractor progress. Earned value analysis.	Construction
Firm cost plan	Based on contractor's tender. Tender reductions if required. Interim valuations and change management.	Construction
Final account	Contractor's tender plus extra work. Final valuation of all approved work.	End of defects liability

From a client's point of view no cost increases are welcome, which do not produce extra value for the client. Danger areas are:

- Misreading the market tender price
- Buildability problems inducing contractor delays
- Poor planning creating an unpredictable cash flow
- Poor tender documentation inducing additional claims
- Abortive work due to late changes or instructions
- Claims due to disruption or late information or client risks

The best way is to anticipate problems before they occur, so that avoiding rather than corrective action can be taken. For example, if cladding chosen is found to cost 30 per cent more than the scheme cost plan, the first reaction is to find a cheaper one to contain price escalation. If it is cost checked at the point of choice then it may be possible to look at a reduction in another element of the building which has less effect on the value, reduce the amount of cladding, or re-orientate the building to make a cheaper cladding possible.

Feed forward allows a value decision based on all client objectives and not just cost. For example, alternative cladding may have more costs in use, such as energy loss, cost more to fit, wear out sooner, delay the contract or be less environmentally acceptable. Cuts in other areas may cause equal problems in which case the budget may need to be exceeded or some leeway on completion time may help to reduce costs also.

Cost control measures

There are a number of cost control measures which may be used by the contractor and they break down into systems based on budget reconciliation and systems based on cumulative cost compared with cumulative earned value.

1   Measures to control budget levels in cost planning. Typical feedback systems are:
    - cost value reconciliation based on the monthly or stage valuations process, which measures the level of spend with the value of the same priced activities, for example, bill of quantities items inclusive of variations
    - cost centre control which is a more sophisticated system and allocates all expenditure to a cost breakdown by activity, for example subcontract package or trade, and also may also break down into elements such as labour, material and overheads.

2   Cash flow which measures the scheduled value for all costs showing this as:

- a monthly or cumulative cost curve of scheduled value against time as in Figure 4.11 with a curve for the actual value earned according to the project valuations. This is often called *earned value*. In an ideal world this would be the same as the scheduled value, but assumes exact compliance with programme on time and in sequence
- a reconciled actual expenditure curve which discounts expenditure not yet valued (prepayments and retention), or value gained not yet paid for (accruals). This then makes it fully comparable with the valuation figures and should differ by the margin expected on the project.

Figure 4.11 shows a graphical representation of a client that pays the contractor monthly shown by the black lines under the dotted curve. These have been cumulated to give the dashed cumulative value curve and compared with the cumulative costs of the contractor in the solid line. The contractor is currently making a small margin.

There are two forms of feedback which can also lead to a feed forward prediction based on the current rate of spend.

1. *Cost slippage* measures the difference between the actual expenditure and the actual value. This indicates whether the margin is currently being achieved if gross figures (including retention and any other valuation deductibles) are used.

The cost performance index (CPI)

= actual value × 100 per cent actual cost

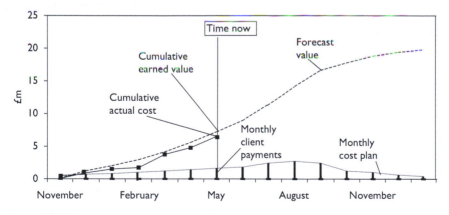

Figure 4.11 Earned value cash flow chart (from contractor's viewpoint)

For example, if the actual value is £2.5m and the actual cost is £2.2m and the margin to be achieved is 10 per cent, then we have

$$\text{CPI} \;=\; \frac{2.5 \times 100 \text{ per cent}}{2.2} \;=\; 113.6 \text{ per cent}$$

Therefore the margin required is being exceeded at the current rate of remuneration. All variation needs to be properly reconciled with the scheduled amounts.

2. *Programme slippage* measures the difference between the scheduled value and the actual value, indicating whether the project is running behind or in front according to the expenditure. This is quite a coarse form of control as cost is not directly connected to unit production. So material price differences and delivery times may distort the true picture. However it is a leading indicator and a tool for prediction of completion at current rate of spend. The schedule performance indicator (SPI) = scheduled value/actual value. The formula for calculating time taken at the rate indicated by the spending rate is

Predicted time period

= (scheduled value/actual value) × total scheduled time

For example, if the schedule value on week six is noted as £3m and the actual value is £2.5m and scheduled time is 25 weeks then we have:

$$\text{Predicted time} \;=\; \frac{3 \times 25}{2.5} \;=\; 30 \text{ weeks}$$

Therefore the project is predicted to be five weeks behind schedule if the spend rate continues as it is. This is related to the contractor's account, but may provide good information for the client.

Reporting systems to control costs for the client

There are three stages for cost reporting which are:

- design development cost checking in the project definition stage
- confirmation of commercial price and budget when tenders arrive
- cash flow control measures during the construction phase.

The client's budget is not finite and any movements need to be planned and managed. At least 80 per cent of the costs are decided in the design development stage so this becomes a critical stage to control.

In the case of the client, cash flow management is critical. They need to know dates for financial commitments so funds are made available to pay. Sudden changes in work sequence which speed up payments need to be agreed and signalled early to allow arrangements to be made. It is likely that the inexperienced client in particular will latch on to early estimates of cost and will find it much harder to adjust to the pattern of payment to meet early procurement of equipment and materials. Clients may also wish to make changes in scope to manage their budget and to respond to changes in external factors which will change the early cost estimates.

According to the CIOB Code of Practice *cost reporting systems* need to provide the established project cost to date, the anticipated final cost of the project which should be fixed within reason, the future cash flow and any risks of expanding costs should be reported and any potential savings. The last three fit in with the idea of feed forward and provide room for proper budget reordering so that the final cost is contained.

Reporting needs to deal with two stages to assist the client's financial planning. They will need to know the last date they can change their minds (commitment) and they will need to know approximately how much they have to have each month to pay the bills and what the financial implications of any scope changes are.

Figure 4.12 indicates the typical cumulative cash flow and the progressive nature of the client's commitment. The costs will change but it can be managed to make savings to cover escalation, so that the final budget is unaffected. In Figure 4.13 the contractor cost curve has been added to show the relationship of earned value (payment to contractor by client) and the development of contractor's profit.

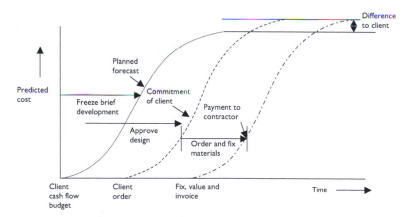

Figure 4.12 Connection between *client* cash flow forecast, commitment and payment

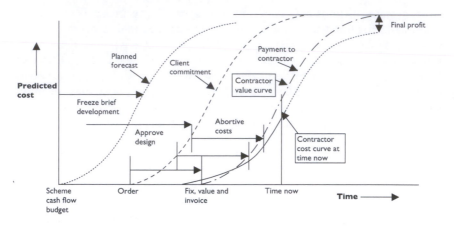

Figure 4.13 Relationship of *contractor* cash flow and earned value with client

Inflation and the possession of fresh information may have a significant effect on expanding the budget and a proper contingency should be allowed for inflation and the client advised of the danger of delaying outside the original time frame.

Summary of cost control

The principles for budget and cash flow control are:

- Time spent at design stage on identifying risks to costs, so specific rather than general contingency allowed, for example, ground investigation, option appraisals to reduce risk
- Contain design inflation in a VFM framework
- Integrate design and construction risk to one or co-ordinating party if possible
- Work together to give accurate advance costing information for any changes to the budget, so decisions are made on basis of knowledge of expenditure, share savings
- Transfer risk to contractor to cover reasonable unforeseen circumstances, but risk they are unused to managing will be budgeted expensively
- Give reliable information on the distribution of the payment cash flow schedule so that funding is easily available
- Consider life cycle costing if possible

Feedback cost control systems need to react quickly enough to give the client information so that they make decisions on the basis of cost. Where accounts are made transparent to the client as in partnering then information about costs and profit are easily available. If they are not available, as in traditional procurement, then feed forward systems, where costs of a possible change are calculated ahead of the time when go ahead is required, have the same effect, but close collaboration is needed to be aware of the commitment dates. Procurement and contract types should take account of the degree of flexibility that is required by the client. NEC contracts have a compensation event warning system which triggers a cost estimate for change for approval before go ahead.

Procurement principles

Procurement of the design and construction work is about reconciling the client's objectives for the project with the particular characteristics of the procurement system chosen. Each of the procurement options has implications for risk allocation, project organisation and sequence. There are many tables which have been prepared which match the attributes of a procurement option to the client's wishes, and there is one in the CIOB Code of Practice which has 17 attributes measured against four main categories of procurement. These categories of traditional, design and build, management contracting and construction management are by no means comprehensive as they poorly cover clients who wish to shed more risk (e.g. PFI or prime contract) or less risk (e.g. measured term or cost plus contracts). They do however give a range of risk options. The purpose in this chapter is to look at the underlying principles of the procurement of design and construction services rather than the detail of a particular system which may in any case be adapted for the project.

The principles which govern procurement type are:

- Degree of client involvement, priorities and client objectives
- Risk allocation client side versus supply side
- Balance of competition and negotiation
- Tendering process (e.g. single or two-stage, open versus selective, closed or negotiated)
- Whether framework agreements or repeat projects required
- Degree of integration required
- Degree of collaborative practice and partnering
- Controlling life cycle costs facilities in or out

The main procurement types are matched with the priorities of the client. These features may be met in more than one of the standardised procurement types or contract forms as shown in Table 4.3. Alternatively a tailor-made procurement could be developed amending contract conditions to suit. This is not usually recommended, because of the difficulty of working with conditions which are untested.

Table 4.3 Client priorities linked with procurement features

Client priorities	Example of procurement features which suit
Project time and speed of mobilisation	Overlap design and construction e.g. management fee
Single point responsibility	Join design and build e.g. design and build, prime contract, or PFI
Degree of client control over design	Independent designer e.g. traditional
Cost certainty	Single responsibility e.g. design and build (D&B) or PFI
Ability to make late changes	Transparent pricing e.g. New Engineering Contract (NEC)
Early involvement of contractor	Management fee e.g. design and construct or construction mgmt
Reduce confrontation	Partner agreements e.g. use of NEC or prime contract
Facilities contracted out	Design, build, finance, operate e.g. PFI

The choice of procurement system is a major issue in meeting the client's requirements and it is important to make sure that they understand the implications of the different approaches. Case study 4.5 rationalises the choice of design and management approach on a large city centre mixed development breaking down into several projects, masterminded by the main developer.

A table to help proper selection of procurement systems can be found in the CIOB CPPMC.[9] A similar table can also be seen in the CIRIA guide on managing project change.[10] Chapter 8 will consider procurement from a risk and value management perspective.

Contracts

Standard contract forms can be linked with certain types of procurement, but there are further choices here in the contract conditions. The NEC contract has been designed to give an alternative approach to the JCT forms with an emphasis on more collaboration. The design and build contract links best with this form and it makes particular allowance for the contractor and the client to develop close relationships, by amending the contract programme (not finish date) regularly to suit the latest information available, by making provision for open accounting of changes, where change events are signalled early, and by keeping the accounting transparent with agreed prices before changes in scope are agreed. The language of the contract is

Case study 4.5 Multi-project single site procurement

This is an eight-year mixed use project to develop a prime 16 acre derelict brown field site worth £300m in the centre of a provincial city. The site is very sensitive and planning permission has only been agreed with very stringent conditions to maintain site lines and to properly connect the city centre to the waterside and to ensure full public access. The proposals are for leisure, retail, commercial and residential use. The latter is core to the private developer in terms of cash flow and profit. The development links directly to a highly popular leisure sector and a significant commercial venture already in place on the site. The site is a desirable location, which faces onto the old harbour waterside.

The developer does not employ an extensive project team, but seeks instead to outsource the tactical project management. The first phase is procured by three D&B contracts for commercial, residential and infrastructure projects and an in-house management of the residential construction and marketing. On each of the D&B contracts there is an outsourced team for the client, consisting of a cost consultant, an architect and a project manager. These will report to the programme director.

Due to the sensitivity of the site to the conservation of the city character and heightened public interest a separate development director has concentrated their efforts on marketing later stages, maintaining good relationships with the public and community and obtaining planning permission.

The programme director has taken on the implementation and detailed marketing stages of the current developments and maintaining the master programme for parallel and follow on projects. Separate projects need to comply with the overall constraints of the new infrastructure and synchronise with environmental clean up for part of the site. Programming needed to respond to this and the logistical access which affect the methodology and sequence of separately let contracts. The residential developments are key to the cash flow for the developer who needs to control the heavy outlay of building costs by bringing to sale premium price residences in this prime position. Programming needs a flexible response to the market forces which will determine demand and the speed of completion of the residential units in particular and the commercial, retail and leisure real estate which are let to tenants and may be sold on as investments for fund holders, or put in the hands of management companies.

The design has been spread out between architectural/engineering practices to reflect specialities and inject variety. Leisure, commercial, and residential blocks have been allocated and the master planners have retained far more detailed involvement in one area. The process is managed by phasing and allocation of different buildings to different contracts, so that design management is linked to the procurement route chosen and is not developed centrally. The programme director co-ordinates the production control for the developer. Figure 4.14 shows the emerging organisation structure.

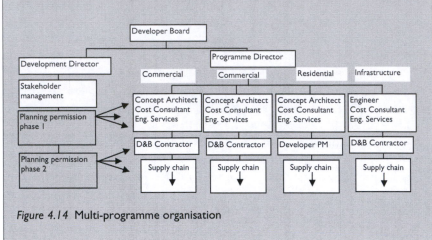

Figure 4.14 Multi-programme organisation

much more open and is designed to encourage more trust between the parties. This is a contract routinely used by many public clients and has advantages for developing collaboration, providing a single point of contact for the client, although the client may also choose to appoint an executive project manager.

The FIDIC international contract gives a suite of contracts to suit different procurements, but provides greater protection for contractors working internationally. PPC2000 partnering contract gives a broad based contract to support partnering teams. Case study 4.6 gives an example of the use of the more flexible NEC contract.

Quality planning and satisfaction

Maylor distinguishes between conformance and performance planning.[11] This may be described as the difference between quality assurance (QA) and quality improvement. In practice this means defining an effective system

Case study 4.6 Contract flexibility

The client was a Hospital Trust requiring a medical training centre which was to be funded by a commercial mortgage with strict budget targets. This was a medium sized contract and was built on several floors in a tight city centre site within reach of other hospital buildings. It was planned to fund the mortgage repayments by letting out areas of the building to related commercial activities who would pay a market rent. The Trust decided on the use of the NEC contract conditions and let the contract on a design and build basis, novating the outline design architect to the successful design and build contractor. An ideal commercial tenant needed to move in early (who was to provide specialist medical training facilities) and, together with the pressure for space in the hospital, the contractor was asked if they could finish the top floor of the building early to enable this to happen, but not to increase the project budget.

Solution: Because of the transparency of the budget it was possible to work out an affordable solution, which cut down the fitting out of one floor to create early access to the top floor, by commissioning the lift early and fencing it off from the rest of the site. Fire escape access was also approved by doing the same to one of the two fire escape stairs. Some contractor usage of the lift to lower floors was agreed.

which is transparently attained by sufficient checks and balances (formal systems are audited) without suffocating the service or the delivery. An effective system is much harder and requires a consideration of the expectations and the perceptions of the customer and other stakeholders. In bespoke construction projects, this should lead to a dedicated system adaptation using the project constraints and client requirements, making the system dynamic.

A non-dynamic project based QA system renders it at least frustrating and at worst harmful to the efforts of project staff operating within them. Accountability should be assigned at operational levels for conformity and at project director level for effectiveness. This will be achieved better on larger projects by using a responsibility matrix which identifies all members of the delivery team as well as specialist co-ordinators and makes them responsible for specific tasks, which are co-ordinated so that no job is overlapped or left uncovered.

Quality also needs to be associated with specific customer requirements so that the commissioning and testing of the building is fully connected with this strategy planning stage of ensuring that the finished building is fit for purpose and all systems work. This is not a static compliance with customer specifications, but a proactive process in checking that the specification which is being quality assured fulfils requirements and maybe improves the status quo by increasing the durability or suitability. Too often buildings are failing because requirements are not interpreted fully. The strategy for defining and controlling the quality is to maintain close relationships with the client and to manage the stakeholders in accordance with their influence and power. This confirms the need to define the brief properly and to develop value.

Figure 4.15 illustrates the relationship of the quality systems with meeting the customer's requirements. It is clear that there are a range of quality outcomes and if perceptions are equal to expectations there will be satisfaction. Maister's law states:[12]

satisfaction = perception − expectation

If there is a negative difference between perception and expectation then there will be dissatisfaction. If the difference is positive then there will be customer delight. The promises which are made in terms of how and when (service) are also strongly related to satisfaction. A customer may well be happy with a longer contract time or a later start on site if what is delivered is not later than advised. One of the problems with competitive tendering is that it often promises more than is delivered resulting not only in claims, but in dissatisfaction. This is a lose–lose or win–lose situation for contractor and client.

Benchmarking is used as a way of maintaining and improving quality. Key performance indicators (KPIs) have been developed within organisations

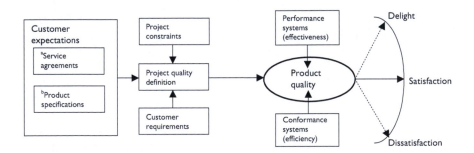

Notes:
*Service is defined as based on stakeholders' expectations and perceptions
*1Product is in conformance with measurable specifications

Figure 4.15 Quality planning and delivery

Source: Based on Maylor (2003: 167)[13]

in order to measure, objectively if possible, the level of quality achieved. They are more easily measured where long term partnerships exist and set uniform parameters. More will be developed on this strategy in the next chapter. Clients like the use of third party accreditation as in ISO 9002 and the National Housing Building Council Buildmark. They relate to these more easily and the industry needs to consider the use of these (see Chapter 13).

Information and communications management

Information management supports transactions and decision making. Decisions may be programmed or non-routine and may be required at strategic, tactical and operational levels of management. Information is specific, but starts with processing raw data into a management information database which supports common decisions. Information systems break down into transaction processing, decision support systems (DSS), executive inform-ation systems (EIS) or ad hoc enquiries. Figure 4.16 shows the hierarchy of these systems in relation to the database and the level of management use, and the context of the decision in proportion to the level of management use.

A database is a collection of structured data. The structure of the data is independent of any particular application.[14] A database management system may draw on organisation and/or external databases. It integrates the different sources and structures data in a format for the outputs which are required in the organisation. Information needs to be timely and appropriate for use in either transaction processing or decision support. A transaction process is routine such as the financial accounting process to record payments and receipts and to present final accounts. Management accounting on the other hand needs budget information to make strategic decisions about investments and project selection.

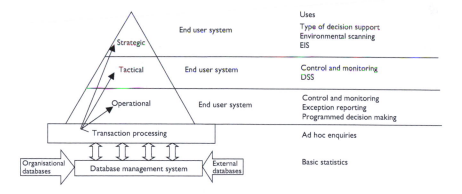

Figure 4.16 Hierarchy of MIS in relation to the database management system

Source: Adapted from Lucey (1991)[15]

Decision support systems are in place to support rather than make decisions and therefore need to be more specific to the end user requirements so that they remain helpful and used. An example of a decision support system relevant to projects would be an update showing the progress critical activities. This would pinpoint where resources are needed most critically to pull the project back onto target.

Certain questions are relevant in the design of DSS such as: What type of decision? Is it for long term or short term? How long have you got to make the decision? How is it integrated with other decision-makers? These questions can be applied for the process of setting up project management information systems (PMIS) which are unique to the project type.

In the case of projects DSS is a key requirement and because of the temporary nature of projects needs to feed back quickly or preferably feed forward to be effective for decision-making.

Project management information systems (PMIS)

PMIS need transaction analysis such as progress reports on programme and budget updates, and decision support such as those supporting risk and value management. Knowledge based systems may be used to support estimating and integrated systems may be used to connect computer aided drawing (CAD), 3D modelling and 'walk throughs', costing and planning systems. Integrated information in its most advanced form is the subject of much research and object orientated models which define generic elements which appear in design, planning, estimating and financial systems are being developed today and appear in systems such as building information modelling (BIM). This however needs a lead in period to make sure that systems and databases can be properly integrated. Currently information may be manual or digitised.

Extranets

Project information is often digitised to create a web based database and distribution called an extranet. This provides access on a controlled basis to all users of the project information. There are many commercial templates for extranets some of which are customised for construction type information. Many users will want to do their own customisation to suit the nature of the project and will create protocols deciding on which user can see what information (e.g. costing). In order to stop overload, certain users will be alerted automatically when information relevant to them is released or amended. See Case study 4.7 for an example.

Advantages of the system are synchronised receipt of information for checking and action and the tracking of approvals. The final approved

Case study 4.7 Example for extranet protocol

For example a services drawing may be uploaded by the consultant that is relevant to the subcontractor. The architect will see it to make sure there are no clashes of services with the fabric and the structural engineer will particularly see services which are in the ground or pass through structures such as walls and floor. The main contractor/project manager will see it to formally accept it and issue it formally. All drawings would be seen by the project manager. However this drawing will not be relevant to other subcontractors or consultants unless it affects finishings or landscaping. The architect will have a central role in ensuring it meets the requirements of the client. Responses to such drawings may be cost and time related if they represent a change to the specification or the scope or are later than requested.

drawings need to be clear by making them the only ones available. The integration of these drawings with others would benefit from private meetings or restricted viewing where major concept issues are to be resolved in primary parties.

Typical information that is stored as well as drawings are change orders, quotations, valuations and cost control reports, time schedules (programmes), health and safety files and maintenance manuals, risk assessments and registers, incident reports, purchase and stock control, invoices, meeting minutes and action plans, short term and Gantt charts and critical path diagrams, quality plans, earned value analysis, progress reports and anything else that can be scanned or digitised.

Saving paper depends upon the size of the drawings/document so they can be read comfortably on screen and how many are required in hard copy on site. Mobile devices may be used to store and check information which is not a drawing or part of a drawing at the workface. Pictures and 3D/4D portrayal may be vital to better understanding and are only available on screen. Responses to queries are easier and quicker and may reduce delays and increase productivity/quality. Fast and reliable delivery of information depends upon the confidence that the users of information put in the system. For all users to benefit they need the correct equipment to receive information and this may not be cost effective for all to have this, so paper formats need to be distributed. Face to face communication at meetings, toolbox talks and supervision are still needed for the workface and supervision team, but also to build up professional team confidence.

Table 4.4 PMIS matrix

Project Databases	Functional Database			
	Cost	Time	Quality	Health and safety
Project 1	Price per bed			Accidents/100,000 workers
Project 2		m²/hour		
Project 3			+ 3mm	

PMIS systems

In the PMIS the databases will include cost, time and performance data as well as many other databases to serve the transactional processing such as accounting, inventory, buying, estimating, health and safety etc. Table 4.4 shows a simplistic structure for a PMIS.

Information may be hard or soft. Hard properties are those that can be defined, measured or assessed in a tangible way, for example, the cost and production outputs of various expansion options against the market forecast. This may include the attachment of various levels of uncertainty. Soft properties are more imprecise and may depend on individual taste and value. For example, for a company moving to a new HQ out of the central business district, it may be necessary to consider the incentives that will be required to move staff away from city centre facilities. The PMIS should adjust outputs to match the client's business case parameters. For example the client might want the price per bed rather than the price per square metre.

Socio-technical

These systems are a way of combining the technical hard data of production outputs with the soft data, which include the expectations, aspirations and value systems of the organisation's workforce. A knowledge of both constitutes a more holistic decision making process.

Expert systems

Sometimes known as knowledge based systems (KBS), these are computer based systems designed for making more complex decisions more accessible to non-experts. They are built up by rationalising the steps made by an expert to take the non-expert through the decision structure based on the answers given. They can build in 'what if' scenarios as well to provide instant sensitivity analysis. Solutions are derived by their link with a database such as estimating price book. A price book on disc is not a KBS, nor is a computational system such as software that generates a bill of quantities. The user is not surrendering their judgement, but is using the system to enhance judgement and improve decision making. Case study 4.8 shows the use of a knowledge based system.

Case study 4.8 A knowledge based system for project cost

This expert system interacts with the user to determine project cost and works through maintaining a knowledge database. With an input for the geographical and relational location of the project, the floor area, the quality of the finished building, the degree of external works, the nature of the building use, the type of design and the structural form and the programme time available if critical, it will calculate average cost.

It then assesses an updated database of costs such as the Building Cost Information Service (BCIS) in order to give a range of current construction prices. It automatically adds statutory and professional fees and allows a land price to be added to give total project estimate for that location, building type and size etc. A report is generated outlining constraints, elemental cost breakdown and uncertainties in particular circumstances.

It is not an econometric model so depends on the significance of the current and updated projects used in the database accessed for its accuracy. A further sensitivity analysis might be necessary for the effect of interest rate changes, inflation, local factors and non-standard specifications.

It does not take direct account of market supply and demand which might affect the competitiveness of tender prices or the procurement system used, but an allowance can be made. The tenders may also be adjusted manually for soft factors such as maintaining client bases, partnership agreements, client preferences and prestige.

Grant regimes and tax breaks may be fed into this type of database if they are locational, but not if they are negotiated.

Executive information systems (EIS)

These provide selected summary information for director level. In the project context these need to be standardised across projects for comparison. Key features should be easy to use, give rapid access to the data that feeds into the report, provide data analysis – trend and ratio calculations and a quality presentation – and be interesting and understandable, for example use of graphs and colour.

The project manager also needs to provide data to the client in a format suitable for them to monitor their business plan. For example if a change has been requested in the scope of works then the effect of this change could be worked out by the client in terms relevant to their own EIS, prior to approval. Issues such as client commitment, cash flow market considerations and production targets may or may not override extra cost.

Conclusion

Project strategy is critically considered at the feasibility stage of the life cycle as well as its ongoing development and implementation throughout the implementation stages. Typically there is a lot of uncertainty even after feasibility studies are complete and the overlapping strategy stage identifies, prioritises and allocates risk where it can be best managed. Further discussion on evaluating and responding to risk optimising value systems is dealt with in Chapter 8.

The master plan typically includes a time schedule, a cost plan, a quality plan and a risk assessment which provide frameworks to optimise the client objectives and recognise the project constraints.

Cost planning needs to take into account an on going value exercise to suit the iterative nature of the design and the cycle of developing it to its final approved post planning application stage. The initial outline cost plan budget becomes an important marker for the client and market changes, design changes and client changes are often not seen as reasons for an inflated budget which means that value has to be generated and design changes made to suit.

Taking external factors into account is vital to the success in identifying risk. Political, economic, social and technological factors can easily change so it is wise to build in flexibility and contingency factors in to cost and time schedules. Sometimes professionals have often been criticised for their reluctance to give advice that might seem unpopular, so that clients are unprepared for steep price rises, which exceed budgets and cause expensive late changes.

A strategy for managing the programme time at this strategic level of detail is more dependent on reaching critical milestones than it is in calculating durations and the planning needs to be iterative and flexible to allow for this. A strategic view at this stage involves being aware of the linkages between the design, construction, procurement, commissioning and client fit out stages. In a climate of 'time is money':

- A fast track strategy would allow a certain amount of overlap between these stages dependent on the nature of the work with impacts on the level of risk. The design management needs to be very tight to reach a series of critical package milestones. Commitment to early packages orders will mean early involvement of a construction manager.

- A fast build strategy will focus on the need to complete each of these stages sequentially and squeeze the timing of each of the stages so that there is a proper handover between. This has more risk in resource management as the critical path of the project becomes more squeezed.
- A procurement strategy sets the constraints for the delivery sequence of the work.
- A commissioning programme formalises the sequence for the end stages of testing and client fitting out and occupation. The beginning of commissioning is the end of construction for each holistic phase and so commissioning needs to be considered first. Quality depends on proper testing and a holistic approach to the whole building in order to hand over with zero serious defects.

The quality planning process is a strategy for meeting clients' requirements and expectations and should address service quality as well as product quality if perceptions are not to fall short of expectations. This is best achieved by close relationships with the client through the continued value management process. The quality improvement process has a place both in the formal post project evaluation (Chapter 14), and also in the development of a project organisational culture, where lessons are continuously being learnt and simultaneously being applied. The basis for this is covered in Chapter 6.

A project information system needs to be integrated across the supply team and keep the client informed in a format which integrates with their own reporting and monitoring systems for asset management and the business. One of the developing issues is the compatibility of CAD/BIM and document distribution systems so that information can be received electronically and simultaneously in order to integrate say design and construction approaches. This helps to change the culture of fragmented information systems that are separate for the project team and client reporting. The sharing of project accounts transparently with the client is becoming an important part of the partnering culture. BIM is discussed in Chapter 3 because of its role in project definition.

The setting up of structures which promote teamwork and integration of the project packages is critical for good communications, responsible health and safety, and creating if possible a blame free culture. An enlightened strategy here would be the bringing together of the design and construction at an earlier stage. Chapter 6 on organisation and culture and Chapter 7 on engineering the psycho-productive environment take this further.

A project handbook may be useful in order to integrate the various responsibilities and duties. A good example of a handbook framework is given in the CIOB Code of Practice.[16]

References

1 Burke R. (2003) *Project Management: Planning and Control Techniques*. 4th edn. New York, Wiley, p.32.
2 Project Management Institute (2008) *A Guide to the Project Management BoK*. Project Management Standards Committee. Newtown, USA.
3 International Standards Organisation (2003) *ISO 10006. Quality Management. Guidelines to Quality in Project Management*. London, ISO Standards Bookshop.
4 Reiss G. (1997) *Project Management Demystified*. 2nd edn. Abingdon, Taylor and Francis.
5 Smith N.J. (Ed.) (2008) *Engineering Project Management*. 3rd edn. Oxford, Blackwell Science.
6 Edwards L. (1995) *Practical Risk Management for the Construction Industry*. London, Thomas Telford.
7 To read more about this see D. McGregor (1960) *The Human Side of Enterprise* (Classic). (Reproduced 2006; New York, McGraw Hill).
8 Ballard G. and Howell G.J. (1997) *Improving the Reliability of Planning: Understanding the Last Planner Technique*. Available online at: http://www.ce.berkeley.edu/~tommelein/LastPlanner.html (accessed 12 November 2011).
9 CIOB (2010) *Code of Practice for Project Management in Construction and Development*. 4th edn. Oxford, Blackwells, p.170.
10 CIRIA (2001) *Publication 556 Managing Project Change A Best Practice Guide*. London, CIRIA.
11 Maylor M. (2003) *Project Management*. 3rd edn. Harlow, Pearson Educational.
12 Maister D.H. (1993) *Managing the Professional Service Firm*. New York, Free Press.
13 Maylor M. (2003) *Project Management*. 3rd edn. Harlow, Pearson Educational, p.167.
14 British Computer Society (no date) *BCS Glossary of Computing and ICT*. 12th edn. Swindon, UK, BCS.
15 Lucey T. (1991) *Management Information Systems*. 5th edn. London, DP Publications, p.264.
16 CIOB (2010) *Code of Practice for Project Management in Construction and Development*. 4th edn. Oxford, Blackwells.

Chapter 5

Lean construction and benchmarking

The construction of the project is the most interesting phase when work actually starts on site and some highly visible progress is made in one of the last major stages of the project life cycle. Construction however can start too soon because there is an over eagerness to proceed with the demonstration of progress on the project and to make reassuring promises about the programme completion. It is vital though to have properly completed the planning stages for production, so that value can be optimised in the construction process and necessary resource planning allowing for lead-in time can be in place. Construction also needs to prove that it is value for money by having efficient systems and continuous improvements in its outputs. Construction companies, to remain competitive, need to eliminate waste, benchmark their performance and keep themselves abreast of the expectations of their clients following recent reports in the industry. This chapter concentrates on the lean performance of construction and the potential for continuous improvements. Case studies will be used for best practice.

The chapter objectives are:

- to review the performance of the UK construction industry in general and its strengths and weaknesses
- to apply the principles of lean production to construction
- to understand the principles and the potential of benchmarking and measurement techniques
- to assess construction performance, using case studies
- construction procurement and programming
- Appendix – the critical path method and its use.

The performance of the construction industry

Construction is under pressure today to 'improve its act', but there is a great variation in performance on construction projects. Many clients have had good experiences, but equally many have been let down on promises that they believed to be binding and have seen extended programmes, suffered cost overruns and experienced quality problems. The structure of the industry is fragmented between design and production, which has often meant that the client has been confronted with separate design and production teams and sometimes a plethora of planning advisors. It has impeded an integrated approach to delivering value to the client and prevented openness and honesty because of a lack of trust between the separate teams, limited accessibility to the client and extended communication channels. The structure has often led to a confrontational and contractual approach. The industry's products are perceived to be too expensive and it is taking longer to finish building projects than leading practice in other industries. Construction is a significant part of national output and productivity. Ruddock measures construction gross domestic product between 5.58 and 5.95 per cent internationally with the larger percentage in the smaller countries.[1]

Construction projects can be breathtaking in their size and impact, but the question is, does success mean an instant iconic landmark, a world beating dimension or an effective and efficient construction? Projects such as the Petronias Towers, Sydney Opera House or the Millennium Dome and Terminal 5 at Heathrow Airport have become headline news, not for their feats of groundbreaking engineering and architectural innovation which they possess, but for going over budget and programme. Case study 5.1 looks at the credentials of the Channel Tunnel.

UK construction

The UK's GDP is around 5 per cent, which is similar to other developed countries. It had a comparable level of GDP per capita performance with France and Germany, with the USA being higher.[2] The Latham Report (1994) succeeded in providing a snapshot of industry performance that fell short of the productivity achieved in other industries.[3] The industry was challenged to meet tough improvement targets in a five-year period of 30 per cent productivity,[4] and to improve its image. Latham proposed better integration of the team including the client, fair payments, better risk management and value engineering, more technology to improve information flow and life cycle costing and standardisation of components. However, he also saw the need to move away from lowest price tendering procedures to value based tendering systems with early contractor involvement in order to deliver best value.

The Egan Report (1998) follows on from Latham and suggests that good construction is very good, but many clients have a poor view of construction performance because of failures in the system.[5] This shows up in the fact that over half of all projects either finish late or over budget, or are of inferior quality. The Egan target was to reduce construction prices by 10 per cent

Case study 5.1   Great structures

The 'Chunnel' is a joint Anglo-French venture first mooted in Napoleonic times and is, at 31 miles (50km), one of the longest tunnels in the world, joining Calais in France and Folkestone in England under the sea. It was built in three years at a cost of £21 billion, which is 700 times more expensive than the Golden Gate Bridge in San Francisco. It used tunnel boring machines two football pitches in length which bored 150m per week on average × 7.6m diameter in one day, hundreds of feet below the seabed. These tunnel borers started either side of the English Channel and met halfway, exactly lining up. It is in fact three tunnels with an east- and west-bound tunnel for a high speed rail link taking trains at 100 miles per hour; a central tunnel which acts as an escape route; and a service tunnel, and there are cross-over tunnels in between. It opened in 1994 and in the first five years, in spite of some operating problems, it took 28 million passengers and £12 million tonnes of freight. There have been three fires and, in 2008, 32 passengers caught in a train fire were all safely evacuated through the central tunnel with only minor injuries.

year on year by improvements in the process and to raise the profit margins of contractors at the same time, as failures in the competitive process were diagnosed as driving waste into the system and lowering the motivation of contractors who were in sectors outside housing and development and working on very low margins. Many of these principles are picked up in the later Construction Matters Report as beneficial to better industry efficiency and bringing about more sustainable construction.[6]

Wolstenholme's (2009) analysis of 11 years of benchmarking revealed construction projects coming in on time and cost was still a 50/50 chance, but in general there was an average improvement of 5 per cent over 10 years.[7] There were better results for projects committed to best practice known as demonstration projects. These showed a clear progressive improvement in time and cost reductions, though at 50 per cent of the rate needed to keep up with the Egan 10 per cent year on year target. Generally, profitability, health and safety performance, environmental targets and productivity (turnover per head) did increase as per the Egan targets.

The Modernising Construction Report (2001) identified that 66 per cent of public projects out of 66 came in 50 per cent over budget or more and two-thirds exceeded their programme date by 63 per cent.[8] Cain (2004) further compiled some evidence for waste in the industry using the breakdown of 50 per cent labour, 40 per cent materials and 10 per cent profits and overheads.[9] He identifies there was an average 40 per cent labour productivity and a 70 per cent materials usage in the UK industry.[10] Current best practice does not exceed 50 per cent labour productivity. He argues that Latham's 30 per cent cost saving could be made by getting labour up to 70 per cent efficiency and reducing materials wastage to 4 per cent. These do seem possible and give a focus as well as breakdown for eliminating production waste over a period of time.

Defining the terms of productivity

Productivity is an overarching principle to improve outputs for the same inputs or to maintain outputs for less inputs and is generally applied to construction, but has implications for any process including design. Value is described similarly as function divided by cost and is generally applied to the design process to optimise it. Productivity can also be affected by quality management which should be applied to reduce abortive and corrective work. Integrated management of design and construction can be tackled by buildability studies to optimise design and construction.

Olomolaiye *et al.* distinguish between breakthrough in productivity and control, describing both as necessary.[11] Breakthrough uses innovative new approaches to improve systems, whilst control maintains existing systems and aims to make them more efficient. Control is also necessary for the ground-breaking productivity improvements which make a difference to productivity. Innovation requires cultural and business change, management commitment and changes in attitude and this is the area of lean construction. This chapter will concentrate on productivity, lean construction performance, layout and organisation. Chapters 6 and 12 will consider other aspects of cross-project organisation, partnering, leadership and supply chain management that can be improved.

Standard ways of measuring productivity

There are plenty of texts which are devoted to the subject of productivity from the control viewpoint and much has been written about the identification of inefficiencies and the subsequent systems for implementing production improvements since the time of F.W. Taylor in the nineteenth century. Many of these texts have concentrated either on labour efficiencies by improving supply, layout and method or on the delivery of improvements by better technologies and materials. This section summarises the more traditional approaches which mainly come under the heading of control.

Work study – time and method study

As the name implies this tends to look at the production system, labour productivity and the layout of workplace with little reference to the delivery of materials.

Work study is the observation of workers in the context of the workplace starting with the 'scientific' studies commissioned by F.W. Taylor who sought to control the workplace and incentivise production by better conditions and pay, but also by the use of sanctions on those not able to perform. The Mayo studies of behaviour in the Hawthorne Electrical Light Company in the USA (1927–1932) countered some of the carrot–stick based incentives of scientific management by looking at what could be done by giving people responsibility for their output and by providing the right level of workshop environment in terms of breaks and comfort.

Time study provides information for standard outputs using either philosophy. It also identifies unproductive time and tries to eliminate this, categorising idle time, productive time and time spent in supporting activities. The focus is to maximise productive time. Charts and stopwatches record elemental times for completing tasks and their subelements in order to identify better efficiencies. Professional observers are called into the workplace. Measurement of standard output is the basis for improvement.

Activity sampling is a simpler work study technique using the direct managers to observe worker productivity thereby creating a direct link with problem solving improvements which may provide faster results. Observations are usually made throughout the day at regular intervals for a week, because they give a rough and ready percentage of productive and support work time based on the *actual workers* on site. Typical worker efficiency on construction sites is 40–45 per cent.[12] Corrective action is based on reflection on the process, discussions with workers to identify problem areas, and workers' suggestions. This is a mixture of quantitative and qualitative analysis.

Method study is the next logical step to time study and focuses on the improvement of a particular inefficient process using flow charts and identifying an efficient process to eliminate elements that are wasteful. For example, Gilbreth observed techniques to eliminate wasteful action.[13] He reduced the process of laying bricks in a wall from 22 actions to seven actions which he called Therblig(s) (Gilbreth spelt backwards). The method study process challenges existing practice (What? How? Where? When? Who with? Why?) and proceeds to record and examine the reason why, develop alternatives and select, install and maintain. The basic standard flow ASME symbols, attributable to American Association for Mechanical Engineers, are still useful in plotting a process.[14] These are shown in Figure 5.1.

The most common manifestation is an outline process chart. The symbols do not distinguish between obvious non-productive time (delayed concrete

○	Operation	Productive. Something is produced	e.g. mixing concrete
⇨	Transport	Movement of materials or workers	e.g. moving concrete
▽	Storage	Materials stored or not used straight away	e.g. stocking bricks
☐	Inspection	Checking or testing quality	e.g. cube test
D	Delay	Labour or material held up	e.g. concrete lorry delayed

Figure 5.1 ASME standard symbols for flow charts

with a concrete gang waiting) and supportive waste such as sending someone to go back down the ladder empty handed. Rest periods are essential for human input, so there may be a necessary delay unless machines are used. A flow diagram shows the outline process chart superimposed onto a real layout. On a flow process chart distances and time can be noted. There are many other ways of representing actual processes and alternatives. The time taken to carry out the processes in Figure 5.2 is controlled by speed of mixing and transport across site, with probable delays if only one crane.

Transport raw materials	Store sand and cement	Carry material to mixer	Mix concrete	Carry concrete to floor	Lay concrete	Vibrate and level concrete floor	Test concrete cubes	Wait for floor to cure and strengthen concrete

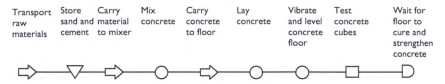

Figure 5.2 Outline process chart – laying concrete floor

If the floor was large there could be substantial savings in manpower and delays by using a concrete pump. In Figure 5.3, automatic mixing, the simultaneous transport of concrete pump with concrete and no transport time for raw materials or concrete on site, lead to vastly improved productivity of a concrete gang laying concrete. This has to be offset against the pump cost and ready mixed concrete. The key issue is to match the controlling rate (laying gang speed) to the delivery of concrete and pumping capacity then no person or machine is wasting time. 'Just in time' (JIT) delivery of concrete will also mean that lorry drivers are not wasting time.

Motivation is a broad subject, but indirect productive action is an important aspect of creating and maintaining the right work conditions and trying to manage change. Work study tends to get filed under a scientific approach, but there is no reason why the intrinsic aspects of human behaviour in regard to work are not aired as was the case in the Hawthorne Studies where a norm of productivity was established which belied attempts to improve it. Other writers have spoken of equity, valency (the worth of the outcome),

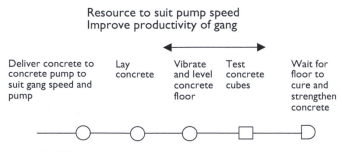

Figure 5.3 Effect of concrete pump

work satisfaction, fulfilment and achievement. Chapter 7 deals with some of these aspects as well as the impact on the working team and the ability or not to synergise with each other to create something that equates to more than the sum of its parts.

There are many books which focus on the tools and rules for worker observation and various means for improving productivity. In construction projects workers are under the control of different organisations with very little direct labour from the co-ordinating contractor with the overall responsibility for delivering to time and cost. General contractors have a responsibility to co-ordinate between different contractors in terms of efficient delivery of materials, layouts and shared resources such as cranes and sequencing of workplaces. Their overall programme determines the rhythm of production and the resourcing of production in terms of the durations allowed.

Specialists supply their labour on a price basis that is competitive where other factors such as market forces may affect price. Productivity depends on the rate of working, the availability of work and the matching of space to size of workplace. It also depends on the balance of sequential trades, for example in a beam, the concrete, the formwork and reinforcement fixing, which are specialist trades, need to work at the same rate. Availability of labour may slow down specialist completion targets. Specialists may be efficient, but other organisations or shared plant they depend on may not be, so these need to be co-ordinated. Their resourcing needs are critical in proportion to whether they are on or near to the critical path of production. Critical path refers to whether their delay impacts on the delay of the whole project or the section only. This balance is a professional choice by an experienced person and several pieces of work by the same specialist input may be balanced by moving gangs around. Certain shared resources such as forklift truck or a tower crane capacity can be critical to avoiding congestion of use. Efficient layouts can also reduce transport time. This leads us on to the important principles of lean production.

Principles of lean and value

The definition of value streams is the sequence of activities required to design, produce and provide a specific good or service along which information, materials and work flows so that the customer demands (pulls) the product.[15] In customer terms it delivers what the customer values efficiently. In construction it is characterised through design, procurement, production planning logistics and construction. Success means front loading the resources in design in order to eliminate waste efficiently in manufacture by planning ahead, concurrent working between the design, manufacture and supplier and working from a reliable database of products, systems and components so that there is a carry over of learned systems into new products and designs.[16]

The five principles of lean proposed by Womack and Jones are:[17]

1   Define the value needed by the customer.
2   Identify the value stream and challenge every step and eliminate those steps that don't add value.
3   Line up and balance the flow to eliminate waiting and stock.
4   Let the customer pull the value so that they guide production.
5   Seek perfection by redefining the process for continuous improvement.

Value focused thinking is a holistic approach to the whole business to bring the best value product to the customer. This means looking at all the processes to bring the product to the customer. In projects it means understanding the customer's business and the wider organisation of the project, not just applying value to the process itself. A value stream map (VSM) for the project would consider purchasing, design, production, accounting, recruitment and market. For specific projects there is a need to incorporate the detail of parallel activities, such as parallel subcontract actions, in the production process and for an awareness of critical path due to integrated sequencing and resource constraints (see Appendix). Flows in materials information and other resources are important to know and integrate to maximise worth and elimination of waste.

Continuous flow is a key activity which tries to balance the connected activities in a process such as bricklaying so that no one is waiting for, say, mortar and bricks and bricks stocks are eliminated by batching deliveries to suit daily production ('just in time'). The customer pull is more subtle in construction, but on a private housing site, houses are produced at the selling rate predicted. That speed determines the manpower for all following trades e.g. brickwork, first fix, plaster, second fix, flooring and painting, and each must run at the same speed, sometimes called line of balance. On less repetitive jobs this might relate to phasing requirements. Production lines run parallel and are interlinked and this is shown through a critical path logic

diagram, a process known as CPM. Successful lean production and, by implication, construction projects have developed their own culture.

Standard symbols developed by ASME and Gilbreth facilitate the
identification of waste, and value is added by reducing or eliminating such
activities as storage, transport, multi-handling and delays. Lean systems seek
to balance the flows between design and construction and client, not just
maximise individual outputs or processes. VSMs are usually shown as flow
charts with an integration of related flows.

F.W. Taylor had great success concentrating more on the production
process (time, motion study and incentive), but VSM is more holistic and
emphasises the importance of information flows, business objectives and
manufacturing lead times. These additional issues are critically important
for construction projects, which need better communication and operate with
many organisations and often in prototype where flows need to be devised
as well as fine-tuned.

Toyota is the leading and most used example of a lean production system.
Toyota chief engineer Ohno would make sure that the workers are involved
in what is known as the Plan, Do, Study and Act cycle (PDSA) developed
by Deming, because he believed that separating planning from doing was a
big waste factor (see Figure 5.4).

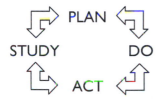

Figure 5.4 Deming's cycle of learning

The Toyota MD Sakichi Toyoda developed the quality and productive
process of Toyota cars, post war, in response to what they saw of waste
built into the large mass production car factories in the USA. What they
saw in the supermarkets where an empty shelf prompted restocking was
considered to be more cost effective in being driven by customer demand
(pull) and not a wasteful push system that built up stock and the need
for excessive storage as seen in car factories. Case study 5.2 shows their
solution.

Other terms which help to develop a defects free culture are coined within
the team concept such as 'the wind in the sails' which refers to the essential
human contribution to the system. There are also various phrases which have
been developed in the Toyota production system to eliminate defects and
correct human error without abortive work:

- *Jidoka* to diagnose and correct human error automatically or stop the process to allow correction thus allowing wider supervision.
- *Poka yoke* which means incorporating failsafe systems to avoid counting, sequencing or product physical attributes errors.

We will study defects free more in the quality management chapter (Chapter 14). Construction is less machine orientated, but could adapt the principles.

Case study 5.2 Toyota production improvements

The Toyota production system (TPS) was developed by Toyota and Ohno their chief engineer to tackle crippling problems which existed in many manufacturing systems. Ohno wanted to build cars to customer order, deliver them instantly and maintain no inventories. It was competitive on three simple principles:

- Eliminate *muda* (waste).
- Avoid *muri* (overburdening people or machines) to stop bottlenecks and breakdowns.
- Enable *mura* (smooth flows) from one process to the next to eliminate waiting and transport.

The system incorporated the idea of production *kanban* at its root which refers to just in time deliveries for the rate of production and withdrawal *kanban* which adjusts production rate to customer demand. Its other route was *kaizen* (continuous improvement); this recognised that improvement was always possible by relentless reflection (*hansei*) and it used its workers to identify and problem solve on areas of waste. In addition to these productivity improvements, quality had to take precedence.

In 1961 every employee in Toyota produced one car per month (10,000 units/month and 10,000 employees). Because of improvements, by 1987 Toyota produced 230,000 to 250,000 units with 45,000 employees, so that every employee produced five cars per month. This improvement is critical to competition, and 'best in class' productivity without reducing quality increases profit margins and profits assuming the same or more sales are possible. This type of manufacturing technique involves radical changes of people, machines and material procurement which is usually dependent on incremental board level decisions in the long term. On independent large projects there is more autonomy which gives potential for faster radical applications and results.

Ohno's seven wastes

The Japanese term *muda* is radical as it identifies unnecessary action and is seen as the actions in production for which the client is unwilling to pay. The Toyota view of traditional production is that there is 5 per cent of pure value adding activity, but there is also essential activity that supports this such as an access scaffold, lifting, walking to workplace, machine downtime, rest periods or stacking material. The remainder is outright waste such as keeping paid workers waiting, throwing away offcuts or rework and sweeping up. With support work there are opportunities for innovative technology and methodology to reduce rest periods and machine maintenance or to eliminate the need for storage and to make fetching and carrying more efficient. Method study is used to identify this and leads to the use of less material and fewer personnel. Ergonomics may also make the worker more comfortable and therefore more efficient. There is no excuse for outright waste but this needs a cultural change in the construction industry which wastes 90m tonnes of waste in the UK alone; 15m tonnes are new unwanted materials. Ohno, the chief engineer at Toyota in Case study 5.3, considered all waste.

Case study 5.3 Cultural development waste

> Classically Ohno observed a worker manning their machine.
>
> Ohno   'How often does your machine break down?'
> Worker 'It never breaks down Ohno san.'
> Ohno   'Well then what do you do all day?'
> Worker 'I watch this machine.'
> Ohno   'All day you watch this machine that never breaks down?'
> Worker 'Yes, this is my job.'
>
> Ohno is reputed to have termed this 'a terrible waste of humanity'. It really demonstrated the potential for the worker to be engaged with the planning system as they had the power to improve it. It demonstrated a waste of not using every worker's thinking capacity more effectively which may also motivate them more.[18]

Ohno identified seven wastes as a way of analysing where work needed to be done to become more efficient. These are well known but they are described below in terms of construction activity:

• Waiting means machines or people are subject to delays because of lack of materials/stock or unbalanced production line processes which are dependent on each other. The solution is to balance resources to make each process matched in length and/or ensure delivery is reliable so that materials are always there. In construction, a series of repetitive trades often follows each other sequentially into each room and each needs to be resourced to take the same time to avoid fast trades waiting.

- Over production means that too much work in progress is done in one batch. In construction it doesn't mean it is not wanted, but it is delivered early so is stored or damaged, or it is processed early so needs temporary protection and workers are laid off or inefficiently diverted whilst others catch up. Capital is tied up in paying for it early. The solution is 'just in time' or *kanban* with a calculated minimum buffer of stock to allow for unexpected problems or interventions. Flow charts help to identify problems of non-value adding storage, protection and damage replacement. Space on site and poor storage conditions make it worse.

- Rework refers to not getting it right first time and could be quality defects from workmanship, wrong or late instructions or drawings. The solution is better preplanning, value management, quality control and feedback, worker accountability and firm change management control. To assume no change of scope in construction is counter productive and flexible planning is effective to keep customers on side and stress levels low.

- Motion waste refers to excess fetching and carrying and poor ergonomics and twisting which are down to layouts and workplace planning and design. Foundation work may force bricklayers to work in cramped conditions that cut down productivity. Filling up trenches with concrete or battering the sides may be more expensive, but speeds work up and saves resource costs. Calculations, method study and preplanning help choices to be made. A standard bill of quantities (BOQ) description does not.

- Over processing refers to over specification and doing more than the customer asks for or wants. A function analysis and a joint customer-designer value appraisal should be a base for brainstorming a 'just ok' design and specification.

- Inventory refers to having to house, stack and multi-handle material whilst it waits to be used. A solution can be worked around 'just in time' which strictly means delivering just enough material for a day's work. Excess material gets in the way like work in progress above, but in a pull system (call off) excess can be controlled better. Some materials are on long leads and so suitable buffers and associated costs can be built in. Supply chain management helps here.

- Conveyance waste is harder to visualise and happens because of unwieldy batches or targets, so the target is to have smaller batches and to design layout carefully. It induces motion, inventory and delay wastes. An example is the use of the tower crane on site when one trade ties up the crane for lifting a large load of formwork plywood for the next two floors of concrete frame. This causes a bottleneck delay for other trades needing to use the crane daily. The stock also takes up floor space and creates blockage, multi-handling and diversion of labour from true value adding activity. A part load with a single lift supplying a day's plywood is likely to be less disruptive.

To these has also been added talent (or resource) to recognise that talent needs to be identified and nurtured where it exists to avoid wasting it. This suggests flexibility in working teams and multi-tasking. Each of the wastes needs to be reduced so for example reducing motion means to stop moving things to storage and to eliminate multi-handling. These wastes mainly emerge from the production flow and can be quantified by value stream mapping the production process. Ohno trained the workers in reducing defects (rework) by giving them the power to stop the line if they received a defective component from upstream of the value stream. Dennis also warns against a 'scavenger hunt' for *muda* which takes away from the more proactive management of planning a process.[19] He calls for awareness of waste so that methodology and worker training are prioritised to incentivise efficient working layouts.

Techniques have been developed such as the 5S system to ensure a logical, efficient, transparent and standardised system of work.

Visual management and the 5S control system

Visual management refers to the use of diagrams to communicate improvements and change to workers. 5S is a tool to improve layouts for better productivity. The five Ss are Sort, Stabilise, Shine, Standardise and Sustain.

- Sort is a way of removing unnecessary tools and materials from workplaces to ensure choice is simple and uncluttered and there is greater effectiveness in use.
- Stabilise is to make sure everything has a place and to identify minimum and maximum stock levels relevant to lead times of materials.
- Shine is to keep clean and to prevent clutter gathering and to fine tune improvements.
- Standardise is to create rules and make sure they are meaningful for everyone to understand.
- Sustain is to monitor performance and challenge existing practice in the light of feedback and performance, but also to protect gains made so that the system does not slip back into waste.

This system has the benefit that it goes beyond worker productivity to look at the constraining factors for the whole team. It also, importantly,

Case study 5.4 Just in time accommodation

A contract for student accommodation in the south-west is made up of elemental flat pack concrete walls or floors or roof units which slot together with windows and services already installed in the units. Accommodation for nearly 2,000 students is multi-storey in several different blocks and time is short. Units are delivered on edge in small numbers on the back of large articulated low loaders. There were 11,000 units to deliver and 300m^2 per day were made at the factory. 400 units could be fixed each day which meant the manufacturer had to stockpile in advance of delivery.[20] The contractor used six 100 tonne cranes. Gangs needed to bolt units together, place floors and structurally screed and put up edge protection for each six person flat. Units were fair face

Figure 5.5 Views of student flats under construction
Acknowledgement to Buchan Concrete, by kind permission

internally and clad by 25 different types of external cladding materials, including brick.

Lorries are loaded in the north and informally wait at a nearby service station on the motorway to take up the buffer of traffic delays. Units are called off to arrive on site in the exact sequence of fixing and are lifted up by mobile crane to four different gangs working simultaneously in separate parts of the site on the structure. Lorries arrive at set intervals to 'their' gang which is supported by a crane which lifts units into position and holds them still fixed. Lorries for different gangs wait in lay-by in the university off site till they can enter a single site access and be processed and driven to position. Gangs stagger their work to receive lorries at different start times and are trained to work to identical targets to maintain staggered deliveries.

This example of JIT requires a high degree of balanced production to keep the lorries flowing at their maximum rate of production. Any unplanned hold ups would have drastic knock on effects on the manufacturer as well as on the progress of the site in terms of sequencing and stocking at the factory. If sequence was changed or restricted to only some of the blocks the long turnaround for curing the slabs would likely delay the new sequence.

encourages ideas from the bottom up which workers can more easily buy in to, such as shadow boards for keeping track of tools. The system puts in place structures such as visible indicators, and charts to pass on lessons learned and to stop slippage. This is particularly important in construction where the teams are changing all the time and a single culture between them needs to be developed. This is illustrated in Case study 5.5. In construction it is much more powerful where a subcontractor can take ownership of the solution.

Just in time (JIT)

It could be argued that JIT is a precursor, along with total quality management, of lean construction, but in this chapter it is regarded as part of the process of achieving lean credentials.

JIT is defined as getting just enough product on hand needed for that day's work. It means that stock levels are reduced and space is saved. Stock levels match the lead time so that delivery does not hold up production e.g. if the bricks can be delivered twice a week then keep two to three days' stock. The number of bricks delivered is dependent on the output rate of the final product e.g. two houses per week. Case study 5.4 gives an example.

The optimisation of the cycle depends on resource levelling (*heijunka*) of the different trades involved so that no trade holds up another. This balance can be achieved by a combination of resource levelling so one bricklayer lays 60 bricks (1m^2) an hour and two bricklayers lay 120 bricks (2m^2). If one plasterer then can cover 16m^2/8 hour day (five bags of plaster) then you need one plasterer and two bricklayers. This will amount to 4,800 bricks per week and 25 bags of plaster, say 30 bags of cement and 2 tonnes of sand. This can be delivered as one small load of bricks and a mixed load of sand and cement. This is efficient as it is sufficient to keep production going and it optimises storage space (that not needed in the same day) to about one-third.

Just in time has many implications as it is a tight control of the material suppliers and needs their support to make it work and to split loads. Alternatively a contractor can use an intermediate logistics centre which can send materials in mixed loads in exchange for discounted bulk deliveries from suppliers to pay for it.

The distinction between bulk materials such as sand and cement and small multiple high value deliveries is important. A bulk low value delivery takes lots of space and should be JIT. A bulk highly engineered delivery such as door sets or kitchen units suits JIT, with some buffer built in to allow for faster fix rates and long lead delays. Small complex deliveries can benefit from initial grouping and labelling by a co-ordinating supplier, such as iron-mongery sets for different door types. This makes sequential delivery much more feasible and saves time, sophisticated storage and retrieval, and loss or damage on site.

Lean construction

Lean construction is defined by Constructing Excellence as 'managing and improving the construction process to profitably deliver what the customer needs'.[21] Lean construction is the adaptation of the principles first pro-pounded by Ohno, of eliminating waste from the *overall* design and production process, to construction. It deals with the optimisation of value streams eliminating waste between activities or stages, i.e. the integrated whole, as well as waste within more discrete parts as might be targeted traditionally in time and method studies, for example, the faster laying of bricks. The implementation of lean therefore requires leadership and organisation.

To adapt to construction means to develop it for project based work. Projects are also the equivalent of several production lines. This means identifying related parts of a construction project, for example, constructing the wall as part of the whole superstructure and not just a micro focus to a particular trade. The 'production lines' are also related and these must be co-ordinated where there are interfaces to give production flow and balance. Koskela's (1992) seminal technical report suggested a new production philosophy for construction based on the Toyota system and suggested that non-value adding activities were not recognised as a problem in traditional control processes and that critical path method needs to eliminate waste as well as sequence activities to get the most logical flow.[22] Techniques such as Last Planner® (see section on lean construction) were further developed by Ballad and Howell to provide a basis for making the traditional conversion logic more productive, by examining constraints and cutting out the waste of rework. They opened the Lean Construction Institute.

Lean construction has been described by Howell as similar to traffic flow where the efficiency for arriving at a destination will depend upon the speed of the slowest vehicle in each lane, but will also be affected by the smoothness of the driving experience as it takes time and wasteful effort to break and accelerate in tailgating.[23] As interventions are highly likely this can be controlled well by the use of intensive preplanning with the use of appropriate buffers between each activity. Small or no buffers (tailgating) increase the likelihood of wasted effort.

The new philosophy is also concerned with client values (business, specification, scope and quality) and the overall project efficiency (time and cost). As such it moves away from the tools of earned value analysis, elemental cost control, work breakdown structure and organisation/task responsibility to looking at an integrated project delivery (IPD) which cuts across a 'pass the baton' approach between different disciplines and the client. To prevent waste there is a necessary greater intensity in the preplanning as we are not tweaking a well oiled system, but defining reliable processes which are applied to a unique project. The following sections deal with specific examples of lean in construction.

Building Down Barriers (UK)

Building Down Barriers was an MOD (Ministry of Defence) sponsored approach to procuring construction projects and successfully managed to make significant savings through a particularly collaborative process based on lean principles called 'prime contracting'. This propounds an integrative partnership working in clusters with early contractor/specialist involvement and providing incentives (pain-gain) to continuously improve process and performance in design and budget (prime cost) and methodology. Prime cost is studied in detail in Chapter 16. Case study 5.5 summarises the level of achievement of better productivity and reduced waste through the use of prime contracts.

The primary principle the Building Down Barriers team recognised was 'compete through superior underlying value' which comprised right first time and developing a positive integrated culture to reduce unnecessary costs whilst improving profits for all parties. Integration in this case meant direct collaboration between the designers and the specialist contractors and manufacturers for continuous improvement to build value into the

Case study 5.5 Building Down Barriers pilot projects

Two pilot projects jointly called Building Down Barriers used more advanced collaborative procurement called prime contracting.[24] One of these was the Andover North project for the MOD, a £40 million project which was finished on time and below budget in October 2002 (see Chapter 16 appendix) with a six year maintenance contract. This resulted in vast improvements in reducing target budget prices by £2 million and running costs by £375,000 per annum. These projects achieved a 20 per cent reduction in construction time, a 15 per cent reduction in steel frame costs, labour efficiency in adding value to the building of 65–70 per cent and rework and materials wastage of only 2 per cent with no reportable accidents and no claims. This was used as a demonstration project in the immediate aftermath of the Egan Report in 1998.[25]

individual projects and to work in clusters of relatively independent elements of the building – frame, ground works etc., and to integrate across packages. Lessons were learnt for future projects. Working together in supply chains is discussed in greater depth in Chapter 12.

Construction Lean Innovation Programme (CLIP)

CLIP is a consultancy set up by the Building Research Establishment (BRE) to work individually with organisations to improve their processes and to make lasting savings in the workplace by changing the culture of their project work. It operates across the construction supply chain from raw materials, components, contracting and design and its case studies demonstrate an average productivity improvement of 50 per cent.[26] The system is based on looking at one key process at a time to bring quick wins which attract commitment from others for the longer term. A manual was produced on the basis of seven case studies. It gave a checklist for success – the need for the support of top management, visual displays to increase communication and discussion, a clear aim and a pilot project that can be measured and recorded, engaging the workforce, and tools developed to help change the established mindset. It helped to have third party CLIP engineers who worked closely with the team to overcome resistance to change, closing the learning loop to put ideas into action and sharing best practice with others. Case study 5.6 examines the process.

The CLIP scheme helps to make a productivity assessment, to introduce improvements, to educate and roll out change management and to develop

Case study 5.6 Mechanical and electrical contractor on six floor new
build office

> The team was tasked to 'reduce the amount of identified waste with a
> focus on productivity and site processes, to demonstrate the improve-
> ments and to establish a measurement system as a base for further
> refinements'.
>
> This project was concentrating on the improvement in productivity
> of the fixing of low level pipework and conduit in which they achieved
> a 13.8 per cent improvement overall. Initially the pipe fitters said they
> could not work any faster, but ultimately they improved productivity
> by 40 per cent. These improvements were mainly derived from better
> layouts as well as reducing standing time in the team of three. They
> initially used logging sheets to identify proportions of productive, non-
> value added (supportive work such as loading) and waste. Observation
> identified bottlenecks, waiting time, cramped working space, poor
> storage and more snags than expected in a type of 5S inspection. A system
> of allowing two of the team to fix pipes, rather than one, and one to
> bend pipes, improved the output. A standard layout for the workplace
> and materials was devised which was also used on all future jobs. Visual
> charts were used to communicate improvements to other teams.
>
> Conclusions were that they let the team suggest the improvements,
> passed on the knowledge locally and got further involvement, and they
> intend to have master classes to roll out improvements across many
> projects.[28]

measurement systems to help maintain improvements. In addition there is a
need for top level leadership and financial commitment. One set back is that
it rarely works to integrate the related wastes of many suppliers on the project
so there is frustration if one bottleneck is removed only to create another.

Trimmed construction (Denmark)

According to Bertelson 'trimmed construction' is characterised by a
framework of viewpoints such as transformation, flow and value generation,
leading to the use of methods such as chaos control, design workshops, Last
Planner®, value management, logistics and process management.[27] This has
been used on 20–30 projects to various degrees with government support
for value generation on public housing projects. Last Planner® is also
supported by the Construction Workers Union who are actively involved.

Agile Construction Initiative (UK)

Agile Construction was an initiative based on the Latham Report in order to improve productivity. It partnered between Bath University and industrial and government partners to 'utilise the lean production management techniques pioneered in the automotive industry'.[29] It has a particular emphasis on the development of benchmarking in an integrated way to provide a platform for the introduction of innovation and the development of the lean concept for the construction sector. Agile's twin strand approach was top down to help improve the process productivity such as procurement for the client and bottom up to build up a database of demonstration projects which used good practice aimed at improving site production processes. Agile research gave senior managers in participating organisations confidence in the claims for productivity and quality improvements so cutting out waste and rework and inducing competitiveness locally and worldwide. ACI has published several case studies which are available for training and also developed work under the Construction Clients Charter which is mainly specific to registered social landlords.

Last Planner®

Last Planner®, developed by Ballard,[30] is a particular practical development of lean construction and has taken some of the paradigms of lean production and added its own. The five elements of Last Planner® are:

1   Master scheduling where milestones and long lead items are identified.
2   Phase pull planning where phases are agreed and conflicts in operation identified.
3   and 4 'Make ready' and weekly work planning which are the two immediate look-ahead resource planning elements to make sure that nothing is going to stop planned work.
5   Learning by measuring progress against plan and looking into cause of failure and improvement.

In other words, it concentrates on progressive planning stages engaging to ensure readiness. It has developed its own terms. The system was set up as a way of countering the delay in feedback systems and getting a 'right first time' approach to increase reliability to 70 per cent. Its research also tested the system to achieve a 90 per cent reliability level.

Collaborative scheduling results in 'make ready' task sheets which emerge from weekly work plans feeding forward to the week's work ahead. A series of promises are extracted from trades' foremen, but only after they have negotiated conditions of satisfaction so that issues for non-compliance can be eliminated. This is monitored by a percentage of progress completed

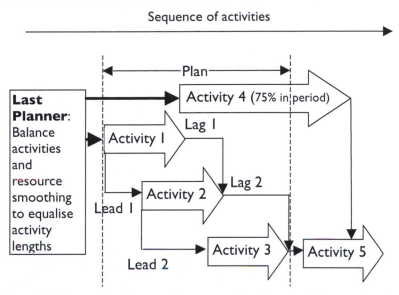

Figure 5.6 Balanced flow for next planning period with an overlapping series of activities

(PPC) as a secondary feedback control. This allows continuous improvement to take place by looking at the reasons for non-compliance. A reasons sheet may be generated to plot the number of occurrences for a particular fault and to provide a Pareto 20/80 analysis to provide focus to the 20 per cent priority items.

Each subsequent activity is constrained by a minimum start to start. The work flow control chart balances information and materials to a network of logically related production activities and helps in determining resource levelling, duration and sequencing for the activities in the weekly detail plan. This is illustrated in Figure 5.6. The weekly period planned for is shown between the dotted lines and sequence flows left to right as the precedence diagram protocol (see Appendix). *Lead time* from the previous activity is shown, determining the amount of overlap. The end of the subsequent activity may also be delayed by the finish to finish *lag time* from the previous activity.

In Figure 5.6 activities on the critical path pass through activities 1, 2, 3, and 5. Activity 4 is not critical. The system shows that leads and lags between activities 1 and 2 are equal and leads and lags between activities 2 and 3 are equal and activity durations need to be equal to get a smooth flow otherwise delays occur waiting for one to finish.

In order to achieve this in the planning process there is a need to get a strong commitment to the rate of work (e.g. m^2/hour/person), look at the available time for completion, dividing this by the total area in metres and the rate/hour and determining how many people/machines are required.

Table 5.1 Detailed method statement to build a plastered wall with resource levelling

Activity	Quantity to lay	Rate to lay (60 bricks/m²)	Time taken for one (hours)	Target time (8 hour day)	Resource levelling (column 4/ column 5)
Laying bricks	12,000#	1m²/hour	2,000	10 days (80hrs)	25 bricklayers
Plastering	750m²	16m²/day	46.8 days	10 days	5 plasterers

Table 5.1 shows how resources need to be levelled to the target time for a component (which may have many more activities) to be completed. Buffer times need to be allowed to take care of unforeseeable delays which are encountered in, say, weather or sickness. These are usually allowed for in the rates (column 2).

The system recognises continuous improvement and has an inbuilt feedback review to learn lessons where things go wrong. The reasons needs hierarchy is prepared. Ballard describes the system change from the traditional push system which plans a 'should be done' into a 'can be done' input culture and this gives an output into a 'will be done'.[31] The traditional system with a lower reliability relied more heavily on time for corrective action. A 'will be done' system has spent this time up front alongside previous production actions so it is saved from the critical path of production.

Difference between lean manufacturing and lean construction

Lean production has invented certain processes that are based on a manufacturing process which is based in a single place where movable products are brought to the work station and passed on to the next process. A solution for lean is to improve the layout of the work stations and the efficiency of the work stations themselves. Koskela lists differentiating characteristics of construction as 'one of a kind', site production and temporary multi-organisation.[32] The latter distinction is the use of many specialist contractors to work together in the site location. Ballard and Howell describe construction as unique because it is dynamic and not a fixed workplace.[33] With construction the product is not movable and is constructed in position on site with 'the stations moving through the whole and adding pieces'. The building site (place) has certain unique characteristics bringing uncertainties such as weather and ground conditions, so it needs a far more responsive system. Rather than a continuously improving fixed serial layout there is a need to plan and co-ordinate parallel activities layouts within a larger programme of events uniquely compiled for the project. The value of a project is related to a specific customer who wishes to live or work in that particular building. Projects vary in these three respects depending on their use – so houses can be batched with plenty of repetition, shopping centres require a general attraction and pipe laying projects may be less multi-organisational.

Waste eradication in this context is a more dynamic learning process as new teams work on a unique process and layout created for the project. Ballard and Howell claim that lean techniques can be applied to much of construction by minimising its peculiarities by, for example, offsite manufacture and the use of standard components so that the dynamic site process is reduced to assembly and testing.[34] However there is a remainder which resists these techniques and fresh ones need to be applied to reduce waste and increase efficiency in these areas. Here they suggest the reduction of fragmentation, for example, between designer and constructor and the more intensive application of preplanning.

Arguments against

Lean construction (LC) is an application from the Japanese lean production and its productivity claims are not without critics. Green puts forward the alternative viewpoint that LC is 'technocratic totalitarianism', believing that closer control in LC sometimes translates into surveillance and exploitation and that the productivity gains may not always be to the benefit of the customer.[35] He criticises what he sees as the evangelistic approach of Egan, and Womack et al.,[36] which is unquestioning about the suitability of LC to different environments (e.g. from manufacturing in Japan). He calls for a balanced appraisal of the evidence. He points out that some concepts of lean are remarkably close to Taylorism which would be a step back in time not a step forward. As a particularly Japanese phenomenon success it is dependent on their more controlled industrial conditions which are not renowned for worker freedom and quality of life. With these warnings in mind, any venture in this area needs to take into account local conditions and ensure worker control, resisting management exploitation.

Off site construction, MMC and prefabrication

Modern methods of construction (MMC) cover prefabricated components built off site and fast build components developed on site such as tunnel form. Prefabrication has developed substantially, but prefabrication essentially turns up in four forms:

- Volumetric which refers to a finished module of accommodation such as a student or hotel room usually fitted out inside and ready to plug in to water, electrical and drainage supplies. The size of the module is dependent on transport and open volumetric can be bolted together. They can often be stacked or rest in a steel framework. Pods are smaller, highly serviced units such as a bathroom or kitchen sitting on traditionally constructed frames and connected up in the same way.

- Panellised construction which can be open or closed is a flat pack system which can be bolted together to make boxes or rooms and includes windows and often with services built into the panels. These usually create the whole structure of the building and can be solid concrete or timber framed. These structural panels will have a height restriction or require bracing and extra frame or in-situ core support. Panels may or may not have external cladding fixed. If they do joints and screeds need to be applied. Precast stairs make up the final component.
- Component form, for example cassettes such as floors or pitched roofs. These often work with prefabricated light steel frames or timber frames. Sub-assemblies can include prefabricated foundations, dormer windows, insulated brickwork or stone insulated panels (SIPS) or even chimneys. There are also prefabricated plumbing and cabling systems designed to snap together quickly.
- Hybrid systems may include elements of panelised and volumetric often used in the way in which pods are utilised within traditional construction.

The type of MMC helps to shorten the construction programme and the less reliable conditions on site.[37] Design needs to be more complete to allow for lead-in times. This brings in specialist suppliers early on following concept design so as to cope with the longer lead time of manufacture and to co-ordinate the structural requirements and to determine sequencing of unit manufacture. Foundations can go on during the manufacture of units and panels, but units need to arrive to suit the sequence as sections of site work are complete. The volumetric has an even bigger impact as its detailed design has to be partly complete before infrastructure can be designed and approvals sought. Figure 5.7 shows the equivalent lead-in and construction times for MMC and traditional construction.

Standardisation, modular co-ordination and case studies e.g. MacDonald and Linden study

Standardisation of construction components is an important aspect of cutting out wasteful redesigning. In the Linden Study which compared work that Bovis Lehrer McGovern were doing in South Carolina and the UK it became apparent that buildings in South Carolina were cheaper because there was much greater standardisation of components, especially the mechanical and electrical systems and the modular division of the building to make this possible. Tailor-made mechanical and electrical systems in the UK made design and manufacture expensive. Other models used are the McDonald's chain of restaurants which use a modular standard construction for their

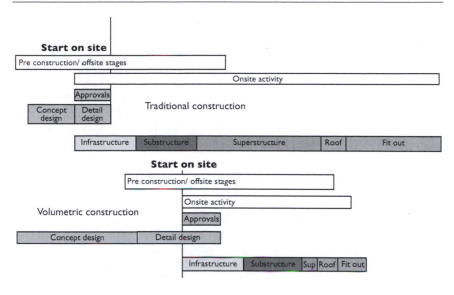

Figure 5.7 Comparison of the lead-in and construction times of traditional and
volumetric

Case study 5.7 Global cost comparisons

Turner Townsend estimate global construction costs in US cities to be
three-quarters of the London costs though these are similar to New
York.[38] These comparisons change sharply with currency fluctuations.
However prices did drop about 10 per cent in the recession from August
2009–2010 for new build. The UK is about the 8th most expensive place
to build in Europe and 16th in the world (August 2010), dropping from
9th in 2009. Large European countries only vary in range 97–107 per
cent of UK prices with Eastern European up to 45 per cent less. Other
sources indicate that US averages are 10 per cent lower than the
UK although they vary greatly between cities. China is about 50 per
cent of UK costs, though Hong Kong is only 10 per cent lower. The
Middle East is broadly similar with Bahrain most expensive at 11 per
cent higher.[39]

drive-through outlets which are erected in four weeks (Egan claims 24 hours for the superstructure). The foundations are simple for a single storey building and drainage and utilities need to be provided and connected to the mains. This saves preliminary costs, but modular offsite components are expensive due to transport, though when the modules are repetitive then costs can be dropped and quality is high. Case study 5.7 compares global costs.

MMC methods can be counted lean if they cut down product waste because of more precise factory production or squeeze waste from the process eliminating abortive rework.

Identifying construction waste in procurement

Construction is also well known to separate different types of activity into silos, shown by its lack of interdisciplinary communications. Traditionally the architect does the design and does not think about construction methodology and the contractor is not prepared to take the responsibility, or appointed early enough, to assess the buildability of design solutions. Supply chains do not share in financial or other incentives which emerge from such thinking, but do take on risk and liability for alternative innovation. Instead the process incentivises a blame culture and a wasteful passing on of risk down the supply chain or back to the designers. When things go wrong, delays or abortive work (waste) has already occurred before the system corrects it.

Non-traditional forms of procurement and collaborative working can help to integrate objectives between client, consultant, contractor and supply chain by helping parties work together for better performance in project completion, value for money, quality satisfaction, and safe and sustainable construction. Some forms of procurement, such as design and build, allow overlap of construction and detail design which may save time and, if properly planned, money. PFI and design build procurement help to integrate risk management by putting risk primarily in the hands of two parties. Construction management gives earlier contractor involvement and increases buildability and the chance of incentivising quality and value improvements at an early stage for meeting client requirements. Proper change control systems can deal with necessary flexibility in all systems.

Latham propounds the use of partnering and more collaborative contracts (e.g. New Engineering Contract (NEC)) to avoid apportioning blame and to concentrate on early warning and problem solving.[40] Figure 5.8 shows different ways of procuring to give more involvement of the contractor in the value process.

Procurement helps to determine value streams in construction as sequences vary as in Figure 5.8. Three options are shown to indicate the progressive moving away from silo effects of traditional procurement, which fragments the parties, to an integrated procurement which makes lean construction become easier so that joint solutions can be applied.

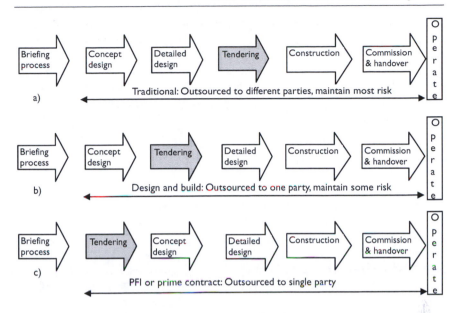

Figure 5.8   Procurement sequencing showing risk distribution in procurement types

Measuring performance

Performance benchmarking

Benchmarking is a method of improving performance in a systematic and logical way by measuring and comparing your performance against others, and then using lessons learnt from the best to make targeted improvements. It involves answering the questions:

*   'Who are my competitors?'
*   'Who performs better?'
*   'Why are they better?'
*   'What actions do we need to take in order to improve our performance?'

Benchmarking needs to compare your performance with another of a known standard. A benchmark is a business standard like Olympic qualifying standards – you need to meet a certain minimum, you are unlikely to be a competitor. The standard is also likely to move upwards in a competitive market. The average of other competitors is the least acceptable standard.

These benchmarks may be based on industry, national or global standards. They may also be compared between organisations in the same sector, same industry or against different industries. The choice depends on the competition which you have. Measuring performance in construction may be based on:

- Internally set targets which are often based on continuous improvement between projects, year to year or as a stated target for best practice in the business.
- Performance of another in the construction industry who are carrying out similar activities, who are the 'best in class'.
- Financial performance, for example by comparing the profitability or turnover growth ratios against others.
- Generic targets across industries which represent excellent management achievements such as the levels of training, business results and customer satisfaction.

The 'best in class' term refers to a leading performance in a company's own defined market place. Measured key performance indicators (KPIs) may be measured against average industry performances, but to be competitive more than average industry level is required. Being in the top 10 per cent of companies is a requirement for best in class.

Continuous improvement of KPIs is the concern to keep up with the best and to offer a better service to the client, in the belief that it produces returns. It involves developing a culture of innovation, target setting, continuous monitoring of standards and reviewed performance. As expected it also recognises the role of the client and the construction/design team working more closely together to produce improvements and adding value. To be successful benchmarking should be focused and a senior manager needs to select critical areas for improvement which will impact on the project as a whole. It also needs challenging goals that are achievable, a willingness to change and persistence in getting there. It is not always going to be possible to improve over one project.

Benchmarks can be applied to various business parameters such as levels of profitability, respect for people and the environmental impact of the company, or to project parameters such as customer satisfaction, predictability of time and cost plan forecasts, number of defects and accidents recorded. It is the latter that we are interested in here and systems have been set in many national construction industries. There is a strong connection with quality management.

Case study 5.8 is a construction best practice profile covering the period over which the two Egan reports have been current and indicates the take up of issues that have become important to competitiveness.

Case study 5.8 Benchmarking – continuous improvement

This is a contractor with a £500m turnover and 40 offices throughout the UK and employs 3,000 people which put it into the large category for contractors. They have adopted the maxim: 'If you look within the sector you are always going to follow. You need to look outside to lead.' This has caused them to make leading edge improvements by looking to solutions which have been adopted by firms in other industries who have made substantial improvements.

Over a period of 12 years they have steadily looked at a series of initiatives that have built upon each other and led to an integrated process, such as prompt project completion commitments, process definition and re-engineering, building an understanding of client requirements and the development of customer service units, standardisation and lean thinking. These have steadily improved and client satisfaction stands at 87 per cent.

With the publication of the Latham and Egan Reports they revisited processes and introduced a programme of lean construction to further remove wasteful activities from the whole value chain. This led to an integrated management system combining quality, health and safety, and environmental management together and individual rewards in meeting Egan targets and they use KPIs to demonstrate this. They believe the key to continuous improvement and integrated construction is:

- Top down commitment – the direct involvement of a director in driving each of the improvement agendas.
- Bottom up workshops and team building – the involvement of all in the simple on-the-job improvements.
- Measurement – use of KPIs to demonstrate where improvement is taking place. They measure likelihood of repeat business, user friendliness of the business, environmental awareness, level of defects.

Critically, the company show a financial improvement in their results in the period 1995–2005 as fast growing profit margins, a continuing increase in turnover and a growing customer satisfaction score, defects above industry average since 2002 and registration for integrated assessment by the BSI for quality, safety, health and the environment management systems. Their accident incident rate has lowered, but they are also testing customer perception of whether they are a safety conscious contractor. They want to extend KPIs to their supply chain.[41]

Benchmarking standards for the UK construction industry

Egan issued a challenge to the UK construction industry by setting up some improvement targets that are effectively benchmarks. These are:

- Decrease defects by 20 per cent per year.
- Reduce accidents by 10 per cent per year.
- Reduce time periods by 10 per cent per year.
- Increase value for money (client gets more for their money) by 10 per cent per year.
- Increase contractor productivity (turnover per head) by 20 per cent per year.
- Increase contractor turnover and profits by 10 per cent per year.

Demonstration projects have given a lead, but a *measurement* system is required to validate the claims of those who say they are giving better value for money. To be sustainable VFM gained through quality improvement must be a win–win situation in which contractors' profits are increased and a client gets more for their money. Innovation and process improvement then is an important part of the objective.

In the UK a benchmarking facility is available based on nationally prepared profiles of performance for the construction industry as a whole and for particular sectors, such as housing and refurbishment, and particular things such as environment. Each is published as a set of graphs annually. Constructing Excellence co-ordinates the collection service for calculating average construction industry KPIs in key areas in their KPIzone.[42] This provides a calculator to rate a company's own KPIs with the industry average. In addition it identifies best practice (demonstration) projects and collects KPIs for them, publishing results which generally have been shown to meet the Egan targets on demonstration but not on other projects. In conjunction with present selection criteria KPIs may be useful predictors for the client and critical marketing tools for the contractor.

International benchmarking schemes

Internationally studies have shown that KPI indicators in several countries' schemes are tested for their competence and ability of the company, project design suitability, and the productivity of the each project and customer satisfaction. PSIBouw identifies:[43]

- National prequalification network in Australia which prequalifies contractors. Many others exist such as the construction quality assessment scheme in Singapore and the New South Wales and Queensland scheme in Australia which also assesses consultants. In Holland the contractors' past performance assessment is used for public projects to amend the lowest price tenders favourably towards performance.

- Performance assessment scoring system (PASS) assesses contractors in the Hong Kong Housing Authority projects on management capability, build quality and maintenance performance. In the UK a voluntary scheme called Construction Line exists.
- Building design appraisal system in Singapore assesses design quality similar to the DQI scheme in the UK (see Chapter 9).
- Calibre is a system for measuring the productivity of the labour resource in the UK.
- Construction Industry Institute (CII) benchmarking and metrics measure the time, cost, quality, safety and productivity of projects of the CII members in the USA, similar to the UK Construction Industry KPI scheme. The Danish construction benchmarking which measures similar areas as the UK and USA scheme is privately funded, but is mandatory for public projects.
- National customer satisfaction survey assesses satisfaction nationally from new house owners which enables a comparison between different housing suppliers in the UK. A similar scheme is operated in Holland called VEH Oplevemonitor.

These show the popularity of benchmarking and its importance particularly for getting on tendering lists for public projects.

Making KPIs effective

Effectiveness is related to meeting set objectives. It is important that there is a wide and measurable range of data for industry comparisons. There are three areas of concern with the use of KPIs:

- There may not be a direct relationship of indicators with performance.
- The past performance may not be repeated for all sorts of reasons on future projects. For this reason they must always be interpreted in present context and not as a magic formula. A standard comparable format is needed to make them useful and give incentives to use them, but where suppliers wish to use their own this is much harder to certify.
- There may be disagreement about which indicators show excellence. Many league tables produce distortion in comparison where companies have different strengths or weaknesses.

Latore et al. have suggested an amended list to the Egan list and connected their results in a causal loop to try and understand how the performances of KPIs interact with each other and to help interpret them to drive project efficiency.[44]

The Construction Industry KPI

The initial 12 Egan construction performance KPIs are averaged for all sectors of the industry and are measured each year. They are also available for different sectors of the industry such as housing, new build, refurbishment and building services. More generic sets are available for the environment, health and safety, and people. They are a different set which apply to construction consultants. Table 5.2 shows the basic set of KPIs.

Crane outlined a comparison report on the demonstration projects that were set up after the Egan Report.[45] These are year on year 2000–2003.

- Client satisfaction with product   81–90 per cent.
- Client satisfaction with service   76–86 per cent.
- Accidents on site dropped   716–428/10,000.
- Construction time improvements   10 per cent saving.
- Profitability ahead at   5.6–7.0 per cent (25 per cent better).

Table 5.2 The construction all industry KPIs

No.	Parameter
1	Client satisfaction – product (questionnaire 1–10)
2	Client satisfaction – service (questionnaire 1–10)
3	No. of defects significant faults at completion range 1–10
4 5	% change in predicted budget – design (start to finish) – construction (start to finish)
6 7	% change in predicted time – design period – construction period
8	% profitability (pre-tax margin)
9	Productivity (£ turnover/person)
10	Accidents (No./100,000 employees)
11	Change in construction cost (cf similar last year)
12	Change in construction time (year on year)

To calculate construction cost change year on year

The following example shows a hypothetical case study for the headline benchmark for the year on year underlying cost saving. It uses data from two similar types of project built 12 months apart and making adjustments to the variable factors of cost inflation, location, size of facility in square metres of floor space and quality level. Agencies such as Building Cost Information Service (BCIS) publish year on year cost inflation, locational variations in cost and ranges of quality prices to help in arriving at the adjustments. In this case an office in different locations of different sizes and underlying quality has been used. Adjusted cost is the comparator. Data is shown in Table 5.3 to calculate construction cost change year on year. The objective is to measure the change in real construction cost, over 12 months, for a particular contractor.

Table 5.3 Case study particulars

CASE STUDY OF TWO OFFICES	Office A 3rd quarter last year	Office B 3rd quarter this year
Raw cost office	£10m	£8m
Size of building	10,000m²	7,500m²
Inflation index [BCIS 1983 =100]	447	478
Quality	100	105 (5 per cent greater)
Location	Outer London 1.03	SW England 0.99

The objective is to measure the change in real construction cost, over 12 months, for a particular contractor taking the data from two similar projects which will be normalised for comparison purposes.

Step 1 Calculate the normalised cost of the two offices

Note, you normalise both buildings for size (per m²) and location (equal to 1.0). You only normalise office B for inflation and quality difference as shown in Table 5.4.

Step 2 Calculate the construction cost indicator by comparing the normalised cost of the two buildings

$$\text{The construction cost indicator} = \frac{B - A}{A}$$

Table 5.4 Case study calculation for adjusted price

Adjustment requirement	Building A	Building B
Adjust for both for **size**	£10m/10,000m² = £1,000/m²	£8m/7,500m² = £1,066.67/m²
Adjust both for *location* of building Outer London 1.03 SW England 0.99	Outer London $\frac{1,000}{1.03}$ = **£970/ m²**	Bristol $\frac{£1,066.67}{0.99}$ = **£1,075.98/m²**
Adjust for *quality* of building Bldg B 5% more quality	High = **£970/m²**	High+5%[normalise by ×100/105] £1,076 × (100/105) = **£1,024.76/m²**
Adjust for *inflation* Bldg A index = 447 Bldg B index = 478	447 = **£970/m²**	478 £1,024.76 × (447/478) = **£958.30/m²**
Normalised cost	= **£970/m²**	= **£958m²/m²**

$$\therefore \text{ Construction cost indicator } = \frac{958 - 970}{970} \times 100\%$$

$$= -1.24\% \quad \text{i.e. better performance}$$

This means that this contractor is becoming more efficient, but only by 1.24 per cent/year, which is less than the Egan target of 10 per cent/year saving.

The above example indicates the calculation of just one of the KPIs – the efficiency of cost – and a similar calculation can be done with time. Care in interpretation is required, as the normalisation exercise does not cope with all the physical or market differences between two projects. To overcome different ground conditions the costs for comparison are often compared above ground. However the role of project management is to reduce extraneous costs by perhaps reducing access problems, orientating the building to minimise shading/heating costs and negotiating efficient Section 106 agreements. Market conditions may also be different, creating competitive tender prices. Comparisons are therefore more reliable where two buildings for a similar client are compared.

The profile of a project is important to be shown across a range of KPIs in order to work out the underlying causes. The 10 Egan KPIs are shown in Figure 5.9.

Figure 5.9 indicates good scores in defects and predictability and service satisfaction, but poor scores in client satisfaction for the product and improvements in cost and quality. Since the defects are low it is more likely

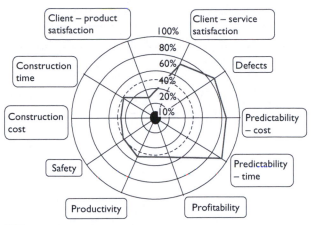

KEY
Each radial arm re-presents a benchmark. The radial zones represent the bench mark score. Those that achieve <50% (below dotted line) are below average and those >50% are above average performance.

Figure 5.9 A spider diagram of the 10 all industry KPIs for a contractor

Source: adapted from Constructing Excellence, KPIzone

that the briefing process has not clearly provided an effective building. If this had occurred then it might also have dragged down the client satisfaction with service of the design party to the contract. Thus client satisfaction surveys may be influenced by a halo effect or the opposite. A balanced scorecard might have more of a weighting on client satisfaction and defects and therefore provide a much better score for this particular contractor.

Construction performance

This section uses case studies to give illustrative examples of performance in action. They are real examples of good practice in the construction stages of the project life cycle. They illustrate areas of logistics, prefabrication, quality control, programming and package management.

Case study 5.9 indicates the close association that there is between the design and construction stages in deciding the most advantageous methodology in running the project. The construction stage tests the effectiveness of the strategy and pre-construction design and planning stages. The project manager is faced with their most complex organisational challenges as the supply chain opens out to a wide range of specialist contractors and interacts with different parts of the design team in the detailed design. The speed and complexity of construction means that production targets need to be clearly stated and closely monitored. Feed forward control is more effective than feedback given by benchmarking measurements.

Case study 5.9 Logistical planning on construction management fit out

The site is a seven floor section of a 12 storey building with double basement and plant room adjustments, for a prestige client in Canary Wharf. At the height of the work 600 people were working on the site on 30 separate packages including a logistics package. The packages were let and managed by a construction manager but each package contractor had a direct contract with the client. A separate fee was payable to the construction manager. Cluster co-ordination was done by a lead specialist contractor for each zone of work such as the ceiling, floor zones and the kitchen and canteen area.

The construction manager developed programmes for each floor broken down into quadrants as a way of co-ordinating the zones and monitoring progress of the work. These programmes became the basis for the trade contractor co-ordination. For communication a trades dated finishing programme was established in order to monitor completion of individual parts of the programme. This provided a latest finishing date for each of the trades in a particular sector. The programme was tight with most of the trades running on two floors and overlapping the previous trade on each floor. A 25 week turnaround for each of the follow-on trades in a 38 week programme was planned.

In order to control quality a mock up of the office finishes was carried out as a package at the start of the contract. One problem area for the quality was the drywall finishes. In some cases these had been redone three times to meet a demanding specification and this had caused problems to delay following trades.

This project was an example of a highly intensive work programme making any hold ups affect the critical path. The hands off nature of the construction management function has only been possible with the setting up of well-prepared packages and a well-administered change control system. As most decisions for significant design change had to be made in New York where the client representative resided, this had the potential to delay the site team if the problems were not anticipated in good time or client changes were frequent. However the collocation of an integrated design and construction management team on site helped in getting design decisions implemented immediately. Giving ownership to the trade contractors by making day-to-day co-ordination their responsibility provided space for more advanced planning and contingency management. It was probable that the pure construction management procurement created by letting out all work including logistics allowed more objective strategic decisions to be made by the construction/project manager, but this probably cost the client more up front.

Case study 5.10 Prefabricated construction

The Superior product team has been set up as a consortium between contractors, designers and client to provide completed corridor units which fit inside a structural steel frame. Each prefabricated unit weighs 6.8 tonnes and is 4m long, 7m wide and 2.5m high. The unit can be double stacked or divided into two narrower corridors of 3.5m wide. It is externally cladded and glazed to provide a finished waterproof product bolted within the on site in-situ steel frame. The units are complete with modular plug and fix heating, electrical and IT services as in Figure 5.10. The steel frame is stand alone and is configured to the route of the corridor which generally links up between two buildings as in Figure 5.10. In order to install the units within the frame, rails are fixed to the frame at the height of the floor of the corridor and units are lifted into the frame by leaving out a section of the top frame structure. Units are lowered onto the rails and pushed along them until they meet up with the last corridor unit and are sealed, bolted together and services joined. This action is repeated for each corridor unit in both directions up to the access holes in the steel frame. Tolerance is critical allowing the steel to be slightly out of true longitudinally, but requiring high tolerances on the rail level and the plumb of the structural columns, to which the units are bolted top and bottom. Once a corridor is completed between the two buildings the services can be hooked up live and the services tested and commissioned.

The production line has been set up in a factory close to site and units are stored for at least the capacity of a night's work. Night time is usually when free access is guaranteed to the building works and is least disruptive to the client's business. A single unit size has been designed to fit on the back of a rigid bed lorry and is similar in size to a large container that is open sided, so needs to be braced for safe transport. A unit is lifted straight off the lorry and into the in-situ

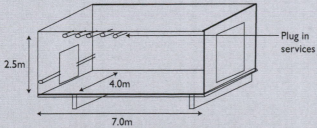

Figure 5.10 Diagrammatic view of a corridor unit

structural steel frame. A multi-skilled production gang receives the unit, manhandles the unit along the rails and installs it. Specialist 'Hiab' equipment is used to load and unload the units, saving the constant hiring of cranage. The fixing rate varied from two to eight units per night shift as a learning curve took place. Approximately two modules were produced per day in the factory.

The installation programme was quite flexible, because there was a need to cope with delays for installation to suit on site progress and client operations. The critical areas of management are caused by the limited storage capability and the need to deliver components to the factory in job lots to suit a flexible delivery schedule. This meant that forward planning was required. This was often achieved by negotiation of extra site visits if stored units were mounting up. To maintain quality and motivation the factory teams and installation teams were interchanged.

Design changes resulted from a better integration of manufacture and installation. Quality improvement led to the re-engineering of brackets in order to make fixing easier and cope with lesser tolerances on site. The factory management stop work on an ad hoc basis when feedback is given in order to give the whole workforce 'one point lessons'. This group learning process ensured equal delivery to all on a check, prevent, repair basis and saved the need for quality inspectors. Another problem was the deflection of the structural steel with the fixing of units in the middle of spans, creating movement after two corridor units were

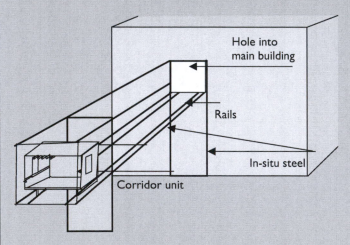

Figure 5.11 Steel framework with unit within

complete and stressing the sealing between them. Both the design of the sealing and the rigidity of the structure were reviewed. Adjustments to the painting methodology were also made to ensure consistency of finish. The manufacturing unit is now in the process of extending its operations to other sites needing the same units and is developing its transport fleet to cope with a longer turnaround.

Case study 5.10 illustrates the way in which a flexible installation programme has been achieved by the coming together of the client, the contractors and the designers to design a particular standardised component (corridor units) to give a fast build scenario to meet a continuing client requirement for limited site access. This could also be applied productively to any type of repeat unit needed in a project, or preferably a series of projects. Another example of this is the new Severn Bridge crossing where heavy deck units were standardised and manufactured on the bank in order to crane directly onto barges for fixing. In addition there is a culture of continuous improvement for the factory production and this has included motivational as well as technical considerations.

Detailed pre planning drawing together designer, main contractor and manufacturer as is evidenced in Case study 5.10 on prefabrication provides a measure of confidence which works well with the very limited access that is afforded on site and the speedy installation that takes place once the units arrive.

Case studies 5.9 and 5.10 indicate the interaction of quality and time constraints in particular. Best practice has been around reducing cost as well at a sustainable profit level for the contractor, which means considering the reduction of project waste, not just in materials, but in the process. Rework and the duplication of management tasks is a waste of resources. This points to better pre planning and better communications. The next sections look at the systems which have been used to control the programming and procurement to give planning flexibility and to add value.

Construction programming and control

It is often perceived that a programme carried out at the beginning of the project with limited information will still be able to predict accurately each detailed activity time. It is also expected that there is a single logic for the assembly of a construction project. Neither of these is true and a progressive system such as that described below allows more respect for the planning process and allows response to current data.

The *overall* programme should be used as a framework with the critical start and finish times of the project and its major subsections. It sets out subcontractor start and finish dates for the procurement of major suppliers and long lead in material deliveries and a basis for reporting to the client. This is the baseline programme and is often prepared using the critical path method which is in the appendix to this chapter.

The *procurement* programme is a co-ordinated programme with the designer in order to get information in time to meet the lead in times of the overall programme. It provides a backward linkage from the date that materials and components and specialists are required on site. Thus if steel was required on site in the eleventh month and there was a lead in time for design and manufacture and delivery of six months then information is required in the third month to give two months to tender and get competitive quotes and place an order.

The *stage* programme provides a quarterly time frame responding to new information and predicting the impact on activities and reprogramming to contain the correct progress within that stage compared with the overall programme. This programme takes into account major changes in scope and looks at programme acceleration or alternative logic and critical path, giving time to agree major alternative methodology and its implications on health and safety and resources.

A *sectional* programme focuses on a particular area of work in more detail so that interacting packages in this location can be identified in more detail. It is likely to relate to a sectional manager's area of responsibility. It might be derived from each stage programme.

The *monthly* programme fine-tunes the stage programme and gains agreement and sign off for subcontractor start and finish dates. It shows the detailed integration of activities on a detailed basis, by discussion with subcontractors. This can be used with the system of interface planning indicated in Figure 5.12. A monthly programme is five weeks long in order to interlink.

The *weekly* meeting provides an opportunity for the site team to meet all the supervisors involved in the current work and to make final adjustments to avoid last minute problems which are anticipated. On a weekly basis there is an opportunity to discuss whether any of the common problems of lack of staff, late deliveries, availability of lifting facilities, inefficient management, lack of information and poor integration of work are likely to occur and to look at the impact of unexpected events on planned activities. This also gives ownership to the planning process and not imposition. This, to be successful, will show the true strength of the team to help each other.

Feed forward control (Chapter 4) is most important with time because of the programme constraints. The importance of tackling slippage on a regular basis to bring the programme back into line cannot be over-emphasised. Feed forward has the same framework, but tries to stop slippage by early intervention. It can be used in cost control where the contract permits and the client requires formalising the cost of design and construction

Case study 5.11 Programme management

The project is a specialist facility for physical research. The contract is worth £80m. The building is circular with many specialist packages and high tolerance requirements. The programme time is 55 weeks for the main building though there are other phases including an office block and enabling works. It involves the use of many specialist package contractors.

In the main project a master programme gives overall coverage of the 55 week construction. It is based on the production needs of the seven key packages that provide 80 per cent of the work. This amounts to 100+ items. There are also phased handover dates for each of the sectors. The contractor monitors this programme weekly and provides a monthly report to the client on each item and the overall progress. The master programme is linked in with the procurement programme in order to determine when packages need to be let. Table 5.5 illustrates the hierarchy of the programming.

Table 5.5 Hierarchy of programme control

Programme type	Programme output
Master programme	Overall report to client Contract start and finish
12 weekly programme	Specialist contractor control and tactical adjustments
6 week stage review	Control and management action
2 weekly specialist co-ordination programme	Co-ordination and work schedules

In order to break down the detail on relevant packages in progress a 12 week rolling programme is produced. This is adjusted to the current progress and also reviewed every six weeks. Tactically it aims to retrieve any programme slippage from the previous 'stage' programme. These are also monitored and presented to the project manager for strategic assessment of the reasons for any problems.

A two weekly programme is also produced and agreed with specialist package contractors to show the immediate work schedules and their co-ordination with other contractors. This provides an awareness of the interfaces and critical path of other contractors as well as an opportunity for *feed forward* with immediate risk and contingency management. The meeting provides an opportunity to discuss any problems and slippage to programme and to work out what impact they have and how the programme may be retrieved.

Case study 5.12 Package management

On the same contract the contractor is responsible for the procurement of specialist subcontractors and their co-ordination and there is a separate contract with lead designers:

- enabling works which include site preparation, roads, external works and piling
- building works which have been divided out into the office block and the physical research facility.

The main contractor tendered competitively on the basis of drawings and specification for both the enabling works consisting of site works and piling which they immediately started. The building works were tendered later and they were also selected. They are responsible for managing the procurement of specialist subcontractors and also have an input into the buildability of the design. They have direct contact with the designers and the client representative. The client is responsible for equipping and testing the facility.

The main specialist packages are piling, ground works, steelwork, concrete superstructure, roof, curtain walling, mechanical and electrical, raised floors and ceilings/partitions, and a procurement programme is used to determine and control the information flow for letting out packages.

The contractor receives and comments on the design drawings before sending them out from the point of view of buildability. The contractor team raised 170 queries from the tender drawings before they were finalised. The main package documentation is sent out on a competitive lump sum basis to approved suppliers on a competitive basis. They receive drawings and a specification prepared by the design team, a scope of works and a programme of works with about 15–20 items, prepared by the contractor procurement team. The specialist contractors provide a schedule of rates and may wish to offer alternative specifications or brands. In the event of tender queries, decisions are distributed to all tenderers.

In the case of the curtain walling the procurement manager went out to five contractors on their approved list and shortlisted two from the tenders received on the basis of selection criteria including price, but not exclusively based on price. In the event these two were not the lowest price because of the specialist nature of the work. This led to further negotiations with the main contractor design manager that allowed a best value final price to be proposed and presented for scrutiny by the project manager and approval by the client. The final signing off of the contract was the responsibility of the principal contractor at project level.

changes before proceeding. Case study 5.11 refers to a planning system used by a major contractor.

The precedence method of critical path programming is introduced in the appendix to this chapter.

This hierarchy of control in the table recognises the dynamic nature of programming in order to retain confidence in the programme as a control mechanism for bringing work back onto programme where there is a slippage. The baseline programme is retained for end and not intermediary control. The short-term programmes remain realistic statements of agreed methodologies and target dates using the latest state of progress and can therefore benefit short-term control and feed forward systems. Case study 5.12 looks at package management.

This process, employed for key subcontractors, was more proactive and provided:

- a basis for transparency for the client
- specialist contractor involvement in the detail design to use their expertise
- more ownership to the supplier with ensuing commitment to quality and time constraints.

Where savings were identified they were passed on to the client on the basis that there was no design liability for the main contractor. Client extras might also be identified at an earlier more cost effective stage. As a general principle handling, access and storage were included in the specialist package, to minimise general site overheads and preliminaries.

Interface planning

The plan should include systems for the proper co-ordination of different subcontractors and designers. Interface planning means:

- Co-ordinating follow on work between different contractors and establishing agreed sequences and co-operative work in areas where many trades have to go on side to side.
- Space co-ordination of designers. For example, different services and structural features in the ceiling spaces or in vertical risers. Many prefabrication packages have to precisely agree service routes, electrical conduits and water pipes to design them into the structural walls and floors before delivery.
- Allocation of work to separate specialist packages must not leave gaps or overlaps.

It is not uncommon for conflicts to arise and a proper system to define the boundaries for responsibility and the hierarchy of authority should be in

Figure 5.12 Workplace flow chart for production control

place. A typical work place planning system concerns a workplace which has sequential procedures of two or more different trades who have to do work which integrated to another trade, such as a room with partition walls followed by first fix carpentry electric and plumbing, followed by plaster, followed by screeds, followed by second fix carpentry electric and plumbing systems, followed by painting, followed by carpets. Especially in the case of painting and carpets there is a need to have a foundation that is acceptable to produce the finished product. The work flow from one trade to another will need the controls shown in Figure 5.12 so that work is both snagged and signed off and then formally inspected by the ingoing trade to accept before starting work.

The main contractor is responsible for dividing the work areas up into suitable areas of work to provide maximum resource smoothing. They will also be responsible for central access scaffolds, conflict management and ensuring that health and safety method statements are available to all affected contractors.

Conclusions

This chapter has looked at the production process with special reference to the need to benchmark the process and make it better. Lean construction is the current terminology for driving waste out of the system. The project manager has a significant responsibility in their unique influence over the design and construction functions by bringing savings that better integrate design into construction and generate possibilities for a more open culture in making 'lean' work. Now clients perceive that service improvements are possible and that more fundamental changes are required to the construction process to give them value for money. This has led to a review of lean production methodology and a more holistic look at systems such as prefabrication, work productivity, defects reduction and indeed more committed leadership and team building to reduce waste in the process.

Closely integrating design and construction is a major change in the normal confrontational approach between designer and contractor and client and contractor. Forces for change that may break this mould and catalyse widespread change in contracting are the threat of greater international competition, more client willingness to develop integrated teams, negotiated contracts, the need to conserve natural resources and the greater awareness of clients (public and private) of the need for controlled competition to meet public expectations for sustainability and investment for the future.

Forces which act against change are business uncertainty and short termism and the culture of a throw-away society with its avid thirst for new facilities and instinct for bargain hunting. Buildings are wanted cheaper and there is not yet a widespread commitment, trust or belief that this can happen to the mutual financial benefit of client and contractor. Where framework and partnering agreements exist there is often mistrust by the second tier supply chain that continuous improvements are brought about by regressive contract conditions between themselves and the managing contractor making long term partnerships less sustainable. More about this will be discussed in Chapter 12.

Productivity improvements are brought about by better planning of the workplace so that work flows are understood better through the value stream map, better layouts and programming are shared with those who will be doing the work. Reducing storage of materials is often forced on a cramped site, but planning deliveries to arrive as they are needed means less work in moving them around and the cost of storing them. Making work more comfortable and benefiting and empowering workers to improve productivity means that many competent minds are at work to improve matters. Prefabrication and offsite manufacture are becoming popular to reduce uncertainty, speed up work and reduce skills required on site and they may also be safer. Extra cost of prefabricated components can be offset by faster times and less fixed costs. Added value is gained in quality and reduced defects.

Productivity is motivated by continuous improvement which is achieved by benchmarking performance to the 'best in class' and generally means that KPIs will need to be kept so that improvements can be measured and business competitiveness may maintained. An average performance is a minimum and should be more. Further management indicators to improve quality and reduce defects, rework and accidents mean that productivity improvements are not thrown away. Getting top management commitment will be necessary to pay for the initial capital expense of better productivity. As much work is carried out by subcontractors the supply chain will also need to be trained in lean construction and brought on board in the planning process so that feed forward systems are allowed to operate on a regular basis throughout the project.

The programming needs to include the design team so that co-ordination of information can be afforded through the procurement programme. The early involvement of the contractor in examining buildability and in making the lead in periods and information required dates clear is always helpful.

Greater respect and long term commitments between client and their contractors are a key to cultural change and the development of trust in the relationships. This does have many good examples for large and especially small contractors who depend critically on repeat business. However as large numbers of clients are one-off clients, ways of ensuring better short term relationships also seem to be critical and it is this area of collaboration which needs to be developed to get away from the 'cowboy' image, the 'Dutch' auctions and claims culture generated by cut throat competition and one-off opportunism. A realistic first cost with an opportunity for reduction needs to be distinguishable from a low 'fake' price. This is an ethical and not a market issue so that savings can be generated on a pain or gain sharing basis, using earlier contractor involvement, value engineering, supply chain innovation and open book accounting.

Appendix – critical path method (CPM)

The appendix looks at the networks which are used to determine the logic of the time scheduling. It assumes familiarity with Gantt or bar charts which show activities as bars against a horizontal time scale. Figure 5.13 indicates a five day activity starting at the beginning of day 1 and finishing at the end of day 5. Figure 5.14 also shows a second two day activity that starts at the beginning of day 6 and finishes at the end of day 7.

An activity indicated in a Gantt chart does not indicate a logical relationship between activities – it is only assumed to be linked to other activities when they have a finish point which coincides with the starting point of a successor. This is imprecise and activities may be connected logically to more than one activity by a vertical line as shown in Figure 5.14.

Figure 5.13 Gantt chart activity

Figure 5.14 Two Gantt chart activities

On complex projects where this is the case a proper indication of logic links may be drawn using the critical path method. There are two main formats – the arrow network and the precedence diagram. The former is sometimes called activity on arrow, because the arrow is the activity and the node is a point in time. The latter is called activity on node because the activity is the node and activities are shown linked by the arrow. Neither has a timescale. The main types of links are:

- An activity starts immediately after the predecessor finishes (e.g. form-work to a beam followed by reinforcement).
- An activity starts after a time lag (e.g. a long trench excavation takes a few days, but drains are laid from the following day).
- An activity can start at the same time as another starts (e.g. when foundation is finished the floor slab formwork and the brick work can start).
- An activity can finish at the same time or at a time lag following the finish of another (e.g. the pipe above can only finish and be tested after the trench has been excavated).

Critical path method (CPM) precedence diagrams

This is a network for linking activities together logically in sequence and analysing the combined durations of the activities and identifying activities with critical time or resource constraints. There are several definitions which are useful.

Activity is an operation or task carried out using a resource and having a duration. In the precedent method they are represented by a box.

```
┌──────────┐
│ Activity │
└──────────┘
```

Figure 5.15 Precedence activity – activity on node

An *event* is a point in time (no duration) and will mark the beginning or the end of an activity. A *link* is a line which links the activities together to make a logical sequence from left to right. It can take one of the forms shown in Figures 5.16 and 5.17. A ladder diagram has both forms in Figure 5.17, restraining the beginning and end of the activity, and is more logically correct than just one of them which leaves a 'dangle', i.e. an open ended finish or start for one of the activities.

This technique can be used to build up a complete network for the logical sequence of various activities. More than one activity can be dependent on another; for example, a concrete beam can depend on the completion of reinforcement and formwork. One activity can also allow two or more to start.

Means Activity A must be complete before Activity B

Figure 5.16 Links between activities

A *lead* or a *lag* links activities together which overlap

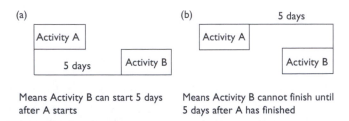

Means Activity B can start 5 days after A starts   Means Activity B cannot finish until 5 days after A has finished

Figure 5.17 A precedence a) lead link, b) lag link

Calculate the critical path. This is the longest path through the network and joins all longest activities together.

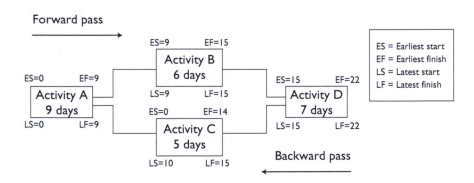

Figure 5.18 Critical path

The longest route through the network (Figure 5.18) is 9 + 6 + 7 = **22 days** which means that a–b–d is the **critical path**. This is called the critical path analysis (CPA). The network can be fully analysed with a forward left to right pass (as above) which gives the ES and EF dates and a backward pass which goes from left back to right to calculate the LF and LS dates. An activity which has the same earliest and latest dates is called critical. C can start one day later in Figure 5.18 without affecting the end date, which means it has float.

The **resources** used in CPM are labour, plant, materials and cash e.g. activity A could be the ground floor walls and have a resource of five bricklayers and if the total cost of the labour plant and materials is £10,000 then this is also a cash (lump) resource. If there are other activities with the same resource which have to be carried out at the same time then the resources are added together to give a total resource. It is much better to float activities with the same resource so that they do not overlap.

Total float is applied to an activity or a milestone date and represents the flexibility to start an activity later without affecting the longest critical route through the project, i.e. C has a one day float from the diagram above because a–c–d is a shorter route than the critical path by one day.

A **critical** activity is one which has no total float, i.e. that is activities A, B and D.

Free float represents the flexibility to start an activity later without affecting any other activity. This is much rarer as it only occurs where an activity is dependent on more than one activity. In this case C also has one day of free float.

Negative float is the period which a critical activity (or an activity which has been made critical) has been delayed. The effect is to lengthen the longest critical route through the project, unless corrective action is taken; for example, if the activity C was delayed by two days it would become supercritical by one day.

A **supercritical activity** is a way of describing an activity with negative float. The only way to avoid delaying the whole project is to reduce the time taken by a supercritical activity by using more resources. It is no good using more resources on another route.

References

1 Ruddock L. (2000) *An International Survey of Macroeconomic and Market Information on the Construction Sector: Issues of Availability and Reliability Research*. London, RICS Research Papers.

2 Blake N., Croot J. and Hastings J. (2004) *Measuring the Competitiveness of the UK Construction Industry*. Experian Business Strategies. UK, DTI, p.2.

3 Latham Sir M. (1994) *Constructing the Team. Final Report of the Government/Industry Review of Procurement and Contractual Arrangements in the UK Construction Industry. Department of the Environment*. London, HMSO.

4 Construction Industry Board WG11 (1995) *Towards 30% Productivity Improvement in Construction*. London, Thomas Telford.

5 Egan J. (1998) *Rethinking Construction: The Report of the Construction Taskforce*. London, DETR.

6 Business and Enterprise Committee (2008) *Construction Matters 9th Report of Session*, Vol 1. London: House of Commons.

7 Wolstenholme A. (2009) *Don't Waste a Good Crisis*. London, Constructing Excellence.

8 NAO (2001) *Modernising Construction, HC87-2000–01*. London, HMSO, January.

9   Cain C.T. (2004) *Profitable Partnering for Lean Construction*. Oxford, Blackwells.
10 Cain C.T. (1997) Research note 14/97 BSRIA.
11 Olomolaiye P., Jayawardine A. and Harris F. (1998) *Construction Productivity Management*. Harlow/Ascot, Addison Wesley Longman/CIOB.
12 Cain C.T. (2004) *Profitable Partnering for Lean Construction*. Oxford, Blackwells.
13 Frank Bunker Gilbreth (1868–1924) was an early supporter of scientific management and led in using motion studies to improve technique and reduce wasted movement.
14 ASME (1947) Adapted from Gilbreth symbols introduced to them in 1921.
15 Business Dictionary (2010). Available online at: http://www.businessdictionary.com/definition/value-stream.html (accessed 16 December 2011).
16 Constructing Excellence (2004) *Lean Construction*. Information Sheet, London, Constructing Excellence.
17 Womack D. and Jones D. (1996) *Lean Thinking: Banish Waste and Create Wealth in your Corporation*. London, Simon and Schuster.
18 Dennis P. (2007) *Lean Production Simplified: A Plain Language Guide to the World's Most Powerful Production System*. 2nd edn. New York, Productivity Press.
19 Dennis P. (2007) *Lean Production Simplified: A Plain Language Guide to the World's Most Powerful Production System*. 2nd edn. New York, Productivity Press.
20 Stansfield K. (2005) 'Precast student flats at UWE, Bristol', *The Structural Engineer*, 6 December.
21 Constructing Excellence (2004) Lean Construction, Fact Sheet, p.1.
22 Koskela L. (1992) *Application of the New Production Philosophy to Construction*, CIFE Technical Report 72. Stanford University, September.
23 Howell G. (1999) 'What is lean construction?' Lean Construction Institute. *Proceedings of Global Lean Construction Conference 26–28 July 1999*, University of California, Berkeley, CA, USA.
24 Holti R., Nicolini D. and Smalley M. (1999) *Building Down Barriers. The Prime Contractor Handbook of Supply Chain*. London, The Tavistock Institute.
25 Constructing Excellence (2002) *Andover North Site Redevelopment Prime Contract*. Demonstration project 1190. London, Constructing Excellence.
26 Constructing Excellence (2003) *Construction Lean Improvement Programme Profit from Process Improvement*. London, Constructing Excellence. DTI and case studies of profit improvement are recorded as a basis for further improvements.
27 Bertelson S. (2002) *Lean Construction in Denmark – A Brief Overview*. Available online at http://www.bertelsen.org/strategisk_r%E5dgivning_aps/pdf/Lean%20Construction%20in%20Denmark.pdf (accessed 19 August 2010).
28 Construction Lean Improvement Programme (2003) *Seven Case Studies*. Watford, BRE.
29 Bath University School of Management (2010) *Agile Construction Initiative*. Bath University. Available online at: http://www.bath.ac.uk/management/larg_agile/overview.html (accessed 19 August 2010).
30 Ballard H.G. (2000) *The Last Planner® System of Production Control*. Unpublished thesis submitted to the Faculty of Engineering The University of Birmingham for PhD. Available online at http://www.leanconstruction.org/pdf/ballard2000-dissertation.pdf (accessed 20 March 2012).

31 Ballard H.G. (2000) *The Last Planner® System of Production Control.* Unpublished thesis submitted to the Faculty of Engineering The University of Birmingham for PhD. Available online at http://www.leanconstruction.org/pdf/ballard2000-dissertation.pdf (accessed 20 March 2012), pp.3–13/14/15.

32 Koskela L. (1997) 'Lean production in construction'. In L. Alarcon (Ed.) *Lean Construction,* Rotterdam, AA Balkerna. E-book Taylor and Francis.

33 Ballard G. and Howell G. (1998) *What Kind of Production is Construction?* Proceedings LGLC 1998. Available online at: leanconstruction.org/pdf/Ballard AndHowell.pdf (accessed 10 March 2012).

34 Ballard G. and Howell G. (1998) *What Kind of Production is Construction?* Proceedings LGLC 1998. Available online at: leanconstruction.org/pdf/BallardAndHowell.pdf (accessed 10 March 2012).

35 Green S.D. (1999) 'The missing arguments of lean construction', *Construction Management and Economics* 1466–433X, 17(2): 33–137.

36 Egan J. (1998) *Rethinking Construction: The Report of the Construction Taskforce;* Womack J.P., Jones D.T. and Roos D. (1990) *The Machine that Changed the World.* New York, Rawson Associates.

37 NHBC Guide (2006) *MMC Guide.* Milton Keynes, NHBC Foundation.

38 Turner Townsend plc (2009) *Global Construction Costs.* Autumn. London, Turner Townsend.

39 Harris E.C. (2010) *Slow Economic Recovery See UK Costs Falling by Further 10%.* E.C. Harris Built Environment Consultancy. Available online at: http://www.echarris.com/reference/news/slow_economic_recovery.aspx (accessed 10 August 2010).

40 Latham Sir M. (1994) *Constructing the Team. Final Report of the Government/ Industry Review of Procurement and Contractual Arrangements in the UK Construction Industry. Department of the Environment.* London, HMSO.

41 Adapted from Constructing Excellence (2001) *Mansell Best Practice Profile.* Available online at: http://www.constructingexcellence.org.uk (accessed 15 October 2004); and Constructing Excellence (2005) *Mansell Turnover Rises on Back of Customer Satisfaction,* Case Study 274. London, Constructing Excellence.

42 Constructing Excellence (2012). Available online at: http://www.kpizone.com.

43 Bakens W., Vries O. and Courtney R. (2005) *PSIBouw International Review of Benchmarking in Construction Roger Courtney Construction Innovations (UK) Ltd.*

44 Latore V., Roberts M. and Riley M. (2010) 'Development of a systems dynamic framework for KPIs to assist project managers' decision making processes', *Reviste de la Construccion,* 9(1): 39–49.

45 Crane (2003) *Crane in Australia – Part Two.* Reported in CIOB International news. Article 4180. CIOB. September.

Chapter 6

Organisation, culture and leadership

With Martyn Jones

This chapter looks at organisational and leadership theory and the development of structures and a culture that can further integrate the project organisation and ensure its effectiveness. This leads to the concept of excellence or world class, which requires effective and ethical leadership. Project teams are usually made up from different functional departments within the same organisation or from other organisations who are brought together to deliver a specific project over a set period of time. The challenge in construction projects is to develop a culture that *integrates* the various organisations involved and ensure that there is a synergy built up in the team that also removes opportunistic and adversarial attitudes. This is even more important with ongoing changes in construction to cope with aggressive economic conditions, address climate change and drive out waste. We examine this trend towards greater emphasis on integrated teams and collaborative relationships.

This culture of greater integration and synergy depends on members of the project team having good relationships with the client and with each other. Clients play a major role in construction projects and get involved in the value management and risk allocation of the project. The structure of a project will vary through the life cycle, but during design and construction a large number of different organisations will be involved and they need to be organised in ways that recognise the uncertainty and yet interdependency that exists in construction projects. They also need to be able to adjust to external and market influences. Chapter 7 is complementary and considers the softer skills of people management.

The objectives of this chapter are to:

- identify different structures under which projects may be run and discuss their impact – traditional and current approaches in construction
- define external factors and their effect on organisational structure
- consider the culture and values of various organisations and consider the factors which are important for creating an effective project culture that meets the needs of the client and motivates the project team

- identify the principles and culture of partnering and integrated project teams
- explain EFQM and project maturity and create a model for world class
- consider the implications of collaboration in making organisation change in projects
- examine the theory and practice of effective and ethical leadership.

Organisation structure

Organisational structures differentiate specialist skills and seek to delegate tasks in a way that ensures coverage and prevents overlap of responsibility. Classically a chain of command is created with unity of command and a regulated span of control of five to six, so that everyone restricts their relationships. Weber developed the term bureaucracy which is hierarchical and permanent with a deep structure and a dependence on procedures.[1] Later research by Woodward indicated that an average span of control for unit production was 23 in more modern business and this leads to fewer layers of authority and better communications.[2] It is also clear that many structures are not so formal in the chain of command, and networking between different departments and levels is critical to modern communications.

Project organisation in particular has cut through the classical principles and set up alternative structures because of the need to work with many different organisations at different levels of authority and to closely co-ordinate interlocking work. Walker maintains that in practice construction projects have three major components influencing organisational structures which are (1) the way the client and project team relate, (2) the way the design team is organised and (3) the integration of the contractor into the process.[3] Figure 1.3 (Chapter 1) represents a simple model which indicates a well integrated relationship with the use of a project manager. However the project manager as the leader or co-ordinator often has other roles and varying authority depending on the contract provisions and can change during the life cycle of the project, for example feasibility to design implementation to construction. The structure and project management role is partly influenced by the procurement and contract type chosen, but also by the degree of collaboration which overlays it. This creates a large number of possible combinations. The key issues for integration of structure to suit the client's priorities and to optimise working relationships with the team chosen. Winch[4] considers project organisations as between two types – bureaucracy and adhocracy,[5] where a flat, flexible or organic structure is used to cope with change.

The organisational structure for a project essentially breaks down into three types: functional; matrix; or project based as shown in Figure 6.1. The matrix organisation arises because team members are accountable to functional and project authority. The matrix organisation may be weak or strong in projects depending on whether the emphasis of authority is with the functional manager (weak matrix as in (a)) or with the project manager (strong matrix as in (b)). Some organisations operate as project organisations, where all activities are project based as in (c). Construction and engineering project teams have traditionally employed a strong seconded matrix with specialist functional roles drawn from professional specialists with pronounced differences in values, attitudes and behaviour.

Of particular interest in the context of this chapter is the degree of authority that the project manager exercises as shown in diagram (d) of Figure 6.1 and therefore the scope of his or her leadership. From (d) this can be seen to increase the more there is a direct link between the line management and the project manager. If they have to use staff from other functional departments or organisations they approximate more to co-ordinators. If these staff are seconded to the project then the authority role is strengthened. Table 6.1 has been built up to indicate the differences and focus for the project manager. Large projects may aspire to a pure project structure where the core project team is full time and the project manager has a more assured authority as members of the team are singular in their focus.

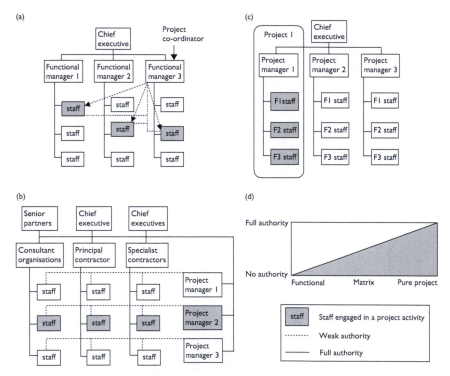

Figure 6.1 Authority: functional, project and matrix organisation types

Table 6.1 Influence of organisational structure type on projects

Project Type	Functional	Matrix	Seconded matrix	Pure project
PM authority	Little or none	Low to moderate	Contractual	High to total
Project personnel involvement	0–25%	15–60%	50–95%	85–100%
Position of PM	Subservient to functional managers	Subservient or equal to functional managers	Leadership and direction	Executive decisions
Title for PM role	Project Co-ordinator/ Project Leader	Project Manager/ Project Officer	Project Manager	Project Manager/ Project Director
Type of project	Minor change to existing product e.g. maintenance upgrading and improvement	Implementing change to work organisation e.g. space planning and implementation	Most types of new build and refurbishment project. Design and construction inputs. Facilities management	Very large construction projects created as business units or singular specialised projects such as IT, roll outs, redecoration and refits
Project manager focus	Diplomatic approach. Integration of functions	Negotiating adequate time allocation. Good communications	Creating an integrated team. Planning and leadership	Productivity and teamwork

There are advantages and disadvantages to each of these organisational structures. For matrix organisation the project personnel have different calls on the use of their time and project managers need to fight for priority use of personnel, making progress slower than expected. In function based projects the project co-ordinator has to be diplomatic if calling on the use of other functionally based personnel who are primarily doing non-project activities and the PM needs to firmly agree resource usage. From a pure project point of view there is good loyalty, but instability for staff as projects come to an end, sometimes causing inefficiency as organisations look for other work and staff get attached to their next job.

Construction industry

Although some staff will be totally committed to the project, for example the project manager, main supervisors, contractor's quantity surveyors (CQS), planners and personal staff, other consultants will spread their time

Case study 6.1 Typical pure project organisation

This project is a new build project worth £26m in the South West for a national company HQ and has a seconded matrix project organisation (Figure 6.2) run from the site. They need to cover most of the roles with project staff who may be a mixture of dedicated or staff shared with another project. Some shorter term staff such as the engineer may be employed through an agency.

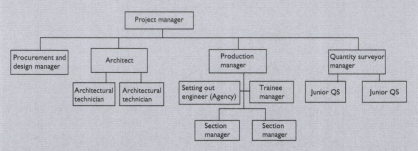

Figure 6.2 Typical organisation structure for intermediate project size

The staff were directly employed by the design and build company to manage the site. In order to ensure that all job responsibilities are covered unambiguously job responsibilities should be allocated to each. In addition some services will be supplied by separate design personnel who are not based on the site. The organisation is an example of a decentralised site with all executive control operating through the project manager who directly reports to the director and has formal accountability for the actions of the company. Because it is a design and build contract there is a much clearer line of authority even though some design personnel are employed from different organisations.

Table 6.2 shows the roles which emerge.

between different projects which their own organisations are committed to. For example a consultant quantity surveyor (PQS) might be committed to valuing work on two projects and an architect or engineer might have two or three design commissions. It is also possible for involvement in the project to be limited to bits or part of the life cycle except for key players such as the job architect. Consultants and subcontractors need to plan for this through the master plan.

Table 6.2 Roles in Case study 6.1

Project manager and health and safety officer	Executive control of design construction, planning and finance and specific accountability for site health and safety
Quantity surveyor manager	Financial control of valuations, change control, and estimating
Junior surveyors	Re-measure and progress on site, procure and check materials
Design and procurement manager	Initiation of design, management of design process, procure and tender supply chain, select subcontractors, method statements
Architect, service engineer and structural engineer	Scheme and detail design, quality inspection and compliance
Production manager	Overall control of production, compliance with BREEAM rating required, health and safety oversight and method statements implementation, quality improvement
Section manager (trainee)	Day to day management of relevant subcontractors or section of work, manage dimensional checks and order bulk materials, safety inspections and records. Technical queries. Access requirements
Setting out engineer	Setting out grid, co-ordinate stations and level benchmarks, checking critical dimensions and levels after setting out by steelwork, bricklayers, cladding and ground workers. Check base, floor and roof levels, plumbing in lift shaft and main elevations and columns. Setting out of drains, roads and car parks and level checks

Generally it is assumed that large construction projects have pure project organisations, but due to the specialisation of organisations in the industry and the division between design and assembly, the authority structure is likely to approximate more to matrix. Normally functional staff are employed by separate organisations and the project manager buys in specialist skills for the project. As they deal with organisations they are managing a supply chain, which is subtly different from employing a person because the organisation is more autonomous in deciding how that service might be supplied, conforming to a performance contract. In practice project managers prefer to focus on a contact person in the organisation to ensure team building and motivation is passed on through the specialist organisation. Main contractor packages are called tier one, but tier two structures might be formed from a sub subcontractor and so on. Chapter 12 on supply chain management explores the challenges of this interdependent network of relationships and hierarchies.

The most common form of organisation in construction projects can be described as the seconded matrix organisation (Case study 6.1) as the most direct project personnel spend all their time on one or more projects and integration of the team is recognised as an important element. Projects in the spend range of £5m plus require significant project organisation.

Centralisation of authority

The degree of centralisation is very much dependent on the amount of authority given to the project manager. In *centralisation* a contractor will try and impress procedures and process on the project set up. Major decisions will be made by the contract or construction manager and contractual issues and supply chain will be appointed and paid through them. Day to day co-ordination with the supply chain and the project team will take place at site level (SM). Smaller and medium sized projects are likely to be centralised to gain from economy of scale. The culture is likely to be less well defined and akin to the culture of the largest organisation in the project. The client will critically communicate through the project manager, which is likely to be the design leader or the design and build contractor.

Decentralisation is a delegation of project control and supply chain pro-curement to project level and allows a more autonomous culture to be developed and a closer understanding between the project team members who, although dealing with different levels of project management, will benefit from a closer knit relationship with all decisions being made at project level. In this structure a strong leader is required who can handle decisions on the construction and design streams and provide reporting of all direct to the client. The client's main point of contact is the project manager who will distribute all instructions and be responsible for the flow of informa-tion between the project team, although the client may hold contracts with individual members of the project team. This elevates the authority and seniority of the project manager. This is a common situation for large projects in construction. Decentralisation is compared with centralisation in Figure 6.3. Case study 6.2 shows a typical organisation for a large project that altered its policy to take account of the environment of other projects the developer had.

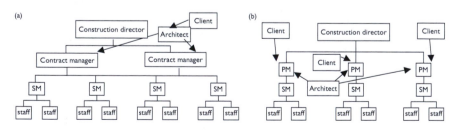

Figure 6.3 (a) Centralisation, (b) Decentralisation

Case study 6.2 Major development projects' organisation structure

A large developer in the West Country have developed their portfolio of building land to take account of what was becoming available and what would be granted planning permission to take on mainly brown field development. On a marketing perspective they noted the rising demand (and prices) for 'smart residential' property in city centres and the planning authority's desire to see past industrial sites in their city centres decontaminated and reused for residential and leisure. Over a period of time they have moved from smaller green field sites to large longer term mixed use schemes. This required the use of standardised footprint designs with a housing mix to suit the local housing market, to more comprehensive schemes which needed a master plan to satisfy the planning authority. This depended on building and attracting tenants or buyers for commercial and leisure and even industrial use and sequencing the whole to suit the general constraints of the market, but one of the major planning conditions was to assure social provision and employment and balance mixed use development to enhance city life. This meant a dynamic plan and sequencing of building to suit the market.

One such site is a £300 million budget over a build period of four years with cash breakeven eight years after the initial start of decontamination and regeneration works on a messy industrial site around obsolete docks. The preparation works included clearing away a disused power station and electrical substations, demolishing chemical factories and remodelling the dock land area to maximise the amenity to the new residential and leisure facilities being provided. In its place seven to nine storey flats were to be constructed adjacent to the dock area, a new marina and basin, the creation of park and wild life areas, luxury house plots and higher density town houses, together with a new transport exchange, shopping and factory facilities and the possible reinstatement of a railway link. The traditional well defined housing scheme had become a regeneration plan in its own right and mainly at the developer's cost.

Structure

This has led them to move away from the typical centralised structure required for the small to medium sized housing scheme, where sites are serviced with site managers calling off bulk purchase and using standard house designs. These were replaced by a suite of operational and executive site offices on the 'muddy side' and a smart sales centre to sell the houses segregated fully from the site traffic. Two directors were housed alongside the site staff to provide strategic direction for the future ongoing planning, the production and design strategy and the sales and marketing. The project operated as a business in its own right with direct accountability to the MD.

Subsequent to this case study research the company decided to centralise the directors away from the project to give them a multiple project portfolio due to the go ahead of a similar sized project in the same area. This indicates the dynamic nature of project organisation to suit the varying business circumstances, which in this case was taking the opportunity to capitalise on personnel experience by broadening a portfolio. The sales office and site offices remained, but the strategic control of the project was made more remote as a culture had become established. It is important in the design of project organisations to allow for changes during the life cycle of the project.

Although we have looked at some case studies to illustrate project arrangements, many of the influences for the organisational structure and culture of projects come from the market and wider environments in which they operate. These external influences are now considered.

Mapping the project environment

Projects and project teams do not exist in a vacuum. They exist in a wider environment or system which includes governments, competitors, suppliers and customers. They are affected by legal, economic, social, political, cultural, technological and environmental forces. The main elements of the wider environment for projects and organisations are shown in Figure 6.4.

The four elements of the external environment have been discussed in Chapter 3. Awareness of sustainable factors also has a significant and growing impact in maintaining a longer-term outlook for organisations and projects. These overlap with social and economic factors in their feasibility and the decision to introduce them.

So why has it become so important for organisations and project teams to look beyond their boundaries? Why is it seen as being so vital to project execution and long-term success? The main reason is change. It is now commonly acknowledged that this wider environment within which organisations and projects exist has been changing significantly since the 1980s when the world entered a new era. Whether this new era is referred to as

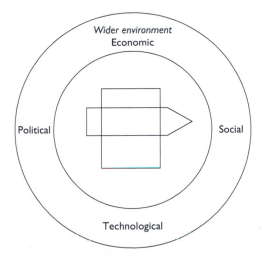

Figure 6.4 Elements of the wider environment

the post-industrial society, or the information and communication paradigm, or the postmodern age, or post-Fordism, what is clear is that the wider environment has been changing significantly during the past 30 years or so. This period of substantial change continues as the world adjusts to globalisation, energy crises, the decline in many commodities, the credit crunch and the emergence of the BRIC countries (Brazil, Russia, India and China), and the downturn in developed countries. It is argued that this postmodernist view of the world within which organisations and projects exist presents a distinct departure from what went before. Clegg has identified clear distinctions between the modernist and postmodernist organisational forms.[6] These are summarised in Table 6.3.

Table 6.3 Modernist and postmodernist comparison

	Modernist	*Postmodernist*
Structure	Rigid bureaucracies	Flexible networks
Consumption	Mass markets	Niche markets
Technology	Technological determinism	Technological choice
Jobs	Differentiated, demarcated and deskilled Highly de-differentiated	De-demarcated and multi-skilled
Employment	Centralised and standardised	Complex and fragmentary relations

What impact has this significantly different approach had on organisations and projects? Linstead argues that under postmodernism hierarchies of merit, legitimacy and authority have given way to networks, partnerships and organisational structures that are constantly shifting, more fluid and with a stronger social nature.[7] On the one hand, this has rendered the management of projects more challenging and yet, on the other hand easier, as individuals, groups and organisations have attempted to adopt these more flexible structures and synergistic processes.

Daft argues that postmodern organisations need more flexible and de-centralised organisational structures with more fuzzy boundaries internally between departments and externally with other organisations.[8] In such organisations leaders are facilitators communicating and interacting more informally and symbolically. In this approach, control is exercised through self-regulation, decision making is inclusive and governance based on egalitarian and democratic principles. For a number of organisation theorists post-modernism has dominated management approaches and is having a major effect on the nature and functioning of organisations, including projects. It might be argued that this shift to postmodernism has been more pronounced in manufacturing as organisations have grappled with moving from the rigidity of mass production to more flexible mass customisation than in project-based industries and organisations that have always had to adopt more flexible and responsive approaches.

What is clear, however, is that with these shifts in the wider environment, today's project leaders need a much greater strategic awareness of the environment within which their projects take place. The importance of mapping and understanding a project's environment is discussed in the following section.

The organisational context

The project environment includes everything external to it. Maps or models are a good means of developing awareness and understanding of the wider environment and identifying what forces are at play. Developing an understanding of the external environment is a tall order, but this can be done by constructing a simple map. The map starts with the project as shown in Figure 6.5.

The next step is to determine where to map the project in relation to the organisation. This is often problematic. Do the projects take place within the organisation or outside it, or do they straddle the boundaries of the organisation in some way? The main situations are shown in Figure 6.6.

Take the case of the project within the organisation. Here the project sponsor will be from within the organisation. The project manager will normally select his or her team from within the sponsoring organisation with the aim of making them a cohesive whole and ensuring that the interests of

Figure 6.5 The project

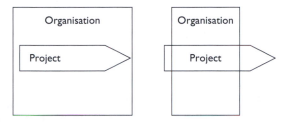

Figure 6.6 Relationships between organisations and projects

all the stakeholders are balanced. This teambuilding is a fundamental role of most project managers irrespective of the situation. Where the project is within a single firm or organisation, the challenge is to organise people from different specialisms or functional departments (for example, marketing, finance or engineering) into an effective project team. There are a number of organisational structures that have been compared in Table 6.1. The next situation is where a single organisation interacts with and plays some part in the project. In many contexts, such as in construction for example, this view is too simplistic, however, as it is common for a number of independent organisations to be involved. In the case of even a modestly sized construction project this may amount to scores of autonomous yet interconnected and interdependent organisations playing some part in the project process, as shown in Figure 6.7.

The challenge for the manager of projects involving a number of independent organisations is to bring together individuals from different

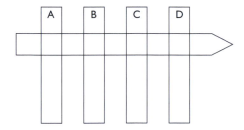

Figure 6.7 A project involving a number of independent organisations

organisations or firms (rather than different functions or specialisms within the same firm) and form them into an effective project team. The firms involved are autonomous firms, which are organisationally independent outside the project. It is likely that each firm or organisation will have conflicting aims and objectives, different cultures and varying degrees of allegiance or commitment to the project. Effective project management offers a means of unifying overall processes and presenting opportunities for participants to place the demands of the project alongside or even above those of their own enterprise. This complex kind of matrix structure raises two significant management issues: managing in the context and culture of the firm or organisation, and managing in the context and culture of projects.

A further complexity arises where, as is common in the case of construction with its short-term approaches to procurement, projects are often undertaken by an amalgam of firms which may well change from project to project. This can give rise to all sorts of problems as the transitional nature of the relationships between the organisations creates a potential for conflict between the needs of each firm and the project.

Figure 6.7 has also greatly simplified the situation because it relates to a single project. In practice, a portfolio of projects may be involved as shown in Figure 6.8.

Again this is a simplified view as it implies that the projects shown are all being undertaken by the same firms or organisations. In reality, this may not be the case as there will often be different mixes of firms involved for each project each with their own structures, cultures and behaviours.

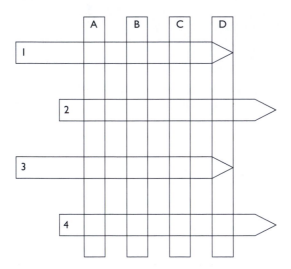

Figure 6.8 Complex matrix of firms and projects

Case study 6.3 provides an inside view of how one developer organised multiple projects on an ongoing development.

In this more complex amalgam of firms, developing a greater understanding of organisational culture and behaviour and the nature of the relationships that exist between firms is seen as being increasingly important in ensuring project outcomes and indeed raising the performance of the whole project process.

Case study 6.3 Multi-project mixed use development

The project takes place over five years on a prime 16 acre derelict brown field site worth £350m in the centre of a provincial city. The site is very sensitive and planning permission has only been agreed with very stringent conditions to maintain site lines and to properly connect the city centre to the waterside and to ensure full public access. The proposals are for leisure, retail, commercial and residential use. The latter is core to the private developer in terms of cash flow and profit. The development links directly to a highly popular leisure sector and a significant commercial venture already in place on the site. The site is in a desirable location, which faces onto the old harbour waterside. A master plan is in place.

The developer does not employ an extensive project team, but seeks instead to outsource the design and tactical project management. The first phase is procured by three design and build contracts. The second phase is a residential contract for the developer built on their behalf, but let directly by them. On each of these contracts there is an outsourced team for the client, consisting of a cost consultant, an architect and a project manager. These will report to the programme director who co-ordinates the whole. Due to the sensitivity of the site to the conservation of the city character and heightened public interest a further development director has concentrated their efforts in key marketing of the subsequent phases of the development and bringing them through the planning process and maintaining good relationships with the public and community. The construction director has taken on the implementation and later marketing stages of the development and is maintaining the master programme for parallel and serial projects. Parallel contracts have impacted on each other and needed to comply with the overall constraints of the site. This has led to a loose structural organisation, which also needed to have a flexible response to the market forces which will determine the speed of completion of the development and the logistical access and environmental constraints which affect the methodology and sequence of separately let contracts.

The design has been spread out between architectural/engineering practices to reflect specialities and inject variety. Leisure, commercial and residential blocks have been allocated and the master planners have been retained for more detailed involvement in one area. The process is managed by phasing and allocation of different buildings to different contracts, so that design management is linked to the procurement route chosen and is not developed centrally. The programme director is the one main player who co-ordinates the production design for the developer. Figure 6.9 indicates the emerging organisation structure.

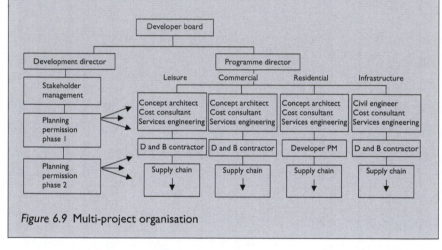

Figure 6.9 Multi-project organisation

Corporate culture and behaviour

Corporate culture consists of an organisation's norms, values, rules of conduct, management style, priorities, beliefs and behaviours. The current fascination of business with organisational culture began in the 1980s with the work of writers such as Peters and Waterman,[9] although earlier writers such as Blake and Mouton had argued the link between culture and excellence.[10] Silverman contended that organisations are societies in miniature and can therefore be expected to show evidence of their own cultural characteristics.[11] However, culture does not spring up automatically and fully formed in response to management strategies. Allaire and Firsirotu argued that it is the product of a number of different influences: the ambient society's values and characteristics; the organisation's history and past leadership; and factors such as industry and technology.[12] The main determinants of business culture are shown in Figure 6.10.

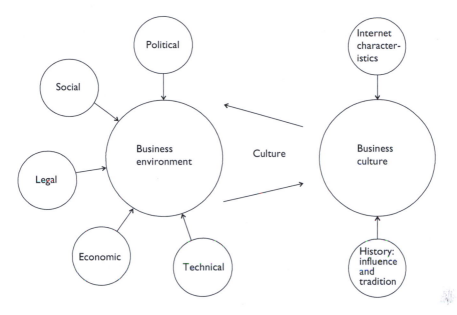

Figure 6.10 The relationship between business environment and business culture

This shows that an organisation's culture is influenced by a combination of its own inherent characteristics and the elements of the external environment identified earlier. Culture is a term that is increasingly used but poorly understood. This is because an organisation may have a number of cultures within it and that even a detailed deconstruction would provide an inaccurate picture of the totality of the organisation.

Harris and Moran have identified a number of determinants of cultural differences:[13]

- Associations – the various groups with which an individual may be associated and this would include disciplines.
- Economy – the type of economy affects the way individuals conduct themselves at work, how they feel about achievements and their loyalty to their employer.
- Education – the types and amounts of educational opportunities.
- Health – recognises the impact of the health of workers on productivity and effectiveness of the organisation.
- Kinship – the family and its importance in the life of the employee.
- Politics – the political system impacts on the organisation and the individual worker.
- Recreation – recognises the role of leisure time in the life of a worker and his/her family.

- Religion – in certain countries this can be the most important cultural variable related not only to the workplace but also to the daily lives of people.

These factors must be taken into account by leaders and managers in their style of management and leadership. Fordist and Taylorist approaches favoured coercion where functional devices and authority are used to force the individual to carry out a particular task and in a non-thinking way. With the shift to post-Fordism there is more emphasis on co-operation and the education of the individual as to the reason why it is in their interests to perform at a high level. This is the focus of the humanistic movement.

Handy has usefully grouped organisations into four cultural types which help to identify the different characteristics, but these are too simplistic for organisations to have an exact fit and indeed it is possible that several cultures will emerge and this is a particular issue between the 'office' and the project.[14] Many small businesses are run as power cultures. In many projects and for much of the project cycle, the role culture prevails as the tasks are routine. Projects run in this way may be effective in undertaking the task but often at the expense of the project team and the stakeholders. The role approach has an emphasis on structure, which although it allocates tasks and responsibilities clearly, they can be bureaucratic and obstructive to change. The task cultures place more emphasis on the team.

International considerations

Hofstede noted four factors which he noticed as having a different emphasis on strategy in different countries and these are identified in Figure 6.11.[15] These refer to the difference in work related values. Long term outlook has been added as an additional dimension covering certain social and political approaches and recognising different emphasis on programme importance.

Although these may be disputed, and others have come up with slightly different dimensions, Hofstede measured a sliding scale for each dimension which broadly recognised some of the critical differences. These dimensions could have strongly different impacts on decision making and negotiations so a prevailing culture at managerial level may be operating in a project reflecting both the client's decisions and the project team's strategy where they are influential advisors to the client. For a multicultural approach to work some of the subliminal issues must be realised and agreed approaches understood. All the dimensions are stereotypical, but produce a base for discussion.

Other obvious things are language, religious, legal, educational and political differences which are important to respect in everyday project transactions. These things affect the motivational climate and communications of a project. There is also a protocol difference which reflects in how

Hofstede characteristic	Those with more	Those with less
Power distance	France, Spain, HK & Iran	Germany, Italy, Australia, USA, GB
Uncertainty avoidance	France, Spain, Germany, Latin	GB, USA, Canada, Australia, Ireland
Individualism	USA, France, Spain, GB	Portugal, HK, India, Greece
Masculinity	USA, Italy, GB, Germany, Japan	Netherlands, Scandinavia
Long/short term outlook	China, Japan, HK	Russia, W. Africa, Indonesia, France

Figure 6.11 Hofstede's cultural dimensions. HK = Hong Kong; GB = Great Britain

much you make personal relationships first or whether you divorce business from personal relationships more and show a caring attitude towards people's personal interests and family. Language may be the same, but the use of different colloquialism, for example 'apple of my eye', may cause misunderstanding. Multicultural sensitivity gained by getting to know each other's culture will help project relationships to avoid cultural collision. The dangers of not doing so can be summed up as mistrust including stereotyping, miscommunications and process difficulties in failing to agree what constitutes an agreement.

Project culture

Project managers need to be aware of culture in managing projects, particularly where different cultures are involved. Weak matrix management often fails to produce the anticipated results because the role culture of the organisation is overlaid with the task-centred culture of the project. In relation to quality, the procedures associated with systems are often applied less effectively in task, person and power cultures, but are most effective in a role culture with its formalised and rule-based approach.

The authority of the project manager varies as in Figure 6.1 from more of a co-ordinative role to one that is controlling. Winch refers to them as a 'temporary coalition of firms' and these are brought together for a set time and purpose.[16] Within that framework culture can be harnessed to produce synergies and to avoid conflict between different ways of doing.

Hopkins *et al.* however talk of accepted subcultures which can be tolerated because they do not upset the overall solidarity of the project or challenge its underlying values.[17] The issue of trust is important for efficiency and has been quoted as a reason for partnering[18] in order to reduce harmful adversarial relationships in construction projects.[19] Lack of trust between client and contractor because of late changes, claims for extras and agreeing payments, or lack of trust between consultant and contractor over efficiency and buildability and between contractor and supply chain over payments and quality are traditional and need to be managed out to improve relationships. For more on trust see Chapter 12.

Construction project cultures

Initially we might automatically assume that a construction project best fits into the task culture in Table 6.4 because a project is task orientated. This is not unreasonable, but different project situations can achieve characteristics that bend them towards roles rather than teams. A contractor's organisation can be quite large with a strong matrix culture between different functions such as quantity surveyors, accountants, project management, plant supply and planning. A strong functional influence can result in the role culture of meetings, reporting procedures and approval forms. The advantages and disadvantages of Handy's four types are shown in Table 6.4 and assessed for construction projects.

The construction project has the involvement of several organisations in the supply chain and these may conflict culturally, or be actively organised by the project manager into a project culture, sometimes called a virtual organisation. Many small organisations contribute to projects and they may exhibit a strong person culture and this may lead to conflict. Fragmentation of project tasks may take place due to the very different culture of learning there is between the different professions of engineering, architecture and construction management, the legal profession and accountants. This can be attributed to the clash of different organisational culture types.

A small architectural or an engineering organisation is likely to be a partnership enshrining a culture of entrepreneurship or a loosely coupled organisation of separate stars. Different educational backgrounds between architects, engineers, quantity surveyors and managers may cause conflict in the project team, but can be harnessed to complement each other if there is respect. There are also many organisations together in a project with their own developed values. In international projects there is sometimes a communication problem where non-verbal signals are misunderstood and there are different approaches to a problem, priorities or interpretations of words. Behaviour may be misunderstood where people have been upset or unwittingly put down. In these cases there has to be an evolved, internationally acceptable culture which may require unlearning set ways to get better teamwork and engagement.

Dubois and Gadde point out that there are both tight and loose couplings in supply chains.[20] Tight because of the direct effect of late delivery or poor performance on the other parties and loose because of long lead times and intermediaries. Loosely coupled supply chains are co-ordinated with a subsequent reduced control of key aspects such as programme and budget. Within firms they are likely to operate on more than one project and these may have different cultures for the reasons given above. A culture may be built up but is difficult without a commitment to supply chain integration.

Loyalties are given to the client in terms of repeating work with the client and the project culture comes second if the client becomes part of the project

Table 6.4 Benefits and concerns for Handy's cultural categories

Culture type	Comments and example	Advantages	Disadvantages
Culture of personal power (Power)	Frequently centred on a single personality in the spirit of entrepreneurship. The culture is based on a personality. e.g. Property development or a small contracting organisation's projects	Very focused objectives with the potential for creativity and fast growth	Control is operated from the centre and some may feel strait-jacketed if they don't catch the vision
Culture of inter-dependence (Person)	A loose coupled organisation of equals. Gain from shared common facilities. Some shared aims & interests. Culture is agreed or existing common one. e.g. Design partnership or the culture developed amongst the professional team	Management and responsibilities are decentralised, allows flexibility and a forum for networking and recognition. All individuals are stars	Synergy is difficult to achieve. One tries to take over. Lack of loyalty, break ups or low retention rate
Culture of position authority (Role)	Bureaucratic depending on procedures and rules and formalised structures. The culture is imposed. e.g. Large longer running project where there is a formal hierarchy. A PRINCE2 or a Gateway Review defines quite formal sign offs	Stable, predictable and everyone knows where they are	Lack of flexibility and innovation
Culture of team building (Task)	Aligns to the matrix organisation. Focus on getting the job done, seconded personnel. The culture is delegated to working groups. e.g. Normal construction projects where teams emerge with their own synergies and develop from their growing knowledge of each other. PM co-ordination skills to gain a common objective and manage interfaces	Satisfying relationships and efficient working organisations formed and reformed to suit needs	Short lived, hard to produce economies of scale and to develop expertise in depth. Competition for available resources

Adapted from Handy (1993)[21]

team. Individual agreements may create competition in the team and a lack of transparency and sharing of information. If it is correctly focused and controlled, a strong partnering culture could be built up with repeat work or client partnerships over several projects, but value for money and reduced risk should be in the forefront of the motivation. An ethical and professional approach needs to be reciprocated in order to nurture it and build up trust and respect, but in general this is like Herzberg's motivating factors – they need to be more than basic professional conduct as slippage works the other way.

As we have already seen, projects and organisations do not exist in a vacuum. They exist in a much wider system, which includes governments, competitors, suppliers and customers. They are also affected and shaped by legal, economic, social and technological forces. It is this wider external environment which facilitates, constrains and threatens activities – and, of course, provides opportunities.

The external environment, strategic issues and forces for change

Clearly organisations and project teams have less influence over the wider environment than their market environment. Day-to-day actions for running the organisation and projects are, therefore, more likely to be governed by the market environment than by the wider environment. Elements of the market environment such as competitors and suppliers will, however, be influenced by the wider environmental forces.

The market environment

Figure 6.12 shows that projects and firms exist within a market environment.

As can be seen, the map of the market environment comprises a number of elements including customers and clients, the labour market, wholesalers and retailers, suppliers, trade unions, professional bodies and other groups and associations. What must be borne in mind is that a firm's ability to make profits is largely determined by the structure of the market in which it operates. Clearly some markets and industries are more profitable than others.

A major feature of most market and sectors of the economy is increasing competition. This means it is important that organisations are aware of their competitors. They need to know who their key competitors are. This is important because their actions can have many consequences for an organisation. Porter identifies five forces that determine the level of competition and underlying profit opportunities within an industry.[22] Consider how these forces apply within the industry and how they affect organisations and projects.

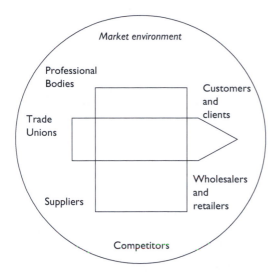

Figure 6.12 Elements of the market environment

Industries are highly competitive when there are many competitors or a small number of equal sized companies, with little to differentiate between products and services, fixed costs are high, economies of scale are important, and industry growth is static or low. Porter's five forces are:

- A highly competitive industry will reduce profitability as firms keep prices low to be competitive.
- The threat of substitutes. Most industries compete directly or indirectly with other industries, which offer substitute products or services.
- Ease of new entry to the industry. Where entry barriers are low an industry's profit potential will be limited because new entrants can enter the market and compete.
- The power of buyers. Powerful buyers can increase their profits by negotiating lower prices, improved quality, better service and delivery from their suppliers.
- The power of suppliers. In some circumstances where there are a small number of suppliers, the suppliers may be able to make excessive profits by forcing customers to pay high prices.

Awareness of these factors within an industry enables organisations and project teams to forecast, recognise and respond to changes in the market environment. A firm and project teams can position themselves to respond to threats and constraints created by the structure and nature of the industry in which it operates.

The role of partnering

From the early 1990s a significantly different form of organisational imperative began to emerge which came to be known as partnering. Partnering was identified by the Latham Report (1994) as 'a way forward in dealing with the conflict in many construction projects',[23] and the CIOB 2010 is more analytical pinpointing its benefit as 'a set of actions by which risk can be distributed and conflict minimised'. Partnering is put in a business context by Bennett and Jayes:

> a management approach used by two or more organisations to achieve specific business objectives by maximising the effectiveness of each participant's resources. It requires that the parties work together in an open and trusting relationship based on mutual objectives, an agreed method of problem resolution and an active search for continuous measurable improvements.[24]

Partnering is based on building up trust and is normally brokered by the signing of a voluntary pact. Essentially it is a risk sharing approach and is not directly connected with any particular organisational structure, but rather with a more open culture and effective leadership to ensure that the trust is reciprocal and respected. Partnering potentially gives savings arising from cultural change, which improves collaborative activity based on co-operation and reduced conflict that cuts out costly contractual procedures. Strategic repeat work with the same client helps this, but is not essential. Trust appears to be a central concept to alliances and partnerships in mainstream organisational studies literature and partnering, but comes at a cost and this may well be associated with the risk of trust breaking down between the parties, because trust requires them 'to be vulnerable to the actions of another party based on the expectation that the trustee will perform actions important to the trustor, irrespective of the ability to monitor or control that party'.[25] This risk needs to be factored against the savings. Certain measures have been developed to seal the trust commitment and reduce the risk of opportunism including, perhaps not surprisingly, the use of a contract! It is important though to recognise there are different levels of trust and some see the transaction of partnership purely as co-operation which is a less costly form of trust because the stakes are reciprocated and controlled. In a more stringent climate of recession it is likely that partnering is going to be seen as anti-competitive by the client, who may perceive that a short term gain to get more for less is to take advantage of a 'buyer's market' and go out to tender to test the market. This type of action, if genuine, can test the fear that long term relationships are too cosy and exploit the client, but if instigated against the agreement terms, it can break down a supplier's trust who has genuinely invested heavily in the partnership. Alternatively this criticism can be countered by the use of reliable benchmarking (Chapter 5) to measure improvements that are independent enough to be trusted by all sides.

Partnering model

The key objectives of partnering according to Bennett and Peace are mutual objectives, continuous improvement and empowered decentralised decision making based on the building up of project teams as indicated in Figure 6.13.[26] The decision making would normally be based on a partnering charter which would describe the principles, attitudes and ideals of the partnership which would promote collaborative working. The mutual objectives are outputs which are expected to be gained by the partnership and will be justified by saving, say, up to 50 per cent of the cost of traditional contracting on a series of projects. The mutual objectives will provide other targets such as reduced programme time, better health and safety, innovative solutions, better profit levels and fewer defects as the team seek to gain win–win solutions rather than zero sum outputs where one party gains at the expense of another. The synergy of mutual objectives must be proven and comes through hard won collaboration and not by wishful thinking. Teamwork is the principle delivery platform as it encourages open communications and transparency, more flexibility, knowledge management and the establishment of workshops. Other tools such as extranets or BIM can be less expensively set up if the team is committed over several projects with the same client

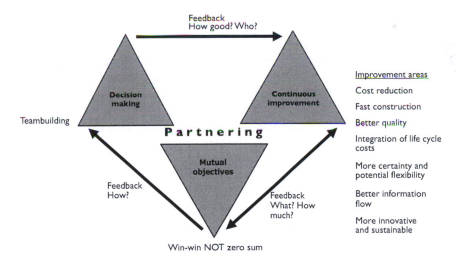

Figure 6.13 The key elements of partnering
Based on CIOB (2010)[27]

and feedback can be used to incrementally improve systems and catalyse innovation to encourage.

Partnering tools

Partnering contracts do exist but essentially it is a culture rather than an organisational structure. There are many tools and procedures to build up benefits for partnering. In many cases partnering uses open book accounting to motivate partners to support each other in that costs and profits are pooled transparently and also there are incentives for suggesting improvements which do not come out of traditional adversarial contracts. It also makes it possible to determine the basis for any profit sharing through either a:

• joint venture partnership between key suppliers
• risk sharing 'pain-gain', 'loss-profit' incentive scheme between supplier and client.

The former incurs single accounting for work done and making payments in proportion only to proportion of work complete, for example, completing foundations triggers payment. This encourages the supply chain to work together to productive working. 'Pain-gain' agreement means that the client is prepared to share any savings made on an agreed target budget, on the basis that the supplier partner is prepared to share any overrun of budget. The proportion of gain:loss is agreed in the contract.

Continuous improvement and value savings are often achieved over several projects in what is called strategic partnering or a framework agreement, where a small number of partners share a repeat client's workload and keep each other competitive. A common 'experienced' client approach is to look at reducing the equivalent project tender for successive jobs by 5 per cent based on identified production or process waste cutting. This can be done by using lessons learnt and applying these to the next project. If both sides are to be motivated then the efficiency saving needs to be greater than the saving to the client and contractor profitability does not simply drop. If this works and the suppliers are willing then the criticism of profitable partnering, that value is lost due to the lack of competition, is countered.

These methods represent the 'hard nose' of partnering and are not essential to the gains that can be made in normal collaborative working. To get repeat work with 'valued clients' never used to be called partnering and yet it is a basic tenet of good business practice to increase certainty of success and cut down extensive tendering costs. Most prefer a negotiated approach due to the costs of tendering. Doing less competitive tendering is witnessed by the behaviour of many large contractors reducing their interest in single stage open or selective competitive tendering.

Partnering at its best can induce real value into a project and share the benefits of savings (up to 10 per cent is claimed for single projects), but it can be one sided where a dominant client induces lower prices *out of the profits* of the suppliers for the promise of keeping them busy. Real savings of up to 30 per cent (see Case study 6.4) are possible over a gestation period and are made by identifying and cutting out waste in the process, reducing the learning curve of subsequent projects, gaining a better knowledge of the customer's real value system, encouraging supply chain synergy and possibly bulk buying power.

Case study 6.4 BAA Frameworks[28]

BAA has had many successful frameworks where proven savings have been made for what have been repetitive contracts. The BAA pavement team, working on runways and airport aprons, carried out 50 projects over five years worth an average of £2.5m starting in the late 1990s. During this time the team was able to reduce construction costs by 30 per cent and total time by 30 per cent. Safety performance improved and staff productivity improved to give value added far over the industry average. Other benefits from the partnership were the client's confidence in being able to gain value for money and a benchmarking process allowing measurements against the UK and work abroad.

In 2008 a new procurement manager was appointed who closed these frameworks down in order to induce more competition on major building projects over £25m which were to be let. This has potential for:

- bringing in new blood and innovation
- ensuring that other contractors and consultants who improved their processes more get a chance to bid for work
- allowing a broader range of procurement and contract which may induce efficiency for specific conditions.

Criticisms that serial partnering does not allow the client to gain from the market are countered where the client is able to market test and to compare against benchmark industry performance. Marketing testing may also help to refresh frameworks which are not working by opening partnership terms and bringing in new contractors from time to time.

Frameworks can be valuable in letting consultancy contracts so that repeat work design does not start from scratch and standardisation is developed to save fees. Frameworks may have disadvantages if they stop new ideas or do not eliminate poor performers.

Customer focus

The longer-term relationships associated with strategic partnering allow the development of a greater focus on the needs of customers. Peters and Waterman argue that excellent companies really do get close to their customers, while others merely talk about it.[29] The customer dictates product, quantity, quality and service. The best organisations are alleged to go to extreme lengths to achieve quality, service and reliability. There is no part of the business that is closed to customers. In fact, many excellent companies claim to get their best ideas for new products and services from listening intently and regularly to their customers. Such companies are more driven by their direct orientation to the customers rather than by technology or by a desire to be the lowest-cost producer/provider. This is how Peters (1992) describes the lack of customer focus in many types of organisation:

> Look through clear eyes and you'll find that almost all enterprises –
> hospitals, manufacturers, banks – are organised around, and for the
> convenience of the 'production function'. The hospital is chiefly con-
> cocted to support doctors, surgery and lab work. Manufacturers are
> fashioned to maximise factory efficiency. The bank's scheme is largely
> the by-product of 'best backroom (operations) practice'. I'm not arguing
> there's no benefit to customers from these practices. The patient
> generally gets well, the car or zipper usually works, or the bank account
> is serviced. And enterprises do reach out, sporadically, to customers –
> holding focus groups, providing toll free numbers to enhance customer
> dialogue, offering 'customer care' training to staff. But how many build
> the entire logic of the firm around the flow of the customer through the
> A to Z process of experiencing the organisation? Answer: darn few![30]

Over the past few decades or so, more and more companies have sought to change their external relationships by developing closer and more harmonious links with their customers and also their suppliers. These changes are taking place due to a growing climate of opinion that customers and suppliers working co-operatively must be more beneficial than more traditional adversarial purchasing relationships.[31] Construction organisations can be seen as having advantages in terms of developing customer focus over organisations in many other sectors of the economy because their products are normally bespoke for a specific client with a specific need.

Case study 6.5 indicates a large project which has responded to their customers' major requirements by creating their own strategic business unit. This is a specific way of giving customer focus and financially protecting the parent company in the event of uncontrolled loss on mega projects. It is also a way of responding to the risks in the wider external environment.

Case study 6.5 Strategic business units

Mega projects where there is a spend of many millions per month are often run as special projects and are allocated as strategic business units with their own profit centre with full accountability. These will be determined where they represent a significant proportion of the turnover of the company. One such £300m project for a government organisation was built in two years by a PFI consortium led by one of the large contractors in the UK. The project involved the moving into new buildings of an almost complete government department on a phased basis, decanting staff from existing buildings on the same site. The construction works included moving major power lines, putting in new roads and diverting traffic flows, releasing land for development, and the design, construction and fitting out of high technology facilities. During construction the cash flow went up to £22m/month at the peak of the contract when 1,200 people were working on the site. At the start of the project and to suit phased handover the speed was much reduced to deal with site and client constraints and this required careful management of contract personnel.

There was a 150 strong project management team, which was part of a joint venture company set up to deal with the design, construction and management of a large supply chain. This company will also continue to manage the facilities for the next 25–30 years of its operation. Major car parks, recruitment teams, training facilities and health and safety systems were created to cope with the workforce and visitors to the site. The project organisation is divided up into procurement management, design management, financial management, sectional management of the construction, and facilities management. A senior project director provided business and strategic leadership supported by a small team of project managers and sectional construction managers. In additional a general manager of the joint venture company was responsible for the strategic level contact with the client throughout the life cycle of its operation.

A unique logistics network was designed into the new building to receive supplies on the non-secure side of the building, where they were vetted, and to then to deliver them by train to various parts of the basement where they were taken by lift to the relevant section requiring them. Security is often a major issue on projects and a security reception was set up to provide identification and escort, and give health and safety induction for all visitors and new workers on to the site. Fire escape routes with fire points and segregated pedestrian and plant routes were designated to back up the health and safety and security requirements on a large site.

Although this project is one of only very few in its size category it indicates the importance of creating a fully competent integrated team for a large business undertaking in a very short period of time to meet. The client required evidence of a world-class organisation who were able to meet the exacting requirements of a very tight programme, budget and phased changeover programme to move staff, whilst maintaining security.

Experienced customers are becoming more discerning in their requirements for getting quality and have looked for more effective methods for choosing between contractors. The following model is one that has been used widely to identify those who are able to give the best and most sustainable service.

World-class performance

Clearly, extremely high customer satisfaction is one measure of what makes a top performing or world-class company. Studies show that a typical profile of world-class performers within the post-Fordist era would also include the following characteristics: strong leadership; motivated employees; a strong and/or rapidly growing market share; highly admired by peer group companies and society at large; and business results that place it in the upper quartile of shareholder value. Achieving business excellence demands the following:

- organisational learning
- farsighted, committed and involved leaders
- a clear understanding of the company's critical success factors
- unambiguous direction setting
- flexible and responsive process management
- people with relevant knowledge and skills
- constant search for improving the ways things are done.

Self-assessment

Self-assessment has become a popular approach for organisations as they seek to assess their performance. It is a way of looking at how a company, organisation, functional department or project is performing. It enables organisations and project teams to look right across all their activities, set a stake in the ground to represent performance to date and determine what is now needed to make improvements in performance.

One way to address the challenges in creating high performance organisations is the European Foundation for Quality Management (EFQM). The Foundation is in the tradition of the American Malcolm Baldrige Award and was initiated by the European Commission and 14 European multi-national organisations in 1988. The essence of the approach is the EFQM model,

Case study 6.6 European use of EFQM

In the Netherlands many health care organisations apply the EFQM Model. In addition to improvement projects, peer review of professional practices, accreditation and certification, the EFQM approach is used mainly as a framework for quality management and as a conceptualisation for organisational excellence. The Dutch National Institute for Quality (the Instituut Nederlandse Kwaliteit) delivers training and supports self-assessment and runs the Dutch quality award programme. Two specific guidelines for health care organisations, positioning and improving, and self-assessment, have been developed and are used frequently.

which can be used as a self-assessment tool on all levels of an organisation and as an auditing instrument for the Quality Award. The model is reviewed every three years, with a team of over 200 individuals from various organisations. The EFQM model has been used in a wide range of contexts including as can be seen in Case study 6.6.

An overall view of the model is shown in Figure 6.14. It is based on the principle that in order for an organisation or team to succeed, there are a number of key enablers and key results on which it should concentrate if improvement goals are to be achieved. The key enablers by which organisations and project teams judge themselves are:

- how well the organisation or project is led
- how well people are managed
- how far policy and strategy are developed and implemented by leaders and people
- how well resources and processes are managed and developed.

The key results areas by which the EFQM model then measures successes are:

- how far the organisation or project team satisfies its customers
- how well motivated and committed is the workforce
- how the local and national community outside the organisation or project views its activities in terms of its contribution to society; audits key business results including profit, return on capital employed, shareholder earnings and achieving budgets.

The EFQM model is used as a basis for self-assessment, an exercise in which an organisation or project team grades itself against the nine criteria.

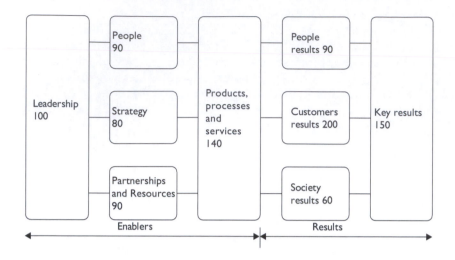

Figure 6.14 EFQM model for business excellence © EFQM. The EFQM Excellence Model is a registered trademark

This exercise helps organisations to identify current strengths and areas for improvement against strategic goals. This gap analysis then facilitates definition and prioritisation of improvement plans to achieve sustainable growth and enhanced performance.

Organisations can score up to 1,000 points on their performance. The significance of customer satisfaction is demonstrated by a score of up to 200 points for customer results. This reflects the importance of providing customers with a product or service which delights them. The best companies in Europe are scoring around 750 points. A score of 500 points is extremely good, and would equate with one of the best in the UK.

Often used as a diagnostic tool, the model takes a holistic view to enable organisations, regardless of size or sector to:

- assess where they are, helping them to understand their key strengths and potential gaps in performance across the nine criteria
- provide a common vocabulary and way of thinking about the organisation that facilitates the effective communication of ideas, both within and outside the organisation
- integrate existing and planned initiatives, removing duplication and identifying gaps.

There are several ways in which self-assessment may be carried out. The most usual are for the organisation to hold a self-assessment workshop, or complete a questionnaire. For a more detailed assessment, and to gain recognition for their achievements, the organisation can produce a written award style application report, against EFQM guidelines, which is then assessed by trained assessors. A low-cost method of assessment now available is the use of computer based questionnaires. In deciding whether self-assessment would be useful for an organisation or team it is important to look at the team characteristics.

As a general rule self-assessment has been shown to work well when an organisation:

- has a clear set of customers
- knows who its key suppliers are
- has a clear set of products/services
- is able to measure its own results
- is run by a management team.

What are the benefits of self-assessment?

Organisations which have used the EFQM model do show a commitment and enthusiasm to continue to give a better and better service to their customers, and meet the needs of all stakeholders. Case studies 6.7 and 6.8 illustrate this.

Case study 6.8 is a summary of the achievements of a small business EQA winner.

Case study 6.7 Highways Agency capability assessment

The agency responsible for highways in England has developed, over a number of years, a Capability Assessment Toolkit (CAT)[32] to identify the competence and capability of major suppliers and select those most likely to deliver best value solutions and services. It recognises world-class performance and the contributions made by their suppliers in the operation, maintenance and improvement of England's road network. Their CAT3 assessment is largely based on the principles set out in the EFQM Excellence Model.

The CAT 3 indicators are:

- Direction and leadership
- Strategy and planning
- People
- Partnerships
- Processes and resources

The scoring factors are:

- Substance – how substantial is what suppliers do?
- Clarity – how clear is what they do?
- Quality – how good is what they do?
- Value added – how much value is added by what they do?

Suppliers are placed in one of four performance bands: early days/ limited; operationally effective; strategically valuable; and external differentiator. The agency is in the process of developing a new assessment tool – the Strategic Alignment Review Toolkit. It retains all the categories from the previous tool while introducing new and challenging indicators in the area of corporate social responsibility.

Case study 6.8 EFQM RBD 50

RBD-50 is a Ukrainian road building SME that has won the EFQM quality award for the qualities of enthusiasm and commitment mentioned above as well as its excellent score against the EFQM model. The firm employs 182 employees who own 72 per cent of the company, with a further 27 per cent ownership by the families of employees. It has survived in a difficult economic climate and still managed to show continuous improvement in all of the elements of the EFQM model. It has maintained its position as the most productive player in a competitive road building industry in the last four years and it improved its workload of 250 per cent whilst improving its profits by 15 times. Its main focus on people and change reflects this and it has consistently scored 90 per cent satisfaction in each of 25 areas under the headings of production and overall image. It was also awarded a 100 per cent score in the categories of willingness to recommend to others, innovation, communication and accessibility so it is well liked by its customers. The company has shown a consistent improvement in performance and is well appreciated by its customers.

The main tool for self-assessment in the case of the UK construction industry is the Construction Industry Key Performance Indicators (KPIs), which emerged from Sir John Egan's (1998) 'Rethinking Construction' Report. See the second part of Chapter 5 for more on this. Case study 6.9 illustrates their use.

Case study 6.9 The use of KPIs in a contractor organisation

Digesting early results of its key performance indicators (KPIs) has made a UK contractor rethink its priorities. Feedback from 40 projects showed promising performance in health and safety, staff relations, understanding clients' needs, keeping to cost targets, and quality of product and service. But lower scores for keeping to programme, defects resolution and supply chain management highlighted a gap in their nine-region structure. Results identified they needed to procure via more strategic alliances with fewer suppliers who are keen to collaborate. A new national procurement manager role was created to make this happen.

There are more sceptical views of the value of the Construction KPIs. Beatham *et al.* argue that the effectiveness of their use has been limited.[33] Most construction KPIs, they argue, are used post event, missing opportunities to improve, and their results are not validated and thus are open to interpretation. They suggest that KPIs are being used within the industry as a marketing tool, rather than as an integral part of improvement.

A study in the early 1990s, by the CBI and DTI, examined 120 of the UK's most successful companies selected on the basis of their profitability, growth and recognised standards in their sector. One of the characteristics distinguishing the leading edge or world class from the run-of-the-mill was their strong leaders who champion change, set targets, and are open with customers and suppliers. A further characteristic is that they constantly learn from others and are committed to innovation and the continuous search to introduce new products and services by exploiting new technologies and ways of securing a competitive edge. It is therefore important to consider the role of the leader in projects in more detail.

Learning projects

Learning refers to the various processes by which skill and knowledge are acquired by individuals, organisations and project teams and is increasingly being recognised as key to innovation and learning.[34] They argue that one of the main reasons for this increasing interest in learning is the growing pressure on organisations to respond to the challenges of the rapid pace of change in markets, the wider environment and increasing globalisation. The challenges and opportunities for change demand innovation, the success of which is dependent upon effective learning and leadership. The greater the pressure for change the more transformational the leadership and the deeper the learning required as shown in Figure 6.15.

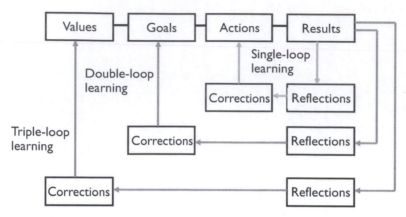

Adaptive or single-loop learning
Reconstructive or double-loop learning
Process or triple-loop learning – creation of a new framework of meaning

Figure 6.15 Scope of innovation and learning

There are three main types of learning which reflect the degree of certainty in which organisations or project teams are operating. Single-loop learning corresponds to an era of relative stability, predictability and certainty. With greater uncertainty and less predictability in the market and wider environments then double-loop or the deeper learning associated with behavioural adaptations becomes necessary. In this situation organisational norms and values need to be questioned and new priorities set which may well lead to a modification of the organisation's goals and indeed values. Triple-loop learning is the deepest form of learning as its main objective is to learn about the process of learning itself. When an organisation learns to learn then it can continuously transform itself to meet the challenges of ongoing and complex change.

Leadership

Issues of leadership and associated power lie at the core of group life in a variety of contexts. Even the most informal groups typically have some form of leadership within their organisation. It can be argued that there are three main forces shaping the process of leadership – flatter matrix forms of organisation, the growing number of alliances and informal networks, and people's changing expectations and values.

The literature shows that there have been four main 'waves' or 'generations' of theory relating to leadership:

- Trait theories
- Behavioural theories
- Contingency theories
- Transformational theories

The earliest approach to leadership was the *'traits'* approach. It was aligned to the notion that leaders are born and not made. It emerged mainly in the context of the military. This view is now contentious and highly dubious. Maylor argues that great leaders in all spheres of human endeavour have developed skills and attributes to the point needed for the task at hand.[35] Both of these are teachable and although intelligence is one of the few characteristics that cannot be taught, this has rarely been a constraint on success.

In the 1950s and 1960s the dominant thinking shifted from leaders to leadership. These approaches have emphasised the functional or group approach, where leadership *styles* are grouped within a behavioural category. The four main styles that have been identified are:

- Concern for task: leaders emphasise the achievement of concrete objectives.
- Concern for people: leaders look upon their followers as people and not simply units of production or a means to an end.
- Directive leadership: leaders take decisions for others and expect subordinates to follow instructions.
- Participative leadership: leaders try to share decision making with others.

More specifically, Hersey and Blanchard identified four different leadership styles for particular situations:[36]

- Telling (high task/low relationship behaviour). This style or approach is characterised by leaders giving a great deal of direction to subordinates and by paying considerable attention to defining roles and goals.
- Selling (high task/high relationship behaviour). Although the direction is given by the leader there is an attempt at encouraging people to 'buy in' to the task. This is sometimes associated with the 'coaching' approach.
- Participating (high relationship/low task behaviour). Leaders share decision making with followers and the main role of the leader is to communicate and facilitate.
- Delegating (low relationship/low task behaviour). The leader identifies the problem, issue or task to be addressed but the responsibility for the response is given to followers.

As Sadler argues, there are inconsistencies in these studies of the effectiveness of different styles of leadership.[37] It is difficult to establish if style of leadership is significant in enabling one group to work better than another. One criticism is that the researchers did not take into account the context or setting in which the style was adopted.

This led to the view that the styles leaders can adopt are far more influenced by those they are working with and the environment within which they are operating. This view places a premium on leaders who are able to develop the ability to work in different ways to match different situations or settings. This gave rise to the *contingency* approach. This can be seen as appropriate in the context of construction's diverse projects and project cultures. Fiedler argued that the effectiveness of leaders depends on two interacting factors: leadership style and the extent to which the situation gives the leader control and influence.[38] Three factors are seen as important in the contingency approach:

- The relationship between the leaders and followers. If leaders are liked and respected they are more likely to have the support of others.
- The nature of the task. If the task is clearly spelled out as to goals, methods and standards of performance then it is more likely that leaders can exert influence.
- Position power. If powers are conferred on the leader for the purpose of getting the job done then this may well increase the influence of the leader.

Case study 6.10 is a transcript of what people felt about their leaders and represents *different* approaches.

The emergence of contingency theory reinforced the view that there is no single recipe for successful leadership. The change associated with the emergence of a new dominant paradigm from the 1980s onwards gave rise to the theory of *transformational* leadership. Hence, this form of leadership is the most recent and is the most change and people focused of the approaches. It has permeated many sectors of the economy, including construction, in recent decades. Burns argued that it was possible to distinguish between transactional and transforming leaders.[39] The former approach their followers with an eye to trading one thing for another. The latter are visionary leaders who seek to appeal to their followers' better nature and move them toward higher and more universal needs and purposes. In other words the leader is seen as an agent of change. Again, this view resonates with the management of projects, especially as most projects are designed to bring about change.

Leader as opposed to manager behaviours need to appeal to follower motives and be interpersonally oriented. Leader behaviours include the formation and articulation of a collective vision, infusing the organisation with values, and motivating followers to exceptional performance by appealing to their values, emotions and self concepts. From this it can be seen that

Case study 6.10 What it feels like to be led

The first transcript indicates a follower's description of some of the best leaders he has worked with in his career to date.

> Over the years I have reported to many managers. Those who have also been leaders have had a certain something about them. Some of them have been charismatic, perhaps not in the same league as Nelson Mandela, but they did have this knack of making you feel optimistic about goals, no matter how tough those goals might be. Some have not been what you would call charismatic but their personal qualities still made them effective leaders. These leaders were extremely clear about what we needed to achieve as a team and then took every opportunity to make sure we understood why our goals were so important and that we were really on board. These leaders were genuine and you knew they would follow through with what they said. They were consistent in achieving what was important in terms of targets and values. Their values influenced everything they did. Their standards of teamwork and behaviour were very high and they applied exactly the same standards to themselves too.

Here is the transcript of another interview contrasting two quite different leadership styles she has experienced.

> My previous manager was . . . tough. You were left feeling that you were just one of her resources to get her job done. My new manager is tough as well but in a different way. She doesn't tell us what to do but does explain why it needs to be done. She doesn't just delegate jobs to us and leaves us to it, but gives each individual just the right amount of support. Her feedback is different too. My previous manager would really let you know when something wasn't right and only every now and then would you get some text-book type praise. My new manager helps you reflect on what you've done and what you learnt from it. It's almost as if she's not just our leader but also one of our resources to help us get our job done. Productivity, team spirit and job satisfaction have never been higher.

The next transcript shows how leaders need to shape their approach to the individual and the circumstances.

> I wasn't sure about our new manager at first. He seemed a bit of a chameleon. At times he would be very relaxed and at other times he would be very strict. Sometimes he would consult with us and at other times he would be very autocratic. But ... then ... the more I thought about it I began to see he was actually being very consistent about what was important, what we had to achieve and how to achieve it and so on. He'd only be strict with people who were negative and who undermined others. He would only be autocratic when we were dwelling on difficulties instead of creating solutions. He'd cut through the negative stuff and take charge. I realised that he had very strong principles about our goals, about us as individuals, and about how we worked as a team. These principles were being upheld through different behaviour depending on the situation. This underlying consistency was one of the qualities that made you really respect him.
>
> Try and identify the underlying approach of each of these witness statements.

managers provide the intellectual inputs necessary for organisations or projects to perform effectively whereas leaders set the direction for organisations and projects and appeal to ideological values, motives and self-perceptions of followers. The relationship between change, innovation, learning and leading is shown in Figure 6.16.

Ethical leaders

Ethical leadership is the belief that a leader does not only have to be successful, but has to be able to provide exemplar principles for good and fair practice. Professional institutions eclipse some of this in professional conduct rules and espouse neutral advice, confidentiality, honesty and integrity, updated knowledge and neutral advice. However as a leader there is the need to be able to explain and justify principles in fair and equitable terms to all parties as a visible example for those they are leading and would likely influence in the same direction. This presumes an experience of upholding standards they are espousing as well as a clear knowledge of what is expected and acceptable and a desire to know what is going on behind the scenes. It is also important that leaders receive support from organisations that also have ethical standards. Principles that emerge are respect, trust, social responsibility, sustainability, fair trading, safe working, fair pay and

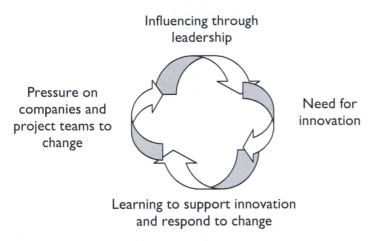

Influencing through
leadership

Pressure on
companies and
project teams to
change

Need for
innovation

Learning to support innovation
and respond to change

Figure 6.16 Relationship between change, innovation, learning and leading

employment terms, career development with long term rather than short term
decision making. It is not just relating to a legal compliance. Attitudes are
powerful drivers of actions and law has been quoted as being a reflection of
public tolerance and views. Corrupt actions and bribery are often interpreted
at a different level. There are cultural decisions to make which may require
discretion to respect diversity.

Ethical leadership is not an easy path because in the short term it may
not bring about the most profitable outcome and it requires a degree of
transparency and honesty so that others can be more critical of leader actions.
Organisations sometimes have dual standards which provide differing views
for employees and bend to economic circumstances, market forces and
organisational mergers. It may not be popular to expose unfair practices
carried on by others and fair trading means sharing profits more equally.
There is a lot more support for these actions because emerging economies
are demanding more equal treatment and bribery legislation has become
stronger internationally and domestically, reinforcing drives for a wider
consideration of fair practice. More is said about the need to develop trust
in Chapter 12. Decentralised construction project leadership on major
projects gives more scope for making an impact and developing a strong
ethical format where there is a clear influential leadership. There is also an
opportunity (and risk) to show that it can work in the success of the project
overall.

The commercial advantage for ethical leadership is a clear direction which
provides common objectives and gives the opportunity for a productive
process and collaborative integrated approach. To be successful it needs all
the instincts of entrepreneurial leadership to see opportunities for continuous

improvement which would be commercially beneficial. An ethical project needs to be seen as a good business. There is always more to learn.

Cultural influences on leaders

It must be borne in mind that there are cultural constraints on what leader behaviours can be used and their effectiveness. Most prevailing theories of leadership have a definite North American cultural orientation: an individualistic rather than collectivist approach; self rather than duty and loyalty; rules and procedures rather than norms; and rationality rather than aesthetics, religion or superstition. There are, however, many cultures that do not share the assumptions on which much of the theory is based and this has given rise to an increasing amount of research geared towards understanding leading across different cultures.

Cultural norms are standards of conduct or acceptable behaviour in any given culture.[40] Different cultures have somewhat different priorities, and their leadership reflects their priorities. This reflects patterns of socialisation but it also likely reflects innate biological differences and pre-dispositions. At a physiological level, males and females differ in ways that have broad behavioural implications. Females seem to be attuned to more diffuse internal sensory and external social stimuli while males seem more attuned to and more focused on external objective stimuli. This suggests that men and women may reason and make decisions in somewhat different ways. Men seem to be concerned more with abstract 'truth' and appear to make decisions using objective reasoning. Women, on the other hand, seem to be more concerned with interpersonal 'fairness' and appear to process information and make decisions using subjective reasoning in a more empathic, personal and feeling way. Denfeld Wood argues that leadership is made up of the task-oriented 'masculine' side and the relationship-based 'feminine' side.[41] Increasingly, these two aspects – task and relationships – are seen to be fundamental dimensions of leadership. He goes on to identify the contrasting features of the two approaches including:

- autocratic versus democratic leadership styles
- production oriented versus employee oriented behaviour
- directive versus participative leader behaviour
- reflections of Theory X (negative and directive) versus Theory Y (positive and facilitative)
- decisiveness versus harmony
- abstract truth versus interpersonal fairness.

In Western society the masculine principle is more prestigious and helps explain why most managers, male or female, think of political and military figures when asked to identify leaders. With the shift to post-Fordism we have seen some movement from the male to the female aspects of leadership. This is reflected in the components and associated traits of emotional competence. The first three focus on self-management:

- Self-awareness: accurate self-assessment, emotional awareness and self-confidence.
- Self-regulation: innovation, adaptability, conscientiousness, trustworthiness and self-control.
- Motivation: optimism, commitment, initiative, and achievement drive.

The remaining two focus on relationships:

- Empathy: developing others, service orientation, political awareness, diversity, active listening and understanding others.
- Social skills: communication, influence, conflict management, leadership, bond building, collaboration, co-operation and team building.

Transformational leadership

Van Maurik argues that the need for transformational leadership has arisen as a result of the more sophisticated demands made of leaders in recent years.[42] These include the shift from Taylorism and Fordism to post-Fordism or postmodernism, increasing globalisation and the imperatives of environmental sustainability. He identifies three broad areas of thinking within the theory of transformational leadership including team leadership (e.g. Belbin),[43] the leader as a catalyst for change (e.g. Covey)[44] and the leader as a strategic visionary (e.g. Senge).[45] However, as argued by Wright, it is impossible to say how affective transformational is with any degree of certainty.[46]

The charismatic leader and the transformational leader can have many similarities, in that the transformational leader may well be charismatic. Their main difference is in their basic focus. Whereas the transformational leader has a basic focus on transforming the organisation and, quite possibly, their followers, the charismatic leader may not want to change anything.

Having charisma is often seen as an important element of leadership but it is a difficult quality to understand. If they are well-intentioned towards others, they can elevate and transform an entire group or organisation. If they are selfish and Machiavellian, they can create cults within the group and exploit their followers. The other possible weakness is that their self-belief may be so high they can come to believe that they are infallible, and hence lead their followers to disaster, even when they have received adequate warning from others. The self-belief can also lead them into psychotic narcissism, where their self-absorption or need for admiration and worship can lead to the alienation of their followers. They may also be intolerant of challengers and their perceived irreplaceableness (intentional or otherwise) can mean that there are no successors when they leave.

Charismatic leadership is most suited for short-term projects, and also for projects that require energy and talent to affect transformational change.

Although the theories and models of leadership examined above can be useful in thinking about which leadership approach to adopt, there are also a number of issues associated with these models including:

- their general and overly simplistic nature – they are neither specific enough in relation to context and culture nor sophisticated enough to represent the true nature of such a complex concept
- their North American bias – there are cultural difference that need to be taken into account in other settings; for example, some cultures are more individualistic than others
- their development, which has been largely by men – evidence suggests that women have leadership styles that are more nurturing, caring and sensitive
- their focus mainly on the relationship between managers and immediate subordinates with insufficient regard for group structure, politics or symbols.

What makes an effective leader?

Van Knippenberg and Hogg argue that leadership has three essential components.[47] First, the would-be leader must convince the other members of the group to regard that person as a credible and legitimate source of influence. This recognises the significance of the status of the leader. Hollander and Julian maintain that the leader's legitimacy flows from the perception that the leader is competent enough to help the group attain its goals and trustworthy enough to remain loyal to collective interests and objectives.[48] Once a person has gained the legitimacy of leadership status, he or she needs to develop relationships with followers that motivate and enable them to act to attain collective goals. This recognises that leaders do not accomplish tasks on their own and that in understanding effective leadership it is necessary to recognise the characteristics of effective followership. This includes ability and requisite competencies that must be recognised by the leader as well as the followers but also the motivation to expend effort to perform at high levels. Finally, the leader needs to mobilise and direct the efforts of the group to make the most effective use of the combined resources of the group in task accomplishment.

O'Neil believes leadership is about:[49]

- providing meaning and purpose
- focusing on the right things to do
- structuring the environment to achieve the organisation's goals
- getting others to do what you want
- motivating people to get things done willingly
- enabling others to take responsibility

- empowering others to do what they think is right
- helping people to feel less fearful and more confident
- developing, sustaining and changing the culture.

Case studies 6.11 and 6.12 illustrate the transformational leadership shown by two practitioners in the UK construction industry.

Case study 6.11 Transformational leadership

A chartered builder, chartered surveyor, chartered environmentalist and corporate construction surveyor has over 25 years' managerial experience in the construction, civil engineering, house building and mixed-use development businesses. Following on from his success with a contractor delivering a large number of complex projects he worked as an executive for a property developer. His role involved establishing and managing a number of major mixed-use developments across the UK. He is currently construction director for a developer investing in sustainable regeneration. His company has been referred to by the United Nations as 'The world's first responsible real estate fund'. Its unique combination of fund, development, investment and asset management expertise is its fundamental differentiator, driving successful sustainable urban regeneration.

As well as showing transformational leadership within his company and its regeneration projects, his contribution and commitment to sustainable development has resulted in him representing the development industry on the Developers Forum for Sustainability in the South West of England, which influences sustainable and environmental issues in the region. He also contributes to the United Nations' environmental research programme as a member of their College of Observers. He is currently a board member of the Construction Clients' Group and provides leadership for a Constructing Excellence Club, which he was involved in establishing and running as Chairman for three years. He has regularly commented in the media on urban regeneration and sustainable developments and has spoken on a number of occasions on these and other issues including value management, project delivery, collaborative working and supply chain integration. Schemes for which he has been responsible have been recognised through the award of the following: 'Best Partnership with the Supply Chain', 'Most Challenging Regeneration Project In the UK', 'National Homebuilder Award – Best Brownfield Development', and 'What House Award – Best Exterior Design'.

Case study 6.12 Transformational leadership 2

The honorary knighthood bestowed on the chairman and chief executive of a major construction company was in recognition of his services to the construction industry in the United Kingdom. It acknowledges his outstanding contribution to the industry spanning more than 30 years. Under his unique style of leadership the company is now an important internationally focused engineering enterprise, responsible for some of the world's most iconic buildings.

He has been a long-standing champion of the government's construction industry improvement plan, demonstrating committed leadership, a focus on the customer and a quality-driven agenda. The group's early adoption of the latest construction methodology, use of lean construction methods and offsite manufacturing had helped eliminate waste and maximise efficiency. Coupled with his commitment to training people the company serves as an exemplar for the rest of the industry.

Following a successful career in the military, he assumed responsibilities for the management of the estates for two leading UK universities culminating in the planning, development and maintenance of an estate of 660,000m² of built space. He also had shared responsibility for a £105m property investment portfolio. Later, as director of construction for OGC HM Treasury, he had the responsibility for leading changes to government construction procurement including the development of whole-life procurement policies and mechanisms, supply/demand management, fair payments in the industry, the promulgation of best practice, and the development of policies to overcome barriers to the adoption of more effective and efficient procurement. He continues to influence construction as it adopts more sustainable policies and the education of the next generation of construction professionals.

Project leadership

Maylor argues that the project leader has a responsibility to the organisation and the team members to ensure that they are provided with high levels of motivation in the new challenging environment of projects.[50] Buttrick argues that leading in a project is different from leading within an organisation.[51] In line management, the manager or supervisor has the power and authority to instruct a person in his/her duties. Most likely they should have it but often do not. They have to deliver the project using a more subtle power base more rooted in the commitment of the team than in the directive of the

project manager. Teamwork and team spirit is important in line management. It could be argued that it is even more important in projects. Reasons
for this include the short time available for 'forming and norming' behaviours
and to optimise performance and the fact that many team members are not
dedicated to the project as they have other duties to attend to.

The project manager must be the leading player in creating and fostering
a team spirit and enrolling the commitment of the project participants.
Factors contributing to this are clear communication, realistic work plans
and targets and well defined roles and responsibilities.

Morris points out the increasing emphasis on entrepreneurship, leadership
and championing in projects from the 1970s onwards. He predicted that in
the twenty-first century, frequent change will become even more pervasive.
Social, economic, demographic and environmental pressures will grow. More
democracies will mean more political change. Technology and communications will become even more global. He claims:

> Projects need strong, experienced people to drive them forward and lead
> those involved. Not only must the project be efficiently adminis-
> tered, there should be a high standard of leadership so that people, both
> within and outside, accept its goals and work enthusiastically towards
> its realisation. Management drive of an extraordinary order may be
> necessary to get the project moving and to produce results of outstanding
> quality on time and within budget. To assure the necessary resources
> and support, the project may need championing both within the
> sponsoring organisations and externally, within the community.[52]

CIOB leadership survey

According to the CIOB, the issue of leadership in construction has been
debated for some time. They argue that the industry's need to rise to
increasing social, economic and environmental challenges has fuelled this
debate and called into question the industry's ability to create leaders who
can inspire and affect real change. Research undertaken by the CIOB was
based on a sample of 655 construction industry professionals.[53] The results
suggest that 'good communication skills', 'strategic vision', 'understanding
of the business' and the 'ability to get results' are the most important traits
for an effective leader in the construction industry.

The respondents when asked why individuals are considered to be the
greatest leaders of all time identified 'vision', 'integrity' and 'communication
skills' to be important traits but 'charisma' and 'inspirational' were also
ranked highly. 'Soft' leadership skills appear to be less prevalent in the
construction industry although when asked about their strongest leadership
qualities the respondents identified 'integrity', 'ability to listen' and
'understanding' as their top returns. The results show that the small sample

of women respondents were looking for these 'softer' skills in construction industry leaders. In particular, they rated 'integrity' and 'open to change' as key leadership qualities.

Respondents identified 'new experiences' as the most valuable way to improve their leadership ability with education, skills, qualifications and training also identified as being important. The main barriers to effective leadership were 'lack of opportunity' and 'organisational culture'. The respondents were asked whether they believed that adequate leadership is present in four main areas of the construction industry: health and safety, business ethics, sustainability, and education and training. The results show that there was a higher level of leadership at company level in relation to health and safety, business ethics, sustainability, and education and training when compared to that at project, UK and international levels. It could be inferred that there is a greater need to address the lack of leadership at project, UK and international levels.

What is an integrated project team (IPT)?

The integrated project team is a term used by the Egan Report, but is also referred to in the PMI BoK as integration management. The Strategic Forum for Construction (SFC)[54] define a fully integrated collaborative team as:

- a single team, including the client, focused on a common set of goals and objectives delivering benefit for all concerned
- a team so seamless, that it appears to operate as if it were a company in its own right
- a team, with no apparent boundaries, in which all the members have the same opportunity to contribute and all the skills and capabilities on offer can be utilised to maximum effect.

An integrated team seeks to break down the traditional barriers between design and construction and between the client and the project team.

IPTs break down functional/organisational boundaries by using cross-functional/organisational teams. Running projects in functional parts with co-ordination between them slows down progress, produces less satisfactory results and increases the likelihood of errors. Lateral co-operation is better than hierarchical communications, which can become bureaucratic and slow down the contract. Increasingly, this involves taking people out of their functional or organisation locations and grouping them in project work team spaces (known as collocation). The recent trend has been to tip the balance of power towards the project and away from the organisation.

When selecting the members of the project team SFC argue that special attention should be paid to enthusiasm and commitment, team attitude and communication skills as well as the technical attributes such as experience and technical qualifications. They recognise that team members often have differing and conflicting objectives and see the project manager as playing a key role in overcoming conflict arising from these different objectives. The project manager should aim to create an environment in which the team member can achieve personal as well as project goals. This means using a problem-solving and no-blame culture where issues are identified, communicated and tackled early in the process.

Some work has been done to evaluate leadership by embracing the principles of an integrated project team that has been formed on the basis of competitive tendering[55] and does not have the culture or track record for integrated team working, and tough committed decisions have to be made to gain greater value. Leadership is having the courage to keep the belief when:

- fighting off the cynics that believe lowest cost is achieved by competitive tendering
- you know that the collaborative relationship that you have with the contractor is not reciprocated through the supply chain
- the agreed time has to be extended to ensure that value is achieved
- the supply chain's price comes in well above the agreed budget
- the supply chain puts more effort in defending why the price has gone up rather than meeting the agreed budget.

Leadership in this context needs to be generous and involved and means:

- building and inspiring a team that understands and is able to deliver the project ideals
- being happy for each member of the supply chain to make a profit on your project
- being prepared to work harder on your project to ensure that innovation can flourish and greater value can be achieved by more collaborative working.

The main challenges facing leaders in construction include exploiting market opportunities and addressing issues through: influencing others, creating high-performance teams, managing intra- and inter-organisational change and innovation, and shaping corporate, project and supply chain cultures.

Conclusion

As project management becomes more established as a management approach, the development of project leadership as a career, with its own promotional path and professional recognition, will be an emerging feature of progressive organisations. Perhaps, in the future, only those who can develop the wide range of skills and knowledge to meet these challenges will be eligible to call themselves project leader. Traditionally construction has moved from full control of the workforce to a network of different contractor, specialist managerial and design organisations, which will have personnel who work for more than one project.

The influence of external factors such as market place competition, customers and the political, economic, social and technological factors has particular impact on how the project is run, as it has a unique timing, location and contract conditions which pass different risks to the project. The competition will influence the behaviour of organisations contributing and may restrain a full co-operative environment. Management style and the authority position and power of the project manager will influence the way the communications occur within the structure and the degree of integration that can take place. Porter's model indicates how major suppliers influence prices, but other stakeholders such as the community may also exert their influence on the project and gain a greater say or force additional accountability.

Experienced clients may superimpose certain structures which they believe are critical such as value management, user group meetings and conflict panels. The introduction of customer account servicing, framework agreements, environmental constraints, incentives, supply chain agreements and partnering is driven by these customers, who will also favour the use of specific contract conditions that transfer different risks to contractors. Contractors may respond defensively to pass risk on, use additional insurance and withhold payment to suppliers, or positively, to create strategic business units, develop trust, abolish retention, create single bank accounts and offer innovation. The culture depends on the preferences of the project team and the client and the culture will be an adjunct of past work experiences and inherent characteristics such as the education, personality and aspirations.

World-class or 'best in class' performance is a stated aim of companies who wish to continue improving performance and will use measurements to benchmark their performance against leading best practice across industries. This requires strong leadership to find out how customers can be satisfied better, to champion change, to set tough objectives that have mechanisms for implementation as well as measurement, to train and develop personnel and to connect all this to improving the 'bottom line' by the elimination of waste in the system. Project leadership needs to recognise the importance of project 'culture making' on communications and motivation. They need to be able to adapt to more numerous multi-national projects where the danger of misunderstandings is great when different cultures come together with different perceptions and expectations of their personal and organisational objectives. It is also important to define the characteristics of ethical and transformational leadership. Dealing with diversity becomes a major consideration with respect for local conditions and ways of doing things.

Leadership definitions have been changing to take account of the need to include the role of champions for change and this better describes the more transformational approach of the project leadership to inspire as well as to provide many management functions. This is especially so as leadership in projects applies to many different contributing organisations in the supply chain as well as the project manager. A leader will develop their own style, but need to adapt it to different circumstances for effect. Political awareness with a small 'p' may also help this approach and help awareness of harnessing indirect stakeholders' support, as well as managing the visible team. Project leadership works with the use of persuasive and cognitive power (relationships) and the need to continuously develop knowledge in a learning organisation, which fits well with the dynamic nature of a project.

Construction project organisations have diversified their leadership with the spread in the use of different procurement types and the challenge to find appropriate organisational forms that provide a framework for an integrated approach. Leadership is needed throughout the supply chain that discourages the traditional adversarial approach. This leads to the concept of an integrated project team that is working together and has time to develop a more tightly defined culture, in a loosely coupled context where organisations build up their trust in each other and seek to work as a single entity to squeeze out more value from the project.

The next chapter looks at the way forward in engineering such an environment and reviews the tricky areas of conflict, negotiation and communications. Is an integrated team realistic and can it provide the solid basis for improvements in an adversarial industry that has lost productivity?

References

1 Weber M. (1947) *The Theory of Social and Economic Organization*, translated by A.M. Henderson and Talcott Parsons. New York, The Free Press and the Falcon's Bring Press.
2 Woodward J. (1980) *Industrial Organisation: Theory and Practice.* 2nd edn. Oxford: Oxford University Press.
3 Walker A. (2007) *Project Management in Construction.* 5th edn. Oxford: Blackwell.
4 Winch G. (2010) *Managing Construction Projects: An Information Processing Approach.* 2nd edn. Oxford, Blackwell Publishing.
5 Tofler A. (1970) *Future Shock.* New York, Random House.
6 Clegg S. (1990) *Modern Organizations: Organization Studies in the Postmodern World.* London, Sage Publications.
7 Linstead, S. (1993) 'From postmodern anthropology to deconstructive ethnography', *Human Relations*, 46(1): 97–120.
8 Daft, R.L. (1998) *Organisational Theory and Design.* 6th edn. Southwestern: Cincinnati, OH, USA.

9 Peters T.J. and Waterman R.H. (1982) *In Search of Excellence: Lessons from America's Best-Run Companies*. London, Harper and Rowe.
10 Blake R.R. and Mouton J.S. (1969) 'Organisational change by design', in *Scientific Methods*. Austin, Texas.
11 Silverman D. (1970) *The Theory of Organisations*. London, Heinemann.
12 Allaire Y. and Firsirotu M.E. (1984) 'Theories of organisational culture', *Organisation Studies*, 5(3): 193–226.
13 Harris P. and Moran R. (1987) *Managing Cultural Differences*. 2nd edn. Houston, Gulf Publishing.
14 Handy C. (1993) *Understanding Organisations*. Harmondsworth, Penguin.
15 Hofstede G. (1991) *Cultures and Organisations: Software of the Mind*. Beverley Hill, Sage.
16 Winch G. (2002) *Managing Construction Projects*. Oxford, Blackwell Publishing.
17 Hopkins S., Hopkins W. and Mallette P. (2005) *Aligning Subcultures for Competitive Advantage: A Strategic Change Approach*. New York, Basic Books.
18 Latham Sir M. (1994) *Constructing the Team. Final Report of the Government/ Industry Review of Procurement and Contractual Arrangements in the UK Construction Industry*. Department of the Environment. London, HMSO.
19 Walker A. (2007) *Project Management in Construction*. 5th edn. Oxford, Blackwell.
20 Dubois A. and Gadde L. (2002) 'The construction industry as a loosely coupled system: implications for productivity and innovation', *Construction Economics and Management*, 20: 621–631.
21 Handy C. (1993) *Understanding Organisations*. Harmondsworth, Penguin.
22 Porter M. (1980) *Competitive Strategy*. New York, Free Press.
23 Latham Sir M. (1994) *Constructing the Team. Final Report of the Government/ Industry Review of Procurement and Contractual Arrangements in the UK Construction Industry*. Department of the Environment. London, HMSO.
24 Bennett J. and Jayes S. (2006) *Trusting the Team*. The Reading Construction Forum. (They have developed this principle to what they call second Generation partnering in a later book *The Seven Pillars of Partnering* (1998) which discusses the new demands of clients and the requirement for a higher level of trust and multi-level partnering. More of the latter is discussed in Chapter 12.)
25 Johnson-George C. and Swap W. (1982) 'Measurement of specific personal trust: construction and validation of a scale to assess trust in a specific other', *Journal of Personality and Social Psychology*, 431302-17. In A. Walker (2003) *Management of Construction Projects*. 4th edn. Oxford, Blackwell Publishing.
26 Bennett J. and Peace S. (2006) *A Code of Practice for Partnering*. CIOB, Oxford, Butterworth-Heinemann.
27 CIOB (2010) *Code of Practice for Project Management in Construction and Development*. 4th edn. Oxford, Blackwells.
28 Construction Best Practice Programme (2000) The Pavements Team Case Study project 64, M4i.
29 Peters T.J. and Waterman R.H. (1982) *In Search of Excellence: Lessons from America's Best-Run Companies*. London, Harper and Rowe.
30 Peters T. (1992) *Liberation Management*. London, Pan Books in association with Macmillan, pp.740–741.
31 See Hines P. (1994) *Creating World-Class Suppliers: Unlocking Mutual Competitive Advantage*. London, Pitman; Lamming R. (1993) *Beyond Partnership: Strategies for Innovation and Lean Supply*. Hemel Hempstead, Prentice-Hall.

32 Highways Agency CAT3 Content guidance 0806. Available online at: http://www.highways.gov.uk/business/documents/CAT3_content_guidance.pdf (accessed 29 December 2011)

33 Beatham S., Anumba C., Thorpe T. and Hedges I. (2004) 'KPIs: a critical appraisal of their use in construction', *Benchmarking: An International Journal*, 11(1): 93–117.

34 Jones J.M. and Saad M. (2003) *Managing Innovation in Construction*. London, Thomas Telford.

35 Maylor H. (2003) *Project Management*. 4th edn. Harlow, Pearson Education.

36 Hersey P. and Blanchard K.H. (1977) *The Management of Organisational Behaviour*. Upper Saddle River NJ, Prentice Hall.

37 Sadler P. (1997) *Leadership*. London, Kogan Page.

38 Fiedler F.E. (1997) 'Situational control and dynamic theory of leadership'. In K. Grint (Ed.) (1997) *Leadership. Classical, Contemporary and Critical Approaches*. Oxford, Oxford University Press.

39 Burns J.M. (1978) *Leadership*. New York: Harper Collins.

40 Cooper G.L. and Argyris C. (1998) *The Blackwell Encyclopaedia of Management*. Oxford, Blackwell Publishing.

41 Denfeld Wood (1997) *Mastering Management*. London, Financial Times/Pitman Publishing.

42 Van Maurik J. (2001) *Writers on Leadership*. London, Penguin Books.

43 Belbin M. (1993) *Team Roles at Work*. Oxford, Butterworth-Heinemann.

44 Covey S. (2004) *The Seven Habits of Highly Effective People*. Salt Lake City, Utah, US, Franklin Covey Company.

45 Senge P.M. (1990) *The Fifth Discipline: The Art and Practice of the Learning Organisation*. London, Century Business.

46 Wright P. (1996) *Managerial Leadership*. London, Routledge.

47 van Knippenberg D. and Hogg M. (Eds) (2003) *Leadership and Power*. London, Sage Publications.

48 Hollander E.P. and Julian J.W. (1969) 'Contemporary trends in the analysis of leadership perceptions', *Psychological Bulletin*, 71: 387–97.

49 O'Neil B. (2000) *Test Your Leadership Skills*. Institute of Management. London, Hodder and Stoughton.

50 Maylor H. (2003) *Project Management*. 4th edn. Harlow, Pearson Education.

51 Buttrick R. (1997) *The Project Workout*. London, FT/Pitman Publishing.

52 Morris P.W. (1994) *The Management of Projects*. London, Thomas Telford, p.255.

53 CIOB (2008) Leadership in the Construction Industry. Available online at: http://www.ciob.org.uk/sites/ciob.org.uk/files/Leadership%20in%20the%20Construction%20Industry%202008.pdf (accessed 29 December 2011).

54 Strategic Forum for Construction (2003) *Integrated Toolkit*. Available online at: http://www.strategicforum.org.uk (accessed 12 December 2011).

55 Constructing Excellence (2004) Focus Group on leadership in the supply chain. Workshop.

Chapter 7

Engineering the psycho-productive environment

At least 50 per cent of project management is dealing with people and building relationships and by definition we are interested in managing these relationships to the benefit of the project. In the life of the project the aim should be to achieve better productivity, but also to look at long term objectives such as retention of staff, building up customer relationships which provide mutual benefits and attracting the best talent for future projects in order to gain a competitive edge through an attractive and effective working atmosphere.

This chapter will look at various strategies that would promote project effectiveness and efficiency by managing people well. Traditional leadership will be evaluated and some new approaches will also be discussed in the light of best practice. We shall consider communication, leadership, conflict and aggression, selling and persuasion, teambuilding, problem solving, innovation, negotiation and team selection. With a lot of interest and generic research in this area it is important to make specific applications for the management of construction projects. As it is a complex subject the chapter will give a broad overview in order to provide a holistic approach to possible changes that could take place.

The main objectives of the chapter are:

- defining the work of groups and construction teams
- developing the role of motivation and leadership in team building particularly
- understanding the communication process and some techniques which have been developed to make communication more effective
- identifying and managing conflict and stress situations that occur in construction that specifically hit productivity
- defining the concept of psycho-engineering and identifying ways that teams can critically improve productivity with the use of Belbin and Myers Briggs and other indicators in determining behaviour
- reviewing the use of negotiation and other forms of bargaining in resolving conflict
- the role of teams in construction project management problem solving.

People need to be motivated and led which is more than organising them, involving relationships to be established, skills to be developed and the delegation and empowerment of their natural abilities. Leading and motivating can be direct from the front, but in many cases provides a capability more indirectly, especially in professional teams, so that others are empowered to make decisions. Flashman considers there is no use in trying to control people by the use of KPIs.[1] It is better to bring them onside and share the responsibility for the project objectives with them. A business needs to measure its ability, but its past work rate and efficiency is not a motivator on its own and can disenfranchise and demotivate people by taking away their responsibility away to know the vision.

Teamwork

It is important to differentiate between group work and teamwork. Groups are defined as a number of people working alongside each other, connected together by social relationships (Forsyth)[2] but perhaps having little control over their combined outputs. He describes group structures as the norms, roles and stable patterns of relationship among the members of the group. Most work is a group based activity, but not necessarily team based. Groups can be informal as well as formal, cutting across the accepted authority structures. Several groups such as trade unions, professional clubs and work teams may work in parallel with some, but not all the same members. A lot has been written on the characteristics of groups starting with Mayo's Hawthorne experiments (1924–1932) and by writers such as Lewin, Likert, Argyris, Tuckman and Argyle on informal groups.[3] These cover such things as group types, cohesiveness, communications, roles, role conflict and group behaviour. Lewin stresses the interdependence of group members and their commitment coming from the 'inter-dependence of fate' because people realise they are in the same boat. Allcorn makes a difference between the intentional group and the other groups.[4] He describes the intentional group as non-defensive where members accept roles and status which are relevant to the objectives of the group and all members are responsible. He contrasts this against the defensive nature of other groups. Groups will often have several possibly conflicting goals which are then expressed as tensions between members. This is called a competitive group and individual goals are hidden or seen as different. A co-operative group seeks to build up some transparency and their goals are similar.[5]

Typical groups in construction are the monthly site meeting inclusive of the client and the project health and safety group which feeds back on improvements to construction safety. The effectiveness of groups can be an important factor in our level of satisfaction.

A team goes a stage further, where the members are committed to common aims and are mutually responsible for their outcome.[6] They tend to be more autonomous and set up their own terms of reference often to suit unique clients. Members of the team regard themselves as belonging and having a generic or technical role. It is the latter which is more aligned to effective project working. However there is a need to measure the amount of teamwork and the conditions in which it may flourish. In construction, teams are often formed from members in different organisations. Typically the construction project team refers to the amalgam of the design and construction professionals who continue to run the project and have fine-tuned time, cost and quality objectives and depend closely on the performance of each other. Production teams may be strengthened by competition by pitting the outputs of teams against each other or against target performance.

The level of teamworking could be measured by the impact measured on the overall project by a member dropping out – team effectiveness should be adversely affected. A team could be improved if a member who is not a team worker is pushed out. A *group* of people will not notice the positive or negative effect in the same way, as relationships are less important. What concerns the project manager is to optimise working relationships in a team and this comes from experience and intuition, but will be helped by an understanding of individual behaviour and interactions among members. Different people react in different ways to stimuli and different mixes of people have a different dynamic that may also be affected by the context and working environment of the project.

Team development

Pre-occupation with the fine-tuning of *ideal* procedures and procurement systems is a setting and not a cure for the optimum delivery of projects. The people environment can be productively engineered to achieve synergistic behaviour and attitudes which benefit productivity.

Tuckman's well-known development of team maturity by passing through the first four stages in Figure 7.1 suggests the need for a team to move *past* a norming or neutral stage to use the synergy of the team to perform.[7] It is still a difficult lesson to learn as many teams have stuck at less than potential because they have not got to know or trust each other. Critical things that affect this are the quality of the communications, the leadership, the conflict resolution systems and the contractual environment.

Construction has a particular challenge as project teams are often only brought together for short periods and the membership, speed and nature of the task are constantly changing at the different project stages. At the completion of the project the divorcing stage is often added to Tuckman's four, when members of the team usually proceed to new non-related project teams. This stage is characterised by a demotivating effect as members become partially concerned with their next project or their search for employment.

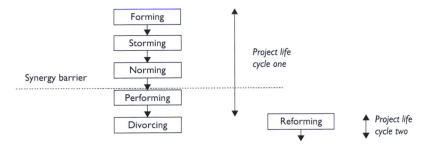

Figure 7.1 Adaptation of Tuckman's four stages of team development

Project teams are often fragmented because of the break up of the project stages into client briefing and concept teams, design teams, construction teams and maintenance teams. Because of their common status, qualifications and professional standing they become strongly cohesive and competitive to other sub groups, creating exclusion or rivalry and developing different sub objectives which become removed from the primary client requirements. This is a disadvantage if a project manager does not take a life cycle view of project integration.

Walker distinguishes between 'group think' where individuals spend an inordinate amount of time agreeing with each other and 'team think' where there is a willingness to talk through the issues, to be creative and to encourage divergent views.[8] Gunning and Harking also emphasise the additional role of the project manager in building the team by 'managing their interactions and satisfying their ego needs' that will also require training in breaking down gatekeeping or defensive behaviour and reducing uncertainty and mediating between conflicting demands.[9] They remind us that there is still productive benefit for meeting face to face and that the need for these skills is not past history.

Team leadership and motivation

Adair's well-known model indicates three aspects of leadership which are achieving the task, managing the team and managing the individuals, as shown in Figure 7.2. The area shown as leadership issues is also critical. In this chapter the main discussion is around the team and individual issues which lead to effective project leadership, although it is acknowledged that leadership also concentrates on the co-ordination of the different tasks or disciplines that members of the team are able to contribute. The aim is to redress the balance and to consider the potential for managing the different human and cultural issues which are brought less consciously into the construction project team. The starting point is to attain client and project objectives.

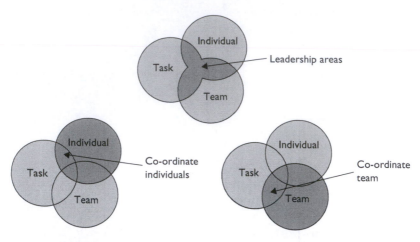

Figure 7.2 Project leadership
Source: adapted from Adair (1979)[10]

The existence of teams does not automatically indicate better productivity. There is a chance that project teamworking may contribute to a less profitable outcome and worsening relationships for the client and the various businesses involved if they don't work together to each other's benefit, if clear insight and leadership is not imposed. The project manager is a team member as well as a team leader and the client may also have a similar dual role. Various sub teams may exist in project funding and business management, design, construction and commissioning also. Each of these teams will need to be led, the interfaces managed and the overall direction established. This leadership role is linked by the CIC to the ability to:

*   Know how the construction industry operates and the roles and responsibilities of the parties. This is likely to require technical knowledge from both the client and supply side of the business.
*   Build or recruit a team inclusive of the relevant skills and make the relevant appointments confident that responsibilities are covered and working relationships are smooth.
*   Establish a communication system which is open and honest and engenders trust in the team in order to maintain and develop effective working relationships.
*   Use motivation and team building theory to enhance the effectiveness of key project individuals and stakeholders.
*   Manage team knowledge and project interfaces to maximise the proper planning and execution of the project to meet project and client objectives.[11]

These abilities represent mainly people skills, but in the context of construction knowledge and understanding. One other is suggested, which takes into account the uniqueness of each construction project context:

- Facilitate appraisal and training regimes which encourage the development of a defined project culture.

Cornick and Mather suggest that there are four main cultures that the leadership evolves through.[12] *Directing* where very clear instructions come from the leader; *coaching* where the leader prescribes advice and exercises to bring on more autonomy to inexperienced members of the team; *supporting or facilitating* where the leader provides resources and an open door for consultation if required to experienced members; *delegating* where the leader is able to brief the team, agree tasks and become a member of the team. Partington refers to the difference between:

- transactional leadership as working within the explicit and implicit contracts to achieve the task
- transformational leadership, which is the charisma needed to change the status quo and unite people to the achievement of a new vision or 'higher purpose'.[13]

In the latter situation the team is most productive and a good degree of trust, respect and understanding has been developed. The output should exceed the sum of the parts and room for further improvements should be planned. There is also a sense of supporting each other so that bottlenecks around the leader can be eliminated by interchanging roles. More team leaders may also be created.

The motivation aspects of management have been well documented, but the main motivational principles in the workplace can be categorised as economic reward or other extrinsic incentives, intrinsic satisfaction gained from the work and social relationships connected with the workplace. In the context of project teams any of the three types of motivation might be applied. At professional level it is almost certain though that there needs to be a degree of satisfaction from the work itself and where there is a close performing team this is also likely to provide satisfaction and social stimuli. Care should be taken with the notion that a happy team is a productive team as it might just be comfortable and complacent or into group think. From the individual's point of view the psychological contract, which consists of the expectations between the individual and their employer, and between them and the project leader, should be balanced. Frustration of the balance of personal and organised goals can be demotivating; on the other hand problem solving any issues leading to desired goals leads to constructive behaviour. Maslow introduced a theory of satisfying needs and self actualisation, and Herzberg modified this by distinguishing between hygiene or basic level factors and those that motivated, introducing the idea of dissatisfiers.[14]

McClelland described needs in terms of achievements.[15] Other process theories such as Vroom, Porter and Lawler, Adams and Locke looked at motivation as being dependent on the dynamic relationship between several factors such as expectancy, effort, rewards and performance.[16] In other words if the credibility of these factors is poor, for example the effort expended did not match the reward offered, then frustration and demotivation set in. It is difficult to 'put money on' any one of these theories, but the instinctive style of the leader should be used to be able to analyse factors that may be causing problems for the individual or by extension, the team. It is also clear that individuals may be motivated in different ways and there needs to be a different treatment of team members to get the best from them. Equally important is the principle of equity within a closely knit team who will see themselves as needing equal treatment. It is in this context that valuing differences may help to build on the different strengths and manage conflicts and stress.

Conflict and stress

Conflict can be defined as a difference of opinion or a difference in objectives. Some conflict is inevitable and in a healthy competitive form can be beneficial in improving productivity. The very nature of construction creates contractual parties which by definition have differing objectives. In traditional procurement, a predictable conflict is created between the two parties to the contract, but also between design and construction. This is conflict which is an inescapable consequence of construction trading relationships. Other conflict which is dysfunctional – actions which have gone outside of the functional and have significant impact – is best resolved in order to stop escalation and the negative results of retaliation. Sommerville and Stocks measure conflict on a three-dimensional parameter model.[17] This would grade complexity by the level of certainty, functionality and significance. In particular a significant, dysfunctional conflict with a high level of uncertainty will be disruptive and requires attention. Other conflict may be insignificant, or within experience.

Stress is a natural reaction which normally enhances performance and may be used as a management tool, but prolonged stress or excessive levels of stress can produce adverse physical or psychological symptoms, which have the opposite effect. The Health and Safety Executive (HSE) defines stress as 'the adverse reaction people have to excessive pressure or other types of demand placed on them'.[18] Individually stressful situations like dysfunctional conflict may be created, which create demotivation or absence. In the UK it is estimated by the HSE that 9.8 million days were lost in 2009/10 because of work related stress,[19] which is almost one-third of the days off work due to all illness and injury. The average time off work for each incident was 22.6 days. The concept of stress hardiness as a measure of the effect of stress on different individuals is well known, but is also as a way of preparing people to react more positively to stressful situations by not taking them personally, seeing them as an opportunity to learn and to grow. It may be possible to increase stress hardiness by means of gaining commitment to work at a higher level than the job conditions, taking the challenge offered by the difficult situation and controlling the situation such as not expecting to win everything. In addition, to do this a project manager may wish to introduce organisational help to others by teaching positive coping strategies, making stress counselling available anonymously, training or matching individuals in teams.

In construction there is a machismo which has propounded 'as norm' certain levels of stress. This norm is being questioned as exclusive practices that, far from inculcating toughness and durable productivity, are distracting workers from the work in hand, nurturing unnecessary safety and health risks and creating regressive conditions which put off some from joining the industry. This may also be affecting the ability to develop creative alternative technology and methods designed to increase value for money and nurture other softer approaches which may be more productive and inclusive. Plenty can be done to deliver a non-threatening culture of continuous improvement and help construction to remain competitive by asking the question 'What are the main areas of conflict that are adversely affecting construction projects and how can they be resolved?'

Our central thesis then for productive working is to psycho-engineer the environment by providing individual and team motivation, awareness of the need to resolve dysfunctional conflict and to reduce prolonged stressful situations. The Belbin team roles and Myers Briggs personality indicator are just two methods that have been used to measure attributes and then to use this information to match individuals in teams and also to allow them to have the information to self-develop. These will be discussed later in this chapter in dealing with conflict. The following section identifies the particular causes of conflict on construction projects.

Conflict and the life cycle

Thamhain and Wilemon collected data on the frequency and magnitude of conflicts found in a range of projects to produce a measure of conflict intensity for different categories of conflict.[20] This allows the possibility of heading off conflict at an early stage. Their findings categorise conflict under seven general conflict categories of project priorities (e.g. the use of a tower crane by many operatives), project procedures, technical problems, staffing needs, costing issues, schedule commitments and personality conflict.

Table 7.1 Main conflict causes compared with project life cycle

Project life cycle stage	Conflict cause (significantly exceeding average conflict intensity)
Project formation	Priorities Procedures
Pre-construction (build up phase including design)*	Schedules commitments Priorities
Construction programme	Schedule commitments Technical issues
Finishing stages	Schedule commitments Personality conflicts (on average, but significantly more than other phases)

Source: Thamhain and Wilemon (1975). Adapted from a report in Meredith and Mantel[21]

These were measured by the degree of average intensity for each conflict type at each stage of the project life cycle. All conflict types were present in each life cycle stage, but some went above the average conflict intensity significantly as shown in Table 7.1.

Table 7.1 clearly shows the importance of managing time, cost and other schedule commitments throughout the executive stages of the life cycle, but differentiates project procedures, project priorities and technical issues in the first three stages. We must also differentiate between the implication of each conflict type for each life cycle stage. The conflict over priorities indicated by the solutions at the formation stage are making decisions in the strategic planning, whilst at the build up stage it is the allocation of resources in support of these priorities. Schedule conflicts develop from commitments in the formation stage to the breakdown of packages in the build up stage, slippage at construction phases and reallocation and completion at the final stages. Personality clash is listed as under average intensity in all stages, but is worst in the frantic finishing stages of the project. However by implication interpersonal skills are needed to resolve the conflicts that are occurring as a prime solution. Meredith and Mantel list perfectionism, demotivation and conflict as being the main behavioural problems facing the project manager.[22] Others have done similar work for construction projects and have shown how conflict type and intensity varies between projects with different procurement types.

Meredith and Mantel have gone further and identified three generic categories of the reasons for conflict which are:

• different goals and expectations of individuals or organisation
• uncertainty about authority to make decisions
• interpersonal conflict.

They identified the main parties in conflict as project or senior management versus client, and project management versus senior management. Further conflict takes place between contractors and between project management and contractors. The matrix nature of supply chains means that resource priority conflicts result from contractor involvement in several projects. Unless this is managed by the project manager there are different goals and expectations for project based staff giving uncertainty about whether project or employer authority influences their decisions the most.

To resolve conflicts of authority and objectives, a project manager is likely to use soft management skills, so that *project* loyalty is engendered. Disputes between client and project about schedule, cost and time objectives and possibly about authority in some of the technological decisions need to be made clear at the start. These are connected to the degree of involvement the client has in the project team and the procurement method employed. In more recent years there has been a push towards partnering objectives where the more experienced client is more fully involved in technological decisions that increase value. It is important to note that partnering is unlikely to lower the conflicts encountered, which are an attribute of the contracting conditions, but as attitudes are changed so the intensity of conflict should be lowered and more productive working conditions experienced especially in the area of personality conflict. They also give incentives for contractors to be involved earlier and to make suggestions for technical improvements. Partnering also involves soft management skills and encourages approaches which produce a 'win–win' situation and tries to leave behind a competitive win–lose scenario resulting in residual resentments, but care needs to be exercised where one party is a dominant size or may exercise undue influence on the agreed terms of the pact.[23]

Personality conflicts may appear at first to be the source of all conflict. Meredith and Mantel suggest that these are mostly created as a subset of technical conflict, the methods used to implement project results or the approach to problem solving. To resolve them care is required to determine underlying problems first and personal animosity may well be resolved as a result. A knowledge of personality type and behavioural strengths and weaknesses will also be key to the solution.

Principles of negotiation and conflict resolution

In a traditional approach to construction conflict it has been established that many disputes are put off only to resurface at the final account stage and this is because of the heavy investment by all the parties in the project. If the project were not to complete especially by the time the implementation stage comes around then all parties would incur heavy losses. Conflict resolution at these stages, as Meredith and Mantel point out, is based on 'allowing the conflict to be settled without irreparable harm to the project's primary objectives'

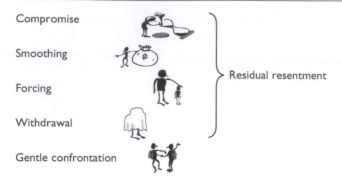

Compromise

Smoothing

Forcing

Withdrawal

Gentle confrontation

Residual resentment

Figure 7.3 Different ways to resolve conflict

i.e. finishing the building on time and to cost.[24] Other disputes which involve determination of contract, unfinished buildings or happen, but are not the norm in spite of what the press would have us believe. Escalated disputes causing failed objectives are not the main point of this section. Collective bargaining may have some similar characteristics. There are five generic ways of approaching dispute resolution and Figure 7.3 gives the essence of each. Although it points to one method, are all the other methods all bad?

In forcing, a dominant partner insists on a 'win–lose' solution which a weaker partner is not in a position to refuse. In compromising, two positions of last resort are established and a halfway point is agreed, commonly applied to financial agreements. In smoothing, persuasion is used in order to make it comfortable to accept a win–lose solution, often a short-term solution with long term problems. In withdrawal neither side refuses to admit that there is a conflict or a conflict is put aside for a time. Where withdrawal is one sided it may be similar to forcing in that psychological pressure forces one party not to face up to the implications of disagreement. Finally gentle confrontation is the opposite to withdrawal, where both sides agree that more effort should be taken to get to the root of the problem to bring about a 'win–win' solution.

The technique of principled negotiation developed by Ury and Fisher is one of gentle persuasion and is a useful one for a project with an open culture, seeking to attain a 'win–win' solution.[25] The four steps are set out below:

1   Separate the people from the problem.
2   Focus on interests and not positions.
3   Before trying to reach agreement, invent options for mutual gain.
4   Insist on using objective criteria.

A typical construction problem on poor information access may have created abortive work because a contractor did not receive their copy of the revised drawing on time. Tempers are running high with accusations flying in both directions and tools have been downed until 'some one sorts this mess out!'

Step one means sitting down around the negotiation table and allowing an element of time. In order to calm down emotions that may be running high it is important to define the actual problem and collect facts. This encourages each party to work on the problem rather than the feelings. Neutral facilitators are useful here.

Step two could establish that one side feels that this has happened too often and they are expecting it to happen again. The other side has particular concerns about the knock on effect on the programme, because it will highlight responsibility without one sided blame and refutation. The matter of abortive work will follow. In focusing on positions pre-judgements have already been made. 'I can't possibly finish that in the time allocated' becomes a barrier if there are heavy liquidated damages. In focusing on the main interest, which is to finish on time without demotivation, an agreement is possible by removing fears.

Step three is the creative one and involves lateral thinking. A more watertight system to avoid this and related problems in the future is clearly the first focus of the negotiation. If more than one viable solution is proffered then the other party is offered the dignity of choice. 'You must remove this architect from the job' may be countered by a separate proposal 'They are staying, but they will complete some design health and safety training'. Other solutions for mutual gain may emerge by comparing solutions from both sides and taking the best from each. This is not the same as compromise as residual resentments arise.

Step four looks at the principles to be used in coming to a financial agreement. It is clear that financial compromise is often the starting point for thawed relationships, though paying for abortive work is not a solution that deals with the cause.

Conflict and communication

Information supply is key to the successful construction project. Any delays disrupt the whole production process. Information supply is a fundamental prerequisite to raising productivity.[26]

CIRIA indicates the need for systematic written communications (or a joint project extranet) on larger projects in order to bring about a co-ordinated understanding of requirements across a broad range of organisations and individuals.[27] Typical communication problems are no allowance for geographically distant partners, poor communications in the early stages, more changes being made because right first time is not achieved, objectives that are divergent. There are also particular issues for communication and conflict in the use of fast track construction where construction is started at an earlier stage by progressively finishing areas of the detail design. It is most important that the design team have fully considered the interfaces between the detail interface of different packages. Particularly helpful are:

- face-to-face communication and the confirmation at meetings of action taken and to be taken
- a record of formal approvals and changes to approvals and briefs
- open and honest communication to take place, particularly when discussing changes which so often produce dispute. Collocation, or use of IT tools for 'virtual collocation and better visualisation
- regular open workshops, routine debates and a good ideas notice board
- a well differentiated and thought through circulation of material, with easy access to further material and involvement of the supply chain in all or any of the above.

Construction Productivity Network (CPN) have also compared the benefits for more overlapping models of communication between design and manufacture indicating earlier involvement of the contractor to increase communication and again to give more chance of 'right first time'.[28]

Communications through meetings

Project teams depend to a significant extent on meetings to develop the design and to progress construction. They include different members to suit the subject of the meeting. We shall concentrate on construction team meetings which bring together the designer, contractor and subcontractors in different combinations. Meetings are often site based during the construction phase enabling immediate problem solving as required. During the design phase there is more need to co-ordinate CAD models or drawings and to include costing and buildability as required. Integrated construction projects include the client in more meetings. Design involvement remains important in the construction phase, but the emphasis transfers to buildability, quality, progress monitoring and problem solving. The meetings may be chaired by an overall project manager, the contractor or the lead designer, for example the architect, or even rotated depending on the emphasis. Change and variation orders are important aspects as there are aspects of redesign and cost checks which need to be agreed. In this respect the client is more involved. In a non-partnering situation the team is often locked together in a situation where there is a given team with little opportunity to choose the personalities allocated. The types of meeting that take place are:

- Design meetings to resolve the client's approval and design integration, for example structural and spatial
- Pre-start meetings between the estimator and the construction team
- Handover meetings from the designer to the contractor to receive drawings and clarify briefing

- Progress meetings to monitor progress and assess impact including the client
- On going technical meetings to ascertain design issues
- Valuation meetings and final account to ascertain interim payments between QSs
- Subcontractor resource and progress meetings
- Subcontractor interim payment meetings
- Health and safety forums for incident and quality improvement

All of these meetings need special attention in order to co-ordinate, avoiding duplication and ensuring efficient use of members' time. Meetings cost time

Case study 7.1 Exhibition centre communications

According to one case study by Foley and Macmillan,[29] construction team meetings on a large exhibition centre project which employed a separate project manager broke down into design technical meetings, construction progress and problem solving. They discovered that the interactions were dominated by the project manager, contractor and the architect and that the other five members of the project team played a secondary role and some of the issues were tangential to them in the technical and progress meetings. However it was different in the problem solving meeting where there was a much more even contribution. The main protagonists, especially the contractor, were multilingual in their communications and communicated well with different players. These team meetings were highly focused and kept to strict lines of communication which were recognised by the team members. For example progress was focused through a contractor progress report which provided standard information, but also highlighted issues for discussion. This made the contractor a centre of attention for most conversations. The technical meeting had similar members, but a single communication channel dominated between the architect and the contractor with the project manager as mediator. The problem solving meeting had similar members, but had a looser agenda with a lot of the interaction passing through the client as their approval was more important to the outcome of the solution. The project manager played the role of mediator. The contractor played a representative role for the subcontractor involved who was there, but backed up the contractor. The agenda was more long term and the communication protocol less rigid.

and money and need to be cost efficient. Many companies are looking at energy efficient solutions involving party calls and video links for single issue meetings to avoid unnecessary journeys.

Case study 7.1 shows how communications can be analysed, but also suggests that there are elements that could be improved so that meetings are productive and more members involved. Meetings are effective in giving systematic information, but having redundant members is not cost effective.

A communication model

The model in Figure 7.4 illustrates the basic elements of communication which are:

- a sender
- a transmitter that chooses a medium such as speech, document or phone
- a receiver.

The difficulties that may be experienced are indicated as distortion, or noise, but other psychological factors are in play as well. It is likely that communication in the construction industry suffers from some of the built in prejudices that exist between different disciplines in the built environment and may also be influenced by the effect of varying education styles, for example between the architect's studio, the engineer's laboratory and a manager's office. Individual situations and personal circumstances may affect the climate of relationships and clearly subconscious signals, often called body language, play a more powerful role than words themselves. Many different media can be used and the correct one should be used for the type of message sent and the impact required. More than one medium may reinforce the message and also provide feedback to the sender as to how accurately the message was received.

Essentially it is the responsibility of the sender to choose an appropriate medium and to make reasonable checks that a message is received and understood accurately. When using non-verbal language different cultures may interpret these in different ways. For example, a 'put off' may become a 'come on!'

Clarity of language, the inconsistency in action compared with words, non-verbal signs, past experience and the relationship that exists between the sender and receiving audience create noise. The 'non-verbals' speak louder than words; in spoken and face-to-face communication the tone and the noted facial expressions or involuntary body movements provide an opportunity for rich interpretation. For example, folded arms may be taken as a sign of keeping uninvolved, but can also indicate a desire for privacy. Covering your mouth whilst listening can indicate a lack of belief in what is being said. Facial expressions and tone make words like 'no' mean entirely

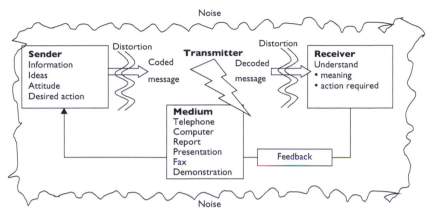

Figure 7.4 A general communication system

Adapted from Lucey (1991)[30]

different things. Responsive 'non-verbals' are also likely to influence what you say and the way you use language.

Perception is an important part of communications and this concept is well introduced by Mullins.[31] In communications it refers to the unique way in which an individual views and interprets the world. This means that we may all get a different message from the same memorandum or instruction and it may be influenced by our personality, our attitude, our reaction to the current environment or our past conditioning. To avoid this in key messages some sort of feedback should be encouraged on receipt. Meetings are an opportunity for interaction and getting an equal message across to everyone.

The project manager could encourage communications by introducing a project culture and an inspired and inclusive communication system and encouraging respect and attention to making ourselves understood by others. This is harder if each new project has a new team, or if the same team works for a different client. It can even be difficult if assumptions about previous behaviour are wrong with the same client and team in a different environment and location. In other words the dynamics of the situation vary on each occasion and there is only a limited project period to develop an understood culture. It is logical then to carry over a consistency which is connected with the client or the project manager or company style and to provide some training and feedback in the early stages of the project. A feedback loop in the system would be an essential requirement in project communications where drawings have to be checked and approved, receipts of information acknowledged and timescales placed on responses such as technical queries for ambiguous, missing or inaccurate information.

Noise is the element that impacts on communication as an external element. In simplest physiological form a loud noise outside the meeting window means that only two-thirds of the words are heard or a headache affects concentration. In a more subtle form it is referring to the other things that are going on for the receiver or sender outside the project. An unresolved domestic row is likely to loom more importantly than accurate interpretation of work documents. Work on another project which is going wrong may affect the tolerance of a subcontractor negotiating price changes on the current project. Severe time constraints may provide problems in ensuring safe environments.

Modern forms of communication have often accentuated the noise part of the model because of the pervasive nature of modern technology. Powerful search machines allow us to collect remote information easily and quickly. The text and the social media such as Twitter, Facebook, blogs and internet publishing give senders options to universalise information instantaneously. We can also restrict communication to what we want others to hear or receive and we can choose an inner circle of contacts (Twitter) that have similar objectives, or a focused business circle to market our company information. Gatekeeping may be an important part of filtering large amounts of information so that what is needed is helpful to the business. However there is a danger that a virtual, but unreal world picture is received, which has excluded vital information, due to our biased world. Feedback has become fundamental to our value of services. Modern communications are thus concentrated on sharpening the relevance of the message so that it is not only received, but it is understood amongst the noise of many other messages. In a project context this means using an attractive or unavoidable media to ensure relevant action. Case study 7.2 communicates H&S in different ways.

Engineering information for projects is often tailor made and is unique to the project. It needs to be well understood by parties with different background education and it has to be precise and accurate and understood unambiguously and the media is critical to its meaning and effectiveness. 2D drawings have been a traditional code for this type of information, but the industry has been slow in co-ordinating information and seriously miscommunicated although CAD and some 3D technology is quite common. It is challenged to use more advanced models such as building information modelling, which forces designers and constructors to work in the same model and not to run parallel ones for the design, costing and time control functions. These systems are projected to be more user friendly in communications to the client with better 'what if?' planning.

Extranets have tried to make common information more consistent and accessible, but have had limited success due to the problems of standardising different digitisation protocols where every project has a unique project team mix which requires standardisation of the technology for quite often short

Case study 7.2 Health and safety information

H&S statistics are expressed in injuries per 100,000 people. Injuries have gone down by 50 per cent in the last 10 years in construction. Injuries in construction still need to go down because construction has four times as many fatal accidents as the average for all industries in 2010. It also has twice as many major injuries as manufacturing.[32] Fifty per cent of all fatal construction accidents are falls from height and objects falling off heights.

Communication methods used in modern construction:

1 H&S induction and training is given to all new workers on site and most large sites require 100 per cent of the workforce to carry a CSCS card which proves they have passed a test in H&S relevant to their trade.
2 Managers go on three day health and safety courses each year and inspect sites daily.
3 Workers on some sites are given a red card similar to football if they seriously contravene H&S rules potentially putting others at risk and have to retrain if they show a smaller contravention of the rules.
4 H&S forums encourage workers to report unsafe working which could be improved.
5 Some companies have a culture of 'don't walk by' encouraging all workers, visitors and passers-by to report unsafe events.
6 Some companies have their own inspectors to shut sites down for significant incidents or accidents to make things safe before reopening.
7 Others show public boards to indicate the number of man days which have occurred accident free out of the project total so far.
8 Some use incident free or zero harm slogans in order to inculcate a new culture.
9 Some use cognitive methods (behavioural H&S) to develop a culture of respect for preventative methods.
10 Almost all sites demand a mandatory wearing of PPE.

Should this information stop accidents or dangerous incidents in construction? Why does it/doesn't it?

This information does not stop accidents, even though it points to a focus for preventing accidents. Egan called for a 20 per cent per year fall in accidents in 1998 and this rate of reduction has not been met. It would have brought the number of accidents down to 249 in five years and 81 in 10 years.

periods. The cost for many small organisations is prohibitive and two tier systems of traditional and new media can lead to mistakes and poor communication and erode the efficiencies of one system. Perception of what is important on receiving last amounts of information that the media can deliver to you are overwhelming on a large or complex project and a gatekeeper is required to send selective information to different users on a need to know basis. There is also a difference between information only and information which is tagged with a desired priority of action. Some may feel that they are being manipulated by what information they don't see.

Psycho-engineering the team

Psycho-engineering in this context refers to the knowledgeable use of psychometric methods and motivation to build a robust team which is then able to form good working relationships and outputs that have been productively enhanced by the team. It is well known that teams may introduce waste (witness four work persons watching one person dig a hole), but a new team needs to create a synergy that is more than the sum of its parts to arguably justify its existence.

A construction project team is formed by appointing a number of different consultants, the project manager, the contractors' managers and possibly the client. The resulting organisations choose suitable members on the basis of availability and experience. It is possible that a client may specifically request a named individual, but this is not the norm so the team is not usually selected on the basis of personality, or with any other psychometric tests that are common to standard recruitment. The job of team building by the project manager is vital especially in respect to working relationships. What can a project manager do if there is the wrong mix of team roles? How can the project manager deal with a personality clash, given that they have limited influence on the choice of personnel?

Belbin team roles

The Belbin hypothesis was based on observation of the interaction of management teams during training exercises at Henley College of Management by Meredith Belbin.[33] He noticed that teams that were stuffed full of clever people were not helpful to the solution of problems. He also noted that where a team was allocating roles and complementing each other, they were usually better than those teams where team members had similar roles and were competing against each other. The roles that he identified as important after further research connected with the smooth maintenance of the team. They are explained in Figure 7.5. In order to assess these reliably he developed a self-assessment inventory designed to measure behavioural characteristics to help him to identify the role that each individual would find themselves most comfortable with. In reality individuals are comfortable with

Plant	Creative and able to produce ideas
Resource investigator	Operates a good network of contacts and good at resourcing the ideas
Shaper	Able to whip up enthusiasm and to motivate the team
Chairman	Co-ordinator and listener. Good at drawing people out and summarising ideas for agreement. Mature and confident
Team worker	Willing to support and develop other people's ideas and encourage working together towards common goals. Co-operative and diplomatic
Implementer	Disciplined application of the team aims and objectives. Conservative, efficient and practical
Monitor/evaluator	Able to objectively appraise ideas and to keep control on critical parameters, for example, cost. Strategic and discerning
Completer/Finisher	Good at finishing off and ensuring the final details are in place to get and maintain a working solution. Painstaking and conscientious
Specialist	The starting point for many technical teams, relevant where specialist knowledge is required. Not really a generic role

Figure 7.5 Belbin roles

two or three roles and can cover roles where there is not a strong team representation of the role. This can work well in smaller teams. Later it was also recommended that a 360° assessment was needed with employer and peers to provide a better result.[34]

Belbin identified nine roles, some have changed names over the years, but have not changed in character. These are shown in Figure 7.5.

These roles have been consolidated into action, social and thinking roles:

- Action roles – implementer, shaper and completer-finisher
- Social roles – co-ordinator, resource investigator and team worker
- Thinking roles – monitor evaluator, plant and specialist

Belbin also recognised that the team needed to determine a common goal. The main problems occurred when the there was an overlap of two or more dominant team roles such as the shaper and the plant as these members were less likely to tolerate other ideas, or being 'shaped' themselves, unless the team could be subdivided. A set of pure team workers without any leaders may also be ineffective, but in the absence of team workers and a chairman the team was tetchy, indecisive and fragmented. The omission of a finisher meant that in complex problems solutions often missed out on vital ingredients and without monitors there was a common problem with schedule overrun. Over time Belbin remixed the teams to prove a startling improvement in the group dynamics which concurred with his theory.

The Belbin inventory can be applied by trying to pick and mix a team based on the inventory results as Belbin did to optimise performance. However if the team is already there the team can be made *aware* of its weaknesses and agreement to cover the roles as discussed above. This awareness of the team working, together with a desire to work together, can be as powerful as having natural role players. Roles are reinforced by training and practice. In addition the occasional or regular use of an outside facilitator or a creative workshop may reinforce these roles if they are lacking. The main reasons for removing persons from the team are if:

- not a team player
- a deeply entrenched personality clash, which has been unresolved
- a failure to accept an open culture
- unwillingness to submit to certain team parameters and agreement.

The above may be described as removing oneself from the team and there are occasions when a client will request that a core member of the project team is moved on. Project disciplines are relevant in construction projects, but functional contribution has already been covered by appointment. However by themselves they do not add to the synergy of the team, or overcome human problems.

It is also considered that roles for any one person are divided into *preferred* roles which can be comfortably and strongly performed, *manageable* roles which can be played if required by the team and *least preferred* roles which require a considerable effort by the individual and may lead to weaknesses in the team. From this a team distributes the generic roles in a way which is complementary for their strongest position. Team roles can change from job to job and subsequent behavioural testing might show a change, which would swop manageable and preferred roles.[35]

Myers Briggs

Another approach is based on Jung's psychology of in-built personality type. A template has been developed by Isobel Myers and Katherine Briggs (her mother) to provide an ethical approach to the identification of personality preferences in four dimensions. In-built preferences are polarised on each of the four scales of:

- extroversion (E) – introversion (I)
- sensing (S) – intuition (N)
- thinking (T) – feeling (F)
- judging (J) – perceiving (P).

A self-assessment questionnaire has been developed over a period of 50 years and has a high reliability factor. The combination of letters differentiates 16 foundational personality types (the Myers Briggs Type Indicator or MBTI™) indicating a starting point for an individual. Used ethically they can point towards strengths and weaknesses and can guide individual choice and help to identify areas of personal development. Ethically they should not be used

ISTJ	ISFJ	INFJ	INTJ
ISTP	ISFP	INFP	INTP
ESTP	ESFP	ENFP	ENTP
ESTJ	ESFJ	ENFJ	ENTJ

Figure 7.6 Myers Briggs type table. Myers Briggs Type Indicator is a registered Trademark of Consulting Psychologists Press Inc.

for selection as they do not define behaviour. Behaviour is a function of additional factors, such as environmental conditioning, experience and choice. If it is used in teams for problem solving a mixture of personality types in teams should complement each other to arrive at better solutions. This gives a much richer understanding than Belbin about why clashes may occur and how they may be avoided between types. One of the more powerful applications for a project manager is to use an understanding of someone else's type to adapt his/her own natural preferences to improve individual communication with them and to tailor development programmes to suit learning styles. This also has implications for improving the outcomes of negotiation in an open culture. Personality type does not explain individual motivation, but it can give clues to understanding individuals' motivation.

Figure 7.6 indicates the 16 types possible using a letter from each pole of the four dimensions. These combinations represent the foundational preferences of an individual and do not seek to stereotype individuals as Jung acknowledges a natural desire for the development of the personality to master other areas which are not immediate preferences. An example of this from the table is that the eight types represented in the first and last rows, which have thinking rather than feeling preference, may as project leaders see the benefits of motivation which recognises the use of harmony (F) rather than step by step logic (T) in the management of the project team. Case study 7.3 indicates a typical type.

Jung recognises the role of environmental conditioning to individual personalities. For example, socially there is need for expressing oneself publicly and this is not the natural preference of the eight types at the top of the table in Figure 7.6 as they gain their energy from within and not from others. This produces a particular twist in that those with an 'I' preference unconsciously present themselves publicly as something they are not and their preferences are easily misinterpreted by others. This stated, it is possible to interpret the behaviour of different types with some degree of training. It is dangerous to guess the type from the behaviour! Environmental conditioning and conscious personal development provide depth and uniqueness to individuals' personalities and contrary to popular opinion has the opposite effect to stereotyping. MBTI™ can be used for developing team

Case study 7.3 Typical characteristics of a common Myers Briggs type

An ESTJ is the most common type found in Western cultures and it may be interpreted letter by letter.

E   indicates the outgoing style of the individual with a tendency to have a wide circle of friends and contacts; together with J it makes for quite forceful planning.

S   indicates a strong use of the five senses and provides a practical down to earth approach based in the here and now evidence with a likely pragmatic problem solving style confident in the straightforward problems within ability. They will need the presence of an N to generate new problem ideas.

T   denotes a thinking logical approach to issues presenting and is likely to be based on strong rules, precedence or business case which together with J is likely to require step by step planning.

J   is a more systematic target based approach with strong desire for hitting planned targets, which may appear as unnecessary pressure to some others.

The dominant and seen characteristic is the thinking one. The S side will support this with a strong desire to celebrate the success of the project.

Approximately this adds up to a loyal, conservative, practical individual, with a logical step by step approach and the ability to close out projects and make effective and systematic decisions. In many cases they make strong up front leadership material in a practical traditional industry like construction. Their limitations will be the tendency to over control, a resistance to change where it 'ain't broke' with a focused rather than a creative outlook. The type makes a good leader, but they may improve by training to develop more creative approaches and team building skills. A deputy may be chosen who has complementary skills that allows creative work together to make a team.

working, ensuring balanced problem solving, improving communications and resolving conflict. It is often used for identifying leadership style, but it is not a measure of good or bad leadership.

Behaviour can also be altered by severe stress and there is a possibility to cross over from one's preference to one's opposite in these circumstances. This can be recognised when a team member is perceived as acting out of character. For example, a normally placid ISTJ will take on the extroverted characteristics of an ENFP and will move from present to future possibilities in assertive fashion warning of doom and gloom. It may also be apparent when certain personality types are under stress when asked to go through major organisational change.

Myers Briggs and conflict

It might be thought that by having opposite types working together that conflict might result with a natural instinct by the leader to keep certain types apart, but this is not the case if progress is to be made in existing teams in standing in each other's shoes. The key is to use the clear vision and blind spots of different personalities and to resolve conflict by an understanding of conflict generators for that particular personality type and what they need from others and in the longer term promote personal development. The most dangerous situation is when the team is working under significant stress. In this situation, what you see is not what you always get.

The research done by Killen and Murphy has indicated that the two last letters contribute most to conflict.[36] They call these the conflict pair. In the fourth dimension judging types have most difficulty adapting to creating space for decisions to be made and vice versa the perceiving types have most difficulty seeing why there is such a rush to make decisions and need more time for information to check it out. In the second dimension indicated by the poles of feeling (F) *or* thinking (T) there is a major conflict between the Ts who want to fix what is wrong and the Fs who want to make sure that everyone is heard and respected. This creates four possible combinations, TJ, TP, FJ and FP, each showing up in four of the personality types (see Figure 7.7). The thinking types' objections tend to arise out of hurt pride in a challenge to their authority or trust. The difference between the TJs and TPs is that one wants a defined process and the other wants closure and conflict sorted. Both detach themselves from emotion. The feeling type objections arise when people are not listened to. The difference between the FJs and FPs is that one wants intact relationships and the other wants open exploration and for all to be heard. They both accept the role of emotions.

By mixing types it is possible to get a good mix for solving problems and the model used is to create space using the Ps, add value by using both the T and F approach and bring about closure by using the Js. See Figure 7.7.

Myers Briggs and communication

Communication may be improved by the sender being aware of the receiver's tendencies to understand the message. According to the communication model the ability to choose the right media and to encode the message with reference to the language understood by the receiver would

Figure 7.7 Killen and Murphy conflict pairs
Adapted from Killen and Murphy (2003)[37]

help. Myers Briggs personality types might give further clues to this process where known. According to Brock this is best determined by the use of the Myers Briggs middle letters.[38] This has been researched with reference to effective selling by understanding your buyer and further applied by Allen and Brock to the health service,[39] which has a reputation for poor patient communications. This approach helps to communicate points clearly and persuasively in a negotiation. The technique will be most useful in aiming to understand mutual needs and producing a win–win result. Used as a tool for domination they are not suitable for the sustainable relationships needed in projects.

Myers Briggs and negotiation and problem solving

The Mobius communication model is another application which can be linked with Myers Briggs.[40] This model equates acknowledgement with perceiving, responsibility with sensing and capability with thinking, judging with commitment, possibility with intuition and mutual understanding with feeling. The main reference to communication is the connection of extroversion with blame or praise (faults or strengths first seen with others) contrasted with the connection of introversion with worry or claim (faults or strengths first seen with self). The communication approach for Mobius gives a negotiation and problem solving sequence as follows:

1   Mutual understanding – of likeness and differences; this is designed to create an atmosphere of wellbeing (F).
2   Explore the possibilities – recognise the common ground (N).
3   Commitment – where choices made within the framework of possibilities get firmed up (J).
4   Capabilities – indicating an implementing strategy for deciding on resources and skills (T).
5   Responsibility – this commits the 'who', 'what' and 'when' to the project so that people are involved in doing it (S).

6   Acknowledgement – a monitoring and feedback action which provides a basis for assessing success (P).

It can be seen this is also useable as a win–win negotiation framework and its connection with Myers Briggs suggests that the use of a team of negotiators, each engaged at the relevant part of the process to use their natural preferences, might be beneficial. The first three steps bring the negotiation to agreement and the final steps ensure the efficient working of the agreement before. Applying this to a PFI negotiation indicates the difference between selection of the preferred bidder stage and financial closure.

Appreciative inquiry

In the area of feedback, appreciative inquiry (AI) is a methodology for giving positive feedback by breaking people out of their typically negative mindset and habits. It looks to the concept of learning opportunities rather than the more negative concept of problem solving. The main drive is to motivate a no blame environment, which encourages learning by doing in which mistakes are seen as the tools to teach. Cynics may call it the art of ignoring the disgruntled, but this may be countered where there is a widespread change in culture developing.

The steps include identify, appreciate, inquire, envision, focus, affirm and sustain. These steps allow a neutral discussion of the issues which have the potential to be improved, by looking at what you do well and analysing whether generic processes may be applied to improvement of this and other areas. A brainstorming 'what if' scenario might be useful in the envisioning step, which then needs to identify areas for action and maintenance of better systems.

An example of AI in customer care is to inquire 'Describe a time when you went an extra mile for the customer – what made it possible?' This provides an opportunity to explain something that went well and analyse it for the innovative features that could be reused.

Change and developing a project culture

Part of setting up the project organisation is the desire to influence the nature of relationships. Traditionally relationships operate on a 'them and us' basis between the consultants and the contractors and between the consultants/ contractors and the client. The closed culture is also likely to have infiltrated the supply chain relationships and the contractor organisations themselves. AI might help the latter in the context of overall change. The discussion here relates to the nature of an open culture and how to move away from a culture of closed and regulated relationships.

The CIRIA C556 report describes a culture of openness as a blame free environment, with admission of errors on the basis of agreement on the best way to move forward.[41] It also provides a single system of reporting to client, where a single understanding of the time and cost safety margins is tendered and change is acknowledged by all parties. This system allows the formation of a true team and incentives which encourage waste saving improvements.

The CIC emphasises the need to identify and ensure standards of specific project management competencies that enable continuous improvement and the appointment of a project manager with experience fitted to the particular needs of the client's project.[42] This may be enshrined in an integrated team of people, which they list as the project manager, design manager, construction manager and trade contractor managers. These competencies are listed in 25 areas under the five main heads of:

- strategic
- project control
- technical
- commercial
- organisation and people.

This raises the issue of extending the culture to supply chain relationships (including consultants and construction managers) and the need to integrate the culture through different life cycle stages. It also may put some responsibility on the client to appoint a team that can work together well in an open culture.

It depends upon competent management to keep costs down and an open book tendering system which allows the client to see where costs are incurred in the event of change which realistically is part of the deal in rapidly changing markets. In the case of cost certainty then extras need to be balanced by savings and a claims culture should be avoided by agreeing who takes on the risks and who pays for contingency. The main challenge of an open culture is the sustained commitment for fundamental culture change of the whole supply chain and perhaps some clients' reluctance or inexperience in becoming closely involved. A number of contracts are suitable for this type of contracting, but they need to be used in a spirit of partnership to be effective.

Traditional contracts encourage a closed culture with familiar, but complex rights and conditions which have grown up with modern contracting, putting the client on the 'other side of the fence' and providing them with a 'fait accompli' that asks for specification in return for product and keeps the client away from the project team decision making. This may be what the client wants, but in many cases greater flexibility is required with client choice to work the system knowledgeably and not by trial and error.

The benefits of a closed culture theoretically lie in the experience of professionals in operating the contract, the 'fixed price' where little change is likely and the ability of the project team to proceed without interruption. In reality these advantages only exist in some market conditions and over short gestation and short construction periods. Track record and conditions should be vetted by the client to ascertain that these advantages exist for their project.

The fixed price contract is an erroneous title as changes and redesign are common and may create extra cost. In a traditional contract, even with a bill of quantities, revised prices for change are rarely fully defined prior to go ahead. This makes it difficult for the client to predict final cost and in the case of disagreement at final account one or both of the parties is likely to lose out, because an agreed common path was not taken at the original

Case study 7.4 A best practice of the use of NEC contract by EDF Energy at Heathrow T5

The electrical installation of the £4 billion Heathrow Terminal 5 contract which is worth £72m is being installed by EDF Energy and includes a 365-day, 24-hour maintenance of the occupied airport HV and LV supplies. The design is being carried out simultaneously with construction with a varying element of supplier design. It presents many challenges to a traditional form in providing cost certainty over a long period, with interfaces with a large number of sub packages which are on a supply and install basis, mostly by suppliers who are not used to being contractors also. It was claimed that the NEC family of contracts was chosen because of its virtues of flexibility and clarity and that it embodies the principles of effective communication. Because of the wholesale lack of experience of this type of contract and because of the limited contractor experience a joint residential training course was set up to focus on the practical application of the contract. The project team accepted that the open culture of the ECC contract helped the resolution of conflicts at package interfaces and enabled cost certainty because of the quick settlement of compensation events. The learning curve though steep was achieved more easily because of the simpler structure of the contract, which suited those who had no prior expectations of other contracts.

The conclusions were that the initial difficulties in understanding were more than overcome and offered excellent reporting and visibility. Visibility also enabled contracts and substantial additional works to be carried out at a fair price to the client.[43]

change position. An additional complexity is the role of third party design consultants who may have had a role to play in the creation of additional costs. In short any one of changing requirements, competitive tendering conditions, insufficient time and agreement for the design, information flow, newly formed project teams and less than competent management can spell disaster.

The New Engineering Contract (NEC) is people rather than contract centred as it sets out early warning systems, which allow the two sides to get together and to firm up the effect of change on programme and budget and requires a regular resubmission of an agreed programme and formal approval of revised budgets to proceed. It has also been criticised for its lack of legal clarity if tested in a court of law. Is that good or bad? Case study 7.4 shows effective use of the EEC contract in this context.

Conclusion

The industry has gained in the past from non-bureaucratic relationships which are a characteristic of project work. However, until now the construction industry has only paid lip service to new ways of changing the confrontational culture which has been created by the contractual forms commonly used over the last 60 years. There is a long way to go in catching up with other industries in developing integrated ways of working which present continuing opportunities for reducing waste and adding value through better understanding of the benefits of an open culture. More is needed in engineering: a conducive open environment, moving away from blame and building relationships that recognise the importance of trust and openness. Communication is a key aspect of this and requires effective delivery of many different types of information and the receiver must be allowed to feed back in the model. Conflict avoidance is unlikely, but good communications is a way of understanding the issues and using this to negotiate solutions and avoid it spiralling into dispute. In this respect the project manager is looking for a win–win approach. Where dispute occurs, mediation provides an option to bring in a neutral party to broker a deal. Under the new Construction Act 2011 they have been given more powers to share the costs of the adjudication process and to outlaw late payments to contractors and subcontractors.[44]

There are opportunities also in the area of personality profiling and team building in order to deal with conflict positively and to stop it becoming personal and dysfunctional. This is not another alternative to dispute resolution with additional mediators, but a wholesale change in manager attitudes throughout the supply chain to an attitude which:

- believes in the bottom line benefit of a 'type Y' approach to management
- cultivates the mutual benefit of win–win negotiation and understands the importance of personality in all types of conflict

- convinces clients of the benefits of collaborative type contracts such as NEC in breaking down contractual attitudes and developing where appropriate long term relationships.

There is plenty of scope for improving the quality and value of one-off construction projects by training and motivating the workforce with these attitudes so that clients respond and repeat work is generated by their recommendation. The challenge for implementation is 'up front' investment in training people until you change attitudes.

References

1 Flashman S. (2010) *Seven People Management Skills in Construction Marketing. Building Construction and Built Environment Forum.* Available online at: http://constructionmarketingtips.com/seven-people-management-keys-in-construction-marketing/ (accessed 12 March 2012).
2 Forsyth D. (2006) *Group Dynamics.* Pacific Groves, CA, Brooks Cole.
3 Lewin K. (1948) *Resolving Social Conflicts: Selected Papers on Group Dynamics,* Ed. G. Lewin. New York, Harper and Row; Likert R. (1967) *The Human Organisation. Its Management and Value.* New York, McGraw Hill; Argyris C. (1964) *Integrating the Individual and the Organisation.* New York, Wiley; Tuckman B.W. (1965) 'Development sequence in small groups', *Psychological Bulletin,* 63(6): 384–399; Argyle M. (1989) *The Social Psychology of Work.* 2nd edn. London, Penguin Books.
4 Allcorn S. (1989) *Understanding Groups at Work. Personnel,* 6(8), August: 28–36.
5 Benson J. (2000) *Working More Co-operatively with Groups.* London, Routledge.
6 Hiley A. (2004) 'Transferring to teamwork', *CEBE News* Update issue. 10 June.
7 Tuckman B.W. (1965) 'Developmental sequence in small groups', *Psychological Bulletin,* 63(6): 384–399.
8 Walker A. (2007) *Project Management in Construction.* 5th edn. Oxford, Blackwell Publishing.
9 Gunning J. and Harker F. (2004) 'Building a project organisation'. Chapter in RIBA, *Architect's Handbook of Construction Project Management.* London, RIBA Enterprises.
10 Adair J. (1979) *Action Centred Leadership.* London, Gower Press, p.10. Used with permission of John Adair Foundation.
11 Construction Industry Council (2000) *Construction Project Management Skills.* London, CIC.
12 Cornick T. and Mather J. (1999) *Construction Project Teams: Making Them Work Profitably.* London, Thomas Telford.
13 Partington D. (2003) 'Managing and leading'. In Turner R. (Ed.) *People in Project Management.* London, Gower Publishing.
14 Maslow A.H. (1943) 'A theory of human motivation', *Psychological Review,* 50, July: 370–396; Herzberg F. (1974) *Work and the Nature of Man.* London, Granada Publishing.
15 McClelland D.C. (1988) *Human Motivation.* Cambridge, Cambridge University Press.

16 Vroom V.H. (1964) *Work and Motivation*. New York, Wiley; Porter L. and Lawler E. (1968) *Managerial Attitudes and Performance*. Homewood, RD, Irwin; Adams J.S. (1965) 'Injustice in social exchange'. In L. Berkowitz *Advances in Experimental Social Psychology*. New York, Academic Press; Locke E.A. (1968) 'Towards a theory of task motivation and incentives'. In *Organisation and Human Performance*, 3: 157–189.

17 Sommerville J. and Stocks J. (1992) *Psycho-Engineering the Productive Environment*. Strathclyde, Strathclyde University.

18 HSE (2004) *Work Related Stress – Together We Can Handle It*. Available online at: http://www.hse.gov.uk/stress/index.htm (accessed 30 November 2011).

19 HSE (2011) *Stress Related and Psychological Disorders*. Available online at: http://www.hse.gov.uk/statistics/causdis/stress/days-lost.htm (accessed 26 October 2011).

20 Thamhein H. and Wilemon D. (1975) 'Management in project life cycles', *Sloan Management Review*. Summer. In Meredith J. and Mantel S. (2000) *Project Management*. 4th edn. New York, Wiley, p.166.

21 Thamhein H. and Wilemon D. (1975) 'Management in project life cycles', *Sloan Management Review*. Summer. Adapted from J. Meredith and S. Mantel (2000) *Project Management*. 4th edn. New York, Wiley.

22 Meredith J. and Mantel S. (2008) *Project Management*. 6th edn. New York, Wiley, p.178.

23 See the CBPP/CIB A (1999) *Model Project Pact*. May. Available online at: http://www.constructingexcellence.org.uk/pdf/document/Cibpact.pdf (accessed 12 March 2012).

24 Meredith J. and Mantel S. (2000) *Project Management*. 4th edn. New York, Wiley.

25 Ury W. and Fisher R. (1983) *Getting to Yes*. Harmondsworth, Penguin Books.

26 Construction Productivity Network CPN (1996) *Design Management for Industry*, CPN Workshop Report 21. 17 October.

27 Lazarus D. and Clifton R. (2001) *Managing Project Change. A Best Practice Guide*. C556. London, CIRIA/DTI.

28 Construction Productivity Network (1996) *How Others Manage the Design Process*. CPN Workshop Report 20, January.

29 Foley J. and MacMillan S. (2005) 'Patterns of interaction in construction team meetings', *Codesign* 1(1), March: 19–37.

30 Lucey T. (1991) *Management Information Systems*. 6th edn. Eastleigh, DPP Publications.

31 Mullins L. (2010) *Management and Organisational Behaviour*. Harlow, Pearson Education, p.209.

32 Health and safety in construction has a poor reputation with 550 accidents/100,000, with people off work for more than three days, of which half of these are major injuries. Although accidents have halved over the last 10 years, construction still maintains a higher rate of accidents than most other industries. Construction has a fatal injury rate twice as big as manufacturing and four times as many in industry as a whole, but is not the worst rate.

33 Belbin M. (1981) *Management Teams: Why They Succeed or Fail*. Oxford, Butterworth-Heinmann.

34 Belbin Associates (2009) *Method, Reliability & Validity, Statistics & Research: A Comprehensive Review of Belbin Team Roles*. Henley, Belbin Associates.

35 Belbin Associates (2009) *Method, Reliability & Validity, Statistics & Research: A Comprehensive Review of Belbin Team Roles*. Henley, Belbin Associates.

36 Killen D. and Murphy D. (2003) *Introduction to Type and Conflict*. Palo Alto, CA, CPP.
37 Killen D. and Murphy D. (2003) *Introduction to Type and Conflict*. Palo Alto, CA, CPP.
38 Brock S.A. (1994) *Using Type in Selling. Building Customer Relationships with the Myers Briggs Type Indicator*. Consulting Psychologists Press, California.
39 Allen J. and Brock S.A. (2000) *Health Care Communication Using Personality Type. Patients are Different*. Abingdon, Routledge.
40 Stockton W. (1994) *Integrating the MBTI with The Mobius Model*. Mobius Inc. Available online at: http://www.mobiusmodel.com/model.php (accessed 24 November 2011)
41 Lazarus D. and Clifton R. (2001) *Managing Project Change. A Best Practice Guide*. C556. London, CIRIA/DTI.
42 Construction Industry Council (2000) *Construction Project Management Skills*. London, CIC.
43 CIOB (2004). Contractors struggled with ECC at Terminal 5 but discovered its benefits in the end. 21 July. *CIOB International News*.
44 Part 8 of the Local Democracy, Economic Development and Construction Act 2009 (LDEDC Act 2009), apply to construction and the Construction scheme has been amended.

Chapter 8

Managing risk and value

Risk and value management are strongly connected with each other because enhancing value for the client also means ensuring that risk is minimised or at least planned and managed. An intention to produce best value for the client must also be implemented by identifying and managing risk.

The main objectives of this chapter are to:

- consider the concept of risk and value, people's values and attitudes and their connection with decision making and business planning
- appraise the responses available for dealing with risk
- demonstrate how risk and opportunity are assessed and managed, reduce uncertainty in construction projects and evaluate the tools and techniques
- investigate the influence of procurement routes on managing risk and value and supporting best practice in the appointment of the supply team
- consider guidance for selection and appointment best practice for managing risk and value and deriving an ethical approach
- understand good practice for change management procedures.

Risk and value imperatives

Fifteen years ago the application of risk and value management was very limited in construction projects and the industry has lagged behind others in recognising how critical they are to building client confidence and in meeting their requirements. Traditionally risk control in the construction stage was based on implicit heuristic assumptions and allowed for in the budget or the programme as a pragmatic contingency or, in the case of the contractor's tender, as a risk premium. Value was considered to be lowest tender cost.

As clients have become more sophisticated they have made more demanding targets, which have left less room for manoeuvre. These have led to tighter budgets, more innovative technology and less tolerance of time slippage, which is often connected to more onerous penalties for lack of performance.

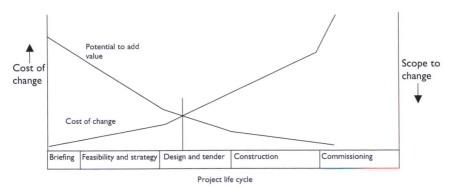

Figure 8.1 The cost of change

In achieving their business clients have expected more innovative technologies and design to achieve more for less (pull). Legislation and fiscal measures have had an equal impact and influenced the values of clients especially in relation to their sustainability goals (push). The process of understanding and achieving client values is important where there is a more complex set of objectives, as in stakeholder interest which influences client values. Stakeholder and external factors cannot be ignored as they impact upon key time, cost and specification goals and may create loss or opportunity. The project manager needs to respond and manage these also in the value and risk process.

In response, the industry has looked at enhancing value for clients and managing their risks better by the use of more advanced techniques. It is now recognised that the scope for changes in design to give significant improvements in value come mainly at the feasibility and strategy stages and that it becomes more difficult as the design is progressed to change previous decisions without disruption and the cost of doing so is greater. This is indicated diagrammatically in Figure 8.1.

Stakeholder and risk management is a part of assessing value by recognising different interests and managing conflict on the basis of priority and influence. Value management workshops are offered to integrate stakeholder value, risk evaluation and site and design constraints with the client's value.

It is difficult to say whether there is more push or pull in the marketing of value and risk management because both sides can gain. Value and risk assessment is an essential part of planning and managing of complex projects, because of the interdependence of external and internal factors and interests, However, strong committed leadership is needed for initial outlays to encourage a level playing field.

Risks assessment makes assumptions about external and internal factors. Risks are first assessed when early feasibility decisions are made and need

Influences on value

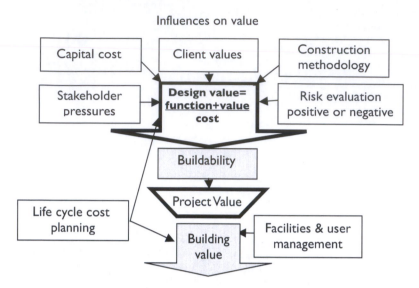

Figure 8.2 The make up of value and influence of risk

to be managed throughout the project life cycle as more information is received and because things change. These require better predictive models to cover the range of probability in response to uncertainty and impact as the project progresses. Simulation may be used to model scenarios where a number of factors may all change. Experience also plays a role and risks accountability should be allocated to those who have experience in the type of project and cultural mix which is essential to interpret a wide range of qualitative and quantitative indicators.

Enhancing design value, as we have seen, is about improving the efficiency of function over cost, and is connected with risk evaluation, through the assessment of alternative methodology and functional value. Risks might be high impact and diminish value where there is more probability. Real cost to the client considers capital and operational costs, by weighing up value minus the cost of risk. This inter dependency of risk and value is recognised as critical as is seen in Figure 8.2. This is not a 'how to' chapter on techniques, but argues the case for a holistic, integrated approach in order to support effective managerial decision making.

Previous chapters have identified risks and their fair allocation amongst the parties to the contract. Aspects of value management and value engineering which specifically consider value as a development of the design process are covered in Chapter 9. This is not a 'how to' chapter on techniques, but looks at the assessment of risk and value and argues the case for a holistic, integrated approach in order to support effective managerial decision making.

Historical approaches to risk and value management[1]

Traditionally risk control in the construction stage was based on implicit heuristic assumptions and allowed for in the budget or the programme as a pragmatic contingency or, in the case of the contractor's tender, as a risk premium. Value was considered to be lowest tender cost.

An over-arching concern of clients in procuring construction is to achieve added value from the project. The value of a construction project is determined by the benefit it creates for the client. Typically this value is measured in monetary terms and therefore the value for money is the primary concern. In the 1960s, when construction cost planning methodology was still in its infancy in the UK, James Nisbet identified value for money as the most difficult of three facets of cost control to achieve. The other two facets of budgetary control and appropriate balance of expenditure over the construction elements are much easier to demonstrate and deliver through effective cost management and cost planning, but value for money requires understanding and evaluation of the client's values as part of defining the project brief. This is no simple matter when the client and/or end user may be a complex organisation or diverse group of stakeholders with disparate interests and requirements.

Value for money in public procurement has been a long-standing principle espoused in HM Treasury procurement guidance; failings in construction procurement have been a long-running criticism by the National Audit Office, particularly in capital projects for the construction of nuclear power stations, motorways, hospitals, the Thames Barrier and the Olympic games. Value for money in government projects has traditionally been measured against cost targets set by reference to historic average costs and by competitive tendering where the contract is awarded to the lowest bidder.

Evaluating the business case for a project implies an assessment of the degree of certainty in the project outcomes. Project uncertainty in terms of completion to the expected project functionality, cost and programme are the main cause of stress for construction project clients; unexpected changes to any of these key parameters are the primary cause of client dissatisfaction with the construction process. The process of gateway reviews at key milestone points is designed to identify and manage project uncertainties, and to allow projects to be deferred or abandoned if project uncertainty is unacceptably high.

The origins of value management can be traced back to the Ersatz movement in Germany when the country was facing the difficulty of regenerating the industrial and commercial infrastructure in a climate of post-war austerity with a desperate shortage of money and materials. The search for added value was not necessarily about lowering cost; value could be added by raising functionality. German industry, exemplified by names such as Mercedes,

BMW and Audi, has demonstrated that the quality and reliability associated with *vorsprung durch teknik* has added value, rather than cut the initial purchase costs of these cars.

Value engineering, 'the organised approach to providing the necessary functions at the lowest cost', owes it origins to Laurence Miles and his post-war work at GEC in the USA. Early development of the techniques of value analysis, function analysis and the VM Job Plan was primarily for US military projects. These techniques, which are characteristic and essential elements of value management methodology, are incorporated in the VM standards of SAVE International, which describes itself as the premier international society devoted to the advancement and promotion of value methodology.

The economic turbulence of the 1970s, triggered by the 1974 and 1979 oil crises and their associated property and construction slumps, stimulated a period of great interest in UK construction of comparative analyses against other countries' practice, most notably of the USA and Japan. In the mid-1980s, the construction management company of Bovis teamed up with Lehrer McGovern in the USA to introduce American techniques to the UK on the Broadgate development, including value engineering, and resulted in innovations such as cladding panels supplied with integral internal finishes and pre-installed engineering services and 'chandeliering' of multiple floors of steel beams into position on one crane lift.

Investigation by Kelly and Male in the early 1980s identified differences in approach between value management in the USA and UK.[2] Central to this was the use in the USA of a value management team to audit a design at scheme design stage. Several states have made this a requirement for verifying value for money in publicly funded construction projects and an example was given of this procedure for prison capital works in the state of New York. The VM audit is undertaken by a multi-disciplinary team under the direction of a VM Team Leader and the team of experts are appointed separately from the design team by the client. The additional fee is often justified on the basis that it will represent some 10 per cent of the savings identified and the savings will be at least 10 per cent of the overall project budget. The difficulty is that these 'savings' are implicitly a criticism of the original design team. Who will pay for and take design responsibility for the required design changes?

By contrast, the UK has tended to promote VM as an additional service provided within the original design team as a way of improving the understanding of the client's value system, business process and the project brief; a way of testing the value for money of the design as it develops and, in the process, a major contribution to team-building and encouraging collaborative interdisciplinary teamwork to improve integration of the design and construction process. Fragmentation of the design and construction process in the UK has been a major criticism of UK government reports for more than 40 years, as evidenced by Emmerson (1962), Banwell (1964), Latham

(1994), Egan (1997) and Fairclough (2002).[3] Since the mid-1990s, the emphasis in public projects has changed to value for money measured against the project business case, where life cycle cost considerations can be as important, and even more important, than initial cost in determining value for money.

Research by Stuart Green in the late 1990s focused on the contribution of VM as a group decision support method, or a method to improve strategic decision making by project teams, which was differentiated from the SMART or management by objectives approach to one which is defined by Connaughton and Green: 'a structured approach to defining what value means to a client in meeting a perceived need by establishing a clear consensus about the project objectives and how they can be achieved'.[4] This gives VM a social value in improving the way the project team work together in defining criteria for project success and then delivering to these criteria and in this respect furthers the argument for the provision of value and risk management services in consort. However Green identified the problem of a lack of an established theoretical underpinning to the softer approach to VM by comparison with the recognition given to the more established quantitative techniques within the SAVE and other existing VM certification schemes. Davis Langdon and Everest presents success as a fusion of managing value, risk and relationships,[5] with the conditions for success being a number of things, but including:

- project objectives that are clearly established and understood by all
- reconciliation of stakeholder needs
- reduced risk
- improved communications and understanding
- innovation and creativity
- enhanced team working
- effective use of resources
- elimination of unnecessary costs and waste
- optimised whole life costs.

Thus effective VM at the strategic project stage can substantially reduce the potential uncertainty in a project and increase the chance of success. Dallas has also argued for the combined management of risk and value to provide an integrated service with the same professional.[6]

Sustainable value and whole life cycle

There will be a need to look at the life cycle costs which include maintenance and operation costs as well as funding costs. It is well known that there is a ballooning relationship of design: construction: operating: business benefit/loss. Romm calculates the proportion of building cost over a 30 year period as 2 per cent design and construction, 6 per cent maintenance and 92 per cent personnel costs (business costs) for running the business.[7] The construction/design and maintenance proportion of business costs (8 per cent) is very similar to Table 4 in Hughes *et al.* who derived average figures for 1999 office blocks, if energy costs are included in the maintenance figures.[8] Wolstenholme also suggested that business costs outweigh early capital costs many times over and therefore small increases in capital cost can provide regular large savings in business costs (see Figure 8.3).[9] Value is recognised by increasing capital cost in order to get a better operating environment which impacts on employee productivity, durable components for less maintenance and energy use to keep running costs low, better layouts for efficient working and competitiveness to impact favourably on market or sales (business costs). Other issues are to fund in attractive ways for cash flow or to locate for additional financial subsidy in low employment areas.

Simple example of whole life cycle cost savings for more capital cost

Use Figure 8.3 for relative costs.

I Standard quality and durability

In the above if construction and design cost was £10m, then

maintenance and energy use is worth $6/2 \times £10m = £30m$, and

business costs are worth $92/2 \times £10m = £460m$

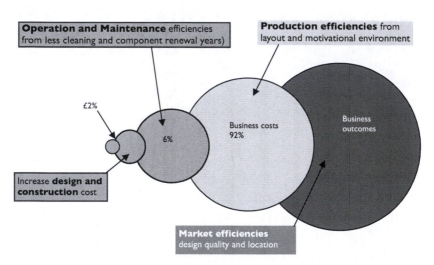

Figure 8.3 Indicative range of cost and value over a building's life cycle
Adapted from Wolstenholme (2009)[10]

2 Better quality, insulation and more durable building

If price increases 50 per cent to £15m for 33 per cent better durability and uses 50 per cent less energy and creates 6 per cent better business efficiency then,

savings of £660,000, £2m and £27.6m = £30.26m
will be made over 30 years (£1m per year)

The savings will pay for the extra £15 million capital cost in *15 years*.

They may also be discounted to net present value (see below).

Whole life costing

Life cycle cost is defined in ISO 15686 part 5 as a 'the cost of an asset, or parts throughout its life cycle, while fulfilling its performance requirements'.

Whole life cycle (WLC) cost for buildings also includes the non-construction costs, externalities (impacts on others) and income.[11] Part of the business plan in commissioning a new building is the consideration of a wider set of costs and incomes which occur over the life of the building.

These costs are greater than the capital cost of building and so may influence the initial expenditure in order to save larger amounts of money later in the maintenance, operation, occupancy and end of life costs as in Figure 8.3. The later costs are usually discounted (net present value) in order to make future costs and income comparable with today. These combined discounted costs and incomes are described as whole life costs. Public procurement clients are insisting on the WLC approach to assess value for money.

Life cycle costing (LCC) is defined by the Royal Institution of Chartered Surveyors (RICS)[12] as the present value of the total cost of an asset over its operating life including:

- initial capital cost including site costs and fees and mitigating grants
- occupation costs, including energy, letting and rates
- operating costs including maintenance, repair and replacement, security, and insurance
- cost or benefit of the eventual disposal of the asset at the end of its life.

The term is relevant to all building clients whether they will retaining the building or not. It helps also to make a universal connection with value management. Developers who are not retaining buildings for their own use will still need to consider the requirements of those who buy or let from them, but often they assess a building on a short undiscounted pay back scale

(Chapter 3). One of the problems with WLC is that apart from public clients the process may be too comprehensive and provide the wrong information for strategic client decision making.

The advantages of WLC are the better evaluation of the business case in considering the whole picture. There are however a number of *barriers* (see Pasquire and Swaffield; Robinson and Kosky)[13] which include:

- lack of reliable data on components and maintenance life and energy consumption before building occupation
- the separation of capital and revenue budget management which stops an integrated approach so that companies want to reduce capital costs to impress their shareholders
- a short term interest by the client who sells or leases the building on
- taxation and fiscal policy encourages allowances on capital expenditure, but this does not include developers for whom it is 'stock in trade'
- business running and maintenance costs can be set off 100 per cent against the corporation tax immediately it is incurred.

Some interesting figures for the relevant importance of the life cycle elements are given by Kelly and Male.[14] See Table 8.1.

Life cycle costing uses a number of techniques to determine costs, mostly connected with DCF techniques. These techniques assess the present value of later operating and maintenance costs and annualise them in what is called an annual equivalent which can be added to the capital cost. Capital allowance tax, depreciation and inflation/replacement costs can also be considered in the model by the use of a sinking fund calculation and years purchase can be used to calculate how much is needed to be invested now to provide an income for an annual yearly cost. There are many textbooks that deal with these calculations in detail and tables are available to support the calculation.[15]

One of the key issues is the reliability of data for the life of components in the building and for the cost of maintenance, energy and insurance. Where these comply to a trend historic figures are available, but step changes in costs can occur creating new taxes and grants such as energy tax, fiscal changes, introducing new statutory requirements such as accessibility for disability, and updating to keep up with new technology imperatives such as IT networking. In any case new buildings may be designed on for a completely new operating regime and internal figures will not be available. Trusting external figures for costs needs to be properly adjusted to take account of unique business and location factors. Other risks are the level of inflation and interest rates and the possibility of the discount rates not being adequate or being out of step with inflation.

This creates the need for a sensitivity analysis on each of (or a combination of) the assumed factors, to understand those which create the greatest

Table 8.1 Life cycle cost elemental breakdown

Relevant cost	Proportion of operating costs	Possible value measures
Maintenance	Offices 12% Homes for the aged 7% Laboratories 30%	Increasing component lives or easy access to replace
Energy and services	20–35%	There is more scope now to tender competitively and special bulk rates may be possible with direct suppliers. Water is not yet truly competitive. Energy saving methods are popular and environmental
Cleaning	10–20%	Access is a key issue in cost such as access to windows. Owners may also choose easy clean options such as self-cleaning glass at greater capital cost
Security	?	The potential for a secure building depends directly on the design. Significant revenue costs in security staff can be saved by the use of passive devices. There is a trade off between insurance premiums and extra security investment. Intrusiveness is a factor for productivity as employees may resent control
Business rate and capital tax allowances	15–45%	The principle determinant for rates is location. Grants, capital allowance or rate relief is available in regeneration areas. Short term relief is possible from councils trying to attract employers

sensitivity to the overall calculation. There is then a need to factor in contingency to allow for some unknowns on the basis of probability. Table 8.1 indicates some of the global ranges of cost for different elements of operating cost and is helpful in arriving at where to concentrate the sensitivity.

The risk is that all parties will continue to work separately and that long term value will not be considered important in a culture of rapidly changing global conditions. So change will depend upon a powerful communication of the long term vision and a consistent methodology for attaining the goal. Wolstenholme suggests these important points:[16]

- Clients to procure for best value with performance specifications to promote innovation.
- Suppliers to take the lead in creating best value, offering attractive competitive collaborative working which integrates lean processes and to continue to move away from systems that promote lowest price and maximum claims.

Definition and evaluation of risk

Risk arises out of uncertainty. Risk in the context of project management is a realistic approach to things that may go wrong on the site and is used in the context of decision making and in answer to the question, 'what happens if . . . ?' Once a risk has been identified and defined it ceases to be a risk and becomes a management problem. In this context it needs to be analysed and a response made – usually to accept the risk, mitigate it, reduce it or transfer it. People reactions are important and the response to the same risk varies according to who is affected, how many and who is responsible for it. This may also lead to the ethical question of acceptable risk. It is important then to classify risks so that appropriate action is taken. It is also important to separate the source of the risk from the consequence of the risk happening:

- Edwards classifies by *source* distinguishing between *external* risk that is beyond project manager control, for example, interest rates; *internal* risk that is caused by the activities on the project, for example, accident damage, design or methodology; and *transmitted* risk which has an impact on others, for example, environmental damage.[17]
- Burke classifies risk as to the degree of uncertainty, connecting it to the level of available *information* about the project. He introduces a risk continuum from total risk to no risk.[18]
- Flanagan and Norman classify *pure* and *speculative* risk. The former typically arises from the possibility of something going wrong and has no potential gain. The latter is connected with a financial, technical or physical choice and has the possibility of loss or gain.[19]
- Donald Rumsfeld famously classifies our level of knowledge varying from *unknown unknowns* to *known unknowns* to *all knowns*.[20]

The consequence and probability of risk happening help us to decide the priority and assessment of the risk, which is discussed later. The unpredictability of the risk is going to lead to less control, or to more sophisticated assessment methods to increase predictability. Many current techniques used in construction projects do not analyse the degree of unpredictability and therefore do not estimate probability correctly. External events tend to be less predictable and their consequences are likely to be greater, so they cost more when they occur outside the planned schedule. They also affect a broader range of activities and are therefore potentially more disruptive. Ignoring or reducing risk by lowering probability of an unknown is foolhardy. A stochastic simulation or sensitivity analysis approach makes sense in this situation.

Risk can be evaluated by quantitative or qualitative methods or both. Quantitative methods allow a tangible value to be attached to the risk and are suited to weighing options such as investment choices or assessing the greatest probability as in the time taken for a programme. These often use tools such as expected monetary value (EMV), project evaluation and review technique (PERT), or Monte Carlo analysis to provide a range of sensitivity. Qualitative is more holistic and provides a systematic evaluation of a range of factors and their interaction. Tools such as decision trees, heuristic analysis and 'fishbone diagrams' provide a logical framework. In both cases some subjectivity is involved in assigning priority or a value for probability or financial impact as in any of the systems mentioned above. Classification then provides a basis for the proper treatment of risk.

The consequence of risk in terms of its likelihood and severity may help to decide what to do with risk. If it is a very frequent risk, but does not create a huge financial impact, it is likely that the risk may be retained and managed better to minimise the risk. If the impact is larger or harder or more expensive to manage, then insuring it provides a way of spreading the cost of the risk. If the impact of the risk is very large such as an environmental disaster and is more than rare, then it may be expensive to insure and a decision may be made to move away from that activity or to specialise in managing it. The case of asbestos removal is an example of the latter case. Another issue in consequences is how closely it is possible to predict the impact and frequency – is it known? What track record is there in its control?

Theory of constraints

Another approach is indicated by Goldratt's theory of constraints (TOC) which provides a framework for identifying critical gatekeeper bottlenecks/risks for focused priority control.[21] It has some similarities with critical path method, but instead of building in buffers for contingency on each activity which he claims are inaccurate and encourage Parkinson's Law to let work expand to the time available, he builds in project buffers at the end of each chain of activity (50 per cent) so that any run over on the activities takes some of the end buffer. He claims it is then more likely that if individual activities are planned and monitored properly then risks which actually occur have a known impact and can be recorded against the project buffer and forward action taken more easily to control erosion beyond the total risk value of the chain allowed. Critical constraints have to be identified. Active management of time consists of working intensively to reduce bottlenecks that are known to slow the whole production chain. On complex projects there is likely to be several 'feed-in' chains of activity which have 'feed-in' buffers before feeding into the critical path.

Insurance and contingency

Risks are often passed on through insurance, but this is an expensive option as someone else is making money out of project uncertainty and all risks are not insurable. Risks passed on to other suppliers often get passed down the line to the lowest point of competence and this means poor controls.

Response is usually in terms of reducing the risk by management, spreading the risk (as in portfolio management between several projects), avoiding the risk, transferring the risk to others, or insuring against it. It is important to note that 'in the uncertain and fast moving climate of projects' there will always be a residual risk which needs to be managed. Transferring risk to someone who has no particular experience in managing that risk is likely to be expensive. Not all things can be insured against including the result of illegal activity and fines. There are many other areas where insurance costs become prohibitive.

Some insurance, especially for third party injury or property damage, is compulsory and is normally taken out by the contractor. Further insurance such as a bank bond may be requested by clients to protect against commercial failure by higher risk contractors. Professional indemnity insurance (PII) is required by professional organisations as a condition of fair trading for its members to provide an ethical and professional service to cover the impact of human error and negligence. Other insurance may be taken out on a voluntary basis, but the cost of insurance and the limitations of compulsory excess and upper limits need to be balanced sensibly with the cost of managing the risk in order to remain competitive. Insurance blurs the boundary between pure and speculative risk. In cases of rare catastrophic loss, and as part of the management solution to limit liability for loss, a subsidiary insurance company known as a captive insurance company may be created.[22]

A contingency sum is normally allocated to a client's budget to cover their liability for risk such as rock instead of soil to excavate, exceptionally inclement weather and other contractual responsibilities and the contractor will build in contingency for labour and material supply problems, inflation where it is a fixed price etc. These sums have to be tightly controlled so as to remain competitive and tightly planning risk reduction reduces unexpected events as a way of providing increased profits and returns. In collaborative contracts risk can be shared in a project insurance policy to incentivise innovative ways of doing things which help to reduce risk further.

Integration

Any risk management system will cost money to administer and this cost also needs to be monitored with the cost of the consequences. If everyone insures for risk it is expensive and one way of saving money is to have a project risk policy with one premium to pay. New ways of working collaboratively that ultimately reduce risk, such as partnering and repeat working, have the potential to reduce costs. Here the key aspect is the building up of trust between partners to value manage and to resolve to eliminate the risk of dysfunctional conflict. It should also provide the opportunity for synergistic team working and of joint insurance coverage in the core team. Unproductive risk transfer may be reduced in the application of better supply chain management and retention can be used on a voluntary and not a compulsory basis. In this case it is likely that the client will insure for the risk at first tier level instead of delegating as in the case of Case study 4.3 (Chapter 4).

Subjective delayed payment can also be outlawed more effectively where teams are prepared to work together better with automatic payment of suppliers, therefore reducing the risk of non-payment. This broadens the application of risk management beyond the application of quantitative techniques, because the measures to work together comprise a managed risk to trust each other. However where it is agreed it is likely to reduce many risks for the project team and the client.

Risk management process

The process of managing risk starts at identification often by brainstorming ideas as well as referring to any standard list of risks. A risk register records the significant risks and tries to categorise them in order to try and assess them. Significance is assessed by evaluating predictability, probability (frequency) and impact. There is then a response to reduce significance by eliminating, reducing risk levels or transferring risk and, finally, managing the residual risk. The process is shown in Figure 8.4.

Figure 8.4 The risk management process

Risk assessment

The process of assessment involves finding the level of significance to provide some guidance to the seriousness of the risk. This is usually carried out on some sort of matrix or scale. Significance is a product of impact and probability. Figure 8.5 is a simple way for assessing a hierarchy of importance. A rating which is 2 or over would indicate the need for investigation. A rating of 3 or over on Figure 8.5 is made out of impact and is multiplied by probability at three levels.

Risks become more risky the more uncertain they are and this can reflect itself in a subjective rating of probability, as indicated in the expected

Impact Probability	Low	Medium	High
Very unlikely	1	2	3
Occasional	2	4	6
Often	3	6	9

Figure 8.5 Rating significant risk

monetary value (EMV) method in Table 8.2. EMV takes each risk and provides a monetary value for impact and reduces it by multiplying by the degree of probability. Monetary value is another way to quantify the impact and probability is rated 0–1. If this is carried out for all items in the risk register a total risk transfer value of £311,500 is achieved by summing all the extended values.

The bottom line shows a total valuation of risk which, if the risk rating is consistently prepared, may also be used to compare the risk of other options and thus mitigate the decision making in option appraisal where others have less risk. Risk assessment in this context particularly varies with the procurement option chosen. Later occupancy or whole life risk impacts will be better shown as discounted as we discussed in whole life costs.

From a ranking point of view it must be remembered that these figures might be subjective in both the financial assessment of the impact and the probability and can therefore be unreliable, or biased according to the compiler's subjectivity. This might be the case where the compilers want a business case to be successful. Objectivity can be built in by forcing compilers to use standard ranges of values for given common situations,[23] or interrogated through a knowledge based system. A project however is unique and an audit of such a method is important where the affordability and value for money are close to the bone. It is, however, an easily understood way

Table 8.2 Risk assessed by monetary value

	Impact 0–1	Probability (£'000)	Extended value = impact x probability (£'000)	Ranking
Client scope change	0.9	200	180	1
Planning qualification	0.4	100	40	2
Planning delay	0.1	200	20	3
Higher cost of construction	0.4	600	24	3
Industrial action	0.05	500	25	3
Latent defect	0.2	100	20	3
Heating failure	0.05	50	2.5	4
Total monetary value			£311.5	

of assessing risk and the project team can own the results and easily explain the impact and probability figures. Probability figures may be backed up by normal distribution and experience, but the final probability should be adjusted for the riskiness as discussed above. For example, predictability is affected by the distance of the event away from the time assessed and the experience of the contractor or client in dealing with that risk.

Risk attitude is important in how the response is managed as we have seen. A defensive attitude to risk is to add a risk premium to the tender, if the risk is sensed to be abnormal to normal business, or to hold retention or to require bonds on a contractor if the contractor is unknown. Risk transfer may be defensive or good business. Reducing risk is part of the business of adding value to the contract. The narrower view involves treating every risk as a pure risk. Speculative risk might provide innovation and value for the project or it might be one risk too far.

The opportunities to change risk attitudes come with integrating risk and value management, so that there is a positive motivation factor over and above the level playing field of hygiene factors. The impact of this is to encourage sensible risk in the area of innovation and to build in systems that properly plan the risk and other issues such as quality and environmental response. Good practice sees risk management as an asset of good business rather than a cost to the overheads.

Risk assessment then needs experience for the project manager to make executive decisions, coloured by their attitude and their ability to influence the client and the project team. In collaborative projects and other negotiated tenders there is a chance to develop a common approach to risk and to share it in the most profitable way.

The risk register is most commonly seen in the form of a list of prioritised risks showing separately a score for probability and a score for impact. There are columns for mitigation, risk responsibility (allocation) and management of residual risk. There may also be a monetary evaluation of the risk. The risks need to be categorised and prioritised. Priority may be shown by using a traffic light system with red for the higher priority risks which must have mitigating action; orange for those risks which must be monitored for the impact of further change to the project; and green for those risks which are low risk or have passed their impact period. To be useful the register must be regularly updated at a progress meeting during design and construction.

Risk attitude

Risk attitude has been classified by Flanagan and Norman as either:

- Risk loving
- Risk averse
- Risk neutral[24]

It is popular to cite bold risk taking as a feature of successful business, but in the context of a project there is a greater tendency to pass risk down the line. The risk allocation established in traditional contracts is shared. Many clients are taking a risk-averse view and pass related building risk down the line to contractors by using different forms of contract. This can be successful where the real estate and technical decision making is also delegated. In PFI, DBFO and BOOT a facilities solution is offered and agreed for a substantial period of time so that the client can concentrate on their core service and/or profit making.

A client who views their building as an investment option will apply a risk premium to their returns, if they feel that they are taking more than normal risk. A client will carry out an investment appraisal to look at the affordability and returns of a real estate investment and the returns which are likely given the market predictions. A risk premium can be applied through a higher discount rate (Chapter 3), which will emphasise the importance attached to a short payback period and uncertainty in the future. This will raise the hurdle return on investment by which a project's viability will be judged.

In a similar way the contractor will lower the tender price at the tender evaluation stage in order to be more competitive for a project, or raise the price to cover the perceived higher risk. 'Risk loving' contractors may actually see an opportunity in the higher risk project to make more money. The danger with a traditional form of construction procurement is that the client perceives the burden of risk has been moved to the contractor on a lump-sum fixed price tender, whilst the contractor perceives that those parts of the building not fully detailed provide the opportunity to bid low and recover costs on later variations.

A risk neutral approach is an approach that could be interpreted as a zero sum in discussing risk allocation to benefit all parties and put risk where it can best be managed. This requires an open attitude and the development of an associated value management approach which in particular makes the early appointment of contractors in the development stage of the contract essential. There is nothing easy about this type of approach; it requires more effort and more early commitment, often at a stage prior to finalising contracts.

The choice of procurement systems introduces its own risk allocation and this will be discussed later. This may also be associated with value management as a process tool.

Risk response and management

Risk response occurs to eliminate, mitigate, deflect or accept the risk,[25] and logically will reflect the cost benefit of the risk management process. Mitigation is action taken to reduce the risk and deflection is action taken to transfer the risk. They are not mutually exclusive, but deflection alone is not a way of reducing the probability. Mitigation may have the effect of reducing probability and impact. Eliminating risks is often not economic, or creates too many other risks. Certain pure risks such as health and safety will have legal or social expectations for mitigating them. Managing residual risk is part of the process of risk management. Clients may:

- retain the risk in which case they have to assess, reduce and manage it
- transfer it so that that someone else can assess it and manage it; this may well appear as a premium in the price for the services provided so this is best done where there is experience of this type of risk otherwise premiums will be expensive or control will be chaotic
- share the risk with another party so both gain or pay for the outcome, but there is more control
- ignore the risk, which is not recommended unless the cost of controlling this risk is more than any possible impact and here a commercial position may be taken as risk management is costly.

Flanigan and Norman refer to certain rules for risk taking such as 'Don't risk a lot for a little, don't take risks purely for reasons of principle or losing face, never risk more than you can afford to lose, always plan ahead, consider the controllable and uncontrollable parts of the risk and consider the odds and what your intuition and experience tells you'.[26] They also refer to the need to gain knowledge about the risk and to remove ignorance. It is in any case unlikely that a risk will be eliminated completely without creating other risks. Ethical and social considerations should be taken into account as well as the business economics in deciding what level of residual risk is acceptable. The reputation of the client and the business is also at risk.

The difference between pure and speculative risk is important in the response. Speculative risk will be accepted in order to take on an opportunity. Success in this case will be in the identification of the risks accepted and mitigation of those that are not acceptable. Here a business may build up its reputation by taking the right risks and managing them successfully.

The transfer of risk by the client is a factor of the procurement method and further risk may be transferred to the contractors in the negotiated contracts so that value for money in allocating the risk to the party best able to manage it is gained. It may be further transferred through passing it down the supply chain (as in asbestos management), by insuring against it or by raising a bond. If the client ultimately wants reliability the use of inclusive maintenance contracts should also be considered.

Contractors, designers and clients will also put different priorities on the same risk. For example the risk of an overrun in budget because of design changes is a risk that will most affect the client in a traditional contract who is likely to pick up the bill for any design and scope changes. Design changes can mean extensive reworking of design for limited remuneration to the designer too. It also affects the designer's reputation if there are too many unexpected changes for which there is little to balance against elsewhere. The contractor will also have to suffer disruption if the scope is changed late with many knock on effects for resource management and abortive work. These disruptions may be absorbed, or a decision made to claim for additional payments. Likewise weather conditions are mainly a contractor risk, but an extension of time affects all parties.

Risk opportunity is connected with speculative risk and this could be illustrated with speculative building developments where there is a chance to make a lot of money where others are not prepared to take the risk to lose it. So you buy land at a low cost without planning permission and sell high when you get the permission and, assuming current land prices, then the risk has been worth while. However if planning is unduly delayed or completely refused, you are likely to not be able to sell the land for what you bought it for and money is lost in the sale and maintenance of the asset. Other opportunities may arise in the tender process for competitive work where you bid low to get the work and keep a good turnover. However if a standard amount of risk goes wrong you are likely to limit profit. In the case of having less than average risk going wrong then opportunity may have been gained which gives opportunities for future work.

Risk management techniques

Many textbooks have been written covering different risk techniques and here it is proposed to summarise the techniques which best fit in with a combined risk and value approach. An example of its application is given, but this is not an exclusive use of the method. Table 8.3 is split into speculative and pure risk methodologies.

Table 8.3 indicates methods used by clients and contractors in assessing their exposure to risk in assessing whether to take on a project and is concerned with the risk of costs including overheads exceeding the returns that may be had from investing the money in bonds or less risky ventures and needs to show economic profit over and above that. It requires entrepreneurial skills over and above the risk management techniques to maximise opportunities, so on their own these techniques might help to compare options or to support a business case or a key market evaluation. Tender adjudication depends heavily on the market conditions and their predictability is often assessed by speculative risk assessment techniques. Major innovative methods and life cycling risks may also be judged to come in the speculative category as they present opportunities which over the extended period of the life cycle are less predictable.

Table 8.4 deals with risk management and assessment techniques during the project.

Table 8.3 Risk assessment methods for speculative risk

Title	Risk approach	Example/application	Connection with VM
Investment analysis	The methods are described in the business case chapter and may be assessed on payback, net present value or through building in a risk premium into the client's 'discount hurdle rate' using IRR or breakeven. More confidence is achieved by also carrying out a sensitivity analysis	Assessment of financial payback associated with balancing capex and opex, with income received from a building over a period of time	Excellent use for assessing financial risks involved in comparative schemes or assessing costs of an alternative maintenance regime for WLC
Sensitivity analysis	This is a mechanism for testing if a small change in one of the expected variables has a disproportionate effect on the overall result. More sophisticated analysis can also be applied such as multiple regression and multi-attribute value theory	Sensitivity analysis can identify the point at which a change in an expected variable value changes the decision to proceed. Thus if the capital cost of energy insulation was to go up by 20 per cent then it might facilitate against the additional insulation by cancelling out discounted savings in heating costs or making the payback too long. Each of the variables can be plotted on a graph of life cycle cost and percentage variation in parameter	This has obvious usage in WLC and may help to establish the evidence for the proposed value for money and the optimum balance between capex and opex
Portfolio management	Portfolio management is diversification of investment to mitigate the risk of loss if one market collapses. By definition it has less potential for premium returns. It is applicable for managing the risk of poor returns and emerges from financial planning. Diversification does not work if the whole industry is affected by adverse conditions such as poor demand for buildings. Niche markets are another variation in looking for opportunities and exploiting them to balance poor returns	Portfolio management may be used by the contractor to spread their risk of loss by taking on a range of projects, so that if work dries up in the leisure market then there are other income earners say in housing. Business wise this can cut both ways as projects are unique and an even wider diversification of skills cuts down efficiency and hence margins may be affected	This is not really a value management tool, but it may be useful to the client in choosing between projects

Table 8.4 Risk assessment methods for risk management during the project

Title	Risk approach	Example/application	Connection with VM
Expected monetary value (EMV)	A risk register is compiled and a value is attached to the consequences of things going wrong. This is illustrated above. A decision needs to be made as to what premium is added to the project	Use in a public sector comparator in a PFI contract and in comparison of different methods of procurement	Alternative schemes can be compared or methodologies to cut down the degree of risk or the probability on the risk register
Decision analysis	A flexible technique for structuring decisions into 'stages' and to apply weighted probabilities for alternative choices. It provides a method which takes into account related decisions and gives flexibility to allow for risk attitude and subjective impressions. Stages can be determined with a means-end chain and there must be a similar number of stages for alternative decision chains	Use to look at probability and cost of say alternative types of river crossings, such as bridge, ferry, tunnel, do nothing. Stages might be 1) which one? what type? 2) when start? how long? 3) how much?	This is very valuable for giving a integrated VM/RM comparison for each of the expected monetary values that have been devised. Discounting can also be applied if there are ongoing costs
Theory of constraints (TOC) Goldratt	Applied to project management. Goldratt based his project application on identifying the weakest link where there is most risk and focusing all management effort on these in what is called the critical chain. This prioritisation is reached after project wide workshops which link to value management and stakeholder management	This has been used holistically to identify major risk of the weakest link in a critical chain of programmed events for major rail infrastructure projects that are logistically demanding and fast track,[27] and to manage it closely on the basis that last past the post is the critical factor for completion of the whole	This method brings value in itself by saving on management time and releasing them for value adding activities
Monte Carlo simulation	Stochastic probability model. Generates sets of random numbers for each of the variables within a range and compares the cumulative outcome (e.g. cost, time) on a frequency basis for the total of each of the random models created. This is most powerful	The PERT model below is a Monte Carlo simulation making a model for project time. The highest frequency of project time estimates is represented as the best estimate of time	Simulation may be limited because it is difficult to quantify functional requirements or clients' values, but it has a role in assessing WLCs

where the number of iterations (simulations) generated reaches a large number, making it statistically significant

Project evaluation and review technique (PERT)	PERT is used on schedule planning. It is a probabilistic model based on an evenly distributed frequency distribution curve with a weighting of most likely, m = 4 optimistic, a = 1 pessimistic, b = 1 The equation Expected time/Cost = $(a + 4m + b)/6$ is built into a computer program based on a range of values for each activity that are usually weighted towards the pessimistic end. There is a level of subjectivity in choosing the range and its distribution around the mean Frequency — Time or cost	Pertmaster have developed a risk analysis which is able to produce a Monte Carlo simulation generating a large number of combinations of activity time or costs within the range for each activity. The most probable overall cost or duration is given together with the range of values for a given standard deviation. Pertmaster generate this on a triangular distribution	This is mainly a scheduling tool, but could be used to test the programme, or cost certainties of different designs
Multiple estimate risk analysis or root mean squared technique	This method is based on calculating a risk free base estimate and an explicit calculation of a risk allowance rather than an arbitrary percentage for contingency just based on experience. The risk allowance is calculated by taking the difference between the maximum (90 per cent probable not to be exceeded) cost of each particular risk and the average likely (50 per cent probability) cost of the risk, taking the square of that difference in cost for each risk, adding the squares for all risks and taking the square root of the total. The result is a base estimate, an average likely estimate including risks and a maximum likely estimate	A project approval process might require the maximum likely estimate and does not exceed the average likely estimate by more than 15 per cent to receive project approval to proceed to the next stage. The method has the benefit of identifying the extent to which risks have a fixed or variable total cost as well as their probability and cost impact	The process of identifying the risk improves project understanding and clarity of the brief. The likelihood of successful delivery is improved by making potential risks more explicit

Table 8.4 shows other methods of risk management which can be used at the project definition and implementation stages of a project in order to differentiate between different alternatives. They have more relevance to the value engineering approach and can also help in managing the project in the cost and time control areas.

The challenge for RM in construction is therefore to challenge and encourage collaborative working, to develop a 'no-blame' culture of openness and honesty and to work within procurement strategies that identify and share risk according to the ability to manage and accept those risks, rather than passing the risk 'parcel' down the line to the weakest member, often the sub-contractor. The difficulty for RM is that much of its techniques are based on mathematical analysis of probability and come from the 'hard' engineering side of the industry. Whilst techniques such as multiple estimate risk analysis have a mathematical rigour, their application and the presentation of their results is often viewed by clients and fellow consultants as confusing and overly complex when they are looking for clear direction and certainty in their project advice. But a similar problem has been experienced by the medical profession in trying to explain risk in surgical treatment, vaccines and susceptibility to diseases. The awareness and understanding of risk needs to be raised in education and professional training and therefore how to develop this 'soft' area of managing culture change with the acceptance of 'hard' engineering based techniques is a fertile area for research.

Another issue is how to achieve the balance between the independent RM facilitator, who may be managing risk in an area outside their particular technical expertise and needs to charge a separate consultancy fee, with the in-house risk manager, where RM is provided as part of a wider project management service. Is RM an additional service to be paid for by the project client or part of the service of successful project delivery by the project design, construction and possibly operation consultants and contractors? Attitude certainly has some bearing on it.

Risk, value and procurement

We have already referred to procurement types in the chapter on strategy. The continuum of risk and value assessment is carried out at feasibility, for example development appraisal, design, tender, construction, commissioning and fitting out, occupancy and life cycle. One of the key factors is the allocation of risk by determining the procurement system to be used. This should match the client's preferences and risk represented by the procurement type is most easily represented by Figure 8.6.

Figure 8.6 and Table 8.5 indicate that the client has a wide choice of risk strategies, but the golden rule is to allocate risk responsibility where it can be managed best. The headline is that clients are looking to deflect risk

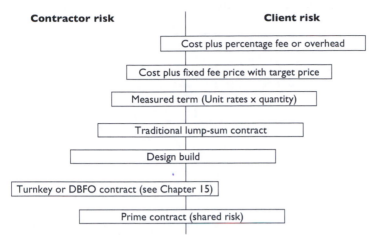

Contractor risk   **Client risk**

Cost plus percentage fee or overhead

Cost plus fixed fee price with target price

Measured term (Unit rates x quantity)

Traditional lump-sum contract

Design build

Turnkey or DBFO contract (see Chapter 15)

Prime contract (shared risk)

Figure 8.6 Procurement and the allocation of risk

Adapted from Flanagan and Norman (1993)[28]

because of their experience of having to pick up the 'tab' on traditional lump sum contracts, but by less involvement this may take away more opportunities to create value. Another option is to use a shared risk approach like prime contract that has a built in pain-gain incentive which is associated with a no blame culture. DBFO contracts often have a chance to negotiate risk at a very early stage of the development life cycle and the relationship is extended (especially in the PFI format over a long post construction occupation phase).

A professional judgement should be made on the tender price. However, the client chooses the degree of risk they take, by the degree to which they require the contractors to guarantee their final price. Changes are inevitable both in detailed design changes and in scope and the risk here is to be managed.

The latter is the reason why some experienced clients have chosen to go with construction management, or design and manage, bringing in contractors early and bringing the whole team round the value management table. Clients have more control over their projects, but take more risk. Design and build has taken most of the risks away from the client as there is a clear contractor responsibility for the whole design and construction process, but risk arises from change. In turnkey projects residual building management risks and capital funding and sometimes operation of the facility risks have been handed over to others. This has not taken away the value management process as it has all happened up front around the negotiation table. Prime contracting represents a risk sharing approach where there is a particular concern for innovation and developing a value solution. This is often

Table 8.5 Table to show comparative financial risk to client of different procurement types

Type of contract	Client financial risk	Measures to control budget
Cost plus	High – any factors Difficult to track expenditure and so control it except by limiting extent	Daywork rates
Measured term	Medium–high Rates control prices, but many rates may not be covered. Work can be limited if budget exceeded	Measured rates + inflation allowance
Traditional lump sum	Medium Quite a wide range of additional rates e.g. design changes, unforeseeable events and scope changes inevitable and push up the 'lump sum'	Contingencies, provisional sums and dayworks
Design and build	Low if no changes Contractor takes on most risks for unforeseen changes and for aspects of design development not to do with scope change	Re-price any scope changes on the basis of known rates
Construction management (Construction management or Management contracting) Design and manage	Medium Specific managed budget and package procurement. Unforeseeable events still applicable, but more integrated design and construction than lump sum so opportunities for initial value management. CM takes no risk, but may have an incentive contract to share savings	Incentive to share savings
DBFO or PFI	Low Best certainty once figure is agreed. The risk is that the negotiated financial close with the preferred bidder may be higher than expected and client is committed. However this figure is regulated for 25 years	Affordable service charged paid yearly by the client, but may limit facility in the long term
NEC contract and partnering Contract (PPC2000)	Medium Better culture for working together to contain budget, but no guarantees against poor planning and unforeseeable events	Use of compensation events to avoid culture of blame. Full access to contractor accounts. Culture of continuous value for money improvements

associated with long term strategic partnering where relationships are built up, synergy established and the learning process is not started all over again for each project. However a client may make different decisions based on the type of project and the degree of control that is perceived necessary. Clients may have a risk avoidance attitude and this may be tied up tighter in the type of contract chosen, or opened up as in the case of the NEC contract and the partnering contract.

Risk behaviour

Risk behaviour is an interesting subject and a lot of research has been done in the area of health and safety risk, for example driving safety and the area of gambling risk, the latter perhaps being more akin to the risk management of projects as a whole. Greenwood has measured a difference between the degree of risk we take personally and that which is taken on on behalf of others.[29] This influences decision making in the area of investment, project choice, the degree of innovation and planning, and the response to external risks.

The context of the risk is very important to the behaviour displayed. Bidding theory identifying the success of 'work hungry' contractors indicates the risk of the lowest price being a loss leader to get into a market or to retain a minimum workload.[30] This in its turn may lead to contractor claims to recoup loss. At the other end of the scale contractors may collude where there is much demand to share out the work at higher prices. Client risk is increased in both of these dysfunctional events for time cost or quality and risk management needs to be aware of the context.

'Group think' indicates a greater willingness to ignore risk in coming to group or committee decisions where an established group converges their members' beliefs by familiarity and starts failing to see the wider picture. For example, the decision to use the nuclear bomb in World War II could have emerged as a 'group think'. It might be thought that projects are unlikely to fall into this category due to the formation of fresh teams, but industry or contract cultures are created that reinforce contractor claims, project overruns and less than excellent performance, creating risk for not meeting project objectives, generally by assumptions that 'there is no other way'.

With health and safety Wilde uses *risk homeostasis theory* (RHT) which suggests that people compensate if there are technological improvements to machinery or environment to experience the same level of risk (target risk) indicating that it is not easy to change someone's risk behaviour without changing their attitude.[31] Familiarity with a task is likely to induce less caution than the approach to innovation. Safety training can for example have a negative effect on behaviour if it induces more confidence and therefore a greater willingness to engage a risk, because of supposed familiarity with

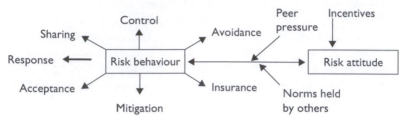

Figure 8.7 Relationship of risk attitude to risk behaviour

the outcome and a loss of fear. Attitudes on the other hand are created and are associated with say a greater willingness to use risk management methods, or to avoid certain things altogether. Their correlation to behaviour can be weak as other factors such as norms and peer/client pressures will have an influence on behaviour. However these attitudes can be iterative with behaviour and make significant changes in behaviour over a period of time, if there is motivation. Reinforcement such as incentives to change or the opportunity to experience benefits can change beliefs/values. If this motivation reaches a critical mass for a project the ingredients are in place to make a change in culture. For projects, ongoing relationships will help to establish norms and positive peer pressure.

Risk attitude is particularly important in how a risk is responded to. Figure 8.7 indicates the influences that attitude puts on behaviours.

Appointing consultants and contractors

Consultants are appointed by the client or the project manager. Traditionally construction consultants are selected on a quality basis for a given fee and contractors are selected on a price basis against a detailed and prescriptive project specification and drawings or pricing document. The trend towards value for money means that both could be considered on a quality and price matrix. The quality issues may be selected by the client, but there is a responsibility on the part of the client to have a selection process that is clear, consistent and complete. This ensures good competition, a basis for a reliable selection process, and builds trust with the service providers. The award of a project to a consultant or contractor needs to be transparent and fair and as such standard rules of appointment and tender are available for public and private tenders. The rules allow for a choice of procurement, but lay down certain basics. In the EU compulsory purchasing rules have been standardised for public works projects in the context of breaking down trading barriers and creating a level playing field. Private clients will benefit from using standard procedures as these will attract competition, but may also make known bespoke rules, which fit their situation.

EU purchasing rules

The EU purchasing rules identify open, restrictive (selective), negotiated or competitive dialogue tenders as three possible approaches to the appointment of contracting or consultancy services.[32] All public tenders over a certain threshold size and need to be advertised in OJEU and are listed on the Tenders Electronic Daily (TED) service to ensure maximum exposure. The open tender is not recommended in construction because of the extensive work involved in preparing tenders and evaluating them. So the restricted method provides outline information about the project and invites expressions of interest. This creates the facility for pre-qualification of the tenderers. The restricted or negotiated tender method can then be operated with a shorter list and tenders invited. Good practice requires that the lists are restricted to no fewer than three up to no more than six. Prices are made on lowest or most economically advantageous tender (MEAT).

In the case of consultants the CIOB recommends no more than two consultants on the short list as these come under part B of the regulations with less jurisdiction.[33] However in any one situation a justification may be given for expanding this list. In negotiated tender the aim is to carry out a second stage to the tender which develops the project value for money. This provides the best of competitive and negotiated systems, by ensuring more involvement of the contractor in the development of the project definition.

The competitive dialogue is normally used in complex projects such as PPP where there is a performance specification that needs to be proposed for client agreement and best value. After client agreement a best offer is made to select the winning bidder. Very little variation is allowed after this in the spirit of competition, but large contracts still need to move to a financial close with a preferred bidder.

Private appointment

Private clients may use the NJCC Conditions for single or two stage tender. There is also a Construction Industry Board Code of Practice for the selection of consultants and main contractors.[34] The DTI Constructionline is an example of a pre-qualified contractor and consultant list.[35]

Consultants are also classed as service providers and are appointed according to the standard industry codes (e.g. RIBA, NEC, RICS or ACE). Selection may be entirely competitive (fee bidding), but it would be unusual not to appoint key consultants with reference to quality factors such as their reputation and track record. This brings in an assessment of their experience and ability in the specifics of the project. Selection techniques may include design competitions, viewing previous work, or proposals presented during an interview with the team expected to provide the expert service. This provides a better forum to assess expertise, compatibility with client objectives and the suitability of consultant experience. Project managers, managing contractors and possibly main contractors also fall into this category.

Clients have to develop their strategy in appointing consultants by deciding:

- Do they want one appointment who contracts with the other consultants and co-ordinates feedback as through a project manager or do they contract with a circle of consultants who will provide them with direct advice that they manage?
- At what level do they delegate authority for decision making?
- Is the appointment for the whole project life cycle, ongoing phases or on a call off basis?
- What scope of duties is required of each consultant and do all the consultant briefs cover all areas without overlapping duties?
- Do they pay consultants on a time basis, lump sum or *ad valorem* – according a percentage of the value of the project? (Clients are asking whether this latter method in simple form does in fact motivate value for money.)
- What level of professional indemnity or collateral do they require? This relates to the level of risk which they carry in the case of mismanagement of the project and poor advice.

The procurement strategy is influenced by these decisions. Other techniques such as partnering, dealing with conflict and integration overlay the procurement method.

Selection process

We have already spoken of the need to have clarity, consistency and completeness in the selection procedure and this means that the project brief and scope must be unambiguous and where there are queries that a process of providing answers available to all parties simultaneously is provided by having regular mailings or a mid-tender briefing session, to deal with queries and emerging issues. Best practice requires that a quality threshold is established especially where a large number of bidders are likely to be interested. This provides the clear criteria for a select tender list and for short listing. According to context (public or private) this may be pre- or post-interest qualification. Quality thresholds may include client values such as a wish to see certain environmentally friendly project policies, the existence of a third party quality assurance scheme, technical suitability, resource availability, financial stability, whole life costing and performance in partnering – these are just a few. It is necessary to be client organisation specific, to reflect client values and past experience.

The price, of course, is important and provides an awarding mechanism once a minimum quality threshold has been achieved. It is obvious that where the quality threshold is only judged after appointment on the basis of price, the appointment is subject to risk of greater conflict as it depends upon contractual interpretation and possibly expensive litigation. In order to choose between price and quality it is important to determine the weighting between the two.

The balance of the WLC factor against capital cost has already been discussed, but is much more dependent on what type of client is going to use the building. Another factor in selection is the cost of tendering, which is often associated with the number of tenderers and the complexity of the tender information asked for. If this includes substantial architectural design elements, such as in design and build, then the cost goes up and contractors will choose to bid more strategically in order to boost their chances to win and recoup more of their tender overhead. Clients on the other hand may induce competitiveness as well as design control by appointing a concept designer who novates to a design build contractor at the scheme/detail design stage.[36] More care in the selection stage is needed to match the design and build contractor and the designer.

Award of contracts

The award of a project on price and quality involves choosing the best value contractor or consultant and needs to have fair and tangible final selection criteria. One way of doing this is to use a balanced scorecard, comparing contractors by providing a weighted score against each of the criteria for the quality scores. The weights must add up to 100. These criteria are relating directly to the tendered information and interview processes and are scored on a percentage and multiplied by the weighting of the criteria. In Table 8.6, if functionality was scored 65 and weighted as 50 per cent amongst three other criteria at 25 per cent, 12.5 per cent, 12.5 per cent respectively, then it would receive a score of 65 × 50 per cent = 33. The others would then score as the table to give an overall quality score out of 100 of 56 for Organisation A.

Table 8.6 provides comparative data which combines the quality and price ratings. Much depends on how you rate the criteria, quality and price to provide a weighted score for the comparative prices. This should be standardised for specific types of project and client values, according to the tightness of the budget. The more criteria used the more likelihood that no criteria will make much difference in the quality scores.

It is therefore important to check out the calculated outcomes with other methods such as a face to face interview and to ascertain the validity of any risks that you have specifically asked the tenderer to price, which are not to be expected. These will give further clues to the validity of a tenderer with

Table 8.6 Illustrative example of a balanced score[37]

Quality weighting 60% Price weighting 40% Quality threshold 55%		Org A			Org B			Org C		
Quality criteria	Criteria weight %	QT reached	Score	Weigh- ted score	QT reached	Score	Weigh- ted score	QT reached	Score	Weigh- ted score
Quality scores										
Functionality	50	yes	65	33	yes	70	35	yes	90	45
Programme	25	yes	45	11	yes	50	13	yes	60	15
Risk manage	12.5	yes	45	6	yes	30	4	yes	60	8
Environment	12.5	yes	50	6	yes	85	11	yes	65	8
Totals	100			56			63			76
Price scores										
Tender price		20,000,000			27,000,000			22,000,000		
Price score (av. £23m = 50)		80			30			60		
Overall Scores										
Quality weighting x quality score		60% x 56 = 33.6			60% x 63 = 37.8			60% x 76 = 45.6		
Price weighting x price score		40% x 80 = 32			40% x 30 = 12			40% x 60 = 24		
Overall score		65			49.8			69.6		
Order of tenders		2			3			1		

experience. You may also differentiate between essentials and desirables in the scoring.

Another way of dealing with dual criteria is to use the two envelope system which means that quality data is evaluated prior to the price data in order to assign additional weight on those contractors who have made excellent quality before the price creates a bias to those with a good track record. This illustrates the importance of changing cultural norms to bring about value for money over the life cycle of the project. A public body will have to work harder to demonstrate objectively that value for money has been achieved where the cheaper short term solution was not value for money.

Notification and conditions of engagement

Most consultants expect to be appointed on standard terms and conditions such as RICS and would price differently for the risk involved in accordance or more for non-standard conditions. The basic conditions should refer to the scope of the work and the responsibilities of the consultant. It is good practice for separate consultants to be aware of each other's conditions and for there to be consistency of style between them and to ensure that the team of consultants can work together.[38] Post appointment meetings need to negotiate the practical outworking of conditions and need to check the PII and/or all risks insurance, bonds, warranties and other formalities, which are part of the risk allocation procedure. Further checklists and detailed information about appointment can be researched, for instance from the CIOB code and the CIRIA guide on competitive value based procurement.[39]

In some cases private clients will be only interested in selective lists based on recommendation, or may wish to further partnering arrangements, which share work between a limited number of framework consultants or contractors. Partnering should be clearly delineated from pre-qualification, which provides opportunities for pre-qualification for each tender, whilst strategic partnering pre-qualifies for a period or series of contracts, to build up collaborative working.

Conclusion

The various tools and techniques that underpin both value and risk management are well established in textbooks and used where collaborative working within an integrated supply chain is established or has been strongly promoted. The traditionally fragmented and rather adversarial nature of the construction industry, particularly with its separation of design and construction, has been both an opportunity and a constraint on the development of VM and RM in the UK. Conversely, using these approaches within project management can also provide a means to encourage and facilitate change. However, both are still largely paid lip service, rather than being applied as distinct project services with their own distinctive and robust methodologies.

Wolstenholme has urged clients to procure for best value with performance specifications and early contractor involvement to promote innovation.[40] He urged suppliers to take the lead in offering best value by offering attractive collaborative working which incorporates lean processes to move away from lowest price and maximum claims. It is also clear that risk management has developed a two edged approach of looking at the opportunities and the threats. This has been clear in the more recent editions of leading risk management guides such as Chapman and Ward.[41] This has helped to bring it closer to the concept of value management and it now makes sense to view the techniques as two sides of the same coin. In considering value, WLC has always been a consideration, but more development of long term procurement relationships such as PFI and maintenance agreements with new build has given the incentive for the use of the technique and better accessibility to reliable life costs. Wider use of benchmarking for operating costs has given confidence in its use for long range decisions, but is still in development. The use of readily available IT has also made more sophisticated techniques such as probability and sensitivity analysis viable, but to better support and not make the decisions.

RM or VM are being seen as a key way of assuring *quality* in construction, so that contractors and consultants are selected on a balanced scorecard of quality and price and risk control and not lowest price to create value breaking claims at a later stage. Value for money is seen as identifying a broader range such as competency of consultants and providers, sustainable and environmental design, ability to finish on time, better life cycle costs and energy consumption, promoting considerate contractors and better and safer products. These can all be built onto a matrix for selecting contractor and consultants able to control risk and value by bringing the client, the designers, contractors and other stakeholders together at an *early* stage of the life cycle.

A higher capital cost due to a better whole life costing may be a better solution. PFI or 'sell and lease back' are examples of solutions that spread the capital cost of building over the life cycle when significant related savings can also be made. The issue of longer term risk management then becomes critical too, for example, in assuring that the cost of financing does not exceed the value gained.

The pitfalls of risk and value management arise out of the late or inappropriate use of these techniques, or the wrong interpretation of client values. Risk evaluation depends upon subjective judgements of impact and probability; value management may have optimistic future savings. A low level of predictability is often ignored as a significant risk, but is generally associated with the unreliability of long range 'guestimate' material. Mechanisms for removing or testing bias for whatever reason (client or consultant) need to be in place, but exclusion because the factors are financially intangible is more dangerous than holistic, systematic, experience based decision making. The latter method can be justified by a constraints analysis which identifies the 'gatekeeper' risks and values which are critical to determining the global risk and value. For example, a 30 per cent cheaper museum is going to be doubly ineffective if the visitor figures realistically will still not support the income to pay the finance and extra running costs which might arise. More fundamental value and risk management is required at the business case stage, perhaps involving the securing of grants, moving location, and cheaper entry charges to ensure patronage.

Two other drivers could also encourage wider use of VM at the earliest strategic stages. These are the increasing push for sustainable development and corporate social responsibility. Both require engaging with a wider stakeholder group, where identifying and agreeing mission statements and strategic goals will require more robust and more transparent decision support mechanisms, including identifying value systems to inform both strategic and more tactical option appraisal to arrive at best value solutions. Developers are increasingly being asked to justify their planning applications in terms of a sustainable development assessment. This requires considering community impact as well as economic and ecological. Clients increasingly want engagement with stakeholders from the wider community, to make their case for sustainable value to the planning authorities.

At the heart of Constructing Excellence is the use of KPIs to measure key quality and value criteria and drive a process of continuous improvement. Continuous improvement is the required benefit to the client as part of the new partnering agreements. Will VM provide the mechanism to facilitate this continuous improvement and will the culture of partnering encourage more 'buy-in' by clients and contractors to the concept of using VM to improve their service, whether or not initiated and paid for by the client? KPIs can follow up in monitoring predictability assumptions made.

On small projects such techniques are costly in terms of person hours and the cost of specialist advice is prohibitive. Here the use of simple but effective heuristics may be a better approach with the direct involvement of stakeholders. This is why EMV has been illustrated. A culture of monitoring and control should also be in place as failure costs have more impact on a smaller budget.

References

1 Acknowledgement for some material in this section from P. Fewings (2005) *Construction Project Management* Edition 1 to Tony Westcott.
2 Kelly J. and Male S. (1993) *Value Management in Design and Construction: The Economic Management of Projects*. London, E & FN Spon.
3 Emmerson (1962) *Survey of the Problems before the Construction Industries*. London, HMSO; Banwell Report (1964) *The Placing and Management of Contracts for Building and Civil Engineering Work*. London, HMSO; Latham M. (1994) *Constructing the Team: Joint Review of Procurement and Contractual Arrangements in the UK Construction Industry*. Final Report. London, HMSO; Egan J. (1998) *Rethinking Construction*. DETR. London, HMSO; Fairclough J. (2002) *Rethinking Construction Innovation and Research, A Review of Government R&D Policies and Practices*. London, DTLR.
4 Connaughton J. and Green S. (1996) *SMART Technique*. London, CIRIA.
5 Davis Langdon and Everest (2003) *Creating the Conditions for Success*. Davis Langdon and Everest. June.
6 Dallas M.F. (2007) *Risk and Value Management*. Oxford, Blackwells.
7 Romm J. (1994) *Lean and Clean Management: How to Boost Profits and Productivity by Reducing Pollution*. New York, Kodasha Amer Inc.
8 Hughes W., Anstel J., Hirst A. and Gruneberg S. (2004) *Exposing the Myth of the 1:5:200 Ratio Relating to Initial Cost, Maintenance and Staffing Costs of Office Buildings*. ARCOM. Available online at: http://www.arcom.ac.uk/publications.html (accessed 12 March 2012).
9 Wolstenholme A. (2009) *Never Waste a Good Crisis*. London, Constructing Excellence.
10 Wolstenholme A. (2009) *Never Waste a Good Crisis*. London, Constructing Excellence, p.25.
11 Martin J. (2009) *Life Cycle Costs and Sustainability*. Presentation to Ecobuild 4 March 2009. London, Building Cost Information Service (BCIS).

12 RICS (1999) *The Surveyors' Construction Handbook – Life Cycle Costing.* London, Royal Institution of Chartered Surveyors.
13 Pasquire C. and Swaffield L. (2002) 'Life cycle/whole life costing'. Chapter in J. Kelly, R. Morledge and S. Wilkinson (Eds) *Best Value in Construction.* London, RICS Foundation/Blackwell Science; Robinson G. and Kosky M. (2000) *Financial Barriers and Recommendations to the Successful Use of Whole Life Costing in Property and Construction.* London, CRISP.
14 Kelly J. and Male S. (1993) *Value Management in Design and Construction: The Economic Management of Projects.* London, E & FN Spon.
15 For example, HM Treasury *The Green Book.* London, TSO.
16 Wolstenholme A. (2009) *Never Waste a Good Crisis.* London, Constructing Excellence, p.22.
17 Edwards L. (1995) *Practical Risk Management in the Construction Industry.* London, Thomas Telford.
18 Burke R. (2003) *Project Management: Planning and Control Techniques.* New York, Wiley.
19 Flanagan R. and Norman G. (1993) *Risk Management and Construction.* Oxford, Blackwell (RICS).
20 Rumsfeld D. (2002) Press briefing, February 12.
21 Goldratt E. and Cox J. (1984) *The Goal.* Great Barrington, North River Press; Goldratt E. (1997) *Critical Chain.* Great Barrington, North River Press.
22 Edwards L. (1993) *Practical Risk Management in the Construction Industry.* London, Thomas Telford.
23 HM Treasury Green Book is an example. London, TSO.
24 Flanagan R. and Norman G. (1993) *Risk Management and Construction.* Oxford, Blackwell (RICS).
25 Burke R. (2004) *Project Management: Planning and Control Techniques.* 4th edn. New York/South Africa, Promotec/Wiley.
26 Flanagan R. and Norman G. (1993) *Risk Management and Construction.* Oxford, Blackwell (RICS), p.24.
27 Gregory A. and Kearney G. (2004) *Restriction Buster.* Project 16(10), 20–22. High Wycombe, APM.
28 Flanagan R. and Norman G. (1993) Risk Management in Construction. Oxford, Blackwell (RICS).
29 Greenwood M. (1998) *Risk Behaviour.* Unpublished Master's Thesis. Bristol, University of the West of England.
30 Harris F. and McCaffer R. (2006) *Modern Construction Management.* 6th edn. Oxford, Wiley Blackwell, p.232.
31 Wilde G.J.S. (1994) *Target Risk: Dealing with the Danger of Death, Disease and Damage in Everyday Decisions.* Canada, PDE Publications.
32 Contractors come under The Public Contract Regulations 2006 and The Utilities Contracts Regulations 2006. Consultants come under The Services Regulations SI 1993/3228.
33 CIOB (2010) *Code of Practice for Project Management in Construction and Development.* 4th edn. Oxford, Blackwells.
34 Construction Industry Board (1995) *Selecting Consultants for the Team: Balancing Quality and Price.* London, CIB.
35 DTI. *Constructionline.* Available online at: http://www.dti.gov.uk/construction line.

36 Novation is the process of the concept architect employed by the client to do the scheme design being transferred to a design contractor as part of the contract conditions to do the detail design.
37 Based on HM Treasury. Procurement Guidance No. 3; Appointment of Contractors and Consultants. No date.
38 CIOB (2010) *Code of Practice for Project Management: Construction and Development*. Oxford, Blackwells, p.167.
39 CIRIA. *Guide 117 Value by Competition: A Guide to Competitive Procurement*. London, CIRIA.
40 Wolstenholme A. (2009) *Never Waste a Good Crisis*. London, Constructing Excellence.
41 Chapman C. and Ward S. (2003) *Project Risk Management*. 2nd edn. Chichester, Wiley.

Chapter 9

Design and value management

This chapter will investigate the design and value management process and the role of any professional in the management of the design stage of the project. This will clearly be of interest to architects and engineers as well as contractors and clients.

The objectives of this chapter are to:

- show basic knowledge of the design process and demonstrate the challenges of improving its efficiency including prefabrication and stand-ardisation
- demonstrate the use of value management in design development and audit and generating skills to identify client value
- generate skills in the flow of design information from the various project parties and manage its integration between design and construction in the supply chain
- set up an effective design change model which recognises the need to reduce waste and in the context of different procurement frameworks
- appreciate the impact of the information technologies on the integration of the project team including client approvals, fast tracking and fast build
- identify and apply the building information modelling technique
- understand the importance of urban design in design quality, sustainability and managing the project stakeholders.

Nature of design

Gray *et al.* define design as 'a process of human interaction and consequently the outcomes contain the interpretations, perceptions and prejudices of the people involved'.[1] Another definition by Hirano defines it more dynamically:

> The learning creative-developmental process can be represented as a spiral. Viewed from the top, a spiral is a moving circle constantly expanding in scope. Viewed from the side upward movement is evident . . . the addition of experience and understanding.[2]

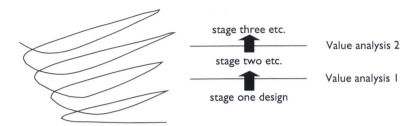

stage three etc.

Value analysis 2

stage two etc.

Value analysis 1

stage one design

Figure 9.1 Iterative design process

Design is an *iterative* process with the designer producing a working solution which is tested and improved or redesigned for further review as greater mutual understanding is achieved in the 'spiral' effect of the process shown in Figure 9.1. This could be related to the traditional stages of the RIBA Plan of Work,[3] or stages in the deepening detail of the life cycle progression.

The role of value analysis brings together client and designer on a regular basis and is a vital part of ensuring design meets need without excess at an early stage and as implemented in the final building. It may be formal (outside facilitator) or informal (design review with the client).

Several related parts of the design could be being developed in parallel subsequent to the concept design such as services, structure and sound. These will impact on each other and need to be co-ordinated. In addition there will be other constraints like planning, statutory requirements and regulatory control such as fire, building control and health and safety. In building projects these are specific to location and context.

Alternative design solutions may have been identified post concept stage partly dependent on location as well as other factors such as access. This method of design is called option appraisal and the choice will be based on client business needs and values as well as efficiency. The two methods are compared in Figure 9.2.

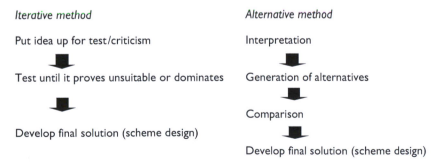

Iterative method

Put idea up for test/criticism

Test until it proves unsuitable or dominates

Develop final solution (scheme design)

Alternative method

Interpretation

Generation of alternatives

Comparison

Develop final solution (scheme design)

Figure 9.2 Comparison of design development methods

Maguire suggests that design comes out of three inputs – the designer, the problem as given and the data about the problem.[4] He admits that not all available data is found or used and also indicates later the constraints represented by the site and the budget/time given.

Engineering design

Gray *et al.* talk about the difference between craft based construction with traditional materials and engineered technology with its tighter quality and fit requirements.[5] Thus a traditional house design is well known by those who have to build. However the use of engineered components such as steel framing or prefabricated non-cavity walls needs much more design co-ordination, research and supervision. This complexity in the detailed design stage increases with the move from standard components to specially developed components and of course with the size of the project. This establishes a need for:

- revision in the concept and scheme in the iterative cycle of design development
- a larger more specialist design team with better communication
- many specialist inputs from designer contractors
- many drawings of a manufacturing detail
- many designed component interfaces.

This stage needs much more management due to the greater co-ordination required whilst the concept and scheme stages are likely to be in the hands of one or two people. It could also be argued that the earlier involvement of the construction manager is appropriate where the methodology for construction is changed significantly. Contractors have specific preferences which are associated with their suppliers and flexibility in design to allow for alternatives may induce value.

Design management

The CIC defines the role of design management as 'ensuring that the right quality of building design information is produced at the right time and conveyed to the right people'. It goes on to say that:

> the increased number of specialist designers and subcontractors contributing to the overall package of design information to practically realise a project has created the need for a defined approach to their direction and co-ordination with a new single point of responsibility. Design management responds to this need and attempts to improve the efficiency of the design process and its integration with the construction process.[6]

Design management is the application of management control in terms of cost, time, quality, health and safety and other issues to the design process to ensure that a good design is also co-ordinated and implemented well, both in construction and in the use of the building. This involves control of information flows between different parties and the proper involvement of client and supply change.

The following are identified in Gray *et al.* and should be good practice:[7]

- Allow the designer time for reflection.
- Choose designers of relevant experience and provide support to problem solve.
- Keep objectives in focus through a framework which allows cycling.
- Provide access to the client and review provision of information.
- Allow time for the design implications of change.

Design management is based on an awareness of the key milestones of the master programme such as value analysis meetings, outline and full planning submission, key procurement dates (supply and contractor), and BAA have a system of process gates which provide joint targets which go beyond design to construction and user acceptance. These gates are client approvals which recognise the coming together of parallel processes (simultaneous engineering), but should not be allowed to become bottlenecks.

Gray *et al.* further list nine essential steps for successful design management:[8]

- Recognise the inherent complexity of design.
- Carefully manage the designer selection process.
- Recognise the changing design leadership role as the design progresses.
- Integrate information supply with construction need.
- Obtain agreement at key decision points.
- Actively manage the integration of contributions.
- Plan at each stage.
- Manage the interfaces.
- Control design development.

Design management at the design stage

The RIBA Plan of Work sequentially only fits the traditional procurement method as covered in JCT traditional contracts, but is acknowledged as a valuable framework for understanding the various stages of outline design, scheme design, detail design and production drawings.[9] Design is often carried out by contractors with different degrees of input depending upon the procurement system, so with this proviso design management can still be overlaid.

The key issue in determining leadership at this stage is the form of procurement. There are two models:

- a lead designer
- a manager with a knowledge of the design process.

The key areas for management need to be defined project by project. Gray *et al.* helpfully identify quality as the primary driver for the management of design and it is this concern which will drive the need to plan in detail, manage the interaction of the design team and ensure the timely production of the information for its use by others to meet targets for client approval, tender and production and to work within the key cost parameters.[10] This is followed closely by the need to ensure that health and safety issues are considered holistically, which does not take the liability away from the individual designer for the safety of their design, but there is an additional role to co-ordinate safe designs.

Design and build procurement is easier to manage due to the executive single point of responsibility of the provider/contractor over design. Design and manage contracts are the same, but in this case there is a consultant responsibility similar to an executive project manager. In construction management there is a tighter client involvement and the design and construction management and contractor inputs need to be reconciled with guidelines, by setting up project guidelines to give the design manager authority.

Case study 9.1 Office fit out

This £27m London office fit out was carried out under construction management procurement. Representatives from each of the contributing consultants were housed in an open plan office with the construction manager. This proved very effective in managing the production stages of the design and dealing with client and design change. The specialist package contractor managers also had open plan office space allocated on the same floor of the building, making their design and management services accessible. Lead contractors were responsible for managing the interfaces of other contractors in a particular zone or area, for example the floor or ceiling. Monthly meetings were held, although some client related decision making was delayed due to the remote location of the client in New York. The project itself has been nominated for a management award, coping with significant client changes without altering agreed targets.

In traditional procurement the architect or engineer has a contractual role in design management, in conjunction with the cost planning role of the quantity surveyor, and liaises closely with the client. It is usual however to appoint, or more informally use, the design contractors for their specialist inputs.

An ideal situation for improving communications and quick decisions is to bring the core design team together in one office, but this collocation has not often been used in the past due to its expense and the detachment of designers away from their core information and their need to be available for other project work. Case study 9.1 gives an example of collocation.

Client values

A project has many stakeholders who may have different values. A client's value system is derived from value chains from each stakeholder group which set the strategic criteria and from a project value chain which develops in tactical detail as it moves through the phases of the project life cycle and may be exposed to different stakeholders in each stage. Various stakeholder interests are related to the sector such as social housing, school or road construction projects.

Understanding the client value system is crucial to successful project management and is central to the function of value management. The essence of a VM study is to question convention or preconceived ideas by a rigorous process of creative and constructive analysis. This is followed by dedicated value management workshops when the client and the project team – designers and constructors if possible – are brought together. It is vital that the VM facilitator encourages participants to share their ideas as equals in a creative process. Decisions may be made more and more by the project team as the life cycle develops, but the client values still remain at the centre of their ongoing approval of distinct stages.

As well as project value Simister also refers to shareholder value based on the distributable returns from the project and customer value which recognises the ability to increase the value.[11] The customer also gains from the reduction of waste from the system. Waste reduction is quite a useful concept in producing goods more competitively for the supplier and the consumer.

Ethically values may be expressed in the sustainability of the design in respecting the ecology, using more renewable resources and fewer nonrenewable ones and reducing the carbon footprint of the building in its construction, for example embedded energy and its use and maintenance. Socially, building design can have an impact on the quality of the work environment for the employees, the availability of employment and the positive effect of the building and its activity on the community.

Economic and market factors are important to clients and the value gained from the business optimisation, the efficient operation of the building and reducing the life cycle costs far exceed the construction cost so if a client is to profit from these efficiencies then design just to reduce construction cost is a lesser issue than business rewards achieved by quick completion or effective operation. A business may also have flexibility in location in profiting from tax breaks from operating in disadvantaged areas where government incentives are available to offset setting up costs or taking advantage of cheaper production abroad.

Value management in design

Value can be defined as cost divided by function and as such value can be maximised by reducing cost and maintaining function, keeping the cost the same and increasing function or a combination of the two, that is, by reducing function and cost or by increasing function and cost disproportionately in favour of the client. To make more than mathematical sense of this we need to consider the worth of function and connect this to clients' values and needs and separate these from the wants which so often confuse the picture when trying to identify an efficient and effective building.

Value can be optimised technically in efficient choice of materials, optimised structural solutions and orientation and level of the building to reduce excavation, shielding and wall to floor ratios. For example an 'elegant' solution on a sloping site may require 'cut and fill' and the building can be multi-storey, long and thin across the slope to minimise the cut and to maximise light into the building and have thin floors with services built in to minimise the storey height. If the building is kept light with steel frame or pre-stressed concrete then the foundations may also be reduced depending upon the ground conditions.

The process and buildability of the building can be optimised for value by efficient control systems, standardised components and familiar detailing so that waste and the construction process are more competitive. Conflict can be reduced in having appropriate contracts and good contractor–client relationships. It is also important that supply chains can be managed by the main contractor efficiently and fairly so that everyone knows what they are doing. The designer and contractor interface are critical and merging the design-assembly function can help in reducing waste. Alternatively an integrated design and construction process can be achieved by early contractor involvement in the design and by building on relationships and team efficiencies from previous working together. Detailed design delegated following the concept design can be passed effectively to the contractor and specialist if information flow is controlled well. The designer can collocate with the contractor or a system such as building information modelling (BIM) can be used which integrates the information flow and standardises information through design, construction and building use.

Stakeholders are internal, external and acting in proxy on behalf of others such as the planning authority. Finding out and managing the needs of external stakeholders have become increasingly important, as planning permission is increasingly difficult to please and pressure groups are more efficient in delaying tactics.

Value management is the more formal recognition of meeting clients' objectives and requirements in the most effective and economic way. It involves creative solutions and a critical needs analysis to test the difference between needs and desirables against properly stated objectives. It takes into account the life cycle costs to promote business efficiency. It is not usually the cheapest capital cost, but may be.

Design economics

Ashworth refers to the term design economics and lists 11 design factors which affect the value of a building without reducing its quality per se:[12] site considerations; building size – small buildings create a worse floor to wall ratio; planning efficiency in reducing circulation space; plan shape in maximising the floor to wall ratio – a square building is most efficient; height – tall buildings have expensive structure; services e.g. lifts, water pumps and fire escape; storey height; groupings of buildings on a site; construction detailing; structural morphology with its impact on spatial design efficiency e.g. large or small span; standardisation of components; and design and prefabrication offsite. All of these factors have to be balanced against the client's preferences to suit function and utility.

Design is an iterative process – recognising the impact of a decision in one area has to be balanced with its impact on another area and so a tall building may be more expensive to design and build, but it will also be efficient if land is expensive.

Value engineering

According to Ashworth value engineering originated from finding substitute materials when there were shortages in supply after the war.[13] GEC found that the alternatives quite often were cheaper or performed better and their use was adopted more permanently as a development of traditional product design. As construction design has developed to rise to new client requirements for quality, but less costly buildings, and a more global perspective has been introduced into the building product market, designers have been forced to look at a wider range of products and methodologies which induce the same functionality for less. Part of value engineering is connected to the concept of lean construction which was discussed in Chapter 5. It tries to eliminate waste in design which is connected to the methodology and waste reduction process in construction. For example prefabricated toilet pods give more quality in design, but also less skilled labour and time on site. They may be more difficult to maintain if things go wrong due to hidden or specialist components.

Value engineering is the next stage of value management in order to consider the more detailed design of the building and generally taking place after more fundamental and unique aspects of the building have been fixed after planning permission has been submitted. The changes made cannot therefore affect the location, aesthetic or the fundamental orientation, height, external appearance and materials of the building. Design changes can be made in the layout, the finishes and the circulation and similar materials may be substituted. It may also make process changes such as the use of standard detailing, prefabrication and quality of materials. It may be possible to omit certain things such as roof lights and to develop alternatives to sustainable technologies and landscaping materials that do not change its appearance. It could offer alternatives for the M&E provisions and finishing materials. For the cynical it is a cost cutting exercise, but where there is flexibility, better research and access to market products previously unknown it gives the chance to better value for money and some manoeuvre in previously desirable but not essential functionality. For example, it may be desirable for flexibility to have all site road specifications which support heavy lorries, but not essential if there is a very specific delivery area.

At the post planning stage it is a good idea to do a further review of value in the light of current information and drawings that become available for the application of detailed design by a contractor. This is done first by the designers taking into account the conditions for planning consent and reviewing the efficiency of any changes. Design drawings are handed over and a third party is able to value engineer the drawings and make suggestions prior to going out to tender. This is an interactive process which is particularly useful for the buildability aspects of design and uses the purchasing knowledge of the general and specialist contractors to obtain value for money with approved changes of specification. For example, a steelwork frame depends upon availability of the designed sections, but other profiles may be designed in if some are not readily available or expensive.

A further review of health and safety of components and methodology may lead to further design changes, for example the use of lighter blocks or a more easily accessible roof. Compliance with building regulations reviews the structural, durability, fire protection, disability access, energy issues and emissions which are essential. Deemed to satisfy solutions may not be the best solution and it may provide better value to develop a solution which can be tested for approval using a system such as Agrément certificates or so called robust solutions. Creative solutions particularly apply to new technologies and proprietary products associated with sustainable construction, and are to be encouraged so they reach maturity and give confidence in use. Transfer of materials used in different contexts such as different climates should be adapted as necessary before adoption.

At this stage it should also be possible to do a review of construction methodology and significant savings may be made overall to the client's business case. One route is to consider the time cost savings of prefabrication of some structural components. Prefabrication is discussed later.

Value engineering needs commitment. Typically a designer does not wish to revisit what they already believe is a robust design which has evolved through the approval and co-ordination of different parties to the design process. The client may be interested if they feel they can reduce their budget targets, without compromising quality, and the contractor will be interested if they are given financial incentives to suggest alternatives or at least stop expensive delays to specified components which are difficult to claw back.

The value management process

The process has developed certain techniques and can be applied at the initial project definition and the post planning stages. Kelly and Male studied the system of value engineering where it developed in North America.[14] There were a number of different approaches to value engineering, but the most common one is the 40 hour multi-disciplinary workshop undertaken (5 days) or in three stages over 3–4 weeks (3–2–2 days so slightly longer). The most common timing was 35 per cent of the way through the design when enough was known about the cost to estimate savings to be made. Savings are claimed to be 5–25 per cent in North America.[15] In the UK they estimate the traditional QS cost planning process makes a probable saving of 5–10 per cent. They argue that the first two stages of functional analysis are not dealt with in the NA system as 35 per cent design is too far advanced and basic project definition issues have been completed. They suggest the NA value engineering process needs to operate earlier in the project life cycle to capitalise on early gains at the feasibility stage and needs existing QS pre-contract cost control procedures in the UK to ensure effective implementation.

Value management workshops typically involve the client whose values are tested, the lead designer and some other key design roles such as architectural, civil/structural, M&E services and environmental who have the technical know how. In design and build they will have the early involvement of the contractor or in traditional procurement they may pay a construction consultant or contractor. It is necessarily chaired by a neutral facilitator who is able to push for honest answers to awkward questions. These questions have to address at different stages according to Kelly and Male:

- What is it? (the objectives)
- What does it do? (considering essential functionality)
- What is it worth? Compared with what it costs?
- What else will do and what does that cost?

This introduces an imperative to consider other alternatives and to be creative. Some techniques have been adapted.

Functional analysis

Functional analysis is a basic process for design in matching the design to the requirements of the client. This process is critical to the briefing process and allows the designer and others to identify what functions are required by the building or infrastructure. Hence the purpose of a dam will be to store water, but may also be to control water flow in the valley and perhaps flooding. In addition a dam provides an opportunity for generating electricity where water flow is processed through a turbine. Secondary uses such as

Table 9.1 Functional analysis levels

K&M's levels	Author comment	Office example
Project task	Identifies client objectives for the building and will have a primary task which can then be broken down into secondary ones. Can question which is primary and eliminate some secondary ones.	Quiet high quality head office building on expensive land in central business area receiving important clients.
Space and arrangement	Considers the breakdown of the building into floors, rooms and corridors and outside spaces in the spatial arrangements which are most efficient to the building functions and are in efficient juxtaposition to other spaces. Client is central to the process.	Multi-storey. Building reception, small meeting rooms and utility rooms on ground floor. Open plan offices broken into departments. Director suites and boardroom on top floor.
Element	Considers the more universal roles for parts of the building fabric and structure such as lifts and floors and walls. Their efficiency is more technical than a client issue. Spatial implications to client.	Fast smart lift to deliver visitors quickly with smart reception areas on floors, high quality curtain wall. Air conditioning. Basic staircase.
Component	Down to the sub-element as a deliverable to site. Thus a wall may include masonry, insulation, window and wallplate components. It considers specific choices which contribute to the spatial and key objectives. It is mainly a technical design decision, but also needs a knowledge of contractor methodology in defining its form and success. Client choices for finishes.	Utility finishes in office areas and panelling in boardroom and director suites. Heating and lighting controls in offices to maintain optimum environment. Tea making and break out rooms. Carpets on raised floors for flexible services and quiet. Cafeteria.

Based on Kelly and Male (1993)[16]

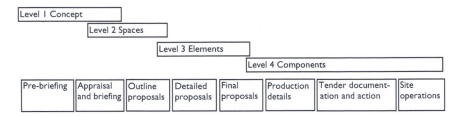

Figure 9.3 Relationship of function analysis to project life cycle

Source: Roughly based on Kelly and Male (1993)[17]

sailing and fishing may be envisaged. In order to produce value for money relative functions must be prioritised. Kelly and Male have defined VM at four levels, shown in Table 9.1. Each have the capacity at different levels of details to optimise the process.

The levels of consideration are also related to the project life cycle as shown in Figure 9.3. The latest Architect's Plan of Work terminology has been added to connect it to the sequence of architect's tasks.[18]

More fundamentally a functional analysis (FA) is used to provide a solution to a problem. It is required to increase production in a factory or to reduce the incident of water shortage in the case of our example above. Male establishes some rules for FA. These are to be clear with the wording by using a clear active verb to precede a noun, for example, 'Provide economical lighting'. Second he advocates first making a functional definition before coming up with a specific solution identifying components. Case study 9.2 continues the example in Table 9.1 for the lighting definition.

Case study 9.2 Functional definition

> 1 Primary level function. Provide economical lighting in narrow floor plate.
> 2 Element level (open office area). The functional definition might be to provide a certain level of lighting on all desk surfaces. This might be provided by natural light by the spatial layout in conjunction with roof lights or windows, or it could be the use of some electric lighting. If it is only occasional working after daylight hours then lamps on desks might be more economic for the few who do. Cuts down heat gain.
> 3 Component level (desk lights). Supply desk lamps to each pair of desks with movement sensors. Basic level cost.

Case study 9.3 VM at the early concept stage

The situation involves an office refurbishment of approximately 2,500m² of office block behind a listed Victorian façade, which has to be retained. The five floors of the building are to be gutted internally and a new steel frame office is to be built inside the retained façade. This façade will have to be retained during demolition and rebuilding; an existing party wall will help to stabilise it during this process, but in this wall there is also a major central chimney as in Figure 9.4(a).

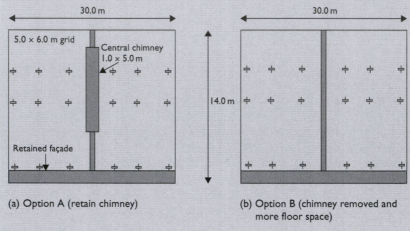

(a) Option A (retain chimney)

(b) Option B (chimney removed and more floor space)

Figure 9.4 Office refurbishment

The structural engineer advises that if the chimney is retained (as in Option A) it will save on demolition costs and provide stability during rebuilding but also reduces the structural bracing required in the new steel frame with a possible saving on the construction programme of one week. The quantity surveyor assesses the construction costs of the two options and estimates that Option B, with the central chimney removed, will cost an extra £15,000 for demolition plus an extra £8,000 for structural bracing to the steelwork. On the basis of construction costs and programme there is therefore a clear argument to recommend Option A, but during the value engineering appraisal the commercial property surveyor raises the issue of the loss of net lettable floor space in Option A.

Option A has some 21m² less net lettable floor area, allowing 300mm for the party wall thickness than Option B and the developer values this at £71,000, based on:

> a rental of £20 per square foot or £220 per m² and a yield of 6.5 per cent
>
> This equates to a year's purchase of 100/6.5 = 15.38:
>
> 15.38 × floor area £220/m² = £71,055.60
>
> After deducting the additional demolition and bracing costs, this leaves a net additional value of £48,000 for taking Option B.
>
> By the same means, we can evaluate the benefit to the client of saving a week on the construction period, which is
>
> 2474m² at £220 per m² per annum/52 weeks = £10,466
>
> Assuming an immediate let, Option B still provides a net added value of some £37,500.[19]

Third he suggests the identification of secondary functions or impacts which are not essential to solving the functional need but are side products of a technical solution. They might be desirable or undesirables so an electric light solution might produce heat and a natural light solution might produce glare and heat. Fourth he suggests looking at the cost/worth relationship where cost is the price paid and worth is the least cost to provide the solution for the function needed. Here an elegant lighting system solves the problem at a greater cost and may have been chosen because it looks nice. If a generic lighting system is chosen is there any fundamental loss to function? Therefore beauty or previous provision would need to be justified in other terms which were primary or reduced harm etc.

The functional analysis allows a lot of choice in how to deliver lighting most efficiently and to look at client values and technical solutions. If the client had strong sustainable development values and specified BREEAM excellent, then an overlay of additional requirements come in the equation such as the heating and cooling solution, the cladding and solar gain and the height of the building. Directors may wish to show off their sustainability in the part of the building they are responsible for.

Functional Analysis System Technique (FAST)

Another technique, developed by Bytheway,[20] is to use the functional levels in answer to the questions 'How?' working from left to right; and 'Why?' working from right to left. This needs to distinguish between a 'task FAST' justifying overall objectives which are primary and secondary and 'technical FAST' which develops the elements and components which serve the sub objectives of keep building warm/cool.

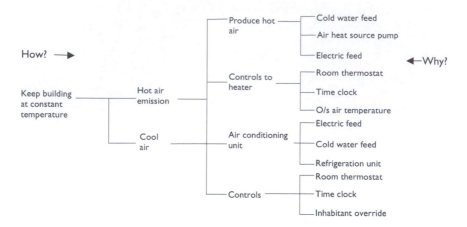

Figure 9.5 The technical FAST example of a system to keep building at constant temperature

Figure 9.5 has a logical design solution to the question 'Why?' Moving from right to left and the value engineering process needs to deal with optimising the components to suit the ultimate objective to heat and cool the room to keep a comfortable environment. In starting at the left hand side and moving to the detail you need to ask the question 'How?' It deals with the technical functional analysis levels 3 and 4.

Case study 9.3 is a more detailed example of the use of value management at the early stages of the project life cycle where there is the biggest scope for developing the brief to build value. It also illustrates how a multi-disciplinary approach to problem-solving can produce more value for money. It describes a situation where spending more produces higher added value by a factor of two or more.

A 'task FAST' has the same format, but concentrates more on level 1 and 2 and the sub tasks for the whole building. An example of a strategic VM study to clarify the project definition was a study undertaken in Limerick for a developer proposing to build a new hotel described in Case study 9.4.

The essence of a VM study is to question convention or preconceived ideas by a rigorous process of creative and constructive analysis. It is vital that the VM facilitator encourages participants to share their ideas as equals in a creative process.

Benefit analysis (BA)

BA is the exercise of checking that required benefits have been delivered in the solution delivered to the client. This needs to be constantly monitored during the design process. It is also a way of comparing in real terms how

Case study 9.4 VM at the early concept stage

> The tax-free industrial area under the Shannon Development Cor-
> poration and the growth of international flights connecting Ireland with
> America had created a strong basis for economic development and
> particularly a strong growth in international tourism and the develop-
> ment of a hotel. As the VM workshop explored the drivers for growth
> in tourism, the importance of an international golf and conference
> facility was realised, to the extent that the project brief became a much
> larger regional and international attraction than a hotel.[21]

expensive that solution was with previous provision. As such it is an audit
on the effectiveness of value management and provides a feedback loop to
check VFM. Clients will have the view that they will have incremental
improvement where creative solutions have been successful. This should be
measured against function and need and not secondary wants. Design and
implementation of the solutions are both important as Figure 9.2 suggests.
It is however possible that change in functional priority has occurred later
during the design and construction process and efforts to incorporate this
may not have optimised the solution.

Value management workshops

To be part of a complete integrated system VM needs to interface with the
project life cycle at key events; Kelly and Male describe a continuum of staged
events, including those in Table 9.2.[22]

The Job Plan in the SAVE toolbox is commonly described as a five-day
workshop, covering the following phases, which have been merged with
Woodhead's proposals for the steps of the initial value management workshop:

- Information – senior management determine metrics. A base case is
  determined where the project team develop an explanation of their
  current thinking about the project (i.e. how they are currently proposing
  to deliver the project and its results). The base case is translated into the
  functions that the system (i.e. the elements of the project) must perform.
  The functional representation frees us from any particular method (or
  solution) so that we can later consider alternative ways of performing
  the functions.
- Creativity – brainstorm different ways of performing the functions that
  leverage value. That is, target specific functions and ask: 'How else can
  we perform this function?'

Table 9.2 Continuum of VM workshops through the life cycle

Description	Duration (days)	Focus	Stage
Pre-brief workshop	0.5 to 1	The strategic brief	Clarifying the business need
Brief workshop	1 to 2	The project brief	After the initial business case has been established
Brief review workshop (Charette)	1 to 3	Examination of the brief already prepared by others (as an alternative to the two workshops above)	At the end of briefing, before concept design
Concept design workshop	1 to 2	Outline design	Before submission of planning application
Detailed design workshop	1 to 3	Final proposed design	Examination of the design by functional elements
Implementation workshop	0.5 to 1	Are recommendations being implemented?	
Contractor's change proposal	1	Buildability studies	After appointment of contractor or in second stage of 2-stage tender with performance incentive
Post construction review	1	Feedback, especially where ongoing programme if capital works	

Based on Kelly and Male (2002)[23]

- Judgement – sort out the ideas on which we are going to spend quality time writing up considered proposals.
- Development – take the considered proposals and create a menu from which we calculate the business results of each scenario and pick the ones we will present to senior management for approval.
- Recommendation – develop our presentation, where senior managers from day one return to hear the team's recommendations and make a decision as to what the new base case should be.

The practical difficulties of timetabling a multi-disciplinary panel of experts, stakeholders and senior management to meet mean that a shorter event is often held incorporating many of the aspects.

VM has had some difficulty in the UK being accepted as a separate discipline. Again, 'Why should the client pay extra for something that the design and construction should be doing anyway?' is still a typical response.

To some extent, value engineering has been seen in the UK as the lifeboat that is brought in to bring a project at tender stage back within budget, and the danger is that participants see this as no more than a multi-disciplinary cost reduction exercise, with little rigour in the focus on value and function analysis, with little creative thinking and the adoption of solutions based on previous experience. Alternatively, VM may be viewed as a contractor-led exercise to improve buildability with a cost-sharing clause added to the contract as an incentive, and yet the textbook approach proclaimed over more than 30 years has been to emphasise the benefit of early strategic VM exercises. An interesting question to pose of companies offering VM services to the construction industry is whether they use VM internally to improve their own product and processes.

Design and construction co-ordination

One of the key areas of design management is the interface between the design and construction stages. Traditionally these are separated so that there is a need for a robust communication system to be employed to ensure that information arrives when and how it is needed. Particular stress is put upon this system when a fast track approach is used requiring construction to take place on certain elements before the whole design is complete. Some of the areas for co-ordination are as follows:

- health and safety risk exposure and hazardous materials
- specially developed or technologically advanced components
- sequencing and timing impacts
- tolerances
- constraints which affect construction methodology
- material lead in times
- available or required mechanisation levels
- prefabrication methodology to maintain benefits and timing
- renewable technologies.

As a contractor is not often available at the early stages, unilateral design action is often taken which tries to predict, or later inform, certain construction methods. Contractors can be aware of the implications for the constraints of the site and the pre-decisions imposed on the methodology when tendering. In an open culture, value may be enhanced where construction management aspects can be dealt with directly with the contractors at an early stage and productivity improved.

Egan (2002) concluded that the UK industry performance in general scored 4/10.[24] The following list indicates some of the main areas which impinge upon design construction co-ordination:

- design waste elimination by engendering efficient methods
- constructability, for example, workable tolerance
- standard components and processes
- prefabrication[25]
- new technologies and research
- change management to allow some flexibility
- benchmarking
- sustainable approaches to reduce whole life cycle costs
- building information management.

However, many opportunities still exist to catch up on better performance, many of them relating to the design process. In this chapter we will
cover standardisation of design and components, design waste issues, some
aspects of prefabrication including the management of new technology,
constructability and change and information management. (We have dealt
with benchmarking.) Case study 9.5 gives an example of design construction
co-ordination.

Design waste and design types

Design waste is created from an inefficient use of resources or by an erroneous interpretation of requirements and has been adapted to cover a more
sustainable approach. Materials can be wasted if they are used inappro-
priately in the wrong position and tolerances are difficult to achieve, or the
material is not fit for purpose, for example window openings do not fit the
brick size so that more bricks are cut and wasted. The use of standardised
components such as standard timber or steel joist lengths for structural spans
is important to reduce waste. Non-standard design and the poor application
of proprietary systems in their interface with other systems, for example
proprietary cladding to proprietary roof join, are likely to require rework
unless they are thought through.

The concept design needs to suit community as well as client requirements.
It can become expensive if it also satisfies design egos, or is not integrative
in meeting the conflicting requirements of stakeholder and community. A
concept design which perfectly meets business requirements and is belatedly
amended in order to deal with planning permission for other community
and stakeholder requirements could be embroiled in expensive compromise
and delay. Different designers need to have a common approach and integrate subconcepts such as fabric, structure, services and landscape early.
Aesthetic elements are important to quality and quality of life, but should
not be given more than their weighting in meeting the client's objectives.

Detailed architectural design is an important process in order to provide
robust details and elegant solutions. It is closely connected with buildability
and the interaction of design, building price and construction. In the case of

Case study 9.5 Design co-ordination on specialist physical research facility

Although it is a traditional contract the contractor was involved in the enabling works and therefore played a forward role in checking the buildability of the drawings.

At the primary level the lead designer was responsible for co-ordinating the different designers who do a full performance design. Full detailed drawings and specification are offered to the principal contractor for scrutiny and tender. He is responsible for assessing the buildability of the drawings and the specialist subcontractor puts in a detailed design which may suggest minor changes to suit economic ways of working. The short listed contractors are also encouraged to fill in question and answer sheets in order to provide acceptable detailed proposals.

The need to change doors in the dry-lined partitions to give a workable tolerance was an example of a change made to benefit the buildability and efficiency of the project package for ceilings and partitions. This would have mutual benefits for the contractor's installation and the client's quality. Value engineering inputs on the design by the contractor are difficult even though the design overlaps the operational phases of the contract as there is no specific brief or system which defines the responsibilities for design liability where design has been adjusted. This makes it difficult to allocate design risks. Incentives to share financial benefits with the client may help to cover risks.

Significantly quicker and cheaper solutions have evolved and good co-operation achieved between contractor, client and architect. The client made use of the availability of the contractor in order to involve them in the design stages of the second phase of the contract. The experience and availability of the enabling contractor on the site was also acknowledged by putting them on the tender list for the second phase construction, which they subsequently won in competition. The informal direct relationship between the client and the contractor project team ended up with a limited formal line of command through the project manager. This was also encouraged by the direct client involvement in the quality inspections. This meant that the project manager has been able to use contractor expertise.

new materials or unusual juxtaposition or shaping of traditional materials or the importation of materials used in different climates and conditions, the waterproofing and durability and health and economy of use need to be thoroughly thought through and tested. Clients may request unusual detailing which they have seen in the design press where the quality and the detail of the finish are critical, for example ceiling details. In this case a mock up for the design should be made to ensure that the requirements are interpreted correctly and understood by all. This allows for the tweaking of the design before all the materials are ordered. The proportion of non-standardised components will raise the tender price.

Spatial planning is about providing comfort and efficiency in use, but the principles of design economics (above) are also important in reducing the use of resources, optimising the wall floor ratios and limiting the area of external weatherproofed walls and roofs. The vertical and horizontal relationships are important.

Sustainable design looks at the choice of materials that have a low embedded carbon and are durable or recyclable. It also considers renewable or low carbon use of energy so as to minimise the use of fuel in use. Short duration materials and constant renewal not only cause disruption, but are wasteful in material use, some of which is non-renewable with finite resources.

Structural design needs to be adequate to take static and dynamic loads and to stop excessive flexing and cracking, but not over specified so that excessive material is used or storey heights are more than otherwise needed for use, wasting space in ceilings for deep members. Control of clear floor spans is important so that structures do not become more expensive than necessary. Foundations which are avoiding other underground structures or obstacles can also become very expensive, and they are often the product of late discovery of these obstacles. Basements are a factor of expensive land prices where space is at a premium. They are also expensive structures and require the retaining of major lateral loads and more robust waterproofing and greater excavations and carting away of materials.

Services design for a comfortable environment may be worth up to 40 per cent of the contract sum and so savings in avoiding over specification in plant need to be weighed against the cost in use. Small proportionate savings will make major contributions to contract sums. Components are often repetitive and careful detailed design should be scrutinised for effective value management. Electrical and mechanical components need to suit the structure and fabric of the building. Sustainable services is a much wider subject in relation to the building architecture to reduce mechanical and electrical building heating and cooling requirements and create natural and renewable fuel alternatives.

Ecological and landscape design needs to match its external environment and 'live' with the existing natural flora and fauna environments. Choices in location, orientation and timescales to suit breeding seasons can help. Waste can be eliminated by harnessing and enhancing these environments to carry out a role in the new facility. Buildings can be shielded by trees for shading and wind exposure. Views can be capitalised by incorporating geomorphic features such as lakes and using natural environments to further wildlife. Water run off can be minimised by using less hard areas and sustainable drainage design.

Civil engineering design needs to provide efficient infrastructure in reducing service runs, optimising highway design harbours and utilities provision and reducing surface run offs to reduce flooding whilst still keeping an economical drain size. Structures have a lot of scope for efficiency in material use where design of some structures is over specified because of unknowns and high factors of safety are built in. For example, there is a tendency to justify larger dams, ever bigger flood defences and huge spans in bridges which are expensive. Power stations and dams and reservoirs are often in out of the way places so do not have to be elegant, but landscaping features are important. Wind turbines, airports, aerial masts and pylons are constantly criticised for ruining areas of outstanding beauty and money is spent fighting planning appeals where people have felt threatened and demeaned in their environment by these structures. There is scope for looking at concrete specifications, natural materials and recycling and fully evaluating the life cycle costs of civil structures, as these structures are enormously expensive.

Prefabrication

Prefabrication is a way of manufacturing larger components offsite in more predictable conditions in order to cut down work onsite where conditions are harsher and less controlled, for example weather conditions. The quality and speed of erection are quoted as two advantages. However this is weighed against a longer pre-delivery period which needs meticulous and early planning and a greater manufacturing cost, though arguably this cost may be offset against less waste and more efficient construction time to produce a quality product. Offsite manufacture (OSM) lends itself to computer aided manufacture (CAM) which is less labour intensive, has standardised components which are not straitjacketed and can be varied in size and a limited range of strengths.

Prefabrication is generally divided into five types which have advantages and disadvantages:

1   Modularised volumetric which means an element such as a room or a toilet cubicle are completely finished in the factory and bolted into position on an existing structure. They can also be structural up to a certain height and stack on each other.

2   Sectional building, which bolt together and are generally structural up to a certain height of building. The limitation is the size that can be transportable by road or rail. This allows for wider rooms and complete buildings such as hospitals where there are a variety of spaces.
3   Pods and other units which fit into an existing structure and connect up to the services such as bathrooms and kitchens.
4   Panel manufacture which makes up structural floors, walls and roofs into a complete building, This is generally structural and is reinforced with structural screeds and bracing and is covered with cladding. It may be in timber frame or concrete panels (Case study 9.6).

Case study 9.6 Panelised construction

1,932 student residences were constructed in 18 months with Buchan pre cast concrete panels. Five courtyards of units consisting of basic building blocks of two six-person flats built off a central stairwell up to six and seven stories. These blocks were joined together in rows to create courtyards enclosing landscape areas at ground floor level with bike stores and reception area for each courtyard.

Panels were delivered in order to create walls and floors which were reinforced with concrete screeds. Windows were already built into the panels and the services were already built into the floors and walls. Panels were inclusive of flats and staircase circulation area. Each panel was designed to suit the block configuration and delivered in order. Five mobile cranes were used to build the panels together on site and only edge protection already attached to the wall panels was required to install the basic structural panels which were bolted together and braced. Structural floor screeds were used to stiffen the structure.

Scaffold was used to fix the 30 different cladding types which included different types of brickwork, cedar panelling, reinforced render and tile hanging. Roofing consisted of steel framed pitched roofs which were not part of the Buchan system and services for each block were provided by a ground floor boiler room externally source and configured into the layout. Internally the walls were fair faced concrete which was painted and concrete precast staircases were fitted. Toilet pods were used for each en suite bedroom.

This was a successful project which fulfilled the needs of the client for a durable design which was also delivered and erected at a faster than usual pace. The accommodation is popular with students and provides a medium priced solution.

Case study 9.7 Component type prefabrication

Durablock is a large light hollow block with an internal insulation and cavity. Blocks are made out of woodchip concrete and cast hollow with internal insulation in those to be used for the wall or floor and can take reinforcement to support beams or lintels; or be filled to create retaining wall strength. Blocks are completely impervious to water and are also more resilient than concrete to explosive blasts so they are being used as shock walls cast in panels and loosely connected. The outside face can be timber clad or colour rendered and a template for a colour render has been developed to realistically give external brick and masonry appearance. The roof would be of simple truss manufacture.

5   Unframed smaller standardised components manufactured in specialist parts which 'bolt together on site such as insulated blocks beams and floor cassettes creating and forming a kit of parts designed to suit client requirements. There are many forms such as Durablock or permanent polystyrene formwork which also insulates the concrete fill on site (Case study 9.7).

There are a number of components that are flexible to use as fast build. Case study 9.7 is one of these.

Standard design solutions

Standard design solutions, previously called system buildings, have a standard approach to the design of buildings with the use of certain standard components. They incorporate flexibility for the layout and facilities required, but try to keep costs down by providing some standard detailing and/or components. They may use a consortium of suppliers or a joint venture between a designer and a manufacturer. They do not require a formal partnership or framework for similar contracts between the client and the provider, but they are effectively a holistic proprietary product available from a single source. Some competition can be built in where the contractor is tendered as an approved supplier. These programmes have potential saving on the design fees and on the construction where the learning curve has already been discounted from the full cost of design and risks are expected to be lower. The older discredited system school buildings was limited on its layout and depended too heavily on a kit of parts which in the event did not have the durability which was required. New systems have been developed and Case study 9.8 is an example.

Case study 9.8 Standardised schools

> Currently a number of solutions have been developed for the schools
> rebuilding programme in the UK. The prices have been predicted up to
> 50 per cent lower than the bespoke programmes that have been used
> to deliver previous new schools and depend upon local partnership, PFI
> or other facility to inform the schools renewal programme. Table 9.3
> indicates the relative merits of four new systems.

From Table 9.3 it is clear that the client for standard design does have
several choices and indeed can induce competition between different systems
together with a review of which system suits best. Either they are based on
existing products such as Tarmac's TermoDeck or they could be based on
standardised designs. Where there is designer and contractor familiarity, then
the learning curve is reduced. Where there is OSM capability there is an
opportunity for improved quality and speed of erection.

New technology

Professionally a designer is expected to keep up with advances in technology and assess them for us in their designs. They also have a professional
liability for anything that goes wrong with the design and this may limit
their innovation. The advantage of innovative technology is that it progresses existing knowledge and is likely to improve it and respond to issues
such as sustainability where simply imposing stiffer requirements, for
example thicker insulation, creates its own problems and more innovation
is required.

However, new technology has a learning curve and also needs more
robust testing and detailed design development to iron out any initial
problems. For example the sealing of buildings to high standards preventing
leakage creates condensation problems and so air treatment systems are
introduced. Other technologies such as high insulation levels require more
consideration to eliminate cold bridges and the use of new materials
may need compatibility testing with traditional or existing ones and the
development of new detailing.

The above solutions represent the melding of new technology with
repeatable construction methods and as they are developed collaboratively
with a willingness to take on liability and with the incentive to corner a particular method and market, innovation is more enthusiastically pursued than
when the design and construction process is completely fragmented.

Table 9.3 Comparison of different standard design school systems[26]

Factors	Edventure	EcoCanopy	NurtureFuture	Learning Barn
Who developed it?	Education consultant Bryanston Square and Biong Arkitektor	Bryden Wood Architects	Cartwright Pickard Architects and Tarmac Building Products	Scott Brownrigg Architects and BAM Construction
How does it work?	Permanent external shell with glulam frame and up to 60m span with plug in modular & panelised components to divide up the internal space	Uses 4m x 4m and 4m x 6m lightweight bolt together concrete waffle panels for floor and wall to create flexible configurations with an exposed timber frame roof which allows quite large spans	Load bearing concrete TBP concrete façade panels incorporating services, windows and TermoDeck which is a hollow concrete deck promoting heating and cooling to pass through the floors using thermal mass. Fair-face products preclude use of ceilings, raised floors or dry-lining	Standardised kit of components including wall panels, windows and door sets developed with BIM Revit in mind. Architects will draw standard component information from model and suit specific sites. BIM allows fully integrated approach
Who has used it?	Liverpool City Council is planning 4 new schools	Ashe Construction has used it on 9 children's centres	Talking to major contractors to take it forward	Used by contractor and architect in Kent
Cost	£1,100/m²	£1,000/m²	£1,400/m²	£1,600/m2
Class leader	High flexibility for space and re-configuration is possible, relatively inexpensively for school uniqueness, or other uses	Low cost solution which is attractive and allows expansion of space for growing school	Factory controlled quality and robust thermo mass of concrete used sustainably with customisable design	Already has full product information allowing early accurate price and reducing risk. Either traditional or offsite delivery
Possible issues	Looks like an airport terminal and relies on staff to configure spaces within standard shell	The timber canopy roof is quite distinctive, but may not be to everyone's taste	Concrete walls internally not flexible once built	Economies of offsite manufacturing depend upon use with multiple schools

Information flow

One of the key areas of co-ordination is the flow of up to date and timely information to each designer and design contractor party and the review of change implications. This requires an efficient change management system.

Design information flows through four levels to production as shown in Figure 9.6.

Production information flow is traditionally controlled by an information required schedule which provides a date based on the contractor's estimation for the lead time plus a contingency from going out to tender and receiving delivery on site. In dealing with technical everyday queries a proforma is used to identify the query and to record a response. A timescale is normally indicated. Alternatively this can be incorporated on an extranet online which alerts the targeted person by email. This provides a record of the exchange and an alert to the relevant person to clarify information. Under contract

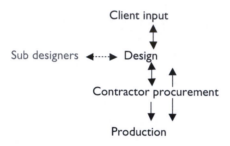

Figure 9.6 Project information flow

Procure	3.10.XX	10.3.XX	17.3.XX	24.3.XX	31.3.XX	7.4.XX
Steelwork contractor	○	△		□	D ▬▬▬	
Roof specialist	○		△		□	D ▬
Windows		○		△	□	D ▬▬

D Delivery
△ Design
○ Order
□ Manufacture

Figure 9.7 Procurement programme for construction work

electronic information is normally acceptable as a formal instruction. A client can be contacted in the same way though they may choose not to subscribe to the extranet.

A procurement schedule programme is helpful. A simple indication of a procurement programme is shown in Figure 9.7. This establishes a framework for main information requirements to allow workload planning by the provider of information.

Design information needs equally critical sequencing and timing for its inputs with the condition that design information from specialist members of the team (consultants and contractors) must be co-ordinated so as not to clash, be incorrect or omit areas. Specialist component information must be made known to appropriate parties, so the information flow is two way and iterative and it needs to incorporate client comments as appropriate. Information in the ground or hidden in the structures of refurbishment projects will be revealed from surveys and later, as construction proceeds.

The Taywood Report on the influence of design on site productivity,[27] written from a contractor's viewpoint, suggests that productivity may be improved if:

- there was a single person responsible for co-ordination
- contractors and specialists were allowed to be involved earlier in the design process
- design is programmed in detail and co-ordinated with construction with regular design reviews and feedback
- construction only started after detail design is adequately advanced.

From a designer's point of view, flexibility is needed for the creative process and time is needed if information is to be correct first time. The nature of fast track projects (overlapping design and construction) contradicts the above, but *fast construction* is a concept to maximise design time at the expense of construction time. This becomes a reality when contractors are taken on early to be involved in the design process so that they can procure in advance in order to preplan the construction phase. For example, prefabrication and simultaneous engineering techniques may be applied to reduce time on site on new build projects with relatively predictable market conditions.

There are opportunities for the client to reduce construction and design time when standardisation is applied and integrated project teams are doing repeat work for the client and can understand the client's business better. These usually involve partnering or framework agreements.

Buildability

Buildability has been termed as the integration of thinking and doing.[28] Buildability is important to the merging of the design process as to what to do and the construction process of how to do it. A contractor will have developed a methodology and a programme in order to carry out the works efficiently. The early involvement of a contractor allows a consultation with the designer from this point of view. Issues that have been dealt with are the material availability.

Change management in design

Change management is an inevitable part of design development and also of client flexibility, both of which are part of an open culture encouraging creative design environments and building in client choice. Design management seeks to control the boundaries of change without closing down the options for later specialist inputs. The RIBA Plan of Work indicates four stages of design – outline, scheme, detail and production. These stages should have defined 'freeze points' which can be agreed with the client by use of a gateway system for approval which fixes change so that the project can consolidate and proceed to the next stage. If this model is used then the following is an example for the boundaries of control:

- *Outline* – client value system is understood and building location, scope, footprint and orientation is established. Outline elevations and layouts based on adjacency plans are proposed and accepted.
- *Scheme* – clear site layout, elevations, plans and cross sections are established with textures and colours of cladding materials so that scheme is suitable for *detailed planning application* often called freezing the brief. Structural and M&E services systems need some definition at this stage especially to assess sustainability.
- *Detail* – internal spatial layouts are finalised and structure and services are firmed up with working detail suitable for tender; no major changes should be envisaged. Specialist contractor inputs are provided. Clarification of detail is given to match type of procurement.
- *Production* – the development of details to suit alternatives and developments on site conditions. In fast track design detailing can overlap to suit package orders.

Managing change also minimises quality defects because it identifies the interdependencies.

The responsibility for the risk of design change is normally shared and may be beneficial. If the design team wishes to make iterative changes which improve the efficacy and efficient delivery of the design this remains their responsibility. If the client wishes to make scope changes or other fundamental changes to their business approach then this remains their responsibility. If a contractor suggests a change or identifies a buildability problem then this means there may be a design problem or a suggestion for an

Case study 9.9 Design development change

The project is a specialist facility for physical research. The building is circular and involves the use of specialist materials, for example heavy concrete. A special company (the client) has been set up for the purpose of running the laboratory and commissioning the works. The contract is inspected by technical staff from the government laboratories and the company has also appointed a project manager who is responsible for vetting the budget and co-ordinating the design and construction sides at a high level. The programme time is 55 weeks + enabling works + a link bridge.

Changes were managed that developed the design. A no blame culture is incorporated to account for up to date knowledge of the project with a clear definition of project scope. In this case packages were tendered either on a bill of quantities (ground works) or on drawings and specifications. Risks need to be properly allocated to parties in the contract and grey areas adjudicated. The Architects Instruction (AI) was to be the official instrument for instructing change. AIs were used to record all changes, whether financial or non-financial, and were also the procedure used for issuing drawings.

However in many cases change is identified by requests for information (RFI) where there is a lack of clarity in the drawings or specification or a clash between them or the BOQ. In these cases a contractual approach is rarely helpful and contingency planning is employed which seeks to focus on the key objectives of operational programme imperatives, which are:

- quality in the sense of fit for use
- overall rather than elemental budgets
- other key client values which may be affected such as sustainability.

This is not an approach which provides a 'contingency sum' as this is unfocused.

In the case study there was a late change in the foundations in order to cope with the ground conditions and this pushed out the programme by three weeks. In order to retrieve the programme loss the drainage layout was revised with main runs outside the foundations so that the drains were taken out of the critical path and other works could be commenced. This meant that key operational programme objectives could be obtained without major budget inflation. For success it depended upon an immediate assessment of the impact of change.

> This shows the close relationship between design, methodology and cost. In order for contractors to cope with their own contingency, they may build in early programme targets to allow for some time contingency if the programme slips. This is becoming less easy where programme times are already tight. Phased handover is another possible option where scope changes have expanded the work, but are not desirable to affect handover of critical sections of the work.

improvement. In all cases the impact on time, cost and quality and other factors are evaluated before the decision to change is confirmed. This needs the co-operation of the client and the design team in managing change and using an early warning system so that the design and construction process is not delayed. However it becomes critical to have a method which controls the process normally in the form of a change management system or protocol. The NEC contract insists on an early warning of problems that occur either because of a change, or because of items that need compensation. An example of a design development change is given in Case study 9.9.

If a design build position is held then the risk for design development lies with the contractor, but the management role on design development does not go away. In open culture procurement incentives for value engineering need to be built in. For example, pain or gain value incentives help to produce benefits for all the team by encouraging innovation and discouraging poor management.

As change is inevitable the key role of change management is to reduce harmful effects of change such as cost inflation, disruption of workflow and expensive time delays which can amount to 10–15 per cent of cost and more. Ming et al. advocated a four factor model in attempting to standardise an effective change management model,[29] which included a:

- change dependency framework to define a standard procedure for managing change in construction projects and understanding dependencies
- change prediction tool to simulate scenarios of likely change and a method for minimising their impact on the planning or the project execution stage
- workflow tool to readjust workflow of teams affected by the change and their interdependency so that productivity is maintained or regained quickly
- knowledge management guide to understand the reason for change and to learn lessons for continuous improvement. Encourages sharing of experience and being explicit about tacit knowledge.

They also distinguished between elective and required changes – the latter where you have no choice to change. If it is unexpected as well, then change management involves adjusting subsequent parts of the programme and budget to incorporate it. There are many direct effects of change such as addition or deletion of work, abortive or rework causing delays or disruption, specification change and reorganisation of schedules. Sun and Howard

Case study 9.10 Scope change

In the development of a postgraduate education centre, a procurement system was set up which required the individual approval of the client of each of the packages as they were let and the use of the NEC form of contract by which compensation events (changes which had cost implications) were advised on an early warning system by either side as soon as they were perceived as a problem. The client had an absolutely fixed budget on the basis of an agreed mortgage and any additional costs had to be offset by a saving so that the project remained in the same budget limit. A contract period was agreed with a single handover.

In the event the client recognised the necessity to occupy part of the floor area at an early stage to decant another department who became 'homeless' due to late completion of another project. It was agreed with the contractor under an early warning compensation event to accelerate completion of the top floor and the lift and one set of stairs so that occupation could take place two months earlier. In order to compensate for acceleration costs and a protected access to the floor a shell and core finish was agreed on part of the ground floor to be fitted out in a separate contract by the eventual tenant of that area. The NEC system also requires a fast turnaround for costing and reprogramming the change prior to approval of the change. The contractor reprogrammed to show a phased handover without affecting the final handover date. Design changes were also necessary, but where these were not a change in scope they became the responsibility of the contractor. One such change involved the resizing of masonry to suit the lift door size. This was disruptive, but mitigated by agreed changes to the lift design.

These types of changes cannot be envisaged, but flexible systems and contractual procedures had been risk managed to meet the stringent restriction of a totally fixed budget and to deal with any contingencies and design changes.

also cite indirect effects as extra communications, loss of productivity because of disruption to workflow, change in cash flow, co-ordination problems, lower workforce morale and loss of float.[30]

Change suggests that there should be some flexibility built into a project, management suggests that this should be well controlled. Change management can be classified as elective and required change.[31] The required change comes as the result of an unforeseen event such as additional concrete in foundations. Case study 9.10 is an example of managing everyday change to benefit the project.

The elective change is a choice and may emerge from the need to do things better. This may be as a result of a value or risk management exercise or as the result of changing circumstances such as market forces. In either case the change should be managed so that the impact of change is well known by the whole team, including the client, and effective choices are made on the basis of full information

Knowledge management is important in making continuous improvements to reduce changes or the impact of change. Often this is done informally by anecdote in problem solving meetings, but the knowledge is not formalised in a way in which it is passed on to a wider audience.

A change management system needs to be incorporated into every project and taking into account project change at the feasibility, design and production stages.

The role of urban design

One of the key issues in a project is the acceptance of its impact on the environment by the community – there are many examples of this. Consider the examples in Table 9.4.

These may be categorised as related to the building itself, the construction, or the proposed process and its intensity. Urban design influences relate to the process of incorporating social values and may influence patterns of land use, transport provision, building design, landscaping, environmental impact, typography and orientation. These all have the ability to enhance the experience for both the occupiers and neighbourhoods of buildings and are normally controlled by the planning authorities, but designed or funded privately. There is often a tension between developers wishing to maximise profit and planners trying to maximise amenity, which results in ugly compromises – more is needed. Case study 9.11 discusses the role of a watchdog in the achievement of good design standards.

These aims hint at the connection of good urban and building design with sustainability and with the true success of built environment projects themselves, if ongoing social value is considered. CABE commissioned a report into the value of urban design to selected commercial projects and found that it adds economic, social and environmental value and does not

Table 9.4 Impact of building on the environment

Project	Impact
Heathrow Terminal 5	Impacting on noise and wildlife habitat
Newbury Bypass	Pollution and wildlife habitat destroyed
Nuclear power stations	Fear of major accidents and health fears
Green belt housing	Foreshortening amenity such as views and congestion

Case study 9.11 Commission for the Built Environment in the UK

The Commission for the Built Environment (CABE) in the UK is now part of the Design Council and has been set up to be an objective gatekeeper for good design in the built environment (see Table 9.5). It has a concern for both the quality of the building or infrastructure design and its standing and sense of place in its environment. Design reviews are used to deliver judgements on elected or key buildings and facilities and to suggest improvements. It has a strong role in the design of open spaces and works with local authorities in particular to train them to understand and encourage good design through the planning approval system and for public architecture. It encourages the exploitation of quality and innovative design and the design of buildings sympathetic and knowledgeable to their context.[32]

Table 9.5 The CABE aims

- campaigning until every child is being educated in a well-designed school and every patient treated in a decent healthcare environment
- fighting alongside the public for greater care and attention in the design and management of our parks, playgrounds, streets and squares
- helping people who are looking to buy or rent a new home by improving the design quality of new houses and neighbourhoods
- building up the evidence that good design creates economic and social value, so that investment in good design is seen as a necessity, not a luxury
- keeping every client of a new building on their toes by demanding design quality in projects of all shapes and sizes in all parts of the country
- working across the country with all those involved in the planning, design, construction and management process so that opportunities for design innovation are always exploited
- thinking ahead to the new demands that will be made of our built environment in 10 or 20 years' time, through drivers such as climate change, technological advance and demographic change

have to cost more or take longer.[33] They found that it also gave high invest-ment returns and open up investment opportunities in successful urban regeneration.

Loosemore *et al.* (2004) believe architects should take a greater interest in urban design issues, because they are in a unique position to develop if not lead and could have a critical impact on the project's success.[34] CABE, who support this, have discovered in a survey of UK local authorities that less than half have an urban designer on their staff to assess planning applications for design quality and only 26 per cent turn down more than 20 projects per year on the basis of poor design.[35] The reasons given by the authorities are that there is lack of policy guidance and support from the government and there is tremendous pressure to accept borderline design which is adequate and yet not inspiring when it is outweighed by the other economic benefits to the community. Further, because of the resource pressures, they have not applied pre-application negotiation to improve a schemes design and help it enhance its context. This raises another issue about the ability of the public planning process to filter out poor design and to consider it on a wider context of urban design.

CABE is trying to get the public more involved in the design and manage-ment of public space and more critical of the standard of new housing developments, by giving them information, getting local authorities more committed, trained and supported by the statutory planning system so that they recognise regeneration and neighbourhood renewal as principal strategic goals.[36] Case study 9.12 gives an example of a sensitive urban site.

Further work by the Audit Commission, in a joint publication with CABE and OGC (2004)[37] in assessing good value in general, outlines six value drivers that include good design and the positive impact on the locality. Included in the 'impact' value driver is the capacity to create place, the experience of the users and visitors in the internal layout and the enjoyment in use of the building, its looks, quality and 'clarity of composition' in the detailing and its ability to present a distinct corporate image. Interestingly another value driver is effective project management, which relates to the efficiency of the delivery process that for public buildings has recently been under the spotlight.

The metrics which they propose for assessing positive impact on the locality are:

- post occupancy evaluation
- design awards
- design quality indicator (a mix of impact delight, build quality and functionality, assessed by client, designers and stakeholders on a six point scale)
- response of the local authority.

Case study 9.12 City centre mixed use development

This £300m, five year mixed use development using brown field space in the centre of a provincial city in the UK was presented for planning permission by a private property developer with particular interest in city prime residential development on brown field land. This was part of a larger scheme by the city council to regenerate the area and regain a valuable waterside site for public enjoyment allowing better usage of the existing docks and access to important historical sites. Any developer would receive partnership money towards the scheme to help excessive infrastructure costs, but would also have to sell the scheme with less parking as a congestion reduction measure.

The scheme was commercially viable, was potentially highly advantageous to the city and was going to clean up an obsolete gas works, but the scheme took nearly four years to get through planning, because of a strong community objection to the first two proposals which gave bad publicity to the developer. The scheme was eventually passed by the dogged determination of the developer to overcome the hostility that had now been engendered and using a neutral third party to redo the Master Plan and to publicly consult to uncover the complex set of variables which might satisfy public objections and still remain viable. Simple suggestions were taken on board by producing better site lines to the cathedral from key viewpoints and by providing more open residential views and wide ranging public access to the waterside and around the residential areas.

This scheme presents an obvious case for urban design that the developers thought they had taken into account with public access and leisure requirements as well as providing access improvements to the road. Two reasons have been put forward for the strong public reaction:

- a sense of ownership of an area of the city which connected up to its history and the need to enhance this with any new development
- a strong lobby provided by the Civic Society to organise the latent feeling which existed and the lost city centre parking which would result.

These metrics seem plausible, but for the reasons discussed above require more resources and education, if they are to be driven by the local authorities. Their value assessment tool chooses weightings for a wide range of criteria, but an example for a court building which has stunning looks and impact and low level maintenance costs still only gets a score which indicates room for improvement. It loses out, of course, on the high weighting areas of financial performance, business effectiveness and delivery efficiency.

Conclusion

It is important to understand the iterative nature of design before you can manage it and give time for reflection and development. Management needs to optimise creativity and design impact in its assessment of value and performance. Design impact will pay back with long term benefits. Design may be managed effectively by a designer or a non-designer who may be the architect, the project manager or the design and build contractor. Efficient information flow between designers and contractors is a key aspect of productivity. This may be enhanced by the earlier involvement of the contractor in the design and the use of a single design and construction co-ordinator.

The management of design is most applicable in controlling cost from outline brief to a detail stage, which is characterised by the co-ordination of many different specialist designers including supplier input. At detail design stage it has a two way flow with the production process which continues to need the input of design. A design manager has a responsibility for ensuring that procured packages have the correct design inputs and produce their design outputs for approval and this crucially means buildability and the elimination of errors to reduce wasteful rework and delays.

The new challenges for integrated design management are:

- Providing a value enhancing service to the client by understanding their business and enabling a co-ordinated approach to the elimination of design waste.
- Optimising creativity in reduced overall project implementation (scheme design and construction) and the use of new technologies to reduce carbon usage and make buildings more liveable.
- Working in the context of sustainability and to co-ordinate the backwards linkage of the building with urban design for true project success taking into account the final users and the social impact of the building, which does not work properly by depending on the statutory planning process alone.
- Developing of teambuilding by the use of excellent communications on complex projects. This can be done through collocation, or through virtual teamwork by standardising electronic documents to remote

positions supported by face-to-face communications wherever possible.
- Getting interest from the client and other stakeholders in the design context so that urban design is enhanced by new buildings which have specifically tried to complement the urban sense of place.

Value management is a key companion to design management and with its potential to release more into the budget and more closely match the client requirements through workshops is certainly a much more integrated approach than a solely client–architect relationship. It has the potential to look at the strategic process as well as the functional components of design and to match them to other life cycle and contextual issues such as urban planning and funding and tax breaks.

References

1 Gray C., Hughes W. and Bennett J. (1994) *Successful Management of Design. A Handbook of Building Design*. Reading: Centre for Strategic Studies in Construction, University of Reading.
2 Hirano T. (2000) 'The development of modern Japanese design'. In V. Magolan and R. Buchanan (Eds) *The Idea of Design: A Design Issues Reader*. Cambridge MA, MIT Press.
3 RIBA (2000) *Plan of Work*. 2nd edn. London, RIBA Enterprises.
4 Maguire R. (1980) 'A conflict between art and life'. In *Architecture for People: Explorations in a New Humane Environment*. New York, Holt, Rinehart and Winston, pp.130–132.
5 Gray C., Hughes W. and Bennett J. (1994) *Successful Management of Design. A Handbook of Building Design*. Reading: Centre for Strategic Studies in Construction, University of Reading.
6 CIC (2000) *Construction Project Management Skills*. London, CIC, p.11.
7 Gray C., Hughes W. and Bennett J. (1994) *Successful Management of Design. A Handbook of Building Design*. Reading: Centre for Strategic Studies in Construction, University of Reading.
8 Gray C., Hughes W. and Bennett J. (1994) *Successful Management of Design. A Handbook of Building Design*. Reading: Centre for Strategic Studies in Construction, University of Reading.
9 Royal Institution of British Architects (2000) *The RIBA Plan of Work*. London, RIBA Enterprises.
10 Gray C., Hughes W. and Bennett J. (1994) *Successful Management of Design. A Handbook of Building Design*. Reading: Centre for Strategic Studies in Construction, University of Reading.
11 Simister S. (2007) 'Managing value'. In R. Turner (Ed.) *Handbook of Project Management*. Aldershot, Gower.
12 Ashworth A. (2010) *Cost Studies of Buildings*. Harlow, Pearson Educational.
13 Ashworth A. (2010) *Cost Studies of Buildings*. Harlow, Pearson Educational, p.473.
14 Kelly J. and Male S. (1993) *Value Management in Design and Construction: The Economic Management of Projects*. London, E&FN Spon.
15 Kelly J. and Male S. (1993) *Value Management in Design and Construction: The Economic Management of Projects*. London, E&FN Spon, p.174.

16 Kelly J. and Male S. (1993) *Value Management in Design and Construction: The Economic Management of Projects*. London, E&FN Spon.

17 Kelly J. and Male S. (1993) *Value Management in Design and Construction: The Economic Management of Projects*. London, E&FN Spon.

18 Lupton S. (2000) *Architect's Plan of Work*. London, RIBA Enterprises.

19 Westcott T. (2005) in P. Fewings (2005) *Construction Project Management: An Integrated Approach*. Abingdon, Taylor and Francis, p.189.

20 Bytheway C.W. (1965) 'Basic Function Determination Technique', *Proceedings of the 5th National Meeting of the Society of American Value Engineers*, 2, 21–3. In J. Kelly, S. Male and G. Drummond (2004) *Value Management in Design and Construction*, Chapter 3. London, E&FN Spon.

21 Westcott T. (2005) in P. Fewings (2005) *Construction Project Management: An Integrated Approach*. Abingdon, Taylor and Francis, pp.190–191.

22 Kelly J. and Male S. (1993) *Value Management in Design and Construction: The Economic Management of Projects*. London, E&FN Spon.

23 Kelly J. and Male S. (2002) Value Management. In J. Kelly, R. Mortledge and S. Wilkinson (2002) *Best Value in Construction*. Oxford, Blackwells, pp.87–94.

24 Egan (2002) *Accelerating Change*. London, Strategic Forum for Construction.

25 Wolstenholme A. (2009) *Never Waste a Good Crisis*. London, Constructing Excellence.

26 Based on Cousins S. (2011) *A New Uniform Construction Manager*. Ascot, CIOB pp.17–20, June; and Tarmac (2011) *Welcome to the Surprising Future of Construction* Building Magazine Supplement. September.

27 Taywood Engineering Limited/DETR (1997) Report No 1303/96/9383. *The Influence of Design on Construction Site Productivity*. A Partners in Technology report, p.18, Table 4.

28 Charles Rich Consultancy (2011) *What is buildability?* Available at: http://www.buildabilityadvice.com/page2.html (accessed 8 August 2011).

29 Ming S., and Sexton M., Aouad G., Fleming A., Seneratne S., Anumba M. (2004) 'Managing changes in construction projects'. *EPSRC Industrial Project Managing Change and Dependency. 2001–2004*. Swindon, Engineering and Physical Sciences Research Council.

30 Sun M. and Howard R. (2004) *Understanding IT in Construction*. London, E&FN Spon.

31 Lazarus D. and Clifton R. (2001) *Managing Project Change. A Best Practice Guide*. C556. London, CIRIA/DTI.

32 CABE (2006) *Design Review: How CABE Evaluates Quality in Architecture and Urban Design*. London, CABE.

33 Bartlett School of Architecture (2007) *The Value of Urban Design*. London, CABE/DETR.

34 Loosemore M., Cox P. and Graus P. (2004) 'Embarking on a project – the role of architects in promoting urban design as the foundation of effective project management'. In *Architects' Handbook of Construction Project Management*. Chapter 6. London, RIBA Enterprises.

35 CABE (2003) *Survey of Design Skills in Local Authorities*. London, CABE.

36 CABE (2011) *Creating Excellent Buildings*. Available at: http://webarchive.nationalarchives.gov.uk/20110118095356/ http://www.cabe.org.uk/buildings (accessed 12 January 2012).

37 Audit Commission, CABE and OGC (2004) *Getting Value for Money from Construction Projects Through Design: How Auditors Can Help*. March, pp.22, 8, 6.

Chapter 10

Project safety, health and the environment

Construction sites have the potential for serious accidents as there are many people close together, many activities are unpredictable because of the dynamic nature of project work and the tolerance towards risk is traditionally quite high, making the frequency and impact of accidents very high. Health conditions are less visible, but just as critical to improving conditions and their link with healthy living environments has given this a higher profile for the construction site conditions, but critically for the designer, for the finished building user and the maintenance team. The mass of legislation designed to reduce accidents has not been as effective as could be hoped and attempts to improve are ongoing. Accident prevention, good management of health and safety and the building environment are the key to improving health and safety and reducing accidents that occur.

The objectives of this chapter are to:

- analyse principles of workplace health and safety legislation
- further the principles of safe and healthy building sites and buildings
- discuss a framework for managing project health and safety through its life cycle for the prevention and improving reduction of accidents
- propose a working system using the Construction (Design and Management) Regulations 2007
- consider the impact of behavioural safety approaches to health and safety
- look at the impact on SME contractors
- provide a critique for effective health and safety planning and risk assessment.

Introduction

The International Labour Organisation have estimated that worldwide there are up to 268 million new cases of work related injuries caused by various types of exposure at the workplace, 160 million new cases of work related illness and at least 60,000 deaths on formal and informal construction sites worldwide.[1] These account for about 17 per cent of the total occupational fatal accidents. Health and safety on construction projects has a notoriously bad record and many countries have a low inspection rate. Construction has a particularly wide range of hazards to health and injury including the requirement for extended heavy physical work in ergonomically difficult positions exposed to the weather, dust, noise and fumes, the use of many corrosive chemicals, regular working at height, the wide use of machinery and power tools in a dynamic environment when conditions change regularly. An EU report produced in 1992 showed a survey of worker perception of work safety in construction was 10–20 per cent worse than other industries, and on EU sites in 2008, injuries were still twice the occupational average. The most common workplace illnesses are cancers from exposure to hazardous substances, musculo skeletal circulatory and respiratory diseases, hearing loss and diseases caused by exposure to pathogens. Low skilled workers in rapidly industrialising countries have a particularly high rate due to the lower skill levels of workers. Accidents level off as there is a shift from construction work to service industries as has happened in Korea, but has got worse in countries such as China and India. The ILO and WHO work together through 70 collaborating centres worldwide to introduce awareness for risk factors and training for primary prevention.

The contruction industry in the UK has improved its accident record in real terms over the last decade but rates are still the third worst after agriculture and mining. Case study 10.1 gives a summary. It is easy to blame a 'cowboy' element for the depressed figures, but it is clear that there is still a major problem. Egan (1998)[2] claimed that accidents can account for 3.6 per cent of project costs and this also gives an economic as well as a

Case study 10.1 Accident figures in the UK 2001–2010

In the UK, the Health and Safety Executive (HSE) indicate there were 50 deaths in 2009/2010 with an average rate of 2.8 fatalities per 100,000 on construction sites in the last five years. This is approximately 2.5 times worse than manufacturing. Major injuries can be added to this and the current rate in the Labour Force Survey (LFS) for all injuries is about 1,200 per 100,000 and for illness is more, at 1,650 per 100,000.[3] Accidents are going down in construction, but not enough. The Construction Industry Advisory Committee (CONIAC 2001) set targets to reduce accidents and incidence of work related ill health by 66 per cent and 50 per cent respectively by 2009/2010 and the number of working days lost from work related injury and ill health by 50 per cent.[4] In this respect the injury rate is down by 34–63 per cent in the different categories over the 10 year period 2001–2010.[5]

social and ethical reason for improving performance. In Wolstenholme's (2009)[6] follow-up report in the UK, only 10 per cent believed that health and safety had improved, although encouragingly 81 per cent felt that they were committed to training and development in better health and safety. Health and safety however was ranked by managers as fourth in importance after commitment to people, sustainability and client leadership.

There are many concerns for poor health caused by exposure in the workplace. The most common is back and other musculo skeletal injury or as an impairment or illness initiated at work. In addition asbestosis and related diseases caused by asbestos exposure are worrying because they have a long gestation as a cause of death. That means that even though asbestos use has been banned for some time in many countries it is a major player with the number of people dying from this disease, for example in the UK, still rising from 200 in 1978 to 400 in 2008 and more getting disablement certificates. Worldwide this figure is 100,000.

Accident and near misses

The HSE has reviewed and reported on the hierarchy of accident types with falls from height listed as the most common serious accident which occurs through falls off or through roofs, scaffolds and ladders, or into excavations. Studies seem to agree that there is a proportion of at least 1:7 between serious reportable and minor, non-reportable accidents and one major report indicates 16 times. Non-reportable accidents need some first aid and there are more than 12 times as many again near misses,[7] producing an iceberg effect shown in Figure 10.1a, where there is a large bulk of accidents which will not be known about, or do not need to be reported. The HSE has shown that 33–50 per cent of accidents are not reported officially by employers

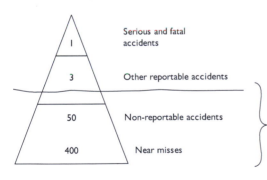
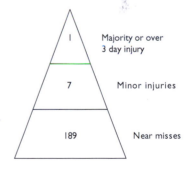

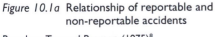

Figure 10.1a Relationship of reportable and non-reportable accidents

Based on Tye and Pearson (1975)[8]

Figure 10.1b Relationship of near misses to reportable accidents

Based on APAU (1997)[9]

when compared with the LFS which is an annual survey direct to the work-force and this is taken into account in Figure 10.1b.

Stress is also a factor where 30–50 per cent of workers in industrialised countries complain of psychological stress and overload. There is a connection here with environmental policy which is also concerned with the impact of construction on the health of others.

There has been a whole raft of legislation which has hit the industry, in Europe generated by directives from the EU.[10] It is important to be aware of the principles of the legislation which were set up in the Health and Safety at Work Act 1974 (HSWA) which underpins much European health and safety legislation. In the UK, the Act set up a framework for more focused non-prescriptive risk related regulations which put the onus on the employer to prove a safe workplace rather than technical contravention. The EU directives have not wavered from these far sighted guidelines.

Principles of modern health and safety legislation and the HSW Act 1974

Prior to 1974 there was a piecemeal industry approach in the UK with quite antiquated and often contradictory legislation. In construction the Factories Act 1890 applied to construction with the lifting, health and welfare and construction operations regulations and to a lesser extent the Offices, Shops and Railway Premises Act 1964 were used as a basis of health and safety on construction sites. In the early 1970s a new committee was set up chaired by Lord Robbins in order to put more emphasis on preventing accidents, and draw the older legislation together.

Integrated guidelines to provide simpler health and safety legislation with parity across all industries were required and the resulting Health and Safety at Work Act 1974 (HSWA) has provided robust forward looking principles which have allowed modern regulations to be produced for all industries under the umbrella principles of the Act. These principles are based on a corporate and personal responsibility and not a prescriptive approach:

- The responsibility of the employer to provide a safe working environment to employees.
- The responsibility of the employee to comply with all reasonable provisions for their own safety and to act responsibly towards others.
- The responsibility of the employer and the owner of premises being accessed by others, to other persons who are not employees.
- The responsibility of designers, suppliers and plant hirers to provide safe products and components.
- The responsibility for employers to prepare a safety policy.

- It provides for criminal as well as civil liability and set up the HSE to implement and enforce the Act and the Health and Safety Commission (HSC) to form policy.

The point of this approach is to encourage a risk assessment method which puts the onus on the employer and building/site owner to adapt and develop measures relevant to their context, rather than a blind adherence to prescriptive measures by Regulation. It also allows, in the requirement, for individual identification of responsible person's accountability at different levels for measures which are sufficient to stop accidents. It is quite clear, for instance, in the Temporary and Mobile Sites Directive, to follow this principle of identifying persons who will assess the risks and set up measures to control and reduce those risks.

In the UK enforcement the HSE has an advisory and prosecuting role in construction through HSWA 1974 (sections 10–14). Enforcement can lead to a notice to rectify specified faults promptly in the form of a:

- warning
- improvement notice
- prohibition notice
- prosecution.

A prohibition notice would shut down the site until suitable action has been taken to make the site safe again. A warning or an improvement notice is for lesser problems and allows work to continue with a requirement for action to be taken within a strict time limit. Prosecution with resulting fines or imprisonment can be imposed in the case of non-compliance to the notice where the employer is proven guilty in court. In the case of serious accidents resulting in death or serious injury an employer or their agent is criminally liable to prosecution. Fines can be quite large and individuals can also be subject to prison sentences in certain serious circumstances.

Other countries have similar occupational health and safety acts which provide a framework for specific construction regulations.

Health and safety policy

All organisations and businesses require a health and safety policy in written form where five or more people are employed. The purpose of the policy is to have a clear statement of commitment to the health and safety of employees and to direct the accountability for safe systems of work to named people. A director or partner must be named with executive responsibility for the policy being carried out. Where the firm is larger there is a separate health and safety manager who is responsible for monitoring health and safety systems, receiving feedback, arranging training and communications and seeking to improve and maintain systems up to date. Smaller companies often share an agency that provides many of these services for them, but must still hold accountability themselves at owner level. The six main areas to address for any organisation writing a policy, including a construction project, are:

- How to prevent accidents.
- What training to ensure employees are competent?
- How do you consult with your employees?
- What emergency procedures enable evacuation in case of fire or explosion?
- What basic systems and responsibilities for maintenance of safe and healthy conditions? For example, responsibilities, risk assessment and checking safety regularly.
- Setting up statutory reporting procedures, first aid and accident book (HSE 2011).[11]

The primary purpose of general policy is to set out an action plan for health and safety.[12] This is shown in Table 10.1.

In order to carry out the policy it is necessary to do a risk assessment asking key questions: What are the hazards? Who might be harmed? What can you do to reduce the risk? Who is responsible to see that it is working?

Organisation of H&S

Overall responsibility for policy and implementation lies with a director or senior partner of the organisation(s) carrying out the work. In a larger organisation they are likely to delegate the development of co-ordinated arrangements to a senior manager to provide continuity and parity across contracts. It is also usual to have safety experts who visit on behalf of the main contractor and subcontractor organisations doing work on site and produce reports for improvements. The preference is for the use of competent employees for health and safety advice, rather than external sources, but this may not be possible in SMEs. Figure 10.2 indicates a typical organisation structure.

On a construction project overall project responsibility will lie with the project manager or other named person who will have health and safety training. He/she is responsible for implementation of the company health and safety policy and co-ordinates the project health and safety plan. He/she monitors the system set up for the project, reports and investigates accidents and near misses and receives feedback from safety representatives and specialist contractors. Under CDM, the principal contractor has special powers to make site safety rules, recognising the co-ordinating function of a main contractor.

Table 10.1 Action planning for the health and safety policy

Main action areas	Application to construction work
Develops safe systems of working on site through *risk assessment* (more below)	Consult those who are doing work for you – construction work is dynamic so each project is assessed and continuously monitored to adapt to conditions. Each employer supplies written risk assessment and principal contractor co-ordinates interfaces, entry, evacuation, reporting and common areas such as access, delivery and excluded areas
Establishes a communication system for informing project workers of health and safety matters	More than paperwork – it is best direct, by using regular inductions and updates and for existing workers and regular toolbox talks. Training courses for managers
Gives details on who will be responsible for health and safety and delegates roles	Primarily the site manager/project manager is responsible and has to delegate specific areas of work for monitoring and checking
Stresses the importance of the co-operation of the employees to using these systems	Set up worker forum which meets monthly or more often on bigger sites
Sets out disciplinary procedures applicable for contravention of safety rules	Stringent indictment for contravention and reward for best practice and improvements

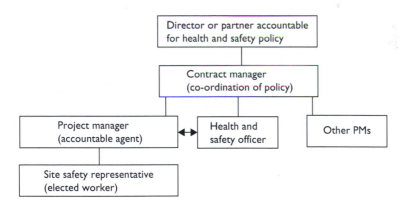

Figure 10.2 Organisation to show responsibilities

Worker involvement

The Health and Safety (Consultation with Employees) Regulations 1996 recommend the appointment of workforce representatives who may be 'ears and eyes' in order to report on unsafe practices on site. This does not take away from the responsibility of all employees and visitors to report dangerous or unsafe structures or practices which they come across. The HSE initiated a campaign to get people involved in their own safety in 2009 based on trust, respect and co-operation.[13] This can centre on the idea of accident prevention, risk assessments and effective inspections. It is less well established in non-unionised or temporary workforces like construction as the unions have taken the role seriously, but it becomes the initiative of the employer to ensure that it happens otherwise. In construction there is a more top down tradition. Each worker is an individual, but engaging a competent and resourceful workforce trained to build in improvements makes sense and removes traditional perceptions that health and safety is 'your problem and a load of bother'. Self-managed teams have a track record in employee engagement in safety. This has led to worker safety advisors who receive training and act as roving safety representatives given time in the same way as a union steward would be for essential duties. The starting point should be the writing of risk assessments by workers aided initially by other trained workers.

The main barriers listed in the RoSPA (2010)[14] research suggested pressure of work, getting everyone together, poor communication, masochistic attitudes amongst workers which had done the job 'that way' for a long time

Case study 10.2 Timber yard worker involvement[15]

A private company which has two timber yards with a turnover of £100m and 400 employees has a H&S and quality officer on both of its sites. It runs a forum for the H&S representatives every six weeks and has an ad hoc safety awareness meeting with the responsible director. It has attempted to move away from 'tick box compliance' towards a change in culture to 'integrate the function into the wider business picture'. They have a standard agenda to cover statutory issues, general H&S, environment, first aid and emergency issues. They raise the specific issues before the meeting to ensure time for consultation with the workforce by the representative. Representatives provide direct feedback to the workforce after the meeting. A small amount of training is given. A health and safety forum operates on an ad hoc basis when a particular issue needs more discussion. There is a H&S suggestion box and a H&S newsletter.

Case study 10.3 Manchester Royal Infirmary[16]

A union specialist was appointed as a site safety officer to advise and train site representatives and to chair the site safety forum so that there was an explicit emphasis on the worker contribution and making it more focused and relevant to site wide measures. These representatives are encouraged to propose amendments which produce a better risk assessment and also to stop work where they think they are ignored. An example of this was the use of a platform at height without proper edge protection, which the operatives insisted on using after being pulled up on the safety aspect. The site representative informed the union site advisor who immediately stopped the work without the sting of management overkill. The power for a concerned operative to stop work in the case of a worry about safety and report their concerns to a supervisor was instituted in the spirit of imminent danger and gave real involvement in decision making.

and management accountants who saw H&S expenditure as expendable. Better management commitment to real communication and feedback for ideas seemed to be a well-run comment for improving existing systems. This would also keep things fresh and regenerate interest in employees. One interesting observation was not to have a system of hierarchical committees where ideas got lost or meaninglessly diluted. There was a need for allowing workers to be involved in decision making in the most effective schemes. Case studies 10.2 and 10.3 illustrate this.

A construction project is a unique workplace and it is compulsory to set up a health and safety plan within the tender documents, which is developed by the chosen contractor during the mobilisation stage in order to set out the strategy and policies, which are relevant to the project's unique circumstances. It is also important to keep this under review as the interfaces between different contractors working on the project are dynamic and so new risks arise. One of the commonest causes of accidents is when the sequence is altered and there is an unplanned and dangerous mix of activities going on alongside each other. For this reason the Construction (Design and Management) Regulations were developed to establish a transparent set of responsibilities for the client, the designer, the principal co-ordinating contractor and the individual contractors to set up clear responsibilities on a multi-organisation temporary worksite such as a construction project.[17] It is important for these to all work together to provide information, to set up training and induction procedures and to co-ordinate the design and construction process. The CDM co-ordinator also co-ordinates the design health and safety.

There are an estimated 200,000 trade union safety representatives in UK industry and organisations and these have a function (not duty) to investigate accidents and near misses and inspect premises where they work or others they have been appointed to represent.[18] They may also consult with and receive information from enforcement officers and require meetings with employers and receive information from the employers such as risk assessments, accident reports and hygiene reports. They are given a basic 10 day union training and represent their members. In construction this representation would be supplied through UCATT and the various specialist unions. Companies have a history of resisting training and workplace inspections as it requires paid time off.

In construction, except on large sites, many are no longer in the union as they work in small disparate groups for specialist contractors. Clearly they miss out on the benefits of safety representation which, according to a TUC survey,[19] was a more important function for the unions than workplace conditions and pay.

Reporting

Reporting of accidents becomes a statutory requirement in most countries. In Europe, under the Reporting of Injuries, Diseases and Dangerous Occurrences Regulations 1995 (RIDDOR),[20] this is whenever there are:

- accidents involving major injury, death or more than 3 days off work[21]
- dangerous occurrences on site which incurred a near miss, even if there was no injury, or incurred damage only
- significant health hazards such as poison, bacteria/virus or pollution
- occurrences of a dangerous disease.

Near misses may cause substantial property damage, but reporting should be against the significant propensity to cause harm such as a collapse of a scaffold. This allows the HSE to investigate accidents and provide accident statistics. It also allows employers to learn from the incident and to improve their systems to mitigate future danger. There is however a serious shortfall of reported injuries which is indicated by the discrepancy between the annual Labour Force Survey (employee response) and RIDDOR (employer reporting). This indicates that only 40 per cent of accidents are reported by employers.

All accidents however small should be recorded in an accident book. They do not have to be reported officially if they are outside RIDDOR, but it is important to review any accidents and reasons which are symptoms to see if there are any patterns which indicate poor practice that can be

Case study 10.4 Tower crane accidents

There have been a number of tower cranes which have collapsed since 2000 and an investigation to see if there is a reason for this has thrown up requirements by the HSE for more qualified inspection particularly of the crane parts which have failed by both the crane hirers and the regular users of equipment. These tightened up the formal requirements under the Lifting Regulations (LOLER).[22] A campaign by *Building* magazine also bought pressure from external stakeholders who were perhaps neighbour to a construction site.

improved. Best practice contractors have a system to study the cause of accidents and to provide new guidelines to eliminate causes with the potential to recur. Case study 10.4 illustrates how guidelines developed for a particular type of accident which was becoming common.

EU directives leading to UK regulations

EU directives which are adapted into UK regulations under the umbrella of HSWA 1974 all have adapted a pattern of risk assessment identifying responsibilities. The way in which this is done is slightly more prescriptive than HSWA.

The Framework Directive (89/391/EEC) for 'measures to encourage improvements in the health and safety of workers at work' led in the UK to the Management of the Health and Safety at Work Regulations 1999 (MHSW).[23] This defines the risk assessment process and prescribes generic measures such as training, planning, health surveillance, organisation and monitoring and escape and the use of workers in risk assessment, which provides principles for a wide range of more specific regulations.

The Framework Directive led also to six daughter directives applying to any workplaces to deal with such areas as hazardous materials, manual handling, biological agents, use of work equipment, noise, electricity at work, protective equipment, equipment safety, visual display screens, pregnant workers, protection of young people at work, safety and health signs and worker safety representation. Figure 10.3 indicates the framework.

Each of these regulations specifies certain hazards and limits in relation to the areas indicated and expects a risk assessment process with control measures to reduce risk, training as necessary and information given out for employees' information. The Temporary or Mobile Construction Sites' Directive (92/57/EEC) led to two UK regulations which apply to construction workplaces in particular; these are:

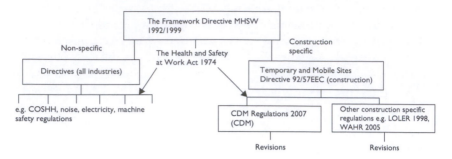

Figure 10.3 The Framework Directive and daughter directives

- Specific, but not exclusive, regulations such as lifting equipment (LOLER 1998) and working at height (Working at Height Regulations 2005- WAHR) and Hard Hat regulations.[24]
- The Construction (Design and Management) Regulations 2007 (CDM) identify prime client responsibilities, the co-ordination of supply chain risk assessments, health and safety plans, general site organisation, designer responsibility for user safety and co-ordination of designer safety measures. The ACOP provides better guidance for designers, coordinators, contractors and clients. CDM part 2 now incorporate the specific measures for construction hazards, health and welfare.

The main difference in construction is the temporary nature of the workplace with unique new hazards and the need to fundamentally consider the context risk. For example, the use of foreign workers for a bricklayer gang has different risks than an English speaking gang, even on a house of the same design. Safety posters will communicate universally if they are more pictorial. In addition the access to another house in the same workplace will have different hazards. There are also physical changes taking place throughout progress of a construction project, which create a dynamic risk environment with key hazards that need communication daily. The use of competent people permanently on the project becomes critical for continuity. The CDM Regulations identify responsibilities for the management of safe environments.

The Lifting Operations and Lifting Equipment Regulations 1998 (LOLER) replace the Construction (Lifting Operations) Regulations 1966. The Provision and Use of Work Equipment Regulations 1998 (PUWER) describe in general terms the safe use of plant and equipment and insist on use only by competent and trained persons. WAHR 2005 take over parts of the old CHSW 1996 by stiffening up the definition of working at height and ensuring proper risk assessment for falls from height and dropping objects which together are by far the worst cause of accidents on construction sites. Fewings refers to the need to reduce working at height and creating a culture of using more prefinished components that reduce working time at height.[25]

CDM part 2 with LOLER, WAHR and PUWER provide a prescriptive framework to deal with specific construction related risks associated with accidents, health and welfare. Typically these are the safety of excavations, falls off or through working places such as scaffolds and roofs, stability of structures, demolition, explosives, falling objects, prevention of drowning, vehicle movements, emergency routes and procedures, fire detection and fire fighting, fresh air to and lighting of work areas, traffic routes, doors and gates and safety in the use of work equipment. Regular inspection and training programmes for competent use are a key part of the operation of these regulations and welfare facilities must be provided and maintained adequate to the size of the workforce. The key issues to check are given in HSE (2009).[26]

The Control of Substances Hazardous to Health Regulations 2002 (COSHH)[27] have been regularly updated to keep up to date with European directives which have become tougher on exposure to biological, dust and fibre, chemical and other corrosive or dangerous emissions hazardous to health, some of which have special regulations like those for asbestos. The principle is to risk assess the use of the substance and to provide safe instructions for use and storage. For example paint with volatile fumes has to be justified against the use of water based substitutions. Questions to consider are:

- Does the product have a danger label? – all hazardous products which affect breathing or are corrosive or let off dust or fumes need one.
- Does your process produce gas, fumes, mist or vapour which are inflammable? – avoid or reduce and provide masks.
- Can the substance harm your skin, or are you likely to ingest fumes or spray or other biologically harmful agents? – avoid or reduce and provide PPE.
- Is harm going to occur because of the way you use it?

Reportable disease includes certain poisonings, some skin diseases such as dermatitis, lung diseases such as mesothelioma and some infections including legionellosis and hepatitis and other conditions such as musculoskeletal conditions. Occupational cancer is often the result of working with substances or materials which are harmful in the short term such as asphyxia or burning, or harmful in the long term such as gradual gassing from urea formaldehyde which builds up harmful fumes in enclosed spaces.

CDM and the management and design responsibilities

CDM 2007 cover the specific management co-ordination, design and client responsibilities to manage health and safety created by the unique fragmented organisation of construction projects. Two *co-ordinators* called the CDM co-ordinator (CDMC) and the principal contractor (PC) are appointed by the client and are responsible for co-ordinating respectively the design and construction health and safety processes. They are not directly responsible for the actions of individuals or other employer liability, but they do have a responsibility for interface management and ensuring good communication between different parties in design and construction. They have to produce a health and safety plan before construction starts incorporating key risks for construction, which is regularly reviewed.

Responsibilities are given to the *designers,* and the *individual contractors* to risk manage their work and to be aware of the effect it has on others, so that they provide risk assessments to the co-ordinators who use these risk assessments to compile a co-ordinated approach both for construction activities and for the building in use. The CDMC may make recommendations for amended design or more information, but does not make themselves liable for unsafe design or methods of installation. This liability remains with the designers and contractors who are considered competent.

The *client* has the prime accountability for health and safety and is responsible for appointing competent co-ordinators and ensuring that a co-ordinated health and safety plan has been developed by the principal contractor before the construction work starts. In practice this is done on the advice of their professional team as the client may not be experienced in construction. The client is liable for not passing on information about hazardous substances or conditions which exist before the site is handed over and they are also responsible for ensuring that sufficient resources are available to cover the costs of sensible health and safety measures. This will be covered by appointing a main contractor who has priced and is committed to proper health and safety measures as indicated in a properly developed contract health and safety plan. The CDMC and others may advise the client when selecting the tender.

The *designers* have responsibilities to produce a safe design and are expected to produce evidence of risk assessments for each stage of the design process. They are liable for accidents or health problems attributable to designs which contribute to unsafe environments for the contractors, the users, the maintenance team and those who demolish buildings.

The designers have responsibilities for safety for building users and for *maintenance workers* which means safe access for repairs, controls and cleaning. Examples of maintenance hazards can include the poor access for cleaning building components working at height or over glass roofs, inadequate or unsafe access to services in the ceiling space, or plant or machinery on roofs, or access around hot machinery or endangering contact with hazardous materials. These issues are properly considered in the scheme design as facilities may feature in the external elevations and in the spatial arrangements for service installations.

For the *user* of the building it means being protected from trip and fall hazards and electrocution and being able to escape safely in the case of fire. Services design also has a role to play in ensuring healthy buildings and preventing build-up of infections such as legionella or general poor conditions sometimes alluded to as sick building syndrome. Buildings which become very hot or fail to induce sufficient air circulation can also incur uncomfortable and unhealthy conditions.

Safe design for the *construction* phase means the proper selection of materials which are safe to use and do not force heavy personal lifting or difficult detailing forcing non-ergonomic working. It requires good information for the hazards and use of hazardous substances and full interpretation of surveys which have identified contaminated ground, unsafe structures or other material hazards such as asbestos. Design can assist in reducing work at height such as prefabricating components so there is less in-situ fixing and in reducing component weights. Safe methodologies are the liability of the contractor and important discussions on buildability can be had when the contractor is on board so that design detailing can be amended to help safe construction.

The *CDMC* is an early design appointment co-ordinating the separated design process. Their responsibility pre-contract is to co-ordinate the health and safety aspects of all the designers and to ensure that documents describing health hazards are fed into the tender documents for pricing ensuring that health and safety hazards are made clear and information is full enough to give 'a level playing field' for each contractor's tender. Later they help in incorporating them into the construction health and safety plan. They also collect and present information from all the designers and contractors, collating all drawings as built, safe operating procedures and use of materials and components guidelines into a health and safety file (HSF).

The *principal contractor* is normally the main contractor who formally manages health and safety co-ordination on site. In practice this requires all contractors to produce risk assessments before they start work, which are integrated or amended by the *PC* to suit the needs of others. Site rules are established to create safe overall conditions, safe common access with escape routes and fire fighting, lifting operations and measures to manage access to areas. They also arrange site wide information, induction and training with updated skills and ensure the wearing of protective clothing. A card indicating a level of competence in the worker's trade or manager's supervisory skills is now compulsory on many sites. Updated trade skills are implicit in safety to include safe ways of working. They need to control access to the site, protect the general public and provide adequate hazard and prevention signs.

Contractors have a responsibility to have safe methodologies for their workers and to others working in the vicinity including the public. As employers they must have a safety policy and provide protective equipment and provide risk assessments for other contractors to see.

Impact of H&S legislation on SMEs

SMEs dominate the construction industry in terms of the number of organisations. They carry out about 60 per cent of the work and that sector has more accidents. They are however notoriously difficult to get on board formally and tend to be non-compliant with significant parts of the legislation such as full PPE, safe scaffold, risk assessments, method statements and paperwork. Research by the HSE has suggested that there are three types of SME philosophies to be found in SMEs which cause them to react differently to health and safety.[28] These are:

- Duckers and divers (DD). They do have some basics in place, but believe they are exempt from 'big site' legislation and will excel in tokenism in order to falsely show compliance. They do not think they have the authority to insist on common measures with others responsible for their own liabilities.
- Confident captains (CC) who believe strongly in their own regime of health and safety and will insist on compliance, but who have a sceptical view of the HSE inspectors who they believe insist on generic conformance, often at the expense of common sense and safety.
- Ex big site conformists (EBSC) who have a remembered regime which they adapt in a more conformist way to their requirements such as inductions, risk assessment and a more uniform use of PPE. They may make some allowances for the site size. Typically they run the larger small site.

In all cases there is an independence of thought with SMEs which moves away from a slavish *'make sites safer'* approach to *'remember it's dangerous'*. This indicates the greater emphasis on the individual's liability that people will learn by their mistakes, though it may be interpreted as moving away from the critical holistic role of the main contractor co-ordinator to stop conflicting H&S arrangements by extensive rules, to the need to make and police simple common arrangements for a few. For both there is a need to cover the longer term 'invisible' health issues.

The research also found that there was an in-built resistance to the HSE inspector who requires conformance to rules rather than applying judgement, experience and common sense arising in what CCs see as compliant procedures which are unsafe (over the top), such as too much PPE so that you lose your balance with say goggles and ear muffs; and where you have to wear gloves you lose your sense of exposure. Alternatively it is perceived as being safe with non-compliance such as wearing trainers rather than boots on a roof for a better grip. The context is also seen important where it is recognised that large sites do not know each other and so do need stronger rules. Some smaller sites are deemed to be safer and rules can be flexible where there is a team used to working with each other. The CC in particular takes responsibility for this type of risk assessment and its control. This type of assessment is not seen as written, but as a perception of who has the final say. The DDs might use token compliance, but are less aware and subject to greater risk if they don't take a leadership role.

As always making these judgement decisions is dependent on a single person or a few people to make it work and when a major accident occurs it may be able to teach people a lesson, but any permanent hurt cannot be undone. The middle line is to consider the barriers to better safety and tackle them. These barriers to accepting the health and safety rules in their entirety are too much paperwork, removing common sense, not dealing with context, ignorance of the actual rules and the reasons for them, and a feeling that the rules are made by others who have experience which is alien to the building experience. Other more uncomfortable barriers are pride, habit, devolution of responsibility, loss of time or money, reluctance to sack or upset people so they won't work for you, unwillingness to take advice from people you don't know, weariness with the subject and the expediency to choose those who are willing to do it the cheapest and fastest way. There is emphasis to learn from others you trust and from your own mistakes and to provide some training to back up belief in your own and experienced tradespersons' awareness.

Health and safety management

Overall health and safety management, including risk assessment, is addressed generically by the HSE.[29] The approach reflects total quality management processes and tries to engender a top management commitment to change and a culture that encourages bottom up participation which is motivated towards improvement. This makes review an important element. Figure 10.4 indicates the HSE's framework for successful management.

The key management issues are a basic policy framework, accountable persons, safe designs, comprehensive planning and organisation, good supervision, segregated or restricted areas, availability of personal protective equipment, careful sequencing of activities which impact on each other and training and induction processes with monitoring and checking. The CDM Regulations 2007 and its Approved Code of Practice (ACOP) are also important for a good project health and safety management system.

Implementation of the system is often harder than developing the system as it depends on developing a better culture of safe working than many construction projects have been able to develop.

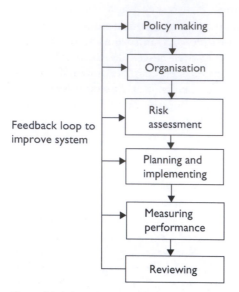

Figure 10.4 A management framework for the management of health and safety
Adapted from HSE (1997).[30] Used with permission

Safe design means information about the hazards of materials specified and revealed in surveys carried out and the risks involved in special methods implied, for example manual handling and contaminated land or asbestos. Supervision cannot be total, but it is connected with knowing what is going on. One way of dealing with this is to require permits to work for particularly hazardous activities such as deep excavations, hot work, confined spaces etc. Other activities will not need to be supervised, but spot checks will need to occur to ensure competent persons are doing their job and that co-ordination with other activities is working.

An effective health and safety culture involves the whole workforce in the planning phase and ensures that there is a commitment for on going training and updating from all the contractors. To achieve this top management support is required so a budget for a quality improvement approach is committed. Some cultures try to remove supervision and replace it with personal responsible behaviour for both assessment and implementation. Some risks must be co-ordinated where many trades are working together in restricted areas and using common access. Some risks may be made worse through unsafe design requirements or poor information about the existing site hazards or the products specified to use.

Risk assessment has been added as a separate phase to planning in order to emphasise the importance of this activity at the start of all planning activities. The concept of risk and its assessment in order to take acceptable action is discussed later. Measuring and review of performance is critical to improvement of safe methods. The review stage is an active stage, which for time limited project work needs to apply at regular intervals and particularly when changes are made in the original planned design or methods.

Liability in health and safety law can emerge from a number of angles, but the chief of these is unsafe design, product safety, negligence in reducing site worker risk, poor site organisation, health impacts which emerge later and public safety.

Culture and behavioural health and safety

There is need to create a more enlightened culture in health and safety management so that H&S is not seen as a necessary extra budget to the company, but as an essential for good business methods which motivates the workforce and is seen as earning money and not leaking it away. HSE claims that a safety culture is built on the 'individual and group values, attitudes, perceptions, competencies and patterns of behaviour that determine the commitment to and the style of proficiency of safety and confidence in the efficacy of preventative measures'.[31]

Project managers require the support of top management to resource a training programme to change attitudes on site and to manage their supply chains to follow suit. Fewings (2010) spoke of the need to create a good awareness culture in the workforce as a whole so that workers are 'eyes open' to the dangers that exist and have been trained in how to mitigate and avoid harm.[32] This would be similar to a mountaineering safety regime where proportionately there are much fewer accidents because awareness and preparation are heightened for what is a known risky environment and which also emphasises the involvement of the skilled worker who is at the workplace.

The CDM 2007 regulations themselves need arrangements for enabling workers to engage in promoting and developing measures to ensure health, safety and welfare and in checking their effectiveness and require worker consultation on matters which may affect their health, safety and welfare.[33]

Construction is never a risk free environment and achieving zero accidents is an attitude of mind which works better if people complement each other and work to the same culture. A safe worker is a knowledgeable worker who wishes to see continuous improvement in best practice. If the project manager can orchestrate a mutual improvement culture in the supply chain then bad habits, short cuts and poor planning will be tackled and a bottom up culture of improvement can be encouraged through worker forums and through spreading a shared learning. This may be seen as a Deeming 'Plan-Do-Check-Act' type cycle where the previous actions are monitored and scrutinised for improvements. Accidents can be seen as a loss of management control, but they may also be opportunities to learn from mistakes so that future practice may be tweaked.

A fresh set of eyes to look at better safety can also help and this may include regular inspections from all levels of management including directors and non-production staff. Case study 10.5 shows good practice.

The example in Case study 10.5 shows a multiple approach on several fronts to involve all the people involved in the contract together. In order to ensure improvements are yielding results it is necessary to use performance standards to measure the progress made. These are likely to be measured on a project in terms of man hours worked without an accident or a dangerous incident report. This has some benefit if it is compared with other projects where there is a good record and can induce a competitive spirit which puts pressure on more poorly performing contractors to up their game. Other KPIs which might impress future clients are accidents per 100,000 people. This can be used corporately to cumulate a company's record as better than average with the national statistics.

Behavioural health and safety

Other company cultures are to develop behavioural challenges. These contrast with refereeing and use a reasoning approach for smaller infringements and look at impact after unsafe incidents that are spotted or reported to challenge the possible outcome if things had gone wrong personally such as personal injury on family life or impact on career if others had been severely injured or killed in the incident and is it worth it? This has often proved effective in amending behaviour where used intensively and consistently – so that a major proportion of the workforce is challenged.

A third culture is 'don't walk by' combined with the idea of zero accidents, to challenge apathy when rules are broken or there is an unsafe environment. The philosophy encourages all workers, and the public and visitors, to report unsafe environments and incidents in a no blame way, so that accidents are prevented by taking immediate avoiding action. All incidents are recorded and analysis done as to the cause to find common ways of avoiding future incidents.

The CIOB has also run a campaign akin to the CCS scheme where visitors to sites are encouraged to report untidy sites where housekeeping is poor and to name and shame the company and the management apathy in not keeping order and control on the site.[34] It argues that a clean site is a safe site and also instils confidence in workers and visitors for their own safety and in working safely. The CSCS scheme encourages all, even regular site visitors, to have a competence test and carry a card to prove their knowledge.

Behavioural methods look to prevention requiring the changing of ingrained bad habits by intensive training and reinforcement of safe habits. It requires a consistent awareness raising by managers and supervisors to communicate the impacts of injuries and poor health on their families and

Case study 10.5 Cultural development of health and safety

On a large general hospital site in a northern county town health and safety was taken seriously in several ways to build up a culture of mutual support and improvement throughout the whole workforce. This included

- Strict rules were set by the managing contractor that all package contractors only started work after attending a project induction session and were handed a site map with accessible areas and access ways and safety rules.
- Toolbox talks were given regularly by different personnel to give variety and to match their interests and expertise. Paid time was allowed for this.
- Management personnel were expected to be involved in touring the site in order to spot poor H&S practice and to suggest alternative improvements.
- Individuals from a wide range of subcontractors were invited onto the workforce forum which was chaired by the health and safety officer, but decisions for implementation were decided democratically so that ideas were valued from all levels and democratically evaluated by that same forum.
- Directors of the managing contractor and of the supply chain contractors would commit to inspections at regular intervals and generate ideas for innovative safe practice.

Refereeing

The health and safety officer of the PC and other specialists also patrolled the site operating a traffic light system which would warn with an amber light for mild infringement/first offence and re-induct at the employing contractors' expense. On a further offence or a gross infringement – putting others and themselves severely at risk – they would receive a red card which would banish them from working on that project and they would need to be replaced. They also operated a green card for an example of good practice and this would build up a points system for individual and contractor award and recognition.

friends endangered on site by careless and non-thinking habits. They also encourage safe conditions such as:

- tidy sites
- automatic wearing of PPE
- exclusion areas planned and set up according to daily activity on a formal permit basis
- good practice systems aired in toolbox talks
- health surveillance.

The language of behavioural safety is that there is always a better way and you are the answer to a safe solution. Management is involved by encouraging workers to take responsibility in their own hands and to report unsafe incidents.

Corporate manslaughter

The corporate manslaughter offence seeks to prosecute the actions of a corporation or similar organisation which has with its policy making and resource allocation caused a serious impact on the individual's action that caused the safety and subsequent death of an individual(s) to whom they had a duty of care. It takes up the concept of a controlling mind or a controlling group and is able to indict an organisation or a senior responsible person who had ultimate responsibility for the actions of their delegates. It is not meant to supersede individual criminal acts where free or negligent actions or omissions caused the death. It is therefore critical to the area of safety where a jury must decide how serious the breach is and how much of a risk of death it posed.[35] The breach that causes the death must be directly attributable to the senior management failure.

The issues that may cause prosecution are a poor or lax approach to H&S and a policy that encouraged the breach or produced tolerance of it. The policy may seriously breach existing enforcement policy guidance without an equivalent coverage. Senior management can include those in the direct line of control as well as those who are responsible for forming organisational policy, but their relevant role must be significant and substantial. The latter is wider than the former 'controlling mind'. It applies to any actions by companies, home or foreign, that operate in the British Isles. Although it can also be applied to government departments it excludes issues to do with wider public policy or military activity or police operations dealing with terrorism or violent disorder. It does apply to construction work. The police will lead the investigation in partnership with the HSE or the relevant regulatory body so both have powers of entry. An individual prosecution can be made in tandem with the corporate one and covers such areas as the HSWA 1974 contraventions (where the HSE are the prosecutors) and gross

Case study 10.6 Manslaughter sentence for ignoring improvement
notices

> The employee who was a roofer from Brighton fell to his death in 2005
> whilst working on a roof without a safety harness or safety net below
> him and he died of severe head injuries. According to the local paper,
> his director who was in charge of operations had already been served
> with two prohibition notices by the health and safety authorities before
> the incident that led to the roofer's death.[36]
>
> The director was jailed for 12 months after being found guilty of
> corporate manslaughter following the death of a roofer who fell through
> a skylight on a site the company was responsible for. The director had
> pleaded not guilty to corporate manslaughter although he admitted
> breaching a general duty to employees at work.
>
> He was also given a £10,000 fine and disqualified from holding a
> director's position for three years. The company was fined £20,000.

negligence manslaughter. The CPS decide whether to prosecute corporate
manslaughter. The level of fines will seldom be below £5,000,000 and will
be ratcheted up to match ability to pay and size of company.

Corporate manslaughter relates to a serious problem because many
companies or clients may have an inadequate policy for health and safety
so that their employees are not trained or inspired to respond to accident
and ill health prevention. Alternatively the company may have an excellent
policy but there is no attempt to allocate the necessary resources and
top management commitment. Updating and reviewing data is part of the
expectation for continuous improvement. Case study 10.6 illustrates a manslaughter case.

In order to avoid manslaughter charges senior managers need to prove
that competent workers were used and that the work was adequately
planned, risk assessed and resourced. Other employees and workers will be
interviewed and the case will be treated like any other manslaughter charge
under formal interview. In the case study there was quite obvious flouting
of advice and no learning from the prohibition notices. Equipment was
missing and fragile roofs are well known danger grounds which need proper
protection and training to avoid more fragile parts.

Product safety and liability

The European position on product liability is related to health and safety
and the damages caused by product defect and produced to a construction

product safety directive initially in 1991.[37] In UK construction this was backed up more specifically with the Construction Product Regulations which emerged from the need to consider the complexity of liability caused by the interactions of products fixed in the building as a whole.[38] The core of the directive refers to the six 'essential requirements' to ensure the product is fit for the purpose intended when fixed in the building structure. These are:

- mechanical resistance and stability
- safety in the case of fire
- hygiene, health and the impact on the environment
- safety in use
- protection against noise
- energy, economy and heat retention.

These requirements carry a broad health and safety brief in all six of the requirements for product quality and put an onus on the manufacturer to incorporate an auditable 'factory production control' QA system or compliance with a harmonised standard, enabling the issue of a conformance certificate. CE labelling is required to encourage correct use and instructions for use once they arrive on the construction site. For example a pre cast concrete lintel with two reinforcement bars in the bottom must be labelled with the top and also with its safe load bearing conditions.

Harmonised standards are part of the drive to use ISO or European (EN) standards rather than national ones. This tries to deal with the hidden problem of products with different standards of safety moving round the EU as well as encouraging the headline 'free movement of goods'. Case study 10.7 illustrates how Europe wide compliance has been achieved for a particular product.

The liability of services providers such as designers and contractors lasts for different periods in different EU states ranging from contractual obligation only to 15 years (Spain). Ten years is quite common though not in the UK. The common denominator for most countries would be a five year period of liability, which the Commission cites as covering 75–80 per cent

Case study 10.7 Door set compliance

The performance of a BS476 part 22 one hour fire for a door is well understood in the UK, but not in Germany. The harmonised standard is BS EN 1634-1:2000 conforming Europe wide for fire resistance. The internal hinge edge of the door carries a label certifying its compliance as FD60 and giving the manufacturer's test certificate number. The test is invalid if the door is not installed to manufacturer's instructions.[39]

of the liabilities, whereas 90 per cent would emerge in a 10 year period. There is no agreement on EU wide liability as yet or on its necessity. Insurance rates may be prohibitive for underwriting particularly long liability periods. Damages are the most common outcomes of product and services liability, but there are cases under the Construction Products Regulations that are open to criminal prosecution.

Risk assessment

Risk assessment is a concept present in the HSW Act 1974, but is more definitively defined in the MHSW Regulations 1999. It is important to define the difference between hazard, harm and significant risk.

A *hazard* is the potential for harm, but the context and the outcome are not defined, for example the use of hot bitumen. If this is used by competent people who have proper protection and the bitumen is not accessible to others this is unlikely to be a significant risk. If it is being used on the roof in windy weather many significant risks emerge.

Harm refers to an injury or longer term health problem. For example, the bitumen might scald skin, cause a health problem through breathing the fumes or get overheated and cause a fire which might put others in danger.

A *significant risk* is the likelihood of harm occurring and follows all the possible harms inherent in a hazard by reference to the severity and probability of that harm occurring. This will be affected by the context, for example near to many workers or few or near to the public who are more vulnerable, or being carried out unsupervised, at height or in bad weather will change the rating of severity and impact. It is these conditions and context that should be addressed as well as the inherent harm of the product or action.

For this reason the risk assessment should not be generic and transferred from site to site, but if possible reference should be made to the methods which are uniquely used by the workmen.

The risk assessment process consists of three steps:

- identify the hazards
- assess their significance, and then
- institute control measures to reduce significant risks.

These arrangements are implemented by the following steps:

- Set up a written plan in order to establish the procedures.
- Train personnel.
- Organise the staff and assign responsibilities and communicate the information to all those affected.

- Monitor and control, making sure the procedures are understood, work and are being carried out.
- Review risk control procedures regularly to make sure they are relevant and effective.

Risk controls

Control of risks is very similar to the system in Chapter 8, but more focused on pure risks because it is unacceptable to injure or kill people or make them ill. However, practically it is necessary to assess the risks and give them priority by giving them:

- Probability – how likely is it they will happen?
- Impact – if they do happen what is the size and seriousness of the impact? Injury is more important than loss and death and permanent maiming are more important than days off whilst an injury is healing. Property damage would be below both.

This could be measured as an overall impact by using a grid as in Figure 10.5.

In a hierarchy of control, risk may be eliminated, reduced, given overall protection against or personal protective equipment provided in this order. It may be possible to eliminate a risk altogether by, for example, eliminating the use of a particular hazardous material or redesigning or banning certain methods of working, but the alternative material or method is also likely to create a level of risk that will need to be assessed. If new materials or untried methods are proposed this is almost certainly the case. One contractor's system advocates a deeper hierarchy which resonates with the different risk types of elimination, substitution, enclosure of hazard, exclusion zone, permit to work, remove, reduce exposure time, training, safe system of work and personal protective equipment.

The key method then is to reduce significant risks to acceptable levels and to be aware of the *residual risk* in managing the workplace. Acceptable risk may also vary with improved knowledge and technology and so the control measures should be reviewed for effectiveness. It is perhaps motivational

Frequency	Probability		
	Low	Moderate	Serious
Seldom	1	2	3
Occasional	2	4	5
Frequent	3	5	6

Figure 10.5 Risk priority grid

to work *towards* zero accidents, and then complacency is controlled by a policy of continuous improvement. A residual risk needs to be assigned a responsible person to ensure it is managed and to ensure it stays at acceptable levels.

Those at risk are those carrying out the work, those who work with them (other workers and managers), visitors to the workplace who are less familiar with the workplace dangers and vulnerable groups such as young people, trespassers on the construction site, the general public and foreign or inexperienced workers. The access to risk of each of these parties needs to be considered separately and not lumped together if proper controls are to be devised. For example, holes which are fenced but not completely covered are likely to be a risk to young trespassers, but not to workers; signs which give written warnings may be misunderstood by immigrant workers newly arrived.

Normally any risk above 2 in Figure 10.5 would be dealt with as a matter of some urgency with some residual protection given. For example a nuclear risk would be seldom and serious so although only a 3 on this system, would be important to deal with urgently. A cable across a doorway would be low severity and high probability and should not be left for long as many are exposed to such a risk and it becomes a critical nuisance. The graph measuring accidents against number of people on site is useful in picking up priority actions. The numbers of people at risk are partly subsumed into the frequency axis, but the risk is heightened if the risk was affecting an unsuspecting public. A secondary population also increases the number of people affected such as other workers/public affected by noise or dust. Case study 10.8 gives such a system.

Research has indicated that many accidents take place because of two normal but un co-ordinated activities clashing to become hazardous. So additional attention is needed to measures for overall site co-ordination and to any likely conflicting activities – in Case study 10.8 the use of a mobile crane to deliver material to a fragile roof would cause additional hazards. Any changes to the programme or the original intentions also require further risk assessment. For example a change of the crane model may affect the exclusion zones. Other things which might be exacerbated are unpredictability, length of exposure, the nature of the risk and the nature of the controls.

Improving the effectiveness of accident prevention

Accident prevention is a primary tool for management effectiveness. The statistics indicate a limited success in reducing accidents so what are the main principles? The HSE suggests that management action could have prevented seven out of 10 construction site deaths i.e. approximately 37 deaths on current figures. Accident prevention arises from the focused application of

Case study 10.8 A risk assessment pro forma

A contractor is likely to develop their own risk assessment systems, which they feel comfortable with and Figure 10.6 is part of a pro forma risk assessment sheet. A separate sheet is used for each activity or system on the site. Critically there is a responsible qualified person to write out the assessment and take responsibility for implementation. In order to ensure supervision is in place for hazardous processes a permit to work certificate needs to be issued.

The graphical effect is useful in picking up priority actions. The scores of probability range from 'unlikely' (2) to 'likely to happen at any moment' (10). The scores of potential outcomes range from 'minor injury' (2) to 'multiple fatality' (10). A description of all the actual risks and the proposed measures to bring the risk to an acceptable level are given; and length of exposure and the type of people at risk are separately noted on the risk assessment sheet. The risk would be heightened if the risk was affecting young people, or an unsuspecting public.

For example, in the case of tower crane operations an accident's outcomes are likely to be classified as serious injuries, because of the height and wide exposure of employees and possibly the public to falling objects, giving a score of 6. Probability of likely to happen occasionally would give a score of 6 (dotted line in Figure 10.6), making it a moderate risk. This score would catalyse actions to reduce this risk to at least the level shown in the diagram. In this case actions to work on reducing

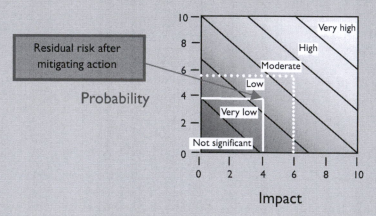

Figure 10.6 Alternative risk priority grid. Reproduced by permission of ISG Pearce

the probability to 4 which is improbable or unlikely brings the risk down to the low level shown. This would be achieved by having strict protocols to supervise loading, test the crane and exclude people from working under loads. It would also be illegal to slew over external property and highways whilst loaded without exclusion zones, because the general public are not inducted into looking out for site dangers, so reducing impact to 4.

measures to areas where accidents happen regularly. As discussed earlier, injuries and fatalities from falling or by being hit by falling objects consistently account for 50–60 per cent of the total. Management and lack of organisation and planning have been consistently blamed by reports on health and safety.

The strategic approach to prevention is to connect good business with safe business mainly espoused by the HSE.[40] The system depends on an adequate culture and commitment from top management to the strategic importance of health and safety management systems. Some companies have committed the resources up front and reaped longer term business benefits by way of reducing costs and getting work on their safe approach reputation. Most clients do not want their reputations sullied by accidents.

It is important too that lessons that can be learned from investigating cause and effect from other accidents and near misses can also be passed on to learn lessons for the future. In particular, individual companies can learn a lot from their own near misses and accidents. The following are some major accident reports identifying particular problems more universally. Examples in Case study 10.9 point to key management failures.

HSE distinguish between active or reactive measurement based on past accidents. One of the key aspects of the active measurement below is the ability to measure performance for preventative action.

Reactive measurement is normally done by the use of KPIs based on past measurement of accidents, ill health, close misses and other evidence such as past accident reports which informs future planning. *Active* measurement measures the standard of planning and organisation procedures and not the failures. It is based on the achievement of objectives and targets for accident prevention and will include the number of health and safety training days and a test of manager and worker knowledge of safe practices, so that the development in competence can be assessed. It will also include the training of trades personnel and managers to keep updated on the current technology. The initial training is to comply with schemes such as Construction Skills Certification Scheme (CSCS) in the UK which requires a certain base level of health and safety relevant to the particular trade and level of supervision

Case study 10.9 Major disaster reports

Buncefield petrol farm. Unknown leaked petrol vapour creating vapour clouds well offsite before igniting into spontaneous fireballs threatened a much wider area with the excessive heat putting unsuspecting public lives at risk. Buncefield Major Incident Investigation Board's final report (2008) blamed poor risk assessment for major explosions and made recommendations for site planning and the containment of large petrol storage facilities.[41] It prosecuted five companies.

Crane failures in the UK. The HSE investigation into tower cranes failures from 2000–2007 resulted in a tower crane register and the notification through the Conventional Tower Crane Regulations (2010) after inspections have been done.[42]

Ferry disaster. Management with its absence of safety policies was 'infected with a disease of sloppiness' according to the Sheen Report (1987) on the Zeebrugge disaster when the Herald of Free Enterprise ferry sank due to open sea doors.[43]

Underground fire. 'London Transport had no system . . . to identify and promptly eliminate hazards' the Fennell Report (1988) stated of the Kings Cross disaster.[44] There were 31 fatalities and 60 injured after a fire started by a cigarette butt on an escalator which it set fire to.

Oil platform explosion. The Cullen Report (1990) on the Piper Alpha disaster concluded 'significant flaws in Occidental's management', where 169 of the 229 people on board the North Sea drilling platform were killed due to routine maintenance error.[45]

Train crash. The Hidden Report (1989) concluded a 'frightening lack of organisation and management' on the Clapham Junction train crash caused by a signalling fault with 35 fatalities and 415 injured.[46]

Construction. An EC Report in 1992 claimed 35 per cent of accidents in construction in Europe have occurred because of design and organisation faults and a further 25 per cent due to the planning and organisation stages.

being practiced. Many larger sites in the UK will not allow workers on the site without CSCS certification or its equivalent.

An important point that HSE makes is that a good track record in health and safety does not necessarily mean that the framework for avoidance in the future is in place. Typical activities in active monitoring equate to preventative maintenance by the use of inspections and replacement of components before failure. A system which is active provides continuous updating to the training and uses toolbox talks and other daily or weekly health and safety activities.

Causal factors

Causal factors may be applied to an accident or serious incident investigation in order to provide preventative information. This is common in industry in examining major disasters such as those listed in Case study 10.9 above. Factors may be regular such as a missing action in maintenance or exceptional such as a unique combination of two factors happening together. It is useful for construction projects to study a range of incidents with all reported accidents and incidents to identify common causes and mitigate against these.

HSE research (2003) into causal factors of construction confirms the particularly strong connection of accidents with poor designer awareness of their impact on construction health and safety.[47] It concluded that 46 per cent of 100 accidents investigated were due to permanent design issues and a further 36 per cent were due to a lack of attention to design in temporary structures.

One of the most common accident places is on the way to the 'workface' including handling materials. The research also suggests that there are many generic site hazards that will continue to occur because of common poor practice on site that needs to be attended to before design change will have a safety benefit. Originating influences such as poor risk management were said to occur in 94 per cent of the accidents in spite of CDM responsibilities. However it found little correlation of accidents to poor weather as generally assumed. This provides conclusions in better risk management and in protected routes to the building. It also recommended the inclusion of health and safety items in the BOQ which were considered to be key aspects in gaining a safer design.

A later study by Donaghy concluded that common causes are poor experience, information and advice deficiencies, poor risk perception, rescheduling of work without planning, equipment operability, space, poor PPE use and tools that don't suit the task and the user.[48] Most of these accidents are preventable.

A prevention focus group study by the HSE (2004)[49] on falls from height (a major accident type in construction) estimates that 50 per cent of accidents could be prevented by the application of risk control measures to just two areas out of seven considered, which were design process (30 per cent) and compliance to safety measures on site (20 per cent). It was found that supervisors tended not to reprimand experienced workers about poor safety practices, either because they believed these people could look after themselves or because of lack of confidence.

A similar influence network for compliance exposed the 'it won't happen to me' attitudes of experienced workers and their failure to assess the scale of the risk even though they were well aware of the hazards. Managers and supervisors leading by example and zero tolerance to the ignoring of safety measures were identified as a key prevention measure for non-compliance. There is a gap in the communication process in that many accidents do happen to experienced workers because they are too confident of their ability and not alert to the combinations of activities or circumstances that can increase the risk of a simple job exponentially.

In construction high risk operations such as scaffold and lifting equipment are inspected daily before use by a manager who has been trained to inspect the basics. Tags are often used to indicate the date and time of inspections. Designed scaffold needs periodic specialist inspection by the designer and a specialist scaffolder. A more extensive check is signed off in a weekly register. Mobile plant and tower cranes need consistent preventative maintenance and regular inspection. Protections of excavations need to be supervised continuously by the use of banks persons, designing side supports and issuing permits for work within the excavations so that people are accompanied and have proper access and help in unexpected collapse. Other confined spaces such as drains also need to be permitted, so that a 'surface worker' can be in attendance. Lifting and manual handling operations all need risk assessment as does being aware of and working with hazardous substances. Carrying out of hot work must be permitted for proper fire precautions. Low risks are dealt with by general inspections. Building up a knowledgeable culture which is aware of common risks and reports them helps the project manager to have a far reaching control.

Reactive measurement and recording can be accused of being too little too late, but if it records and analyses minor injuries and near misses and the patterns that occur, specific areas for improvements may emerge as a proactive action. For example, consistent small injuries in one trade

Case study 10.10 Staircase collapse (hypothetical)

This is a causal analysis of a near miss.

An escalator with a cantilever dogleg was to be installed from a floor one storey above a multi-storey atrium to the floor of the atrium.

On the way down it passed through a free standing media wall providing electronic advertisements to escalator users (see Figure 10.7).

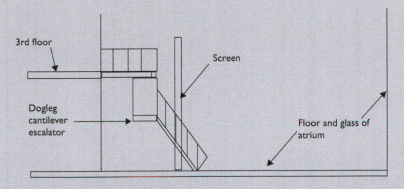

Figure 10.7 Diagrammatic of staircase

Initially the escalator was to be made in one piece and a risk assessment determined a suitable method statement for the fitters. In the event the escalator was redesigned and delivered in bits to get it through existing doors and was bolted together in-situ. Craneage and frames were used to lift and support the parts until they were completely bolted together. At completion the frame was taken away and the equipment commissioned by the manufacturer. During its first major usage the assembly slumped out of line hitting the media screen which crumpled and could have hit escalator passengers. The escalator was immediately shut down. The design or the installation had failed.

The subsequent investigation indicated the shearing of two bolts holding two of the sections together. These were not the specified friction bolts or tightened to the right torque. Potential dangers were the complete collapse of a full escalator with subsequent injury or death. The accident was the result of a workmanship problem, but with a narrow margin of safety on the design. The risk assessment was carried out by the installer who was a subcontractor to the designer. They were unaware of the design parameters and a hurried change in the design to accommodate the access restrictions of a nearly finished building had weakened the design safety checks and created a shortfall.

- What can be learned from this near miss incident?
- Who was primarily to blame? Would anyone have been prosecuted?
- How could this have been avoided?

may point to a shortfall in training or a poor attitude to safe working or inadequate supervision. HSE points out that an observant approach on the behalf of workers with a non-blame, responsive action by management encourages open and honest communication so that improvements are constantly carried out and those affected trained. Observation can pinpoint the unsafe and unhealthy parts of a work process.

In terms of investigation, *underlying causes* are important and these can occur in the:

- person (poor attitudes, behaviour and knowledge)
- organisation (insufficient procedures to ensure a safe working place)
- task (inherent danger in the process or materials/components).

A risk assessment needs to identify the context of the task with the person doing it and not remain generic. The question is not what actually happened, but what is the worst situation that could have happened. It also uses feedback from accidents and near misses to improve processes before the accident or injury occurs and the influence of company practice and culture is important here. Case study 10.10 analyses the causes.

The active model is a feed forward model for prevention with the use of both types of measurements and an aim to prevent accidents by planning ahead and encouraging self-supervision rather than intervention. The culture of improvement is foremost. Case study 10.11 represents this.

Case study 10.11 Plasterboard fixing – improvements

An example of improvement by observation would be the HSE study into the manual handling of plasterboard where there is sustained heavy lifting due to piecework bonus, loading out in confined spaces (e.g. upstairs and around landings), awkward extended neck and body positions in cutting and supporting the board and the repetitive action of hammering nails to fix the boards.[50] It particularly looks at housing practice where there are tight limited spaces. Evidence showed that plasterers were one of the most liable to musculo skeletal disorders (MSD).

The research examines the suitability of manual handling aids and the use of two people to work together to minimise stretching and weight carried. It also suggests ways of improving postures in cutting and supporting boards by avoiding awkward bending and avoiding awkward loading up journeys by, for example, loading through windows off scaffolds or by cutting slots through floorboards to avoid manoeuvring boards round staircases.

Safe design

A designer has responsibility under the CDM Regulations to provide a safe design and to risk assess their design. It applies to all buildings including domestic buildings and demolition requirements of all sizes. The definition of a designer is quite wide and includes temporary works design as well as architectural, many other types of specialist design, as well as those who put constraints upon it such as the cost engineer. Their responsibility extends to buildability and safe products for the construction stage, maintenance safety and user safety including fire safety. It does not include double guessing the contractor's method statement which remains firmly with the contractor on the basis of safety information on materials and existing building hazards. Designers should be supported by client information available from the existing building, for example asbestos. The three stages in the design process are outline design at the feasibility stage, scheme design leading up to a scheme for planning approval and detail design leading up to the tender documentation in traditional procurement. Risk assessment for a healthy design is best at the earlier stages of design as more opportunities are given to design out hazards (the potential to cause harm). It is often connected with sustainable design. The HSE has a code of hazards rated red, amber and green, which prioritises under red the hazards that should be designed out. Amber recommends reducing the hazards as much as possible and green refers to design practices to be positively encouraged.[51] The designers' role then is to look to reduce hazards or to avoid hazards at source. For example, heavy beams can have lifting points designed in to facilitate the use of lifting gear and discourage manhandling. This is in addition to the use of protective equipment. ERIC refers to eliminate, reduce, inform of residual risks and co-ordinate between designs.[52] Many hazards are health related and five common areas affecting construction practice are:

- manual handling
- noise
- vibration
- respiratory, for example asbestosis and asthma related disease
- dermatological complaints.

CIRIA has also produced some guidance notes which support designers.[53] Construction Skills (CS) have written a manual for designers.[54] CS define the broad view of designers which may be the architect or one of the consultants but may also cover the temporary works designer the QS who specify materials or components, the project manager insisting on a particular solution, or a specialist package designer, or the utilities needed to be designed outside the boundary.

Buildings in use

The designer accountability is not for construction alone, but also for a building which is safe to occupy and maintain. This consideration is not new, but the safety of maintenance access has been reinforced and should include access to services in ceiling and roof spaces. The designer should achieve a safe design for construction maintenance and users and for neighbours affected by the building. It should be noted that revisions in the guidance for the CDM Regulations tried to be more specific about the responsibilities of designers. Particular areas of danger are:

- Impact on adjacent buildings and on maintenance of external services.
- Cleaning windows over the top of conservatories.
- Cleaning windows or curtain walling in multi-storey blocks.
- Cleaning the underside and top of glass atriums.
- Cleaning out of gutters.
- Changing light bulbs in tall spaces.
- Moving around on fragile roofs, or near roof lights.
- Moving above fragile ceilings where there are service access points.
- Accessing heavy plant and equipment safely for periodic maintenance.
- Accessing plant in ceiling spaces.
- Safe fire exits.

The creation of an *inclusive environment* is important for the health and safety as well as compliance with the requirements of the Equality Act 2010 for those with special needs. It considers various needs such as mobility issues, partial sighting, hearing loss and dyslexia and provides where possible access and provision without making anyone a special case. This might include raised Braille on signage, tactile paving or flooring for crossings and entrances, contrasting colour schemes and furniture, wheelchair reachable handles and switches and fire doors that are easily manoeuvred or automatically openable, that everybody uses. The Equality Act 2010 prefers a universal communication suitable for all needs. If this is applied to fire escape then the system needs to include notices for fire escape routes and to provide alarms. Those with sight problems and in wheelchairs will need help, but ideally escape routes should be the same for those of all mobility. Fewer special cases means less confusion or frustration for users and this is a safer culture. The user will also have training for clearing a building in the case of fire.

The *CDMC* has a role at handover stage to produce the HSF which provides information about how systems can be operated safely and how to get the best out of the building for health. This has often resulted in criticism about the effectiveness of this role, as they have no liability for the adequate safe functioning of the building in use, just a responsibility to ensure that documents about working parts and product safe use are received. This role is dependent on the co-operation of other parties to provide comprehensive information and for the maintenance teams to access and use it. There is a need to ensure training of maintenance teams and users so that they understand the building and documentation is easily accessible for reference in the handover of the building. Currently there is no requirement for training or for any quality of provision of the file, although they are required to agree the format with the client and let the designers and contractors know what information is required early on in the contract (ACOP para 108). The client has the responsibility to incentivise the designer and contractor to provide full and timely information i.e. as soon as available (ACOP para 109). The main purpose of the file is to ensure that accurate information is available to guide the H&S of future work. It may be combined with other maintenance information.

Typical information contained in the file is residual hazards such as stabilised asbestos, contaminated land, lead paint, information on structural principles such as pre- or post-tensioned beams, location and nature of underground services and cables and 'as built' drawings of the structure, its plant and equipment, and the means of safe access to plant rooms, service ducts and voids. Also any information for the demolishing of buildings or the replacement of equipment is included. These relevant parts of the file need to be passed on to users and to leaseholders by the client or to future owners.

Safe construction

The key thing is that a health and safety plan is available before work starts and that this is developed throughout the on going design and construction process. It should be available with the planned method and risk assessment, and should have strategies to be effectively implemented and co-ordinated, taking into account all the human factors and the practicality of ensuring good quality safe systems. The plan should be tailored to each project and should be focused and easy to read giving a dated plan of risk assessments to be prepared, by whom and not including irrelevant material. A safe system includes induction, training and communications with the direct supervisors and the supply chain and also the monitoring and updating of the plan. The issues of competence, inspection, culture, controlled access, training and communications are discussed. The plan may include site rules and the appropriate direction of non-compliant contractors. Visitors must be inducted and the general public must be especially protected. Even site trespassers have rights of compensation so security is important.

Competence

A key issue in the maintenance of health and safety standards is the achievement of competent workers. Competence is the ability, the experience and motivation of the individual. This will apply in the recruitment procedures exercised by all contractors, but should not be exclusive of those who are willing to be trained and this will incur temporary greater supervision and feedback. The client has a responsibility to appoint competent co-ordinators, but the principal contractor goes on to ensure contractor competency.

Schemes for ensuring management and operative competency are important and there has also been a move by the HSE to target special areas of concern and make themselves available as a training resource during this target period. Health and safety training is part of the competency certificates, for example as shown in Case study 10.12. This is a necessary addition to site induction, which highlights known hazards, but does not specifically test competency and understanding or provide updates.

Case study 10.12 CSCS cards

Many large contractors now require all workers to have passed a test of competence that is associated with a series of compulsory training sessions including health and safety. In the UK competence is recognised by issuing a Construction Skills Certification Scheme (CSCS) card. Critically this is required of all managers and operatives representing any of the trade organisations on the site. The card is specific to the trade or management level and also confirms that the holder has gone through appropriate health and safety awareness training. The cards are graded and recognise competence equal to NVQ level 2 or completed apprenticeship for the trades, NVQ level 3 for supervisors and NVQ level 4 or a professional standing for managers. This important association of competence with health and safety gives a better indication of an individual than a signed disclaimer. Site based health and safety induction is still relevant for specific site hazards.

There are also affiliated schemes covering specialist trades and plant qualified plant operators. GasSafe is a typical scheme for ensuring the competence of works installing gas installations in the UK.

Managers need to remain updated if they are to exercise an effective role. Information is passed on to suppliers via their supervisors. The regular development of a health and safety culture is achieved by keeping health and safety in the spotlight through toolbox talks trucks and prizes for examples of safe working. This also has a direct connection with management and shows their commitment and specific contractor based training such as safe driving of forklift vehicles.

Supply chain

Most production work is now carried out by package contractors and the self-employed. As indicated they are responsible for safe working methods by carrying out a risk assessment and devising control measures to reduce significant risks caused to their employees and eliminating harm that their work may cause to others. The submission of a method statement allows proper consideration by the principal contractor of sequencing activities to avoid dangerous situations caused by a combination of work activities by different contractors. They may also vet the method statement to protect others and to achieve better standards of safety.

Clearly there is a supervisory role for both the specialist contractor and the principal contractor. There is also the likelihood that there are sub subcontractors or self-employed who will need to be controlled further down the supply line and may have their own employees on site. Particular issues to watch are:

- Quality of the risk assessment.
- Receipt of the method statement before the contractor start date.
- Incorporation of changes into the main programme so that method statements are reviewed.
- Proper training and competence update in post.
- Competent self-supervision.
- Adaptation and project changes to method are properly planned for impact on *all* parties.

The unplanned response to change is the cause of many accidents, because change needs a regular review of health and safety that may not be appreciated where risk assessment is carried out on a 'told to do' basis at the beginning of the contract, or on behalf of those who actually do productive work. Case study 10.13 is an example of an effective system.

It could be argued that a quality product is a safely produced one. This has led to many main contractors insisting on CSCS cards for all workers and managers on site in the belief that they will be able to reduce defects and accidents on site. With the use of a wide range of contractors and self-employed, this target depends on being able to influence suppliers to comply with the training and testing requirements that this will entail. It will be much harder for smaller contractors to afford this investment without help, so a more formal supply chain management procedure should mean more intervention. Health and safety training is a requirement of the MHSW Regulations and this gives some leverage for compliance. Some main contractors provide training for their supply chains. Other training is available through schemes such as the Contractors Health and Safety Assessment Scheme (CHAS) which is a cheap scheme for accrediting a proper health and safety policy for smaller contractors.

Case study 10.13 Site safe management system – hospital site

A community hospital site has two areas of construction with different teams working on each. The hospital is a multi-storey building. There are also areas designated for external storage, waste segregation and recycling and offloading. There are quite a few package contractors working on site and there are regular deliveries to different parts of the building as well as to storage areas. Vertical access requires scaffold externally. The site offices are outside, but connected to the site.

The site boundaries are secure with a hoarding and a car parking area beyond the boundaries. Access to the site is gated and automatic access is gained via use of ID cards issued to competent workmen who have completed the induction. Visitors are admitted to offices outside the secure area and non-regular visitors are accompanied on site. There is a segregated and fenced paved walkway leading to a single entry point in the building and where service roads are crossed these are gated, sometimes manned, to provide awareness of a busy crossing. Internal access to work areas in the superstructure is across internal floors or via the scaffold. Temporary stair towers with handrails are used to ascend the scaffold and precast staircases are fixed internally for general access to higher floors. Exclusion zones have been built around delivery points such as hoists, forklift or crane loading platforms. All fixed and mobile scaffold is colour tagged with latest inspection dates (at least weekly). Access to the recycling bin area is specific with clear labelling and a separate lorry access is given to change skips.

Inductions are carried out on a daily basis for all visitors and new workers and a site plan is issued to new workers listing the key site rules. All workers are expected to hold CSCS cards indicating their areas of competence. A board showing updated daily hazards is displayed at the site entrance so that exclusion zones and particular hazards are clear. Fire fighting stations have been placed on all floors against the main access and alternative egress routes. Fire escape routes are marked up as in a completed building and alarm buttons are also provided. Evacuation is controlled by wardens, to an assembly area. Toolbox talks are carried out weekly with a programme of relevant safety and health talks and time is specifically allowed to workers for the talks.

A workers forum meets monthly and provides opportunities for worker led representatives for feedback to management of improvement issues. Inspections are carried out regularly by a company safety officer who visits and provides lists of contraventions and improvements. He/she has the power to shut the site down in the case of serious contravention or to issue improvement notices to be inspected on the next visit. The site manager is responsible for regular inspection and tagging of scaffold and shows the register to the safety officer. Changes to scaffold are on a permit system to the site manager.

The site manager has the ultimate responsibility for the safety of the site and receives and co-ordinates the method statements for each area of the work. They also use their construction management team to police the site rules such as wearing of PPE and keeping the work areas tidy. Movements of vehicles and cranes are aided by banks men and deliveries are booked through the gatekeepers on site. The crane timetable is centralised and subcontractors book up time according to requirements on a first come, first served basis mediated by the site manager the day before. There are gatekeepers on the loading platforms at each level.

The key to this regime is the tight control of movements on the site and the strong policing of site rules such as induction and the competence of workers, updating information, inspection, the use of segregated access, the control of deliveries and the security of the site and its evacuation.

Inspection and supervision human factors

Traditionally workplace inspections are carried out by those who manage the site and have accountability for health and safety plans. They include mandatory scaffold and excavation support inspections, PPE checks, vehicle checks and formal investigations after things have gone wrong. This puts the onus on a management led system where the burden of proof lies heavily on relevant competence or experience and time available to check. Due to other commitments it can either end up in the hands of trainees or become a reactive rather than preventative activity.

Dalton notes the importance of a broader inspection procedure due to the downsizing of the HSE enforcement and inspection role.[55] He describes the new approach as a management system approach, firmly linked to QM and environmental systems. These systems would:

- pull in a wider view on safety including experts (fire officers, insurance assessors), workforce (trade union or safety committee representation)
- note evidence of management involvement and planning
- take a general view and impression of the efficiency and openness of the implementation of the system – is it clean and efficient in appearance, what are the risk assessments like and are they accessible?

- make specific third party inspections and make reports to management with copies to senior managers with recommendations.

Workplace control

A quality product allows every pair of eyes to inspect but also encourages ownership of the problem and a responsible workforce by peer pressure. This means clearly delineating responsibility for health and safety:

- a formal handover between different trades working in an area
- a clear connection between method statements where work overlaps to be programmed to take account of this
- a quality improvement attitude that includes training in health and safety planning
- a drive towards self-supervision to equal or exceed project standards.

HSE put forward a case, shown in the adapted Figure 10.8, for encouraging a culture of improvement by contrasting the polarised scales for a culture for deterioration and for promotion of health.[56] Figure 10.8 indicates the belief that health and safety is not just a drag on the project budget, but that financial benefits accrue from health and safety in terms of productivity to pay for systems that are continuously raising the health and safety standards.

The same publication also discusses the benefits of self-regulation when workers take responsibility and pride in their work. This in turn will offset the costs of close supervision against the costs of training. In a project with many different employers, it will be necessary for all contractors to buy into the same approach otherwise the system will break down around the weak link. This will need to be driven by the project manager and built into the conditions for the procurement of package contractors. They point out that there are certain underlying beliefs that must be understood to make this culture possible. These are:

- All accidents, ill health and incidents are preventable.
- All levels of the site organisation are co-operatively involved in reducing health and safety risks.
- Trade competence includes safe working risk assessments.
- Health and safety is of equal importance to production and quality.
- Competence in health and safety is an essential part of managing construction.

In this context there is a much more fundamental preparation stage than a remote health and safety plan. Inspection and supervision are shared and discussed and workforce representatives are proactive with management. The system is not unsupervised as building up experience is important and

Problems No response = deterioration of health	Benefits Positive response = promotion of health
Ill health and injuries Damage to property Disillusionment Absenteeism and increased liability	More commitment General health and efficiency rising Improved job satisfaction and motivation Reduced absenteeism

Figure 10.8 Spectrum of occupational health

training needs to be on going, but it sets in place a wider accountability. Teamwork can be important in respecting what you can learn from others.

Communications, information and training

One of the most difficult things to achieve is a fail-safe communication system that provides a clear delivery of the requirements of the health and safety plan to the whole supply chain. This must be communicated to supervisors and their workforces. It will be achieved through co-ordination meetings and through regular workforce briefings, posters and signs. Posters and signs are better using graphics so as to cover the international workforce. Communications are also about ensuring that when there are amendments made to method statements to suit design changes these are broadly communicated to all affected parties.

Basic information about evacuation and site wide hazards and procedures is delivered through induction. Site rules need to give a stable environment and be sharply focused so that they are perceived to be relevant. Different employees will have different training needs and access should depend on competency. The requirements of different company health and safety systems and specific client requests for health and safety features need to be co-ordinated. Signposting will conform to international standards and will be mainly graphical so as to cover those, such as migrant workers, who don't understand the language.

Consultation and informal employee reporting is an upward channel of communication, which needs to be received responsively if the system of improvement is to be maintained and new systems are to be introduced.

Protecting the public

The general public remain a very vulnerable feature of construction particularly if they are working or passing close to the site boundaries. They are not issued with PPE, inducted or trained for the particular hazards of construction and so public space should be heavily protected and site boundary control in place to exclude unauthorised entry. They are in particular danger from undermining foundations of adjacent occupied buildings, falling materials, over sailing loads from offloading and lifting, noise, dust and obnoxious fumes and explosion. Deliveries for building materials will increase the traffic and provide extra hazards especially where they are offloaded in the street. Recently a spate of fierce fires in unfinished timber frame buildings, falling tower cranes and collapsing scaffold structures have created major accidents. The number of public accidents has caused injury and death to passers-by. Road works which are temporary and are in or adjacent to pavements or public highways need to be safe in use which means effective barriers, noise attenuation and control of deliveries and traffic flow and speeds. Excluding traffic completely is not smart and transfers risks in terms of congestion elsewhere, so work overnight with temporary road closure or shifts is better public safety. Contractors are also liable for accidents occurring as a result of trespass onto the site.

What is the correct response to protecting the public? The key measures to reduce risk to the public to acceptable levels are:

- Protection by an effective barrier around the site and any other public areas such as road works and drainage connections, including security guards in busy areas. Scaffold fans or gantries to protect from falling debris.
- Information about the hours of work, programme length and activities and irreducible levels of noise or dust.
- Warning and information signs to instruct, make the danger understandable, viewing windows to satisfy curiosity and limit trespass.
- Cut down pollution levels such as dust and emissions, especially exclude any likely to have biological or allergic reactions.
- Control noise as public have no PPE. Use exclusion zones fairly and limit hours.
- Strictly control any toxic or hazardous materials with secure storage and delivery.
- Agree a traffic plan with the local authority for safe egress and access and temporary traffic controls; and provide banksman for reversing vehicles.
- Control lifting and limit any crane slew to site air space except in agreed control procedures. Set lights on tall temporary structures.
- Exclude public access without protective equipment and sign clearly.

These measures are based on total protection, considerate contracting, providing information and reducing nuisance. The Considerate Constructors Scheme has been set up as a voluntary compliance with the following principles.

- Ensure that work is carried out to limit exposure to the public.
- Keep sites and surrounding roads tidy.
- Control deliveries and the blockages they may cause in tight conditions.
- Keep neighbours well informed of what is going on and the measures which have been carried out to minimise the exposure to the public.
- Limit exposure to 'nuisance' activities.

Integrated systems for quality, health, safety

Integrated management systems arise from the overlap and duplication which occurs between health and safety, quality and environmental management specified in the well known OHSAS 18001:1999 Health and Safety Management Systems, ISO 9000:2008 Quality Management and ISO 14001:2000 Environmental Systems which are designed to be compatible with each other. Griffiths and Howarth point out the need for a risk assessed *open* system (recognising external influences), which copes with 'unplanned events' and the appropriateness of the 'whole organisation' approach shared by each of the standard systems.[57] A few organisations on construction have adapted such a system.[58]

Holistic control in an integrated system releases the potential for a better product for the client and better value for money, saving paperwork. The challenge is to keep the system simple and this is usually done by using a single system such as quality and adding on environmental and health and safety modules. To be effective the paperwork must be kept to a minimum and control remains in the hands of production staff accountable for outputs and not through a remote systems manager. The system should also be simplified for small projects. Figure 10.9 shows such a system.

A paper by HSC has suggested that an integrated team approach which is forced upon PFI contracts is something that would benefit the introduction of a better health and safety culture.[59] This method would have advantages

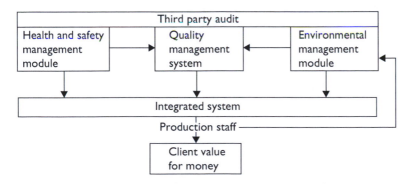

Figure 10.9 Integrated health, safety, environmental and quality system

for risk sharing and the early involvement of the supply chain in the health and safety plan. It is also suggested that a solution which is good for health and safety at this stage can be value engineered so as to be good for cost saving, also moving away from the approach that health and safety costs money.

Conclusion

Health and safety regulation has multiplied in recent years with some encouragement from the EU and continues to do so as headline accident statistics refuse to reduce significantly in construction, for example the increase in tower crane accidents and the need to further tackle falls from height. Because of this, health and safety training on construction sites is a substantial investment and most large contractors require evidence of training to a minimum competency for managers and workers. It is however easy to go through the motions of paperwork and inspections and not develop a rigorous culture of awareness by the entire workforce; that is, that all workers report poor practice and are committed to preventing accidents in the first place. This is difficult given the fragmented nature of the workforce, and the many sub subcontractors in the supply chain make consistent communications difficult. The culture of induction is often substituted for specific training and is not focused to the work area of the individual. It needs to be backed up with specific training, by each contractor represented. This may have to be policed by the main contractor.

The CDM Regulations have provided an integrated framework for the management of the construction project safety, by formally including the client and the designer with the traditional contractor role in health and safety management and introducing the idea of co-ordination. This rightly recognises the process throughout the project life cycle from inception to completion. It has much potential in making sure that health and safety resources are priced in adequate tender prices and risk assessment. Implementing safer systems is critical, if organisations do not understand, or apply risk management co-operatively with others and enforce it in an integrated way between design, project team and facilities management. This is compounded by the problem of macho and complacent attitudes by workers that believe that 'it won't happen to me' and by supervisors and managers who turn a blind eye to shortcuts by specialist trades who have the experience to recognise the hazards, but have not entered the culture of *assessing the scale* of the risk.

Some would argue that SMEs want a more prescriptive system where generic standards become well known and can be used in a directive way to ensure compliance, others that self-regulation is a better way of owning the problem and developing universal awareness and accountability. As SMEs make up a major proportion of organisations on the site, there is a case for more prescription. The CHAS system used in local authorities is a pre-qualification measure for SMEs and recognises training and an organisational commitment to health and safety. Culture is still important, as prescriptive solutions create generic methods to cater for interruptions and changes, which create unforeseen situations that often cause accidents. The culture in prescription is to meet minimum standards rather than to encourage continuous improvement.

Worker engagement is poor in construction due to the fragmented nature of the workforce and the limited trade union representation, though good practice has shown its advantages in reducing accidents especially where workers can be part of the decision making. The benefit of this is more ownership and integrated feedback by the project workforce in what is already perceived to be a dangerous environment. Worker committees need to be given power to change things and to stop unsafe work and a route to feed back to project and employer management. In terms of encouraging self-supervision, more accountability for worker inspection of the workplace needs to be given.

The best practice movement is a way of engendering a sense of pride and competition to reach higher standards and to name and shame those who don't. It can encourage better safety standards and help produce this culture of continuous improvement. This could work if clients recognise their role in insisting on certain standards of qualifying compliance, such as the compulsory use of CSCS cards which evidence minimum standards of worker and site management competence. Top management commitment to safe working improvements and an auditable safety management system which monitors and ensures good implementation by the use of visible site wide KPIs is essential. If these requirements are not part of the selection process in competitive tendering, those that take the shortcuts on health and safety will initially have an unfair financial advantage in a competitive market.

Closer connections have been made between safety, health, environmental and quality systems in the development of integrated management systems which reflect the numerous interfaces between these areas of management control. Here the client has a better chance for a value for money solution where quality (Chapter 13) is linked with health and safety and is seen as good business. This could be illustrated by more 'right first time' and less supervision. Some contractors have health and safety built into their quality assurance systems.

Liability and negligence may be seen as good enforcement tools, but in practice they are seen as inevitable and as an insurable cost to the business. The 1997 HSE study found that non-insurable costs were at least a factor of eight of insurable costs and this suggests that attention to management systems is critical to avoid waste in the system and even future loss in an environment where litigation may become more commonplace.

References

1 ILO (2005) *Number of Work Related Accidents and Illnesses Continue to Rise.* Joint news release with WHO (2005). Available online at: http://www.who.int/ mediacentre/news/releases/2005/pr18/en/index.html (accessed 11 November 2011).

2 Egan Report (1998) *Rethinking Construction.* London, DETR.

3 HSE Statistics (2009/10) Figure 11, p.17. Available online at: http://www.hse. gov.uk/statistics/publications/general.htm (accessed 15 August 2011). This compares worse with the average all industry rate of 0.6 workers per 100,000 which is 4.5 times worse.

4 CONIAC (2001) Health and Safety Summit.

5 HSE Statistics (2011). Available online at: http://www.hse.gov.uk/statistics/ industry/construction/index.htm (accessed 15 August 2011). 4.5 to 2 workers/ 100,000 fatal injuries and 375–275/100,000 workers for major injuries.

6 Wolstenholme A. (2009) *Never Waste a Good Crisis.* London, Constructing Excellence.

7 Tye and Pearson (1974/5) *Management Safety Manual.* British Safety Council 5 Star Health and Safety Management System, in HSE (1991) *Successful Health and Safety Management.* London, HSE.

8 Tye and Pearson (1974/5) *Management Safety Manual.* British Safety Council 5 Star Health and Safety Management System, in HSE (1991) *Successful Health and Safety Management.* London, HSE.

9 APAU (1991) *Successful Health and Safety Management.* London, HSE Publications.

10 Fewings P. and Laycock P. (1994) *Health and Safety in the Workplace.* University of Northumbria Built Environment Group, pp.8–10. (26 safety related directives were adopted in EU countries in the period 1986–1994 alone.)

11 HSE (2011) *Risk Assessment and Policy Template.* Available online at: http://www.hse.gov.uk/simple-health-safety/write.htm (accessed 17 August 2011).

12 Fewings P. and Laycock P. (1994) *Health and Safety in the Workplace.* University of Northumbria Built Environment Group.

13 HSE Scotland (2009) *Safe and Sound at Work: Do Your Bit.* London, HSE.

14 Fidderman H. and McDonnell K. (2010) *Worker Involvement in Health and Safety: What Works? A Report to the HSE.* 21 May. Birmingham, RoSPA.

15 Fidderman H. and McDonnell K. (2010) *Worker Involvement in Health and Safety: What Works? A Report to the HSE.* 21 May. Birmingham, RoSPA, Case Study 10 summary.

16 HSE Case Study UCATT North-West, *HSE Worker Involvement Website.* Available online at http://www.hse.gov.uk/involvement/casestudies/ucatt.htm (accessed 14 November 2011).

17 UK SI 2007:320 The Construction (Design and Management) Regulations. April.

18 UK SI 1977:500 Safety Representation and Safety Committees Regulations. October.

19 TUC (1995) Commissioned National Opinion Poll to 1,002 people; 98 per cent felt they had the right to be represented by a trade union on health and safety.

20 SI 3163 RIDDOR 1995 Reporting of Injuries, Diseases and Dangerous Occurrences Regulations (1995).

21 Possible change to 5 days in the UK to match self-assessment of sickness rules.

22 UK SI 2307 (1998) Lifting Operations and Lifting Equipment Regulations.

23 MHSW 1999 now incorporates special sections for young people and expectant mothers. It expects that where possible employers should use competent

employees to carry out assessment of risk in order to give them ownership. Part 3 of the Fire Precautions (Workplace) Regulations have also been incorporated for escape measures.

24 SI 1989 2209 Construction (Head Protection) Regulations.
25 Fewings P. (2010) 'Working at height and roofwork'. Chapter 14 in C. McCeenan and D. Aloke (2010) *ICE Manual of Health and Safety in Construction*. London ICE/Thomas Telford.
26 HSE (2009) *Working with Substances Hazardous to Health* INDG136. Available online at: http://www.hse.gov.uk/pubns/indg136.pdf (accessed 18 March 2012).
27 SI 2002 2677 The Control of Substances Hazardous to Health Regulations.
28 Willbourne C. (2009) *Report of Qualitative Research amongst 'Hard to Reach' Small Construction Site Operators* RR719. London, HSE.
29 Accident Advisory and Prevention Unit (1997) *Successful Health and Safety Management*. London, HSE. Revised Edition. (The first edition was issued before the 'six pack' in 1992 and reflected much of the new culture of putting the onus on the employer to propose and develop new systems of work that was started in HSWA 1974.)
30 HSE (1997) *Successful Health and Safety Management*. HSG65 London, HSE Publications.
31 HSE (1997) *Successful Health and Safety Management*. HSG65 London, HSE Publications, p.22.
32 Fewings P. (2010) 'Working at height and roofwork'. Chapter 14 in C. McCeenan and D. Aloke (2010) *ICE Manual of Health and Safety in Construction*. London ICE/Thomas Telford.
33 SI 320/2007 Construction Design and Management Regulations 2007, Regulation 24.
34 CIOB (2004) *Improving Site Conditions: The Construction Manager's Perspective*. Ascot, CIOB. (Based on the 'Change in our Sites' workshop 2003 as a contribution to the improving site conditions campaign by the Strategic Forum for construction.)
35 Crown Prosecution Service (2010) *Corporate Manslaughter Guidelines*. Available online at: http://www.cps.gov.uk/legal/a_to_c/corporate_manslaughter/#a04 (accessed 16 August 2011).
36 Griffiths S. (2009) 'Brighton roofer gets 12 months for manslaughter', *Building Magazine*, 27 January.
37 Regulation EU 305/2011 of the European Parliament and of the Council (2011) laying down harmonised conditions for the marketing of construction products and repealing Council Directive 89/106/EEC.
38 SI 1991 1620 The Construction Products Regulations (revised regulations due to be released in 2013 with better recognition of energy and simpler CE marking for micro enterprises).
39 Fire Safety Centre (2011) *Fire Doors*. Available online at: http://www.firesafe.org.uk/fire-doors/ (accessed 18 August 2011).
40 HSE (1997) *Successful Health and Safety Management*. London, HSE.
41 Major Incident Investigation Board (2008) *Final Report of Buncefield Major Accident Investigation Panel*. Available online at: http://www.buncefield investigation.gov.uk/reports/index.htm (accessed 19 August 2011).
42 HSE (2010) *Notification of Conventional Tower Crane Regulations* INDG47. Available online at: http://www.hse.gov.uk/pubns/books/indg437.htm (accessed 19 August 2011).

43 Sheen Report (1987) *The Merchant Shipping Act 1894, MV Herald of Free Enterprise Report of Court No 8074* (Sheen Report) Department of Transport. London, HMSO.
44 Fennell Report (1988) *Accident at Kings Cross on November 18 1987*. London, HMSO.
45 Cullen Report (1990) *Public Enquiry into the Piper Alpha North Sea Platform Fire of 6 July 1988*. London, HMSO.
46 Hidden Report (1989) *Investigation into the Clapham Junction Railway Accident 12 December 1988*. London, HMSO.
47 HSE (2003) *RR156 Causal Factors in Construction Accidents*. With UMIST and Loughborough Universities. London, HSE.
48 Donaghy R. (2009) *One Death is Too Many. Inquiry into the Underlying Causes of Construction Fatal Accidents*. CM7657 For the Secretary of State for Work and Pensions. London, DWP, p.16.
49 HSE (2004) RR116 *Falls from Height: Prevention and Risk Control Effectiveness*.
50 HSE (2010) *An Investigation into the Use of Plasterboard Manual Handling Aids in the GB Construction Industry and Helping and Hindering the Practicability of their Application*. Health and Safety Laboratory. RR812 HSE.
51 HSE (2004) *CDM Red, Amber and Green Lists*. Available online at: http//:www. hse.gov.uk/construction/cdm/hse-rag.pdf.hse.gov.uk/cdm (accessed 10 December 2011).
52 Watson D. (2006) *Report on the Designer Compliance with the CDM Regulations and Health and Safety Guidance for Designers*. April. Birmingham, WSP for the HSE.
53 CIRIA (2007) *C662 CDM Regulations – Work Sector Guidance for Designers*. 3rd edn. London, CIRIA.
54 Construction Skills (2007) *The Construction (Design and Management) Regulations: Industry Guidance for Designers*. Bircham Newton, Construction Skills.
55 Dalton A.J.P. (1998) *Safety, Health and Environmental Hazards at the Workplace*. London, Cassell.
56 HSE (1997) *Successful Health and Safety Management*. London, HSE.
57 Griffith A. and Howarth T. (2000) *Construction Health and Safety Management*. Harlow, Pearson Education.
58 Bhutto K., Griffiths A. and Stephenson P. (2004) 'Integration of quality, health and safety and environment systems in contractor organisations'. In F. Khosrowshahi (Ed.) *20th Annual ARCOM Conference* 1–3 September 2004. Heriot Watt University, Association of Construction Management, 2: 1211–1220.
59 HSC/CONIAC (2003) *Integrated Teams – Managing the Process*. HSC/M1/2. March.

Chapter 11

Sustainable delivery of construction projects

Buildings and infrastructure are a critical part of the development process and in achieving sustainable development. The sustainable building or structure is one that satisfies environmental, social and economic concerns in a balanced way. There is a need 'to promote sustainable skills and behaviour, inspired by creative and critical ways of thinking, in order to encourage the resolution and management of problems that stand in the way of sustainable development'.[1]

A project manager has influence in co-ordinating the design with the procurement and construction process to achieve the level of sustainability required. Sustainable delivery is critical to ensure that the level envisaged is in fact achieved in use. This will depend upon the efficacy of the design, the quality and method of the construction and the way the structure is used. BREEAM and some certification schemes are now requiring post occupation testing before releasing their final certificates.

The scope of sustainability in construction design and delivery

Sustainable development, of which construction is a part, has been broken down into energy, industrial development, air, pollution and climate change.[2] For construction, sustainability breaks down more recognisably into climate change, zero carbon, recycling and waste reduction, water conservation, energy use reduction, transport strategies, sustainable procurement of materials and services, and biodiversity. On the social side looking after your workforce and providing a safe and healthy working environment and providing effective training with awareness of sustainable construction and design are important. The industry also has to deal with health legacies such as asbestos and de-stressing lifestyles. Clients, designers, contractors and their supply chains all need to work together to develop new solutions and pass on their knowledge to the users of buildings. Government legislation is an important factor to move things forward and initiate new efforts. Cost effectiveness in the short and long term is important to incentivise change

Figure 11.1 Proportionality of total carbon emissions

Adapted from *IGT Low Carbon Construction: Emerging Findings*[3]

for private business and co-ordinated guidelines are needed for public clients. Clients cannot fail to get involved in driving the process if it is to be effective.

It is said that buildings are responsible for 45 per cent of energy consumption,[4] and the break down of this is shown in Figure 11.1. Design and manufacture will both have emissions and will heavily influence total life cycle emissions when buildings are in use and most energy is used.

These are the joint challenges of building and retrofitting efficiently but sustainably:

- Achieving zero carbon building targets and incentivising new technologies at the design stage.
- Innovating technologically in new environmentally friendly materials and collating these together in energy efficient designs and the use of local materials and sustainable transport.
- Tracking and certifying products from sustainable and ethical sources and increasing use of local resources in a time, cost and quality efficient way.
- Making voluntary models such as BREEAM and LEED, Passivhaus and the sustainable code of housing models less bureaucratic for clients and designers so that they are attractive and easy to use and supplement government compulsory measures and are effective into the occupation period also.
- Creating measurement systems that take into account local conditions and context; for example, rural and urban solutions would be different.
- Having equitable planning control requirements on sustainability which ensure an incremental improvement in social, environmental and economic outputs towards a desirable longer term sustainable goal without suppressing entrepreneurial initiative.
- Retrofitting existing building stock to insulate it, make it run efficiently and reduce carbon emissions, supplementing requirements with incentives for domestic owners.
- Developing effective site waste management plans complying with government fiscal policy to reduce waste to landfill and recycle and avoid 'greenwash' as a way of fudging issues.

- Reducing the cost of green buildings so that they are attractive financially and competitive.
- Providing ecologically and socially sound solutions which enhance quality of life at affordable cost and manage our natural environment.
- Encouraging clients with 'loud' CSR policies to put their money where their mouth is and keep robust budgets for sustainable buildings in spite of short term premium costs.

These challenges emerge from the wider world agenda for climate change, but also the desire for living more sustainable lives and ensuring a legacy for our children.[5] The challenges to reduce carbon are spectacular but the change in social attitudes, the emergence of corporate social responsibility (CSR) for larger organisations that are prepared to budget for sustainable buildings in return for longer term payback and the wider use for contractors to run a Considerate Constructors Scheme (CCS) in order to increase BREEAM ratings are encouraging. However there is still a long way to go to reach targets of zero carbon buildings.

The Millennium Development Goal 7 also aimed to 'integrate the principles of sustainable development into country policies and reverse the present trend of loss of natural resources'.[6]

UK government targets

In 2008 the Strategy for Sustainable Construction was issued as a joint government/industry initiative providing overarching targets for the areas of procurement, design, innovation, employee care, better regulation, climate change mitigation and adaptation, water, biodiversity, waste and materials to fight climate change.[7] Targets such as below were agreed:

- 15 per cent reduction in carbon emissions by 2012 for government departments
- reduce water consumption to 130 litres/day/person by 2020
- all projects over £1m to have a biodiversity plan
- 50 per cent reduction of construction waste in four years to 2012.[8]

A government response in June 2011 confirmed all 65 targets for buildings related to climate change and agreed to lead by example as 40 per cent of construction is commissioned by public agencies.[9] Notably these are to reduce waste to landfill to small amounts and to achieve zero carbon construction in 2016 for housing and by 2019 for all other building types. By 2050 they have targeted 80 per cent reduction in all carbon emission in the UK. They will support retrofitting of existing stock and new homes with financial incentives such as feed in tariffs for private renewable energy production as in Case study 11.1, stamp duty relief on the sale of zero carbon homes and green deal incentives for retrofitting the 75 per cent of existing stock likely to be in operation in 2050. These incentives are to reduce the barriers for responsible building owners and to reduce payback periods for capital expense within occupation/ownership averages, including social landlords who own large building portfolios. The green deal incentive is a financing mechanism for energy saving improvement for all homes, which gives reductions for hard-up house holds and spreads repayments for others linked to subsequent electricity and gas use to reduce fuel poverty for vulnerable households. A publicly available standard (PAS 2030) will be used to ensure high standards of product and installation for green deals.

Case study 11.1 Greater Manchester Housing Retrofit Partnership[10]

A pilot of 16,000 social homes out of 260,000 owned by Greater Manchester Council was set up, using the green deal to provide solid wall insulation and efficient condensing boiler systems to make them energy efficient and reduce carbon emissions. There is a potential for creation of 1,800 jobs in construction and a £100m boost to local business in terms of profit. It is designed to act as a catalyst for take up and an example of good practice to investors and other social housing providers.

A new joint Industry and Government Board was created to enhance co-operation and research/innovation and to develop capacity and skills especially in the area of building engineering physics and apprenticeships. This also looks at promoting innovation, leadership and exporting ideas internationally through existing good practice such as the 2012 Olympics. As a leader it favourably met its target to reduce carbon emissions from all of government buildings by 10 per cent during 2010–2011. There will also be a European standard (PAS 2050) for a consistent measurement of sustainability which will replace the energy display certificates (EDCs) which are reputably unreliable and poorly distinguish between different property efficiency and do not account for embodied carbon in the manufacture of materials. There is desire to close the gap between modelled (laboratory conditions) and actual (installed and operated) performance. Figure 11.2 shows where carbon is generated.

The Construction Commitments (2008) signed up to by most large construction organisations and government indicate six key principles for better building.[11] They put sustainability at the heart of design and construction, including the environmental, social and economic factors:

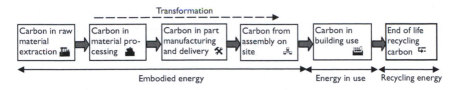

Figure 11.2 Life cycle model of carbon use showing embedded energy

1   Procurement and integration
2   Commitment to people
3   Client leadership
4   Sustainability
5   Design quality
6   Health and safety

Each has a relationship to sustainability; for instance, principle 1 includes ethical sourcing and principles 2 and 6 cover the social sustainability aspects of looking after the workforce. The sustainability section covers the principles of a sustainable construction strategy. It talks as well about 50 per cent waste reduction on site using construction site waste management plans, 25 per cent ethical sourcing and a 20 per cent reduction in water usage by 2012. In health and safety they are looking for improvements in health screening and reduction in injuries where there is a specific focused training of SMEs. These are linked with the sustainable construction strategy. One other important issue is to reduce CO_2 site production emissions by better insulation of temporary buildings, connection to the grid for energy needs and reducing transport miles.

The Wolstenholme Report and sustainability

The Wolstenholme Report (2009)[12] reports on the UK industry progress 10 years after Egan's *Rethinking Construction* and reflects upon opportunities for further sustainability in a recessionary climate. Wolstenholme recognised the need to create business models that integrate the client with the construction process if sustainability is to be tackled effectively. Clients and contractors and designers need to create opportunities for sustainable solutions taking entrepreneurial risk. Main contractors need to move away from pushing risk down the chain to subcontractors and sustain collaborative practice where innovation is encouraged. For clients, they need value based, life cycle costing which supports sustainable innovation and reduces long term running costs. Wolstenholme sees BREEAM excellent rating as a starting point to sustainable delivery of buildings, but a proper understanding of the client's business is also needed to maintain life cycle sustainability.

The positive focus of sustainability has the potential for the public to connect sustainability with construction activity and its capability to generate long term value for the society and the economy. In a recession the industry can benefit from a green led recovery where lower activity gives opportunity for more environmental training and profitable market leadership as the economy recovers. He sees sustainability equally led by the younger generation who have a greater vested interest in the future.

The G4C (generation for collaboration) which consists of early career construction professionals taking jobs in the 1998–2009 era have criticised the industry for failing to create opportunities for sustainable development and a low carbon environment and for lack of opportunities such as skills development in sustainability and too firmly focused on the bottom line. Wolstenholme connects the poor sustainability agenda with missing out on an opportunity to attract the best graduates. Further he criticises the university sector for compounding the problem of un-integrated approaches by the emphasis on discipline training.

He suggests that 'adopting carbon as a unit of currency would be a powerful way to promote the right kind of change for our industry'.[13] Using carbon like this would prioritise sustainable design and construction solutions and could be justified in the business case if decreasing carbon emissions was incentivised by the government by feed in tariffs for renewable energy generated. Client VFM could be gained by avoiding direct government penalties under the carbon reduction commitment where industry has to progressively use less carbon or pay for carbon credits to cover its shortfall in imposed government targets across Europe. Designers will need to work closely with clients on the life cycle payback of energy efficient solutions and not just functionally. More use of innovative technology should bring the price down.

The planning system and sustainability

Some governments now require all planning applications to include a design statement on sustainable aspects of the building which meet the needs of the future. As a key part of this a sustainability assessment is required. In the UK travel plans must show continuous improvement of numbers of employees using alternative transport and force less parking provision in new buildings. In the application local planning authorities require first a design and access statement which gives a reasoned justification of the design, including how it complies with less energy usage, second a sustainable homes or BREEAM assessment to show what level of compliance the design is likely to attain and third an energy assessment to show the CO_2 emissions and a 'what if?' if any renewable energy generation is being offered. The level of attainment proposed might be critical to the granting of the application. With large projects it will be necessary to show an environmental impact assessment to deal with noise, with ecology balance and with pollution. The transport impact assessment is part of this and a Section 255 agreement agrees access improvements to the highway. The requirement for a Section 106 agreement sometimes called 'planning gain' is required of private developers who are gaining profitably from their development in order to contribute to the capital expense of social development of the site such as schools and community recreation provision. These add to the sustainable community target.

In granting planning permission planning authorities wanting harsh requirements need to be aware that they ask the question 'What will be the impact of not having further development?' as well as the obvious question 'What impact will this development have?'[14]

It is quite important that planning is an equitable process and not based on postcode as the system outlined above can be very subjective in what is asked for and what level of sustainable provision is acceptable to that authority. It is also possible that a review of alternatives is a much better option and that this would open up a debate between designer and planner and developer. In the UK there is an opportunity for a pre planning discussion with the local planning officer in order to clarify what is acceptable and likely to be approved. The risk still remains of an incorrect prediction as planning permission is given by elected members.[15]

Private clients' corporate social responsibility (CSR)

A building's sustainability level is guided by the client's policies enshrined in their CSR statement. A project manager will generally claim that they deliver the building according to the client's requirements, but there are occasions when there may be an ethical dilemma between the client's low requirement and the reasonable professional perception to deliver sustainability within good practice and the government timetable. Some of these issues are dealt with at the planning application stage and may return as conditions in the planning permission such as keeping trees or limiting car parking and capping site use. However renewable energy is a specific issue that needs a business case and a payback period. Many clients have high standards clearly stated in the brief that require innovative developments within tight budgets far above minimum compliance.

The client's CSR policy is a statement of their intention to deliver a programme of economic, social and environmental improvements in their business otherwise known as a triple bottom line. The development of new or refurbished assets is a one off for most who are not experienced building developers. This will require some guidance as to various elements of the building layout, fabric or infrastructure which will impact their ability to deliver the policy and meet their KPIs for running the building. Good quality designed environments aim to be motivating and induce productive work.

In addition the capital cost is not a guide to the building life cycle cost and may be inversely proportional to reduce running and maintenance costs. These life cycle costs represent 84 per cent of the cost according to Figure 11.1. The location of a factory or warehouse is also crucial to the transport impact on the environment.

One of the areas of responsibility is to reduce waste and this leads to lean construction as well as a lean business process. In addition recycled products could be used such as the Barclays Bank headquarters in Case study 11.3. The design and assembly process in construction needs to be lean and reduce the capital cost as well as to incorporate an efficient use of the building through the layout and working environment. Sometimes this depends on the right choice of procurement to meet the client's needs. So a client who knows what they want would benefit from a single point of contact. If they want innovative architecture and flexibility in the design process a more traditional form of procurement is useful.

The social issues relate to local employment and being good neighbours to the community. They also ground themselves in the safe working of the employees and the protection of the general public from unsafe incidents. The project manager needs to provide assurance and manage these. Clients often ask for an environmental assessment system to be used. Site sustainable production systems may also be expected as part of the equation for contractors to win work.

Environmental assessment schemes

LEED and BREEAM systems

These two schemes are the leading environmental assessment schemes from the USA and the UK with at least 500 certifications each. They are both adapted for different types of buildings such as offices, retail, industrial, educational, healthcare and homes (BREEEAM: a Code for Sustainable Homes). They have categories in which credits are awarded and these vary slightly though both have:

- energy
- ecology
- materials
- water efficiency
- indoor health/environment.

These translate into four performance ratings with BREEEAM having recently introduced a fifth zero carbon rating of Outstanding. Rating is done by a trained assessor or by the project team which is audited. Both have a design assessment to predict environmental performance, but measure it differently.

Both schemes, post 2008, now have a further mandatory in-use assessment to confirm the expected performance. The latter is to take account of the proper awareness and training of users to ensure optimisation of sustainability in use. Both are used internationally and some buildings have achieved certification in both which has shown slightly higher ratings given by BREEAM, though with in-use assessment this could disappear. The cultures are distinctly differently, but internationally they are sharing features more. BREEAM methods are considered to be quite exacting in the UK and take considerable effort to corroborate evidence. LEED is policed in terms of performance by the country's Green Building Council. BRE International is more academic based than the country based Green Building Council but carries more weight and BREEAM is used substantially more than LEED internationally. Both methods can be used for new construction and existing buildings.

CEEQUAL and Green Globe

CEEQUAL provides an alternative project assessment for civil structures and is an assessment and award scheme which offers UK and Ireland and international fixed term maintenance assessments for civil engineering structures.[16] It is primarily an environmental quality scheme incorporating sustainability. It has 12 categories of assessment which are similar to BREEAM, but also include historic environment. It puts more emphasis on

Case study 11.2 Cross Valley Link Road

The project by the Homes and Community Agency (HCA) secured the future site for 12,000 new social houses near the Nene Valley by installing a new 1.5km link road to join the main A4500 and providing flood attenuation works for the river which was also realigned. The new river bunds also provided a path network in the country park for the enjoyment of locals. The civil works paid attention to reducing flood risk to Northampton and also preserving and enhancing the ecological and local characteristics. The project obtained a top excellent CEEQUAL rating (85.9 per cent) overall.[17] Consultant ecologists worked closely with the local Wildlife Trust to preserve critical species and to transplant valuable hedging and plant life and provide habitats for bats, reptiles and amphibian wildlife. 100 per cent topsoil and subsoil was reused and 1,000 willow cuttings were grown in the on site nursery. The contractors also won a bronze CCS award. The two year project was successfully completed in May 2009.

landscape issues. It was launched in 2004 and by 2008 had had £5bn of assessment.[18] An example of a project CEEQUAL award is the Cross Valley link scheme shown in Case study 11.2.

In gaining a CEEQUAL organisational award it recognises a triple bottom line assessment in respect for people and society, active environmental care and enhancement, ethical operation and responsible concern for environmental impact in the public and stakeholder realm. The award is suitable for client, designer and contractor organisations and built environment funders.

Green Globe is a similar system to LEED and scores a percentage score of perfect for each of the categories. It is offered in Canada, the USA and the UK. It assesses the seven categories of energy reduction, environmental purchasing, sustainable site, water performance, low impact systems and materials, air emissions and occupancy comfort. The system is heavily weighted towards energy efficiency, but it does look at the life cycle cost of green measures and assesses payback so that 'what-if' design scenarios can be compared. It has four different categories: one globe (35–54 per cent weighted average), two globes (55–69 per cent weighted average) and three globes (70–84 per cent weighted average) and four globes (85–100 per cent weighted average). Two globes is therefore equivalent to very good in the BREEAM system. A questionnaire survey is completed which generates a report with percentage scores for each section and suggestions for improvements. To get a rating a third party verifier assesses the score. Case study 11.3 shows two comparison projects assessed by this scheme.

Case study 11.3 Comparison of Green Globe and BREEAM ratings

Two buildings are compared which obtained ratings from these environmental assessment schemes as recorded by the International Facilities Management Association Foundation (IFMAF).[19] Barclays' new HQ in Canary Wharf is a class A glass and masonry multi-storey office building in central London which achieved BREEAM excellent in 2002, and the Walter Cronkite School of Journalism in the Arizona State University is a public building with ground floor retail space, a working TV station, a news room with upper floors with classrooms and offices. It is a six storey iconic glass and masonry structure with sunscreens and a prefabricated lightweight steel structure costing $55m in downtown Phoenix.

Table 11.1 shows a wide variation on the ways in which excellent certification was achieved and it is important to distinguish the business objectives and the improvements made from existing in assessing carbon savings and sustainability improvements.

Table 11.1 Comparison of main sustainability criteria (expanded information from IFMAF)

Sustainable features	Barclays Global HQ London 92,900m²	Walter Cronkite School of Journalism, Phoenix 21,897m²
Rating	BREEAM Excellent	2 Green Globes
Objective	Move office for better working environment and consolidation of office space	To replace existing worn out building and provide a landmark environmentally leading building
Sustainable site	12 offices consolidated into one Direct access to underground, reduced deliveries/week 115 to 7	Brownfield site minimising disturbance to topography and ecology and using naturalised landscape Near light rail station Scored 71%
Energy efficiency	Reduction of CO_2 of 15% per m² with heat recovery, daylight and motion sensors. Five atria on south for thermal buffers and maximum daylight	165% more energy efficient than similar buildings on energy star and saves 273,000kg of CO_2. Energy efficient equipment and low μ value Scored 62%
Water	A green roof to reduce run off	Achieved most stringent water consumption target with water saving devices inside and drought resistant plants to landscape outside. Scored 84%
Materials	Timber sustainable source and no tropical hardwoods, 80% construction waste recycling and reuse furniture and ongoing waste management	Recycling and waste management. Avoid ozone depleting refrigerants. Recycled content in building materials
Indoor environment	No information	Low volatile organic materials and demand control ventilation to separately control high and low pollution areas. Acoustic control Scored 78%

Passivhaus

This concept originating in Germany in the 1990s is about reducing our use of any fuel to heat a house by the radical insulation of the house to stop heat loss and in summer to stop it overheating so that minimal top up heating is achieved. With mechanical ventilation and air tightness it has been proved that the temperature of a house to Passivhaus standards will not fall below 16°C without heating. This concept works in any country for heating and/or cooling needs. Thus to keep a living temperature, for example a 70m² house with gas heating would spend £25/year at 2011 prices, because it only uses

Case study 11.4 The United Welsh Passivhaus[20]

> The project was funded by Welsh Assembly funding for two ultra insu-
> lated detached Passivhaus which consisted of two sustainable designs
> – the three bed larch house and the two bed lime house. Both houses
> were Passivhaus standard in heating requirements, but the former had
> a higher spec. The lime house achieved a BREEAM sustainable code of
> housing rating of level 5 but the larch house achieved a 6. Materials
> were mainly local, although a German Passivhaus window supplier was
> used for the larch house, but subsequently a local supplier tooled up to
> supply this for the lime house.
>
> The houses were contractor built and the cost of the later level 5
> lime house design was £1,467/m² which is only 17 per cent above the
> guidelines for a level 3 house. The larch house was more aesthetic and
> achieved level 6 and was much more costly at £1,787/m². However
> terraced equivalents and less costly materials could bring this price down
> substantially.
>
> The houses were acclaimed exemplar projects to have met the goals
> of low energy usage in a sustainable and acceptable cost way for social
> housing.

15KW/m².[21] The first houses to be constructed to these standards were in
Darmstadt in 1991 which only use 10 per cent of the normal energy for their
size. The standard can be applied to industrial and commercial buildings.
The proposal for Passivhaus is that the 2013 building regulations should
include a deemed to satisfy methodology which is voluntary, but would
ensure low permeability fabric, use high performance windows and low U
values. Case study 11.4 illustrates the use of the Passivhaus system.

Business case for high environmental assessment

Most environmental certification is voluntary with the exception of the Gulf
ERISDAM system which is enshrined in the building code. There is a strong
moral pressure on clients to insist on a certain level of compliance, but a
cost benefit case needs to be made for the business to assess net costs. Energy
efficiencies will be a useful way of showing payback for the capital cost of
incorporating environmental features which cut down on heating, lighting
and fuel costs to reduce energy, but also enhance the quality of life for
employers at work and residents in their homes. An employer will need to
balance the intangibles of better productivity from attracting good clients

and better employees long term, with the net cost and length of payback for a more energy efficient building. Asset values may also be increased. Certification will carry its own costs. There will still be a need for minimum compliance with legislative requirements designed to reduce climate change and meet national and global targets which should be factored out of the comparison costs in the business case. Cases can be made for moving into new buildings or refurbishing existing.

There are many other certification systems available that are used to a lesser or greater extent on a country basis.

Differences between LEED and BREEAM

These schemes have been developed in different parts of the world in the USA and UK respectively. BREEAM includes site production and public transport and LEED has innovation in design. Culturally they are based on different standards and LEED awards points for more car parking and BREEAM for more public transport. BREEAM is based on carbon reduction and LEED is based on US$ cost of running; this makes BREEAM more easy to adapt to an international assessment, but LEED can be recalibrated on different currencies quite easily.

There is a range of ratings from 'pass' to 'outstanding' or 'certified' to 'platinum' with some evidence that BREEAM rates are higher than LEED. BREEAM is very exacting in its standards based on the UK building regulations which make co-ordinated reference to BREEAM ratings. BREEAM have found that by using a bespoke category they have used the categories more flexibly internationally. They have also in some cases recalibrated the weighting to take account of different climates as in the case of Gulf BREEAM where water is much more critical and public transport not so critical to credits. LEED is sometimes seen as a simpler assessment and is more intuitional and BREEAM requires more testing. LEED has adapted internationally by creating local versions in each case and connected to local currency and conditions. Gulacsy has made five areas of comparison which are shown in Table 11.2.[22]

In general it is thought that the dynamic tension between the two systems as they have moved from their home country to the international stage has produced healthy competition for improvement and flexibility. The Greenstar system has been added to show how countries have adapted an assessment system and spread its use internationally.

Internationally the International Facilities Management Association (IFMA) shows that it supports assessment schemes. It lists other schemes internationally (abbreviated name and number of certifications in brackets): France (HQE over 340), Singapore (Green Mark 300), Hong Kong (BEAM 199), Taiwan (EEWH 200), South Korea (GBCS over 120), Japan (CASBEE 80), Germany (DGNB 78), China (Greenstar over 15), Israel (SI-5281 1),

Table 11.2 Comparison of the metrics for LEED, BREEAM and Greenstar

	BREEAM (UK)	LEED (US)	Greenstar (Australia)
Standards	Legislation and best practice e.g. Building Regulations 2006	Based on US ASHRAE, but optional standards	Nine environmental categories
Thresholds	Quantitative e.g. credits	Percentage of whole	Credits and points with weighting on the basis of the region
Measurement	Carbon	US dollars	Australian Dollars
Application	UK and international adaptation voluntary	US and country adaptation voluntary	Australia and country adaptation voluntary
Trained assessor	Required	Ad hoc training of team with assessor support from GBC-USA	GBC-A assessed
Adaptability internationally	Use BRE bespoke version. Widely used internationally	Country GBC adapts standard to suit. Six regional priority bonus credits. '4 other countries adapted'	Versions used also in New Zealand and South Africa
Lead body	Building Research Establishment (BRE) Launched 1990	US Green Building Council (US GBC) Launched 1998	Australian GBC Launched 2003

Adapted and expanded from Gulacsy[23]

Portugal (Lider A 1), India (GRIHA 1). This shows the growing importance, but not the universality, of these systems.

The contribution of sectors of the construction industry

The industry has much scope to reduce carbon use (36 per cent of CO_2 emissions in EU)[24] and increase energy efficiency of buildings in use. Suppliers can take the lead and build on the achievements of much innovation in manufacturing and help to enhance take up of genuine durable products which are locally procured or are transported sustainably so that manufacture and transport carbon is considered equally with cost. In some cases this could lead to fewer technological, locally sourced materials. Where materials are sourced globally robust sustainable sourcing and transport policies can be offered. Figure 11.3 shows the many inputs that feed into the energy saving of a building.

Designers have a large influence in creative solutions which are more energy efficient and towards zero carbon, more socially sustainable in meeting community aspirations, but that are also cost efficient. They will need

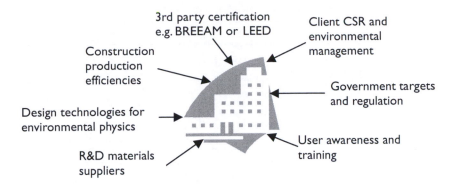

3rd party certification
e.g. BREEAM or LEED

Client CSR and
environmental
management

Construction
production
efficiencies

Government targets
and regulation

Design technologies for
environmental physics

R&D materials
suppliers

User awareness and
training

Figure 11.3 Integrating the energy saving effort in buildings

to enquire into business use to bring life cycle value to the client in balance to the capital cost. All this complicates the discounted net cash flow of future incomes. They will need to work more closely with services designers and ecologists to create more integrated designs and have a better knowledge of sustainable technology as well as ecosystems. In many countries planning authorities are requiring more evidence of sustainable strategies in order to approve design schemes and building regulation requires less energy use. Designers need to design fundamentally to comply with a steadily decreasing level of carbon emission as well as increasing social responsibility. In the UK Part L Building Regulations and Code of sustainable homes have reducing levels to ensure a 2016/2020 zero target. Designers are also encouraged by many clients to achieve recognised levels of environmental assessment accreditation. For example BREEAM and LEED have a wider remit to social and environmental sustainability and give a measured level of attainment. Case study 11.5 has a strong social sustainability rationale.

Private clients enhance their image by advertising their CSR. This often incorporates a drive to carbon planning for new buildings and drives the refurbishment of existing buildings to be more energy efficient and can specify a level of attainment in their new and retrofitted buildings. Clients need to lead the vision for sustainable buildings and facilities as without their commissioning of suitable designs and underpinning of the additional costs low carbon buildings will not be achieved. For users they have scope to use heating and lighting sparingly although technologies for automatic control can be incorporated, but there is still scope for user training and awareness by the industry and an incentive in getting the BREEAM final certificate.

Contractors have a legal duty to control waste and keep carbon emissions down during construction operations and Case study 11.6 indicates a scheme that aims to police this. For example in the UK there is a requirement for site management waste plans and there is a minimum need to design a

Case study 11.5 Sustainable school design to improve learning

In Bristol a 10 year local educational partnership (LEP) was formed to renew school stock in Bristol over a period of time. The first four schools were carried out on a PFI procurement in deprived areas of Bristol in 2007–9 in order to restore confidence in failing educational objectives. Before work started the LEP team consulted with parents, teachers and children to obtain their input so that the schools would be fit for purpose. Bullying, lack of space, vandalism and better IT, more teacher support, better standards of attainment and lack of facilities emerged as key issues. In response the schools were provided with more visibility so that the toilet entrances and washrooms were visible from the staffrooms. The objectives became safer schools, creating active learners, increasing parental engagement, personalised and collaborative learning and 24/7 access to resources in order to transform existing facilities.

Classrooms were made into more flexible spaces and a street was created for better interaction with school zoning and home areas for different cohorts of children where they could see class tutors. The education authority and headmaster were also anxious to have cheaper running costs and a more sustainable school. Biomass boilers were installed and natural ventilation was designed in by incorporating stacks at the back of each of the classrooms which drew in fresh air which was stored in the mass concrete structure by omitting ceilings and providing hard finishings and at night the stale air was drawn out by opening vents in the roof. Rainwater collection tanks were used to flush the toilets. The classroom spaces were also made much more flexible with multi-purpose laboratories and minimum furniture. The ICT system was installed and managed by another PFI and provided automatic registration, administration support and online marking systems and a moneyless system for buying food and lunches as well as plenty of pupil learning. High quality sports facilities were installed which were open to the community after school hours and at weekends. In one school children celebrated their learning with an entrance mural of wishes for their educational outcomes such as 'I wish I could find a cure for cancer', 'I wish for world peace'. Children also had a say in the public art round the school. A further nine secondary schools and academies and two primary schools have been completed.

The results of the design were that pupil attendances went up as bullying went down and many parents who had been sending their children to schools outside the area were prepared to send them locally.

Children feeling safe went up from 30 per cent to 87 per cent. There was better retention of teachers and administrative and they also felt supported by the IT systems and able to track and stop pupil disruption. Vandalism dropped dramatically and self-esteem was raised. Children were happier to come to school with more IT and proud to have the sporting and other leading edge facilities (43 per cent increased to 77 per cent). The headmaster did not to have to worry about rising fuel costs under PFI and energy meters had been visibly installed so that lessons and projects were possible around energy efficiency. Pupil results were 15 per cent improved in just the first year at the first school. Many administrative chores were taken by the contractor's facilities management team, such as a help desk. The community was happy with the new sporting facilities and felt more engaged with the school.

The schools were built on the playing fields and phasing was used without stopping use of the previous schools. The schools obtained a BREEAM 'excellence'. The contractors also found that they were able to make improvements to the three other schools after the lessons learnt on the first one especially in being sure of the BREEAM 'excellence' rating and kept their design and supply chain together. Only 6.6 per cent of construction waste was moved from the sites on average. They felt they had made an impact with the design. They were also to get better bulk buy rates.[25]

house to the relevant level of compliance with Part L of the Building Regulations which reduces energy use. If they are appointed earlier they can also be involved in introducing innovation into the design process and vetting the carbon planning process. In the final stages they can help reduced carbon emission in use of the building by making the building airtight and by handing down easily understood documentation and training to facilities staff in effective use of plant and EMS systems. Contractors are also responsible for choice in procuring materials and in establishing a chain of custody which audits sustainable sources in their supply (see materials sourcing section later). Although it may not be possible to avoid subcontracting, there are opportunities for supply chains to be more tightly managed to ensure compliance with sourcing requirements and to vet contractor design and early involvement to induce innovation. The Considerate Constructors Scheme (CCS) is discussed in Case study 11.6 as an environmental and social way of investing in sustainable delivery of buildings.

Case study 11.6   Considerate Constructors Scheme

The Considerate Constructors Scheme (CCS) is a voluntary scheme for contractors enhancing action towards the environment, the workforce and the general public.[26] CCS awards points for contractors of any size involving educational site visits, giving talks and raising funds for local charities and projects through staff and employee release. It has an eight section 40 point code of: considerate, environment, cleanliness, good neighbour, respectful, safe, responsible and accountable. Sites have to get a minimum of 24 points to comply. Registered sites are monitored and visited and a score of three in each category brings compliance. A score of 35 out of 40 brings a gold award for a construction project. Fifty thousand sites had been monitored in the UK at the end of 2011 and the average score is approximately 32.

   The push to improve environmental scores is to reduce the amount of CO_2 emissions in construction by better insulation of site accommodation, less use of temporary generation of electricity, less fuel usage in site machinery and fewer deliveries and long single driver journeys to work. Sites of course have bigger challenges as many are remote from public transport and workers are casual and itinerant with workers from many different organisations. Site wide education is required to be most effective.

Sustainable procurement

This term is used to cover the procurement of materials and contractors and is covered well in terms of business case, the procurement process and sustainable design, construction, operation and reuse in the OGC sustainability guide.[27] It also considers the three different perspectives of economic, social and environmental.

Business justification

At business justification, consider:

- Economic consideration – consider whole life costing, economic regeneration, future functional use and the investment cost benefit ratio realised.
- Environmental consideration – consider location on a brown or green site, transport infrastructure, biodiversity, energy and water.
- Social aspect – consider stakeholders, heritage and culture and health and safety.

Procurement cycle

The cycle of procurement is considered in Figure 11.4.
During the procurement cycle, consider:

- Economic aspects – ensure documenting on outputs to give a level playing field with a VFM for whole life costing.
- Environmental aspects – assess ethical and sustainable source of materials, the embodied energy, the transportation, the ecology and other environmental aspects of the Construction Product Regulations' six essential requirements.
- Social aspects – assess respect for people, health and safety policy, views and requirements of the stakeholders including community, culture, team sustainability knowledge, competence, commitment and resources, supply chain selection, integration and training for innovation and best practice capability, the safety and health and social aspects of the Construction Product Regulations' six essential requirements.

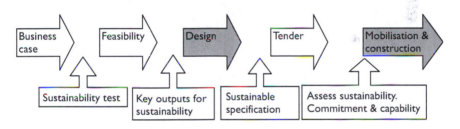

Figure 11.4 Procurement process and sustainability input

These procurement considerations are helpfully broken down to make sure that all angles are considered and to make sure that there is a clear brief, a clear, fair and ethical selection process with maximum freedom for innovation and once on board there is a committed team who are prepared to implement honestly and effectively what they have committed to. Often a sustainable brief is hampered by the budget because of a capital premium for life cycle capability. Requirements need to be feasible so that key ones are not uneconomically cut at a 'cost saving' stage. The procurement process can be expressed graphically as in Figure 11.4. Case study 11.7 looks at the London Olympics procurement.

Ethical materials sourcing

Contractors are responsible for choice in procuring materials and in establishing chain of custody which audits sustainable sources in their supply.

Case study 11.7 The Olympics London 2012

Firms tendering for construction work had to prove their sustainability credentials by complying with the sustainable sourcing code and to be designed to ensure a legacy for the facilities after 2012 and to be useful to the community. They also had to reduce their carbon footprint in delivery by using the Carbon Trust standard. They were required to use energy efficient manufacturers and supply chains. The 'Towards a One Planet' Olympics set out some aspirational goals. The London Sustainability Policy sets out in the details, targets in biodiversity, waste climate change and healthy living for each part of the site. The Olympics Delivery Authority (ODA) oversees the procurement and construction compliance of the plan. The ODA let out 7,000 direct contracts worth £6 billion who will form supply chains with 75,000 subcontractors. It has aspirational targets of zero carbon. There are targets for waste reduction and recycling and legacy. Legacy refers to reuse of buildings for similar or modified purposes; for example, the press centre is to be reused to attract office employment.

This is called responsible sourcing which is an ethos of supply chain management and product stewardship. Sustainable accreditation schemes such as BREEAM, CEEQUAL and LEED put the onus on a responsible supplier to prove the sustainability of the source and this may mean co-ordinating the chain of supply back to the raw material. In effect it is a labelling process for the construction process. One well known and effective scheme is the Forestry Stewardship Council (FSC) certification, which tracks timber supply to its source and provides evidence (or not) for renewable production which generally means protection of rainforests and replanting of timber from commercial forests. Through effective marketing and reliable monitoring this is a respected and well used system which has developed labelling to make sourcing for timber absolutely clear. Client education for level playing fields in bidding is important. There is potential for most other products to have similar systems.

Concrete products are another area where sustainable sourcing is being forced by the taxing of new aggregates and limitations on new quarries. Aggregates also impact the road surfacing industry and recycling is an important consideration. Cement itself depends heavily on quarried limestone and this is a finite product, which also can mar the landscape. Aggregates appear in many other hybrid products. Natural stone is a non-renewable commodity, but can be reused or recycled in products such as reconstituted stone. Concrete structures and products can be crushed on site and used as

Table 11.3 Sustainable material management

Product	Category
Forestry products	Need to manage replanting and rate of cutting. Use FSC certified timber and TRADA reuse schemes
Woodchip and logs fuel	Biomass boilers only if there is a renewable source or plantable woodland, but NO_x emissions need to be cleaned in suitably specified biomass boiler or heating system. See Case study 11.11
Quarry products	e.g. stone, marble, metal ores, tarmacadam and asphalt. Many are registered with BES 6001. Imported stone is not sustainable and local stone has less embedded energy
Fossil fuels	Finite resource, but also heavy carbon omissions. Generally eliminate or reduce use on carbon emissions basis in construction and in building use
Precast concrete products (PCC)	Need to have responsible sourcing policy to avoid substantial fossil fuel use in transporting from afar. These products can be lighter than in-situ concrete and save foundation size. PCC piles can be good but new technology is needed to avoid the noise and vibration nuisance to neighbours. Less waste than in situ, recyclable as road base
Plastic windows	Complex component requiring you to buy from a supplier who has a responsible supply chain or full source checks to be done
Structural steel, cladding, reinforcement, aluminium	Use of ores which are quarried. Steel may be recycled, but process energy intensive. Steel cladding coated and needs more processing. Use of Eco Reinforcement. See Case study 11.8
Glass	May be recycled, but process is energy intensive
Stone	Quarried and recycled product. Also can be used as dust in reconstituted stone and mortars. Imported stone from emerging economies in Asia has potential for unfair conditions and health and safety issues restricting working life to 40 years. Use fair stone standard and win–win scheme
Concrete	Elements of cement and aggregate can be substituted for recyclable products such as grinding of old concrete, iron filings, pulverised fuel ash or recycled plastic fibreglass or coconut fibres. Some processes are quite high energy. Recyclable as road base and hardcore
Excess farm products	Straw stubble can be usefully used in construction. See Case study 11.9

substitute aggregates and steel reinforcement recycled. Most structural steel, aluminium and glass are recyclable but it is an energy heavy process. Table 11.3 indicates issues for a range of materials.

Contractors are key players in the procurement of materials and schemes such as BREEAM and LEED, and contractors gain points if they are able to source sustainably and ethically. In addition non-volatile materials gain points in ensuring that materials with a tendency for 'gassing' do not get used.

Responsible sourcing can be attached to a particular scheme such as the Forestry Stewardship Scheme which provides a chain of custody guarantee for trademarked products where the raw source is established and accredited products and sources are quality controlled – contractors buy these products with full assurance. Alternatively responsible sourcing involves checking back individually through a chain of supply that has a quality control and EMS system to track and assure that sourcing control is in place for each of the suppliers in the chain. The latter is a time consuming and expensive approach especially in the construction industry where such a wide range of products

Case study 11.8 Eco Reinforcement company registration scheme

BRC Ltd obtained a 'good' rating through the BES 6001 scheme for their manufacture of eco steel reinforcement products out of a range 'good'–'excellent'. Eco Reinforcement is a trademark like FSC for responsibly sourced reinforcing steel including recycling from scrap which addresses greenhouse emissions, transport, waste management, employment and skills and environmental stewardship such as water extraction. BRC have 1) third party assessment by BRE and can prove that they have a supply chain entirely based on responsible sourcing that can be traced and 2) a range of environmentally friendly actions such as waste control and reducing carbon emissions and can prove that their business is based on ethical practices. This has enabled them to win contracts and to get onto contractor supplier lists to help their BREEAM rating and sometimes specified by the client. Reinforcement manufacturing with Electric Arc Furnace uses 95 per cent recycled steel but also uses large volumes of water to cool molten metal and more energy. Companies must use stakeholder engagement.

In a sense it is like the scheme that the organic farming community uses where a contravention of one part of the environmental profile certified will invalidate the rating. Eco Reinforcement products and organisations are listed in BRE's 'GreenBookLive' which runs the BES standard.[28]

is used and many components are very complex and have many individual source materials. The scores that are obtained for schemes such as BES 6001 vary on the basis of the percentage of the materials so checked.[29] They break down into the categories of organisational management, supply chain management and environmental and social issues. Scores in compulsory elements must be reached to comply with the standard, but further scores can be gained for exceeding the threshold standard. The majority of companies that have registered as responsible construction products sources are from the aggregate and concrete industries and as such have a relatively limited supply chain. So company registration with BES 6001 is at a relatively young stage.

Social aspects of material production

Ethical materials refers to low carbon materials but it is also concerned with the sustainability of the source such as the labour wages and social conditions in manufacturing and whether the materials are renewable or have finite accessibility such as fossil fuels, hardwood forests farmed with no replanting or methods to ensure future supply. It also can refer to the fair payment for goods at source and the boycotting of products where there are known corruptive influences. In order to control this there are schemes such as the ones above to track sourcing and to accredit methods. Materials of concern in this area either are non-renewable or have the potential for being mismanaged so that they need accreditation. BES 6001 is a framework standard for the responsible sourcing of construction products and covers organisational management, supply chain management and environmental and social requirements. The European project BRC – Building Responsible Competitiveness in the Construction Sector – used a pilot of five countries to set out responsible sourcing in large companies in dealing with their supply chain. There is also a Worldwide Responsible Accredited Production scheme (WRAP)[30] focusing mainly on employment conditions which will affect products supplied to the industry and a portal for responsible supply chain management which provides good access to resources in this field.

Products made in countries utilising child labour are common, but the building industry itself, in some countries, may depend on under age and low wage labour. Bribery is strongly felt in the importation of essential goods into some countries and this may apply to the ethics of international contracts that use many imported products and are prepared to pay out of proportion payments or bribes to get priority for their project's products. International exports in developing countries can help their economies and contribute to sustainability, but must be controlled closely.

An acceptable source is based on their responsible control of ethical and legal operation, effective management systems, sustainable policies encouraged by themselves and their supply chain, their stakeholder engagement, human rights compliance, complaints procedure, health and safety, resource use, site stewardship, efficient water use and waste management, transport impacts, employment records and contribution to local community and built environment. In establishing the points score these points can be used in the materials section of a scheme such as BREEAM.

Localism

Local products are desirable because they reduce the use of fuel and carbon in transportation and provide an additional challenge to the design process. It stems some of the unsustainable impacts of globalism which is economically forcing much of the small scale local production of materials out of business and centralising processing so that bulk material is double transported long distances to and from the processor. For example, the use of local stone can blend well into a design and may cost less if it can be processed in the area. These materials are low carbon. Of course it may have ecological impacts and it may be expensive, but it will also cut down the pollution of major deliveries and possibly create the opportunity for small scale landfill on farms for example. If it is possible it would boost the local economy directly through employment and sale of stone. Balancing this is the need for high quality building products which may have to be processed in centres of excellence as a factor of the design and may be mitigated in some circumstances. An example of a controversial use of local agricultural products is the substitution of local food crops for the growth of oil

Case study 11.9 Local straw bale modular panels

It is worth considering innovation in the localism cause. An example of this is the use of modular straw bale panels called Modcell© which have been developed under patent as framed compacted straw panels, plastered on the external face. These are prefabricated in temporary local factories (farm barns) set up for the purpose of supplying a nearby building site. The straw stubble is harvested for the panels. The large panels are fixed together on site to quickly provide an insulated and waterproof wall which is designed into the structural dimensions and can incorporate windows and doors.

The panels use low tech, local labour and employment. They are likely to use local transport over short distances and they are comparatively cheap to manufacture and easy to use. As a product they have been tested extensively and so quality control is well known and therefore controlled under licence by site staff.

producing plants for fuel. This is sometimes connected with genetically engineered varieties upsetting permanently the ecological and food production balance of the area and in developing countries causing food shortage. A good example is indicated in Case study 11.9.

Carbon reduction and climate change in delivery

Carbon reduction is preferred to energy in today's environment as it represents the impact on climate change more effectively and there is a difference in the carbon produced by fuels or recognised as embodied in various materials. The latter is only measured in terms of manufacturers' production figures and by implication in the green material guides, so if it is not local then there is an additional embodiment of energy to transport it. Energy use in buildings is affected by heat loss and efficiency of heating and cooling and other energy requirements such as lighting. Production requirements and type of building use are considered separately though there are design considerations of layout such as day lighting and shading and thermal mass.

Carbon Reduction Commitment (CRC) started in April 2010 and is designed to encourage companies to good carbon behaviour. It requires companies to baseline their use of carbon in the first year and to progressively reduce their carbon emissions to a target schedule or to pay a penalty.

Meeting more aggressive carbon reduction targets

The key target areas for construction in the UK are for zero carbon in new build housing by 2016 which equates to a level 6 in the code for sustainable housing, a zero carbon for all other new build by 2019 and for public buildings by 2018. One of the key debates is to establish what zero carbon is. Normally it refers to the net emissions that are made to the atmosphere, but these emissions can be offset by:

- generating renewable energy on site and feeding back into the national grid if it is not used
- set off of carbon by planting trees to 'soak up' carbon emissions
- buying carbon credits from others who have come in below zero carbon targets.

In the UK there is a penalty for not reaching climate change targets for reducing carbon output year to year under the Climate Change Act 2008. This is common in Europe and removes the last from the options. The low carbon transition plan 2008–2022 covers the first three budgets for capping carbon emissions enshrined in the Climate Change Act 2008 and this includes new and retrofitting targets. The retrofitting of old stock is to be achieved in the period up to 2050 when the overall target for carbon emissions is to reduce by 80 per cent in the economy as a whole on 1990 emissions.

This plan for construction activities is run by the Innovation and Growth Team (IGT) who are charged with preparing a detailed execution plan as well as encouraging innovative approaches to carbon reduction in buildings. One of the things that the IGT proposes is a requirement to conduct a whole carbon life appraisal factored into the feasibility studies to take account of manufacturing and transporting carbon as well as carbon emissions in the use of the building so that design decisions must factor in location and a consideration whether demolition is necessary due to the high carbon emissions of both demolition and rebuild. In other words there is a need to promote the use of recycling in design as well as in waste management.

The NAO (2007)[31] identifies six environmental impacts for sustainable buildings:

- Reduce energy consumption and associated CO_2 emissions.
- Minimise the use of resources.
- Reduce the release of pollutants.
- Maximise recycling and sustainably sourced materials.
- Promote sustainable materials.
- Conserve or enhance bio diversity.

These are mainly implemented in the UK by the BREEAM and LEED schemes which indicate different levels of achievement and identify different aspects for different types of buildings such as sustainable housing code, education and offices. BREEAM also identifies the social aspects of sustainability which affect buildings in use. Case study 11.10 shows the UK Government performance for their own buildings.

For a project manager, there is a need to cover broader sustainable targets in the feasibility and strategy stage and then to focus on implementation strategies in the design and construction phases of the project life cycle. The client will be involved mainly in defining targets, but their objectives and user characteristics are important in defining the technology which will implement and procure these targets.

Renewable technologies

One of the many ways of gaining credit in environmental assessment is to prove that you will be sourcing energy from renewable sources. This can be expensive as capital costs can be quite high for the outputs. Each technology is carefully evaluated in context as they are likely to be part of the sustainability equation in total. In the UK the government is offering the renewable heat incentive which pays a dividend to any metered energy generation which does not use power from the national grid. RHI helps to

Case study 11.10 **Zero carbon building targets**

The NAO notes that, in their 2006/7 survey,[32] the UK government buildings have yet to successfully implement their own policy which is to require an excellent rating in the BREEAM assessment for new build and at least a very good rating for refurbishment projects. However some projects have aspired to this level to gain tax breaks and kudos. Interestingly the NAO also criticised the BREEAM points collecting approach as allowing even excellent and very good ratings to concentrate on limited aspects of the spectrum of assessment to obtain these scores. They did sympathise with rural projects as having a harder job in attaining higher ratings because of the high drag effects of little public transport compared with quality of environment which cannot be much improved. There has been some improvement since then in the introduction of an outstanding category and more recognition in weightings of the difference rural/urban location makes. More is discussed below.

cut down the payback period of capital plant and/or reduces the cost to the user by giving quarterly or annual cash back. This pays money back per KW produced over a set period of years which can be offset against the capital cost of installation. The feed in tariff is also offered by the government for the generation of renewable energy such as those below where energy is effectively fed back into the national grid by micro site generation. Commercial schemes have some incentives and typical renewable technologies that exist are:

- Central heat and power (CHP) systems which are power plants capturing heat from the generation process to heat water or air and provide heating systems. District systems are most common for the electricity and the heating. They are better powered by non-fossil fuel.
- Biomass boilers which are very efficient wood burners (90 per cent) using logs, woodchips or pellets allowing recycling of trimmings from managed woodland. They need a sustainable timber. Buffer HW tanks are necessary to store off peak production and to boost peak supply. They may also be used for CHP. See also Case study 11.11.
- Ground source or air source heating units which draw latent heat from the air by refrigerating an external source and from the ground by using the latent heat sink in the soil. The former requires either a borehole or a piece of landscaped ground to install a set of water pipes which pick

up the latent heat. Approximately 30–50 per cent fuel is used to run the pumps and fans.

- Wind energy requires the construction of a wind turbine which needs storage capacity in the form of batteries to give consistent supply when calm. They can be an interrupted supply and need backup. Small turbines at low level give minimal supply. Off shore wind is popular.
- Wave turbines, which use the tidal movements and wave currents and have a continuous output.
- Hydro-electric power using gravity fed water to drive turbines, which generate electricity.
- Solar power as a heat source to boost hot water supply/wet heating systems.
- Photovoltaic cells which convert heat to generate electricity, which is then stored in batteries.
- Biogas waste systems which capture methane and other waste gases from sewage and rotting vegetation.

All of these systems have been used on a small scale and there are various targets for their use. In delivering them in construction it is important to take into account the specific type of use by the client, the rate of use, the cycle of peak use. Capital payback may be long, but often the lower running costs make it viable especially where capital grants are available to offset steep initial investment as in Germany.

Renewable energy schemes are sometimes difficult to justify in their capital cost and wind and solar systems often generate below their predictive capacity and need conventional back up for times when there is no wind or sun. The case for zero carbon is usually made with renewable power which neutralises the carbon emissions of conventional back ups by feeding into the national grid.

Reducing waste and recycling more

Reducing waste, that is by optimising design use and wasting less of what is specified, is a primary and necessary approach to sustainable development as it represents a more radical understanding of the limitation of physical resources which can be used in the provision of the built environment including buildings and civil engineering, so that less is better if it was never actually used in the first place. It is radical because taken to its logical conclusion it will challenge our assumptions of usage. A designer would need to challenge a client, do you need it? This is more uncomfortable and affects the supplier's use of electricity.

Recycling is a pragmatic response to the developed environment as we see it and its projection into the future where we try to mitigate the effect of throwing things away by reusing them. It is common sense and also

Case study 11.11 Biomass boiler

Cirencester Agriculture College installed a 220KWh biomass boiler which now heats its library and three other adjacent teaching blocks. The opportunity came when the college needed to change expensive oil fired boilers at the end of their life and reduce its carbon emissions by 34 per cent in compliance with higher education targets.

The boiler, two large heat sink facilities for heat storage and a wood chip storage facility, was installed in a new boiler house and district heating pipes installed in shallow trenches to the adjacent buildings. Grants received to cover some of the extra capital cost made it affordable. With the running costs it is expected to be similar to gas and save substantially on oil fuels which is the only fossil fuel alternative to their country site. With the advent of RHI scheme they will also receive quarterly cash back. Air quality of emissions is good and the high quality boiler is 90 per cent efficient and emits NO_x within prescribed limits, but no carbon. It was necessary to raise the standard flue height to clear the screening effect of nearby trees. The college has considered the possibility of harvesting its own fuel through the farms that it owns. This will be the subject of a further cost benefit analysis. College buildings use less heat in the evenings and they can reduce output to 30 per cent without 'turning off' the boilers which means they can store any excess heat in heat sinks to release and boost in peak demand and to save fuel. The boiler house also includes two back up oil fired boilers for very cold periods and as a back up in case of fuel shortage or boiler maintenance. A new student accommodation block will be connected later to the same circuit as it has an evening peak load and will not increase boiler capacity.

represents economic sense. It challenges our attitudes on a moral level i.e. why are you wasting it?

The 'do you need it?' approach comes within the conception and feasibility stage of the project life cycle and needs to be considered by a client as part of portfolio and programme management in that the answer will affect all current projects as it emerges from organisational strategic planning. Value analysis at its earliest stage is a necessary tool for considering this and as such needs to question normal assumptions of usage and present alternative intrinsic innovative solutions which may impact on company ways of doing things and reducing space requirements.

Design to reduce waste or using it to make use of recycled products, for example reuse of foundations or production recycling emerging from site waste management plans, is evidently limited on what has already been decided to be built, but is still a significant and ethical reduction of the use of new materials. There is an effort for duty of care to ensure structural stability, health and safety, and life cycle consumption and reduce embodied energy in the recycled materials especially where they go off site to be processed. A particularly fruitful consideration is in reducing excavation and its removal from site. This stops noise and nuisance to others in the streets, reduces the use of fuel and economically saves the cost of excavation 'cart away' with associated landfill tax. Excavated soil can be used on site for sound attenuating bunds or as a cut and fill exercise on sloping sites. Top soil can be stockpiled for reuse or sold. Other products such as plasterboard can be recycled. Surface water run off can be reduced by catching to recycle as grey water or by using permeable or less paving so that drainage pipes are smaller (SUDS)[33] and planting trees saves carbon and provides shade and wind breaks. People appreciate the natural environment and work harder. More radically, some research has also been done on the reuse of old foundations by maintaining similar footprints in new buildings. Decontamination is also possible by careful sequencing and planning by 'washing' soils to remove contamination by chemicals, so that soil does not have to move away and virgin hardcore is imported.

Recycling off site can be a problem, because it might not take place, it might have to go long distances for both processing and reuse and it might embody a lot of energy (carbon) in the recycled product. It is sensible in this case to have a unique recycling plan for each project, hence the use of dedicated site waste management plans and the sourcing of local recycling projects. Recycling steel reinforcement or beams after demolition can be very efficient as it is already an established part of the manufacturing of most new steel, but is more efficient after the concrete has been crushed on site. Recycling on site such as the reuse of concrete for aggregates may also be efficient as it reduces the use of non-renewables and the harmful impact of quarrying. It is a design and production issue.

It is clear in recycling that there are conflicting impacts. The recycling of wasted plasterboard to a local manufacturer to reuse presupposes wasteful site use and/or design. Could this have been cut out by standardised design sizes, wider board size choices or more packaging or worker training to reduce damage to the product and to store better? In cutting and carrying and sorting less plasterboard a labour saving is also realised. Delivery miles and waste costs are also saved. If you blow up a 1960s high rise building it will have fewer salvaged materials, but it is less energy to demolish.

Design choice of materials has to balance capital and durability cost against easy breakage and future recycling possibilities and cost of dismantling and reuse. It needs to consider the embodied carbon of recycling as well as in use. In a post stressed structure there might be prohibitive dismantling and reuse costs. Can the building be more sustainably refurbished if the space layouts are flexible for several types of use? Are the costs of retrofitting better than the costs of rebuilding to offset the level of carbon emissions in new build?

Carbon reduction in design and services

Carbon reduction is considered at the design stage and also at the functional analysis. Prior to that strategic planning identifies the impact of expansion and portfolio planning identifies the interconnection between asset management and provision and the outputs of the company. Hence a new office may be reduced in scope and size by 'hot-desking' and encouraging staff to work from home because it reduces the size of the office provision and saves carbon produced in commuting. It will increase home use of electricity and heating. Case study 11.12 shows some example costs of reduction.

Case study 11.12 Asda Stockton retail base case

This case study is an Asda store in Stockton completed in 2008 with a floor area of 9,393m² over two levels. About three-fifths of this is the sales area and two-fifths is occupied by storage including cold storage, staff cafeteria and a bakery. A base case analysed by Target Zero, which was carbon neutral, brought up several interesting comparative findings. In order to be compliant with the Building Regulations 2010 it saved 35 per cent of carbon and saved 0.36 per cent on capital cost and £973,575 on the NPV cost. The store would have saved 51 per cent carbon on the Building Regulations 2006 at an uplift of 0.8 per cent capital cost and 58 per cent carbon at an uplift of 5.2 per cent. On a broader economic, environmental and social scale, the estimated capital uplift compared with the BREEAM ratings is:

Good	0.2 per cent uplift of capital cost.
Very Good	1.8 per cent uplift of capital cost.
Excellent	10.1 per cent uplift of capital cost.

This last figure is far from carbon neutral; the store would require renewable energy such as a wind turbine to comply which would up the cost much more. Location is important in the public transport credits and the ecology which is enhanced and can vary a lot from site to site. A rural site would lose out on improving existing ecology and on good public transport.[34]

Building engineering physics is closely associated with the design of building fabric to encourage lower energy buildings and as such covers the areas of energy loss and thermal performance, air movement, control of moisture, ambient energy in terms of solar heat and light, acoustics, lighting, climate impacts such as wind and heat gain and day lighting, human physiology relating to comfort. These may be controlled naturally by encouraging heat and cooling flows into the fabric and layout or by using heating or cooling systems or a hybrid of the two. Under sustainable design it is preferable to reduce dependency on machinery that delivers high carbon usage.

Under natural systems the exact building requirements need to be modelled and occupant numbers and habits known. Layout, orientation and shading control are critical, and controllable venting and thermal storage need to be introduced, bringing building engineering physics considerations into the structural and architectural concept stage of the design.

Planning consent also requires a statement of intent in reducing carbon transmissions and increasing acceptability of the building form and use in the community. This will involve both social aspects such as transport and stakeholder consultation, as well as a commitment to methods of reducing energy use and/or generating renewables in the environment. The BREEAM provides a complex but still basic assessment of building performance in these areas and generally a commitment to achieving a particular level, for example excellent.

BREEAM assessments

In the BREEAM assessments there are nine different areas of consideration shown in Table 11.4 which cover environmental and social issues.

These nine assessments are minutely broken down and described in a 403 page document. There are also minimum BREEAM credits which are required in each section and a rating is made from a minimum of 30 which is a pass to 70 which is excellent and 85 which is outstanding. A score of very good (55) is a minimum target for a sustainably conscious client. It can be seen in Table 11.4 that the heaviest score is on energy use and health and wellbeing. Reduction of CO_2 emissions has an essential score of six for excellent. We will deal with excellent rating as this is generally considered the top building score economically attainable without significant renewable energy. We shall also be discussing the credits for the management section in the next section. It is often hard to get provenance for the scoring in some areas such as the responsible sourcing and the volatility of the products. There is comparatively little given to a sustainable construction process but getting excellent scores in the Considerate Constructors Scheme is helpful as this is recognised for environmental and social points.

It is however the longer term sustainable design and use of the facility which is more critical to climate change and social benefits and the design

Table 11.4 BREEAM 2008 environmental weighting

BREEAM Section	Weighting %
Management	12
Health & Wellbeing	15
Energy	19
Transport	8
Water	6
Materials	12.5
Waste	7.5
Land Use & Ecology	10
Pollution	10
Total	100%

of the service has a large role to play. The accessibility of public transport, the recognition of good local employment and the proper use of the controls mean that some of the building in use figures do not add up to expectations without good awareness from the users. Training is also necessary in the knowledgeable use of proposals like biomass which may also be seen as polluting by local building inspectors. Case study 11.13 is a good example of some of the problems encountered in meeting a level of compliance.

Case study 11.13 Design and build for further education assessment

This is an example of how a design and build contractor seeking to get excellent on a further education college was able to use the BREEAM rating system for Educational Buildings. Excellent requires a score of 70 credits out of 100 and covers the areas above. The assessment is made at the design stage and following construction. A third party assessor is used to validate.

A plan is prepared which is used to target where the points can be gained in each section and subsection. The particular areas of interest which affect the detailed design and construction are Management section 2, Heat section 1, Materials sections 1, 5 and 6, Waste sections 1 and 2. One third of the points were targeted in these sections. A later in-use assessment will have to meet the points so it is important to have points in hand as it is never certain that the design expectations will be met, due to changes and different interpretations during implementation and use.

A BRE assessor is appointed to calculate and audit the points score; there is continuing communication and assessment. Expected points not awarded may mean future changes. There is a complex requirement for minimum scores in some sections as well as total scores. Some impacts are assessed as an integrated whole.

Materials must be compliant to the Green Guide to Specification (GG) and these give a rating of 1,800 different common specifications based on the life cycle of the material/component called an environmental profile (EP). Manufacturers may wish to commission a tailor made certified EP for their product. EPs do not take into account site location in relation to production location. A favourable EP leads to a favourable GG rating, but aspects of source have to be proven. A software template is used to assure equality of measurement.

The supply chain needs to be able to indicate a chain of custody so that components and major subcomponents show evidence of sustainable sources. There is also a need to prove the sourcing in managing the small quantities of volatile organic material allowed, which may be subject to 'gassing'. To get points for management of the process an environmental management system needs to be in place for both the managing contractor and the supply chain members where there is no formal labelling system. For example a precast concrete supplier must be able to prove an EMS of their compliant sourcing of aggregates. This may mean that aggregate quarrying must be sustainable or have some recycled aggregate to gain a credit. Changes of supplier may take place in order to be able to ensure evidence is available.

Heating and lighting assessment is based on the heat gain and loss and also the natural daylighting. These clash with each other in that small or fewer windows to reduce heat loss in winter will mean that daylighting in the room is lost (room must be shallower). Also solar heat gain in winter will be limited (more heating). Shallow floor plans give a smaller floor to wall ratio which makes it more expensive and uses more material for the same space enclosure. This means a sensible trade off or some anomalies are created due to the rules. On this project windows with lower window sills were used to increase the daylighting and depth of room and a fixed shaded rooflight was allowed to get more light into a restricted area of the room.

Buildings on land liable to flood need to have ground floors and access roads above expected flood levels to not gain negative ratings.

The waste sections meant that a target of 80 per cent recycling was required.

Carbon reduction in production

The Considerate Constructors Scheme is a well established voluntary scheme for constructors to ensure that they are both green and think about their impact on neighbours and their staff health and safety and three other areas. It is mandatory in terms of achieving sufficient points for BREEAM Excellent and impacts on more than one of the management sections.

This deals with the construction and procurement stage of construction and as such only really refers to the management section of BREEAM which covers commissioning, Considerate Constructors Scheme, construction site impacts and building user guide.

Site waste management plans (SWMPs)

SWMPs are specific plans which are made for reducing, recycling and managing the waste on the site. They should be agreed with the client and a specific site manager needs to be responsible for the plan. On restricted sites it is difficult to accommodate recycling of all types of waste. The aim is to help reduce construction waste to zero by 2020. Site waste can be determined in broadly three categories:

- Inert waste means waste that does not break down and is cheaper to send to landfill sites because it does not produce side products like methane. Landfill tax applies at lowest rate.
- Active waste is waste that does break down and is subject to an escalating tax rate which makes dumping progressively expensive.
- Special waste is classed hazardous and as such must go for further diagnosis and special dumping and carrying procedures dependent on the nature of the hazard e.g. asbestos or contaminated soil.

Inert waste such as excavations may also be hazardous as it may be contaminated on industrial sites or places where there have been leakages and this may limit some reuse of excavations. Washing down during construction may also be polluting if chemicals are used, such as façade cleaning and washing out concrete wagons. Wash water may not be accepted in drains or because it will leach into water courses. Landfill taxes are contributed to community funds.

Active waste will need to be licensed and sent to specific tips and this is expensive.

Hazardous waste needs to be treated or dumped at specific tips which can take it. This makes it popular to treat contaminated soil waste on site to neutralise the pollution and to reuse the product on site or alternatively to seal in the contamination so that it will not reach the surface or leach into water courses. There are some unavoidable materials such as asbestos.

The Environment Protection: Duty of Care Regulations 1991 have compulsory procedures to ensure waste is categorised and tracked with proper paperwork to remove and dump waste using licensed tips and contractors. As a part of the sustainable communities programme there is a drive to improve brown field sites as soon as a new development is muted and this is the responsibility of the owner.

Preparation

There is over 90m tonnes of production and demolition waste each year on UK building sites – four times the rate of household waste.[35] 13m tonnes of these are unused materials ordered in error. Nearly one third of all fly tipping in the UK involves construction or demolition waste.

SWMPs are prepared to show where waste reductions can be made and to provide a framework for implementing it so they need to be a little challenging to be effective and not replicate standard achievements. A continuous improvement model should be aimed for. It is a legal requirement in the UK since 2008 that all sites which exceed £300,000 in value must formally sign off a SWMP co-ordinated by the principal contractor.[36] The original voluntary code of practice for SWMPs has been used below and adapted:[37]

1   Appoint an individual to prepare and be responsible to implement the plan before construction work begins. Principal contractor is normal.
2   Identify the types and quantities of waste and when produced according to targets to reduce.
3   Identify management options and rank according to waste treatment hierarchy.
4   Duty of care – use only licensed contractors and set up systems for transfer records.
5   Train staff and subcontractors to identify and segregate waste.
6   Measure actual waste streams to compare with targets.
7   Monitor implementation and be prepare to update plan during use.
8   Put security measures in place to prevent illegal disposal of waste.
9   Feed back for next time.

SWMPs will work if there is good communication up and down the supply chain reaching down the tiers to those that order, produce and install the material. A co-ordinator is appointed who can establish a recycling centre and improve and incorporate different subplans into a master plan in the manner of health and safety method statements. This master plan should be integrated for possible site reuse of materials by others. There should also be incentivised targets to encourage continuous improvements against the plan and monitoring to try and pull back any slippage as once wasted it is difficult to reincarnate. Training can be integrated through toolbox talks to make sure there is a regular reminder of the need and feedback on the benefit. Feedback will mean checking the effectiveness and the cost of waste dumping and offsite recycling. Once the client has signed the plan there is a commitment to reach the targets and a responsibility on behalf of the client to ensure that it happens. Other leaks to the system are to ensure that there is no over ordering and that the buying department is kept up to date with recycling and reuse so that long delivery orders are not abortive and suitable credits are invoiced. Communication is critical in all these areas.

Timely information and change management procedures are important and a client making changes at a late stage needs to consider the impact of abortive costs and waste where the specification changes or work is ripped out. A contractor needs to have a system for following through the impact on existing orders and recycling on the specialist suppliers and contractors.

Fly tipping is a criminal offence in the UK and is subject to large fines or even imprisonment.[38] It means dumping waste on ground not licensed for waste or that type of waste and not checking whether someone has a license for waste disposal.

A site management plan is best accompanied by a checklist and a monitoring form which may look like Figure 11.5.

In setting up recycling a reserved area on site needs to be identified as accessible for skip lorries as shown in Figure 11.6. It is no good having specific skips unless they are labelled with absolute clarity so that materials

Waste type	Quantity (m^3)				
	Reused on site	Reused off site	Recycled on site	Recycled off site	Disposal to landfill
Inert					
Active					
Hazardous Type 1 Type 2 etc					
Totals (m^3)					
Performance score as %					
SWMP targets					

Figure 11.5 A monitoring table to track and target waste material

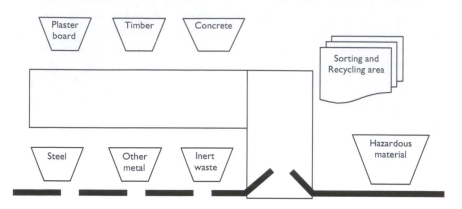

Figure 11.6 Typical waste management layout

can be recycled effectively. Materials also need to be weighed or measured in each bin and recorded. A sorting area and hard standing are required.

Considerate Constructors Scheme

Construction companies also have CSR policies which concentrate on reducing their carbon footprint and reducing transport use of fossil fuels and the energy use in running equipment and heating and lighting welfare and office facilities and temporary lighting in the building. The CCS is outlined in Case study 11.6.

The Considerate Constructors Scheme is a voluntary one which aims to raise the reputation of the contractor in the local community by being better neighbours, but also by being proactive in the area of the environment and health and safety. It has a significant focus on raising awareness and involving the wider workforce in acts of proactive social sustainability in the community, improving health and safety attitude and having a responsible attitude in respecting the neighbours and reducing nuisance in things like cleanliness, dress code, noise, dust, deliveries and public safety and health. The social and community issues include the creation of employment opportunities during construction.

How does it work?

The scheme works on the basis of eight areas for which five points are awarded in each section and above 35 is an excellent score. It is significantly harder to obtain a full score in any section without being innovative and going above standard levels of provision. The scheme was started in the UK in 1997 from a start with 85 sites and four monitors and now has 50,000 registered sites. Many companies now register corporately which ensures that each of its sites is registered and their scheme covers smaller sites without full time supervision. Other countries are showing an interest in the scheme and have their own equivalents.

Companies are registered prior to the site commencement for a small fee. Posters are used at sites to show signed up for commitments to those outside with hotlines to register complaints or praise about non-neighbourly activity. Contractors produce activity plans showing how they specifically intend to attain standards. Assessors come on to site regularly to inspect provision made and provide a formal score and advice for future improvements. Prizes are awarded for exceptional performance at a gold, silver and bronze level citing particular achievements for high scoring sites or companies. Client partners later joined to ensure registration of sites for which they had commissioned the work.[39] The scheme is endorsed by BREEAM and ecohomes and credits are awarded for being a member of it as long as the score of three in all areas is achieved representing an acceptable standard. All public projects are advised to ensure contractors register for the scheme. Case study 11.14 is an example of a winning contractor.

The CCS scheme in Table 11.5 provides basic good practice guidelines and links to source material to help sites in a learning process to do well. The scheme has a lot of potential for expanding good practice competitively even though compliance is at an acceptable level. This encourages continuous improvement.

There are problems with the scheme in that minimal compliance gets the same press as outstanding compliance in a local community and some contractors want the image, but may not put the money up front to make wide ranging improvements. There are many other areas of environmental compliance outside CCS and these sometimes have specific legal methodology, such as transfer of waste requirements. Monitors are sometimes

Case study 11.14 Clayfield construction

In 2010 there were 580 sites selected for gold, silver, bronze or runner up awards. Interestingly the most considerate site in 2010 was a small office accommodation by a local contractor which was recommended for its scrupulous cleanliness on the site and in the surrounding areas and absence of obstructions, tight waste control, top score in health and safety because of additional features, exceptional welfare facilities, active staff and community participation, assistance to a local football club and school, clearing of Japanese knotweed from the area, provision of owl and bat boxes and slow worms.

Table 11.5 Measures for Considerate Constructors Scheme

Categories for action	Possible measures
Considerate	Restrict delivery times, inform scope of works and consult, clear access, visitor arrangements, operative car parking, disability access, personal touch & providing hoardings for visibility
Environment	Reduce waste, respect flora/fauna, spillage bunds, energy saving, stop pollution, segregate & recycle, reduce noise, directional lights, delivery control, reduce cars
Cleanliness	Clean and tidy site, wheel clean and roads, damp dust, clean toilets and showers, food hygiene, no litter, efficient material access, paint hoardings
Good Neighbour	Complaints courtesy and action, 'hotline' information, reduce security alarms, clear and limited hours of work, non obtrusive lighting and deliveries, no radios
Respectful	Screen toilets from public, dress sense especially for local shopping/contact, conduct rules, exclude offensive material, reduce noise, Good welfare, viewing
Safe	Warning signs, effective protective barriers, lighting, walkways & segregation, record accidents/near misses, escape routes & rendezvous, safe material storage. Protection of public.
Responsible	Community participation, be aware of other activities and adjust work to suit, first aid help, information to public services
Accountable	Records of schemes in place, information to the public, operative training an induction, photos etc

wowed by unusual but minimal issues like zebra crossings from the site car park to the mess room, at the expense of more important measures. Poorer contractors may simply lower prices and cut corners on the sustainability up front so that they get the job. However on balance the scheme is successful, especially as it is voluntary.

Environmental Management System (EMS)

EMS is a system that specifies a process for controlling and improving an organisation's environmental performance. These will be applied here in the context of the project management systems. An EMS, as shown in Figure 11.7, provides the ability to deal systematically with implementing policy and complying with complex regulation. It is not compulsory, but gains confidence from clients.

ISO 14001:2004 is a worldwide standard which requires a formal system with a third party monitoring and report.[40] The purpose is to identify what impacts upon the environment in the business and to extend this to the project. There is a need to understand and comply with the laws and regulations that affect these areas and to produce objectives for improvement and to set up a control system to achieve them. Internal and external reviews are

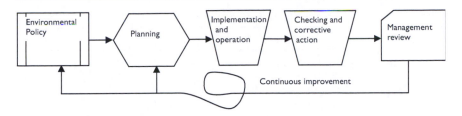

Figure 11.7 EMS process chart
Based on IEMA website[41]

needed to identify improvements. During setting up a registrar is available to advise and an assessor awards (or not) compliance and supports the company in a continual improvement process. Care should be exercised in project work in that a general business registration is the norm and can be set up in general terms not specifically relevant to the project itself. This can lead to a less than hands on approach, unless the company has a substructure to develop each of its projects.

The European Eco-management and Audit Scheme (EMAS) is site based and refers to an existing company environmental policy and also stresses a continuous improvement. The process is identified and quantifiable data on current emissions and environmental effects, waste generated, raw materials utilised and energy and water consumed is generated. This may apply to an office of a building project. These assessments then measure improvements according to targets set against the base date. An EMS is required to set up procedures and monitor these aspects and to ensure the project is kept on track. Communication is an essential factor and motivation will improve if outputs are explained in beneficial terms to meet project requirements or to benefit future business. An EMAS verifier may be involved to verify the company statements, especially in large contractor businesses as shown in Case study 11.15.

Although EMAS is a specific system there are a number of organisations such as BITC and WWF who have other standardised requirements and expectations.

The Green Dragon Environmental Standard is a stepped approach to environmental management systems and audit and has five recognised levels of compliance to get a smaller or less involved organisation along the way to full certification. These are commitment to an EMS, understanding, managing impacts, EMS programme and continuous improvement. In addition BS 8555 has been developed for SMEs and is much less cumbersome to develop.[42] The government IEMA Acorn Scheme develops a simple capacity to manage environmental impacts and can lead on to an ISO 14001 scheme, but there is a separate register in this scheme following an external assessment.[43]

Case study 11.15 Large contractor environment management system[44]

A large contractor score 93 per cent on their assessment to be part of BITC and runs an EMS based on ISO 14001 in the UK and some of their Canadian business, which ensures that they have procedures training and monitoring in place for all their projects. They set targets each year to reduce things such as carbon footprint and use the system to comply with their government set carbon reduction commitment (CRC). They prepare environmental risk plans for each of their projects for which they identify specific mitigating actions and the top five risks are considered to be water pollution, waste management, impacts on wildlife, environmental nuisance offences and their carbon use. They carried out 600 people hours of training to their 4,000 employees through team talks, project discussions and targeted emails and web conferencing. They use an outside organisation as an auditor of their performance. Of their employees 83 per cent think they are a leader in environmental management.

Conclusion

Sustainability is an important part of the delivery of buildings in social, environmental and economic terms, because it can save money in the long term by payback of less energy use building a reputation with desirable clients. Buildings use 40 per cent of the world's energy so there is considerable scope to design buildings which are more efficient. Users and facilities staff need to be trained to use the technology in an educated effective way with predicted life cycle savings which are now tested by environmental assessment systems in use before final rating certificates are issued. These assessment systems are being used internationally across the world. There is still some way to go to make the systems easier to use and to make different ones comparable. The climate and location context where they are used is important and should be acknowledged in the weighting.

Government policy across the world is tuned in with the Rio 1992 carbon reduction commitments and new buildings need to be zero carbon in operation by 2020. In the UK the carbon reduction commitment is a key player in improving company and public department performance and the Strategy for Sustainable Development is a key document for ensuring that construction targets are met. The public sector needs to be an exemplar for these targets which the government has accepted. The planning system is often used to ensure that sustainability and environmental impact have been assessed. The Building Regulations are used to ensure design will comply with minimum requirements, but are prescriptive. In other countries the

Green Building Councils are used to police the requirements of environmental assessment such as LEED.

Designers are driven by the need to reduce carbon, but also by the other issues which come out in environmental management systems. The environmental assessment methods primarily impact designers to gain ratings from the building use, but cover the procurement of materials, carbon emissions, ecology, water use and indoor environment. The LEED EAS seems to be slightly tougher than BREEAM, but is favoured by some organisations because it is simpler and perhaps not so exacting. CEEQUAL is used for civil engineering projects. There is growing confidence in their use, but they could still be clearer on the source of materials.

Ethical sourcing of materials is an important part of social sustainability to ensure renewal and to ensure equitable conditions of production and reward. There is a need to track production and to identify unfair conditions. Sustainable procurement also covers the area of localism and other community factors which favour for example local employment. The project manager has some professional responsibility to ensure that the client is supportive of sustainability and ethical sourcing to make them aware of good practice.

The environmental management systems can be connected to environmental KPIs such as reduced carbon footprint, less water usage and in managing the system which checks and audits this with a third party. Contracting companies value both the ongoing improvements for recycling and waste reduction and the use of systems such as the SWMPs to build up a waste management system. Their CSR is also enhanced by CCS to integrate with neighbours considerately, reduce site production carbon and reduce pollution, noise and dust.

There is much that can be improved in better sourcing and more innovative technology to reach higher environmental assessment grades routinely and with less cost and to make sure that smaller companies are able to understand their role in refurbishment of buildings.

References

1 UNESCO (2005) UNESCO and Sustainable Development. UNESCO. Available online at: http://unesdoc.unesco.org/images/0013/001393/139369e.pdf p.4 (accessed 16 November 2011).

2 UN Department for Economic and Social Affairs (2007) Division of Sustainable Development. Available online at: http://www.un.org/esa/sutdev (accessed 9 July 2010).

3 Adapted from Morrell P. (2010) *IGT Low Carbon Construction Innovation and Growth Team: Emerging Findings.* HM Government Business Innovation and Skills. Note that percentage calculations are preliminary and are available at http://www.bis.gov.uk/constructionigt (accessed 6 August 2010).

4 King D. (2010) *Engineering a Low Carbon Built Environment.* London, Royal Academy of Engineering.

5 Brundtland G. (1987) *Our Common Future.* United Nations Report of the World Commission on Environment and Development, Annex to United Nations General Document A/42/427.

6 United Nations (2001) *The Millennium Development Goal 7*. The Millennium Summit, New York, UN Assembly.

7 HM Government, Dept of Business Enterprise and Regulatory Reform and Strategy for Construction (2008) *Strategy for Sustainable Construction*. London, HMSO. Available online at: http://www.bis.gov.uk/files/file46535.pdf June (accessed 12 December 2011).

8 HM Government in association with Strategic Forum for Construction (2009) *Strategy for Sustainable Construction Progress Report,* 16 September. Available online at: http://www.bis.gov.uk/files/file52843.pdf (accessed 12 December 2011).

9 HM Government (2011) *Low Carbon Construction Action Plan: Government Response to the Low Carbon Construction and Innovation and Growth Team Report*. London, HMSO. Available online at: http://www.bis.gov.uk/assets/biscore/business-sectors/docs/l/11-976-low-carbon-construction-action-plan.pdf (accessed 12 December 2011).

10 HM Government (2011) *Low Carbon Construction Action Plan: Government Response to the Low Carbon Construction and Innovation and Growth Team Report*. London, HMSO. Available online at: http://www.bis.gov.uk/assets/biscore/business-sectors/docs/l/11-976-low-carbon-construction-action-plan.pdf (accessed 12 December 2011), p.41.

11 Strategic Forum for Construction (2008) *The Construction Commitments*. London, Strategic Forum for Construction.

12 Wolstenholme A. (2009) *Never Waste a Good Crisis*. London, Constructing Excellence. (Concludes some progress has been made towards targets for better health and safety, better profitability and productivity, but had failed to deliver on predictability in cost, time and quality and needs to retain the integrated business model with clients involved in the bigger process of the built environment and not regress to competitive tendering in order to focus on issues of sustainability.)

13 Wolstenholme A. (2009) *Never Waste a Good Crisis*. London, Constructing Excellence, p.17.

14 Smith S., Richardson J. and McNab A. (2010) *Towards a More Efficient and Effective Use of Strategic Environmental Assessment and Sustainability Appraisal in Spatial Planning*. UK, Department of Community and Local Planning, in conjunction with Scott Wilson. Available online at: http://www.communities.gov.uk/documents/planningandbuilding/pdf/15130101.pdf (accessed 21 November 2011).

15 For more of these arguments see author's book on ethics: Fewings P. (2008) *Ethics for the Built Environment*. Abingdon, Taylor and Francis, pp.183–184.

16 CEEQUAL (n.d.) *What is CEEQUAL?* Available online at: http://www.ceequal.com/what.htm (accessed 21 November 2011).

17 CEEQUAL (2009) *Awards 2008/9 Client Homes and Community Agency Design Halcrow*. Available online at: http://www.ceequal.com/awards_036.htm (accessed 20 December 2011).

18 Venables R. (2009) *Compendium of CEEQUAL New Civil Engineering 28.05.09*. London, Emap.

19 IFMA Foundation (2010) *Sustainability How-To Guide Series*. Green Building Rating Systems. Available online at: http://www.ifmafoundation.org/documents/public/BuildingRatingSystems.pdf (accessed 21 November 2011).

20 United Welsh Housing Association (n.d.) *United Welsh Passivhaus: Exemplar Case Study*. Wales, Constructing Excellence. Available online at: http://www.cewales.org.uk/cew/wp-content/uploads/Passivhaus-case-study-web.pdf (accessed 12 December 2011).

21 BRE Passivhaus (2011) *The Passivhaus Standard*. Available online at: http://www.passivhaus.org.uk/standard.jsp?id=17 (accessed 21 November 2011).

22 Gulacsy E. (2009) quoted in J. Parker *BREEAM or LEED Strengths and Weaknesses of the Environmental Assessment Methods*. BSRIA.

23 Gulacsy E. (2009) quoted in J. Parker *BREEAM or LEED Strengths and Weaknesses of the Environmental Assessment Methods*. BSRIA.

24 European Union Directive on Energy Performance of Buildings (2002/91/EC).

25 Adapted from information supplied by Bristol Local Education Partnership website. Available online at: http://www.bristollep.co.uk/index.html (accessed 16 November 2011).

26 More information about the Considerate Constructors Scheme is available online at: http://www.considerateconstructorsscheme.org.uk/htm-howtobe/index.html (accessed 19 December 2007).

27 OGC (2007) *Sustainability, Achieving Excellence in Construction Procurement Guide No 11*, pp.10–17. Available online at the National Archives: http://web archive.nationalarchives.gov.uk/20100503135839/   http://www.ogc.gov.uk/documents/CP0016AEGuide11.pdf (accessed 5 December 2011).

28 BRE 6001 (2009) *BRE Environment and Sustainability Standard*. Available online at: http://www.greenbooklive.com/filelibrary/BES_6001_Issue_2_Final.pdf (15 June) (accessed 24 February 2011).

29 BRE 6001 (2009) *BRE Environment and Sustainability Standard*. Available online at: http://www.greenbooklive.com/filelibrary/BES_6001_Issue_2_Final.pdf (15 June) (accessed 24 February 2011).

30 Worldwide Responsible Accredited Production. Available online at: http://www.wrapcompliance.org/en/about-us (accessed 24 August 2011).

31 National Audit Office (2007) *Building for the Future: Sustainable Construction and Refurbishment of the Government Estate*. London, The Stationery Office.

32 NAO (2007) *Building for the Future: Sustainable Construction and Refurbishment on the Government Estate* HC 324 Session 2006–2007. Available online at: http:// www.environmental-auditing.org/Building%20for%20the%20future.pdf (accessed 21 November 2011).

33 Sustainable urban drainage systems.

34 Target Zero (2007) Supermarket base case. Available online at: www.targetzero.info/guidance_reports/summary/supermarket/ (accessed 23 December 2011).

35 Weatherhogg R. (2008) *Waste Regulation: Review and Update*. Available online at: http://www.contaminated-land.org/brmf_members/Richard per cent20 Weatherhogg.pdf (accessed 12 December 2011).

36 HM Government Great Britain (2008) SI 314 Site Waste Management Plans Regulations.

37 DTI (2004) *Voluntary Code of Practice for Site Waste Management Plans*. London, DTI.

38 DirectGov (2010) Available online at: http://www.direct.gov.uk/en/HomeAnd Community/WhereYouLive/Streetcleaninglitterandillegaldumping/DG_10029700 (accessed 5 August 2010).

39 CCS Scheme (2010) *Considerate Constructor's Scheme: An Overview*. Available online at: http://www.ccscheme.org.uk/images/stories/ccs-ltd-section/downloads/overview.pdf (accessed 10 October 2011).

40 ISO 14001 (2004) *Environment Management Systems*. London, BSI.

41 IEMA (2008) *EMS Explained*. Available online at: http://www.iema.net/ems/emsexplained (accessed 6 August 2010).

42 BS 8555 (2003) *Environment Management Scheme*. Guide to the phased implementation of an environmental management system including the use of environmental performance evaluation.

43 IEMA – Institute of Environmental Management and Assessment.

44 Carillion (2010) *Sustainability Report. Environmental Management System*, p.81. Available online at: http://sustainability11.carillionplc.com/downloads/carillion-sr2010-environment.pdf (accessed 21 November 2011).

Chapter 12

Supply chain management in construction

Martyn Jones

The objectives of this chapter are:

- to outline the historical development of supply chain management (SCM) in a number of sectors of the economy including construction and why it has become increasingly important to businesses
- to analyse the key features of SCM and its impact on the way business is conducted between organisations
- to consider how it can be applied to construction projects and supply systems and the rationale for its adoption
- to outline the reasons why its adoption and implementation in construction has been comparatively slow and the main factors inhibiting its take up.

Introduction

SCM provides the means by which the integration of project processes and partners can be extended beyond the boundaries of construction projects to their supply chains. Partnering and SCM with their strong emphasis on integration, improving relationships, the adoption of a process-oriented approach and increasing customer focus, can be seen as an appropriate strategy for improvement in construction. Consequently, SCM has emerged as an innovation to address the problems impeding construction's performance, and to tackle the fundamental issues of the high levels of uncertainty and interdependence in project processes. This was led by more informed private-sector clients who adopted partnering in the early 1990s in their attempt both to increase the degree of integration and collaboration between their preferred consultants and contractors, and to extend this approach downstream to include key specialist and trade subcontractors and suppliers. Increasingly, main contractors are also playing a key role in implementing SCM through a greater integration of both upstream and downstream participants and processes.

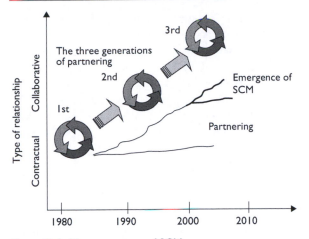

Figure 12.1 The emergence of SCM

Encouraged by improvements brought about through partnering, these pioneers of more collaborative approaches have continued both to develop closer relationships with their partners and to further integrate project and supply chain processes. A number of these clients and their consultant and main contractor partners have extended the adoption of longer term and more collaborative relationships downstream of the main contractor, to include key specialist and trade subcontractors and materials and component suppliers. Using their leverage in their supply chains, frequent users of construction services have been able to successfully make the transition from project-specific partnering, through strategic partnering, and onto SCM as shown in Figure 12.1.

With the encouragement provided from a number of reports into the inadequacies of construction procurement in the public sector, HM Treasury and later the Office for Government Commerce (OGC) advocated partnering and SCM as ways of delivering more value for money. Some public sector clients have responded by attempting to build the new purchaser–supplier relationships associated with SCM into their procurement of construction products and services. An example in the public sector is set by Defence Estates, an agency of the UK's Ministry of Defence. They introduced Prime Contracting, a procurement approach which includes many of the key elements of partnering, TQM and SCM. With its ambitious objectives and scope it aims to promote collaboration through leadership, facilitation, training and incentives, and replace short term, contractually driven, project-by-project, adversarial relationships with long term, multiple-project relationships, based on trust and co-operation. It includes the restructuring and integration of project processes and supply networks with fewer strategic supplier partners. These new relationships incorporate continuous improvement targets to reduce costs, enhance quality, and focus on the whole-life cost and functional performance of buildings.[1]

These and other pioneering initiatives have led to a growing interest in introducing supply chain management (SCM) more widely in construction in order to integrate processes, manage interfaces between organisations in projects and supply systems, reduce uncertainty and increase overall effectiveness in a greater part of the supply chain. However, the construction industry is seen as lagging behind other industries such as the electronics and the automobile industries in implementing SCM.[2] This slower uptake in SCM in construction has been attributed in part to the uniqueness of the industry and its products in particular, which are ultimately produced at the point of consumption. Unlike many other industries, the resources of production are mobile and the product (building or infrastructure) static. This means that the nature of each project shapes the structure of demand and that of construction supply systems. The diversity of end users, clients, buildings and sites, coupled with the need for mobile resources of production, means that construction projects have a wide range of supply systems supplying labour, skills, materials components and sub-assemblies. This myriad of flexible (some would argue fragmented) supply systems has developed to service the demand for a range of technologies and processes associated with the different phases of the project and the different parts, or elements, of the building or infrastructure under construction. These supply systems provide incomplete sets of resources and services which have to be brought together (mainly on a temporary basis) and managed in order to construct the elements of the building or infrastructure and service the work package providers. For each of these elements or work packages the most appropriate supply chains have to be selected, organised, aligned, integrated and managed in a way that best fits with the client's requirements and the overall nature, aims and objectives of the project, and all within a relatively short time span. As well as being shaped by the functional and technological requirements of each building element, these supply chains are further conditioned and shaped by the complex, largely transient linkages and relationships between individuals, business units, firms and the procurement strategies and forms of contract adopted by the main project participant's purchasers, particularly clients and main contractors.

As is the case in other project-based industries, construction practitioners face considerable challenges in managing their supply chains. Construction clients are faced with having to select the most appropriate suppliers from a highly price-competitive and often adversarial supply market. Main contractors operate as both suppliers and buyers having to obtain a regular workload from clients whilst managing large numbers of specialist contractors and materials suppliers who themselves are striving for their own survival and success. Under difficult market conditions this can lead to opportunistic and even adversarial behaviour by organisations within projects and their supply chains.

There is an ongoing debate within the UK construction industry as to how to address and overcome these inherent problems. There have been a number of government and industry-sponsored reports aimed at understanding the nature of construction and increasing its competitiveness. These reports into the state of the industry have identified its flexibility and adaptability but also its largely opportunistic behaviour and fragmented processes. Virtually every report has advocated changes based on more collaborative inter-organisational relationships, more integrated processes, improvements in supplier performance, and greater focus on customer satisfaction.[3] Despite the advice offered by these reports these approaches have failed to penetrate much of the industry where supply chains have remained largely contested, fragmented and highly adversarial with low margins for suppliers.[4]

Some observers have attributed this failure to poor understanding and implementation by construction practitioners. Others argue that it is not implementation that is at fault but the overall approach itself, which does not take into account the diversity of construction projects and the different demand and supply characteristics including operational and commercial differences of the participants. This would suggest that procurers of construction products and services need to be much more aware of what inter-organisational relationships and procurement strategies are most appropriate for specific projects and supply chains and the products and services involved.

As can be seen in Figure 12.2, it is the commissioning client, and their advisors, who identify the need for a new facility or the refurbishment of an existing facility to satisfy the needs of the end users. They, or their advisors, determine the procurement strategy for the project. However, most clients exercise surprisingly little control or management of the intermediate processes of pre-construction and construction. This means the construction supply chain starts and ends with the end user and the commissioning client but with varying degrees of involvement by the client during the project process.

As a result, each part or tier of the project process and supply chain is able to manage their relationships with their suppliers as they wish and this means often in opportunistic ways that suit their own business objectives. This lack of overall management of the supply chain means they can effectively act as 'gatekeepers' to their own networks of suppliers in the next tier. This restricted access to large parts of the supply chain creates considerable challenges for clients, and indeed main contractors, who wish to adopt more collaborative procurement approaches and exercise more influence over their supply chains.

The other significant barrier to the implementation of SCM in construction is its large proportion of infrequent and irregular clients, which

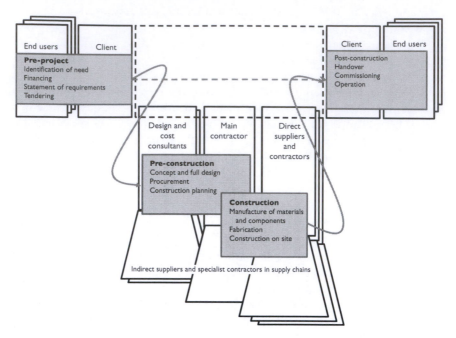

Figure 12.2 Typical representation of construction's process of development

accounts for much of the temporariness and uniqueness of its supply chains. In contrast to manufacturing, projects imply a temporary and transient coalition of a significant number of participants with frequent changes of membership. Although the organisations involved are interdependent they operate through a variety of contractual arrangements and procedures with considerable differences in commitment to the project objectives and improvements to increase efficiency and effectiveness. A relatively small number of major construction clients and main contractors may be classified as regularly having a significant demand for fairly standardised products, services and operating systems. It has been estimated that clients with regular and substantial workloads could account for less than 25 per cent of the total UK market.[5] Blismas *et al.* (2003) argue that multi-projects accounted for less than 30 per cent of contractor output in 1999.[6] The regular nature of their construction demand means that it is possible for frequent users of construction products and services to standardise their procurement process, the design and specification of the elements of their buildings, and work in longer-term collaboration with their key suppliers: in other words to adopt supply chain management (SCM). Other procurers have to consider alternative, shorter-term sourcing options.

Cox *et al.* maintain that the procurers of construction products and services need to understand the nature of demand within their supply chains

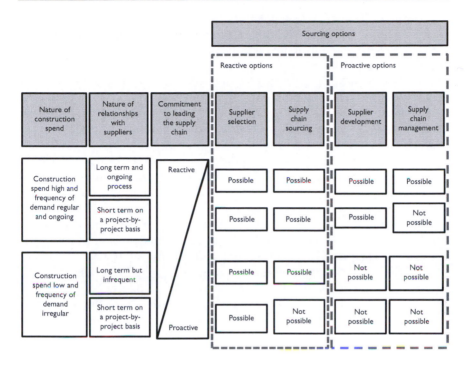

Figure 12.3 A positioning tool for procurers of construction products and services
Based on Cox *et al.* (2006)[7]

and the extent of their influence over their suppliers.[8] They have developed a tool (shown in Figure 12.3) which can be used by the procurers of construction products and services to assess the appropriateness of the various sourcing options based on two main factors: the volume and regularity of their construction spend and the extent to which they wish to influence behaviour in the supply chain.

As can be seen in Figure 12.3, the first column requires procurers to assess and position the nature of their construction spend. The second column requires an assessment and positioning of their relationships with their suppliers. The third column asks them to assess their commitment to co-ordinating and leading improvements in their supply chains. If they have a reactive approach to their supply chains then they will be drawn to either the supplier selection or the supply chain sourcing option. If they have a proactive approach to procurement with a strong commitment to improvement then they will be drawn to either the supplier development or SCM options. In the supplier selection approach the buyer reactively selects products and or services from existing supply offerings made by suppliers operating at the first tier of the supply chain. In the supply chain sourcing approach the buyer reactively selects products and/or services from the existing supply offerings made by suppliers operating at the first tier but having understood the structure of the supply chain through which products and services are created and delivered.

Buyers wishing to be more proactive in the management of their supply chains and encourage continuous improvement may adopt the supplier development approach. Here the buyer works on a continuous and proactive basis with first-tier supplier/s to transform the current trade-off between product or service functionality and overall cost of ownership. Those buyers wishing to be even more proactive in the management of their supply chains may adopt the SCM approach. Here the buyer works continuously and proactively with all supply chain players to transform the current trade-off between product or service functionality and overall cost of ownership.

Encouraged by substantial improvements in performance resulting from the successful implementation of first and second generation partnering (as discussed in Chapter 6) some construction clients moved to the adoption of SCM in the late 1990s. This trend was supported by the Egan Report, which endorsed the use of Lean SCM in construction based on its success in other sectors of the economy. This move to Lean SCM was essentially led by more experienced and informed private-sector clients with the necessary size and regularity of demand to manage their construction requirements in a similar manner to a manufacturing process. They had also successfully adopted first generation partnering during the late 1980s and early 1990s and then second generation partnering in the late 1990s. Unlike partnering, SCM is still often perceived as a concept borrowed from other sectors and which is in an early stage of development in construction. It is also important to recognise that SCM is a complex innovation, which has proved to be difficult to implement and sustain even in sectors such as the automotive industry in which it originated. Given its complexity it is often described as a fifth generation innovation.[9]

As it is a multi-factor process built around close and long-term intra- and inter-organisational relationships, supply chain partners need to take a strategic and long-term approach to SCM. As it is such a strategic innovation, it necessitates strong commitment from the top management of the participating organisations. This means that adopting SCM should not be seen merely as another procurement route. As it is so challenging, it requires continuous learning and is often dependent upon links with, and specialist support from, outside the organisation or project team, particularly in the early stages of its implementation.

The remainder of this chapter reviews the concept of SCM and its application in a number of sectors of the economy including construction projects. In order to fully comprehend how it is being applied it is also necessary to understand the main factors that have been driving and shaping its development in other sectors of the economy. The most important of these factors include the changing wider environment within which business has been conducted and new thinking in relation to management approaches – in the case of the latter, not only within organisations but also externally in relationships with other organisations in supply chains.

Changes in the business environment

Many writers, from a wide range of perspectives, have been arguing for some years that the wider environment is changing significantly and that business has moved to a new model or paradigm. From the late 1970s new technologies, methods of production and ways of doing business began to emerge replacing the Classical approach, which is characterised by 'the horizontal and hierarchical division of labour, the minimisation of human skills and discretion, and attempts to construe organisations as rational-scientific entities'.[10] The key ideas associated with the new emerging paradigm, sometimes referred to as the information and communications paradigm, include:

- A significant shift towards empowerment of people to encourage continuous improvement through learning and innovation.
- A change from dedicated mass-production systems towards more flexible systems that can accommodate a wider range of products, smaller batches and more frequent design changes – hence the 'economies of scale' have increasingly been replaced by 'economies of scope' and greater mass customisation.
- A customer-focused approach with the customer brought into the centre of the business in order to align the organisation with the evolving needs and expectations of customers.
- A move towards the greater integration of processes and systems within companies and between suppliers and customers, in order to align resources more closely to market and customer requirements.
- Developing the appropriate organisational structure and culture to support and sustain learning and innovation and inter-organisational co-operation and integration.

It is the emergence of these new possibilities and challenges associated with this technological-economic paradigm that has been driving organisations – including many in construction – to undertake a fundamental reassessment of the way in which they achieve their objectives by being more innovative, not only within their organisation but also in their increasingly integrated networks of customers and suppliers. Another significant feature of the new paradigm is the growth of outsourcing and the increasing reliance on suppliers and subcontractors to build and maintain competitive advantage. Outsourcing refers to the activity of purchasing goods or services from external sources, as opposed to internal sourcing (either by internal production or by purchasing from a subsidiary of the organisation). This growing reliance on suppliers means that a significant proportion of a company's products or services are obtained from other companies in their supply chains. The implications of this are that unless the quality, cost, design and delivery of inputs from their suppliers are appropriate the final goods or service are unlikely to meet the needs of their end customers.

The main motives for the increase in outsourcing include cost reduction (as external sources can enjoy greater economies of scale) and access to specialist expertise. This allows the outsourcing organisation to concentrate on its core competencies by discarding operations peripheral to their core business. The potential disadvantages of outsourcing include reduced control over the operations involved, fragmentation and concerns over delegating elements of delivery and quality to other organisations. SCM has emerged as an approach to obtain the benefits of outsourcing but also to address its difficulties. Introducing greater co-ordination and integration in the supplier networks allows an organisation to respond as effectively as possible to the needs of its customers and to changes in market and wider environments.

Outsourcing in construction

Outsourcing in construction grew with the increasing complexity of the design and specification of construction products and processes during the early part of the twentieth century. More finely subdivided design and engineering specialisms began to emerge to support the work of the architect. Also general contractors were no longer able to undertake all the work or provide the substantial capital investment needed in the emerging specialisms as many of them required, for example, expensive plant and equipment. In addition, subcontracting has given general and main contractors the flexibility to deal with fluctuations in overall demand for construction products and services and to respond flexibly to the diverse needs of clients and their wide range of construction needs. It has also been argued that the greater use of subcontracting by general contractors during the period between the early 1950s and the early 1970s was a means of circumventing difficulties such as the poor relations that had developed with their workers.[11] Subcontracting either pushed the labour relations problem on to another firm or removed the role of trades unions altogether, for example by employing workers on a labour-only self-employed basis. A further factor was the introduction of Selective Employment Tax in the 1960s (essentially a tax on jobs), which stimulated an acceleration in the use of labour-only workers. All of these developments, without parallel developments in what we would now call SCM, contributed to the fragmentation of the industry, the marginalisation of subcontractors and the disappointing performance of construction. The emergence of SCM in construction from the early 2000s provided parts of the industry with a means of providing more effective co-ordination and integration of its complex supply systems.

Origins of SCM

The term SCM began to be used in the early 1980s in some industries, including the automotive industry, to refer to the management of material flows across functional boundaries *within* organisations. This innovation, coupled with others such as just-in-time (JIT) and total quality management (TQM), often provided substantial improvements within organisations by breaking down barriers between departments and focusing on efficiency in managing core processes. With the increased reliance on external suppliers as a result of increased outsourcing, companies began to extend SCM beyond the boundaries of a single business unit to comprise all those organisations and business units that need to interact in order to deliver a product or service to the end customer.

The theoretical roots of SCM

Similar to partnering, SCM derives from two roots of practically oriented management theory: operations management and partnership philosophies. Within operations management a typical definition of a 'supply chain' as proposed by Christopher is 'a system whose constituent parts include material suppliers, production facilities, distribution and customers linked together via feed forward of materials and feedback flow of information'.[12]

Figure 12.4 illustrates the origin of SCM in the context of operations management. These early internal and increasingly external improvement programmes led to the broader concept of SCM. This means that the roots of SCM include, as illustrated in Figure 12.4:

- purchasing and supply management
- physical distribution management
- logistics management
- materials management.

Figure 12.4 identifies the key management functions involved in the management of the supply chain in the case of a manufacturer of consumer goods. In the case of construction, the organisation might be the main or principal contractor. The second tier customer would be the commissioning client, and the first tier supplier's specialist contractors. As can be seen, purchasing and supply management is concerned with the links with first-tier suppliers. More than merely buying, the aim is to obtain supplies

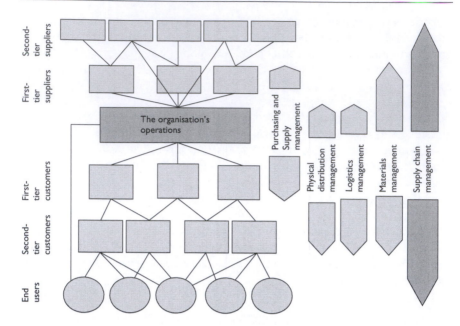

Figure 12.4 Supply chain for consumer goods manufacturer
Derived from Slack *et al.* (2001) and Hill (2000)[13]

according to the five 'rights': the right price, the right quantity, the right schedule, the right quality and from the right source. As discussed earlier, it is now widely accepted that these features contribute significantly to the value offered by organisations in their own products and services to their customers. Establishing deep relationships with a few suppliers or even a single supplier, so-called 'single sourcing', has resulted in improvements in product and service not available in more traditional approaches to purchasing.

Physical distribution management (PDM) refers to managing the flows to first-tier customers. It deals with issues such as the number and location of depots, the type of transport and the scheduling of flows. The skills and investment required in this activity mean it is frequently outsourced to specialists. Logistics is usually seen as an extension of PDM in considering the flow of products to consumers. While PDM considers the best way to deliver to the next tier, logistics seeks to optimise the whole chain, for example to take joint decisions on packaging so that it not only protects the product during transit but also carries useful information, such as a unit for display for retailers. Materials management considers the flows of products both within and outside the organisation including, as a result of globalisation, many sites spread throughout different regions and countries in order to offer high efficiency and flexibility. It involves issues such as purchasing, location of plant and warehouses, stock control, and design of transport systems.

SCM has emerged as a broad concept covering flows within and between organisations along significant parts of the supply chain. It focuses on integrating all these functions, their processes and interfaces. This emergence of SCM has been driven not only by internal pressures to reduce costs and add value, improve efficiency and satisfy customers, but by a number of external factors which characterise the current dominant paradigm such as increasing globalisation, shifts in the nature of competition, systems development, the increasing capabilities and adoption of information and communication technologies (ICT).

Working more closely with suppliers presents considerable challenges. It requires high levels of information sharing, co-operation, increased openness and transparency, which highlights the importance of the role played by ICT and the need for the development of closer relationships between customers and suppliers. The latter gave rise to the partnership philosophies root which is based on greater inter-organisational collaboration to achieve significant mutual benefits. These benefits include the sharing of resources, information, learning and other assets owned by organisations in the supply chain.[14] Effectiveness in adopting partnership approaches is dependent upon creating the appropriate organisational and cultural changes within the partnering organisations.

The partnership philosophies, which arguably have been more influential in shaping SCM in construction than the operations management root, emphasise the importance of the nature of the inter-organisational relationships between the organisations involved in the supply chain. The central concept is about building mutual competitive advantage or 'win–win' outcomes through developing more trusting and collaborative buyer–seller relationships. At one end of the spectrum is the 'arms-length', or 'hands-off' contract for goods or services where a price is agreed for a completely specified, or a standard off-the-shelf, product or service. Apart from agreeing the product or service and the price, the buyer and supplier need know nothing of each other's processes and operations. In contrast to this approach, at the other end of the spectrum, is a more involved and explicitly interdependent model based on a common purpose, which is mutually beneficial to customer and supplier, and which leads to a sharing of profit and risk. This needs to take place in an atmosphere of trust based on sharing of information and knowledge in order to understand issues and problems as they emerge, and devise appropriate solutions. This often requires going to levels of commitment well beyond those associated with traditional approaches to procurement and relationships based on a formal contract.

Definition of supply chain management in construction and other sectors

As SCM is a relatively new concept, especially in construction, it is still in the process of being clearly defined and explained. As well as Christopher's general definition given earlier,[15] Stadtlers[16] definition is closer to the perspective of project managers is 'the task of integrating of organisational units along the project process and its supply chains and coordinating the flow of materials, information and finance in order to fulfil customer demands with the aim of improving the competitiveness of the project and its supply chains as a whole'. According to Constructing Excellence, the term 'supply chain in construction is used to describe the linkage of companies that turns a series of basic materials, products or services into a finished product for the client'.[17] They go on to explain that all construction companies, be they client, main contractor, designer, surveyor, subcontractor or materials supplier, are therefore part of a supply chain. Because of the project-based nature of construction and the way that procurement normally operates, they are usually members of different supply chains on different projects. Each company in the chain has a client or customer – the organisation to which the services are directly provided – but an integrated supply chain will have the objective of understanding and working wholly in the interests of the 'project client'.

Most definitions link SCM with the integration of systems and processes within and between organisations. Accordingly, SCM can be seen as a set of practices aimed at managing and co-ordinating much of a supply chain and, in certain situations, the whole supply chain from raw materials suppliers to the end consumer, or, as in the case of construction projects, the commissioning client and end users.

SCM has become closely associated with delivering key organisational strategies such as health and safety, quality and environmental sustainability. It is also a means of adopting inter-organisational improvement programmes such as lean production with its emphasis on reducing waste and adding value across the entire supply chain.[18] The main objective is to develop greater collaboration, synergy and innovation throughout the whole network of suppliers through a better integration of both upstream and downstream processes.[19] As discussed earlier, this emphasis on co-ordination and integration is strongly dependent on the development of more effective and longer-term relationships between buyers and suppliers with increased trust and commitment.[20] It is about adopting a more holistic approach in order to optimise the overall activities of companies working together in a supply chain to build greater mutual competitive advantage and greater customer focus. However, achieving and sustaining this shift in emphasis presents considerable problems for project-based industries such as construction.

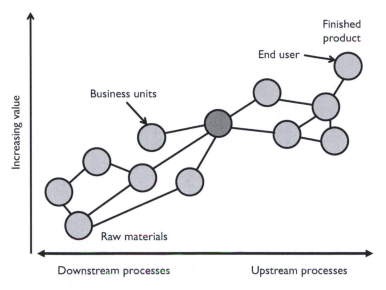

Figure 12.5 Supplier networks

One of the problems is the oversimplification of the context within which SCM is to be applied. In reality, most companies have several if not scores of customers and suppliers. Often, several companies compete for the same customers and have common suppliers. A more realistic picture is more complex, with a multitude or 'spider web' of relationships between customers and suppliers as shown in Figure 12.5.

A number of features of SCM have been identified across sectors of the economy including construction:

- Leadership and co-ordination of the whole (or substantial parts of) the supply chain with the power necessary to apply intense pressure for improvement.
- Long-term and stable relationships between customers and key suppliers.
- Shifts in the concept of what constitutes value and competitive advantage.
- Clearer definition of value and more explicit value chain.
- Mutual advantage through sharing of value created.
- Reorganisation and reshaping of supply networks (e.g. in tiers/clusters).
- Greater selectivity by reducing customer and supplier bases.
- Early (or ongoing) supplier involvement in product and service development.
- Breaking down barriers (internal and external) to improve the management of interfaces and link processes.

- Promotion of team working and collaboration.
- Joint identification, management and more equitable sharing of risk.
- Shift from penalties to rewards for performance improvements.
- Continuous improvement from synergy within the whole supply network.
- Appropriate performance metrics and tools for improvement.
- Joint design, value management, cost development, problem solving and innovation.
- Sharing of information to manage fluctuations in demand for more predictability and provide responsiveness to changes.
- Asset specificity (people, space, equipment) – blurring of the boundaries between organisations and facilitating the exchange of people.

The long list of features listed above is an indication of the complexity of supply chains and SCM, and suggests that implementing SCM requires a number of actions which have to be considered concurrently rather than sequentially. It implies a process involving some degree of experimentation, learning and feedback mechanisms to assess progress. It also needs to be contingent upon the nature of the organisations involved, their present relationships, the volume and continuity of work, the nature of their markets, and the wider environment. Hence the implementation of SCM should not be seen as a linear process and the following features should not be seen as a sequential list of actions for its successful implementation. The main actions in adopting and implementing SCM include agreeing a common purpose to ensure mutual competitive advantage; developing more collaborative inter-organisational relationships; selecting the right partners; reducing and shaping of customer and supplier bases; shifting from inter-firm competition to network competition; performance improvement requirements and management; and having the appropriate governance and the exercise of power.

Agreeing a common purpose to ensure mutual competitive advantage

As with partnering, effective SCM requires a significant level of joint strategy development where key members of a supply chain collectively agree a common purpose and jointly set strategic goals that are mutually beneficial. It is this concept of developing mutual competitive advantage or 'win–win' thinking which is at the heart of successful SCM, and presents a fundamental challenge to its widespread adoption in construction's supply systems. Most failures of SCM are mainly attributable to members of supply chains continuing to compete with each other, behaving opportunistically, not sharing their strategic thinking or agreeing a common goal and strategy. Establishing a common purpose based on mutual advantage requires a collaborative culture and a degree of preparedness and readiness to collaborate within each of the participating organisations. In the first instance a common purpose should focus on developing contractual trust. In other words delivering what is set out in the contracts between members of the supply chain. A common purpose should be based on:

* Reciprocity by which one is contractually obliged and subsequently morally obliged to give something in return for something received.
* Fair rates of exchange between costs and benefits.
* Distributive justice through which all parties receive benefits proportional to their contribution and commitment.

Ensuring mutual benefits is perhaps the biggest challenge in removing hidden agendas and shifting from the often adversarial nature of buyer–supplier relationships to long term and trusting relationships. In this approach, competitive advantage is seen as no longer residing only with a company's own innate capabilities, but also with the effectiveness of the relationships and linkages that the firm can forge with other organisations in the supply chain. The new dynamics of these supply chain relationships is built upon the fundamental idea of sharing both profit and risk. In this context it is no longer appropriate for suppliers and customers to view themselves as independent entities competing with each other but rather as working together to increase the competitiveness of the whole supply chain. This more integrated and collaborative approach, based on longer-term relationships, is illustrated in Case study 12.1.

Developing more collaborative inter-organisational relationships

Evidence shows that, in the appropriate circumstances, long-term agreements can improve mutual understanding, build closer relationships and reduce conflict and transaction costs. Indeed they are seen as necessary for effective SCM. Instead of pursuing short-term contracts characterised by frequent bidding and switching costs and the costs of pursuing claims and resolving conflicts, purchasers and suppliers can direct their energy and efforts toward value-adding activities. Long-term purchase agreements are seen as a prerequisite to achieving ongoing and closer co-operation between purchaser and supplier leading to, for example, early or even ongoing involvement of the supplier in understanding markets and the design of new products, services and processes.

However, building such relationships demands considerable resources over an extended period of time. This can only be justified if there is a sufficient flow of work to warrant this substantial investment. Essentially the buyer and suppliers are undertaking an investment appraisal because

Case study 12.1 ProCure21 frameworks

Following the Egan Report in 1998 the public sector recognised that significant benefits could be realised on capital schemes through collaborative working between clients and their supply chains. One of the key elements of this was to adapt and apply the same principles, practices and processes that made other industries more efficient and successful. The Department of Health in the UK responded to the Egan Report by commissioning a framework to improve the procurement process for publicly funded schemes and create an environment where more value could be realised from collaboration between NHS clients and construction supply chains.

The ProCure21 National Framework was launched in 2003, and following a two-year extension in 2008 came to a conclusion in September 2010. It was one of the first construction frameworks of its kind in the public sector and was used as a basis for similar frameworks in Scotland and Wales.

In terms of benefits to the client, ProCure21 enabled NHS clients to achieve improved levels of performance and value for money than those historically achieved using what is referred to as 'traditional tendering'. Schemes were delivered faster, on time and within budget without affecting functionality or quality. Patients were able to access care faster and NHS clients were able to generate additional revenue. NHS clients were offered cost-certainty and a much-reduced risk of litigation as there has been no litigation on any ProCure21 scheme.

decisions to work more collaboratively carry a cost that is expected to deliver a yield, with risk and power as intervening variables. This is difficult in construction because of the prevalence of irregular and infrequent clients.

Given that SCM is dependent upon 'win–win' thinking permeating the culture of the supply chain, where it is a viable option it can help overcome the often deeply ingrained adversarial attitudes in traditional business relationships. This makes it an appropriate approach to dealing with construction's traditionally adversarial inter-organisational relationships. Indeed, the need for greater integration of project processes and co-operation between project participants has been demonstrated in the succession of reports into the state of the construction industries throughout the world. As discussed in previous chapters, over the past 50 years or so there has been a series of reports into the UK's construction industry and there have been many calls for action to improve its performance and competitiveness. An

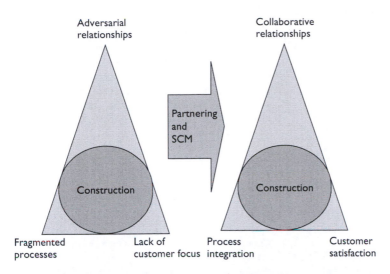

Figure 12.6 The role of SCM in addressing the key problems in construction

analysis of these reports indicates that the problems facing construction can be categorised into three broad areas: poor inter-organisational relationships, insufficient integration of processes, and a lack of customer focus, as shown in Figure 12.6.

The adoption of SCM in parts of the construction industry can be seen as a growing recognition of the importance of specialist contractors and component and materials suppliers in the success of projects. A number of problems in construction's supply chains downstream of the main or general contractor are outlined in Table 12.1.[21]

SCM involves the two-way exchange of information and the building of closer, long term relationships so that a supplier may have a guarantee of the business for several years, rather than having to re-bid for each order or project. In return the supplier is required to hold or indeed reduce prices, improve quality and delivery, introduce more efficient processes to drive out waste, and add more value as defined by the customer, or in the case of construction, the client. The greater trust developed between customer and supplier can lead to the acceptance of open-book accounting, which can also lead to more transparency in relation to true costs and foster more open and honest communication between the members of the supply chain. Longer-term and more collaborative relationships may also encourage the buying company to commit itself to helping the supplier to improve its competencies through supplier development activities, and the sharing of information and key resources.

Table 12.1 The main problems associated with construction

Relationships	Processes	Customer focus
1. Lack of trust leading to opportunistic behaviour and conflict 2. Onerous contract conditions and unfair loading of risk 3. Lack of understanding of the risks involved and their consequences 4. Unfair supplier selection procedures 5. Unfair supplier payment procedures 6. Perceived poor status of specialist contractors 7. Failure to communicate with specialist contractors and to view them as equal project partners	1. Fragmented nature of the design process 2. Inadequate time allocated for the design period 3. Late, poor and incomplete design information lacking specialist contractor input 4. Poor overall planning with inadequate lead in time 5. Fluctuations in demand for the products and services of the specialists 6. Failure to involve specialist contractors early enough in the process	1. Fragmented nature of construction's clients, the construction process and poorly integrated value chain 2. Insufficient focus on internal and external customer requirements 3. Insufficient understanding of specialist contractors' requirements 4. Unclear statements of requirements and ambiguous project information and tender packages

Selecting the right partners

The move to close, long-term and more collaborative customer–supplier relationships increases the importance of choosing the most appropriate partners. It requires both customers and suppliers to have an overall understanding of the nature of their businesses and their links with supply chains. In this complex setting, companies need to realise that some relationships between customers and suppliers will be more important than others in terms of their own competitive advantage. Companies need to consider the wider set of competencies they need when selecting their closest partners. Vollman et al. have identified three different types of competencies which should influence the selection of suppliers and their subsequent development:[22]

- *Distinctive* competencies that provide the purchasing organisation with a unique competitive advantage.
- *Essential* competencies that are vital to the effectiveness of the purchasing organisation at the operational level.
- *Plain* competencies that have no direct effect on the product or service delivered.

Clearly this implies that considerable care needs to be taken when selecting partners especially those who provide distinctive competencies and essential competencies. Greater care in selecting and grading of suppliers in terms of their significance in the success of projects and their competencies has been adopted by many main contractors in construction with a corresponding reduction in the number of suppliers offering distinctive and essential competencies. The approach adopted by a UK-based main contractor in managing their supply chains is outlined in Case study 12.2.

Reducing and shaping of customer and supplier bases

Developing closer customer–supplier relationships requires considerable time and resources. This explains why customers seek to reduce significantly the number of suppliers in their supply chains. In the case of larger organisations this has often meant reducing hundreds or indeed thousands of suppliers down to less than a hundred. In parallel with this, many suppliers have sought to differentiate themselves from their competitors by providing more customised, customer-focused products and services. Again, this normally involves a rationalisation of their traditional customer base. It is common for customers to organise their smaller number of suppliers into a supply chain. A common configuration is a hierarchy of tiers, as shown in Figure 12.7, or some other arrangement, which reflects their importance in the supply chain, their position, and their competencies. This rationalisation of supplier and customer bases allows organisations to focus their efforts on performance improvement with a smaller number of their most significant customers and suppliers.

This hierarchy, which can take the form of a pyramid, demonstrates the tiering commonly found in the early years of the adoption of SCM in the automotive industry. In this example, there are multiple layers or tiers roughly delineated by the size of the firms and their roles in the supply chain. At the apex of the pyramid sits the final assembler who is supplied with sub-assemblies by first tier suppliers. In turn, these first tier companies are supplied by a larger number of second tier suppliers. These second tier suppliers have their own subcontractors who provide them with specialist process abilities. In some instances there may even be fourth and fifth tier suppliers.[23]

Managing the interfaces between the organisations involved

Some construction clients and their advisors are adopting the tiering structure outlined above. In this approach it is the responsibility of the customer tier to organise, communicate and nurture the level below. Thus, the major assembler takes responsibility for the welfare of the first tier suppliers, the first tier that of the second tier firms, and so on down the

Case study 12.2 SCM tiers in a construction firm

A major UK main contractor began re-examining their role and that of their supply chains in construction projects following the publication of the Latham Report (1994) and the Egan Report (1998). This meant challenging the status quo of disjointed supply chains; widespread adversarial practices; inefficient production and procurement processes; the 'suspicion' in which clients and professionals held many main contractors; the lack of openness and honesty about costs and margins; and the need to replace lowest cost tendering with open book principles.

They were successful in winning work under the pioneering NHS Estates' ProCure21 and its successor, the ProCure21+ National Framework. They were one of the six supply chains (principal supply chain partners) selected via OJEU Tender process for capital investment construction schemes across England. Within this framework, an NHS client or joint venture may select a supply chain for a project they wish to undertake without having to go through OJEU procurement themselves. Engagement in the framework provided a catalyst for the main or principal contractor to adopt supply chain management. Existing suppliers and supply chains were reshaped on the basis of the profile of spend with suppliers, key markets and segments, and key selection criteria.

Suppliers were classified as either key, principal, secondary or other key trade partners and were selected on the basis of the level of risk (including design risk) associated with their work package, the influence they have on project delivery and the level and frequency of expenditure. Their principal partner trades are mechanical and electrical installations, structural steelwork, cladding (wall and roof), windows and curtain walling. Their secondary partner trades are groundworks, scaffolding and demolition (on the basis of health and safety considerations), brickwork and roofing (on the basis of programme considerations) and dry lining/plaster, carpentry and painting (on the basis of quality considerations). Other supply chain trades include four single source trades and 26 managed trades which are managed to ensure consistent quality. Suppliers are further classified on the basis of performance into:

- tier 1 (partner/preferred)
- tier 2 (approved)
- tier 3 (qualified
- tier 4 (non-approved).

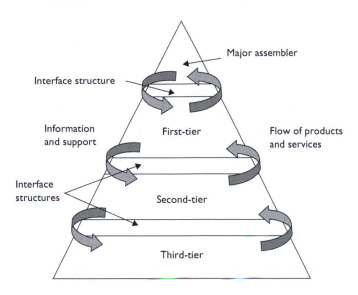

Figure 12.7 A pyramidal supply system

pyramid as shown in Figure 12.7. Given the leading role played by clients in construction, they are at the apex of the pyramid with typically cost and design consultants and the main contractor as their first tier suppliers. The second tier comprises the specialist and trade subcontractors, with their material and component suppliers forming the third tier.

A variation on this structure is where the main or prime contractor occupies the apex of the pyramid. This is rather like the prime contracting approach and cluster structure adopted by Defence Estates. The cluster structure, which is illustrated in Figure 12.8, is an idea developed by

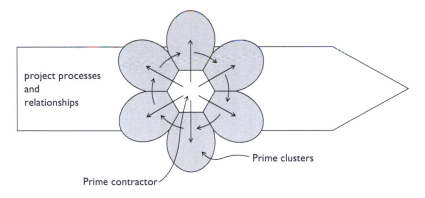

Figure 12.8 Structuring of supply systems using clusters

the Reading Construction Forum and piloted by Defence Estates in the 'Building Down Barriers' project.[24] In this approach a prime contractor is appointed to work with key supply partners, known as cluster leaders, who set the general direction of designing and delivering a significant element of the building – the ground works and substructure, the superstructure, the services, and so on.

This allows the prime contractor to improve in an integrated fashion the process for designing and delivering the overall building as well as the materials and the components that go into its main elements. It means the client has a single contractual relationship with the construction team. The contract is held by the prime contractor, who is usually a main contractor, although the role can be undertaken by other members of the construction team. The team is stable and consists of designers, the prime contractor and suppliers who have a long-term contract with the client for a series of projects. The long-term relationships between the partners facilitate the development of understanding and trust. This leads to improvements in the teams performance from project to project over a period of time and better communication across the interface structures developed as shown in Figure 12.9.

Having assembled and shaped the supply chain in the most appropriate manner, the interfaces between customers and suppliers need to be managed, based on closer inter-organisational relationships. The nature of the interfaces between the business units in the supply chain is a critical element in the success of SCM. Although the management of the interfaces between the organisations is important, the effective management of the relationship requires that each of the organisations has the appropriate *internal* culture,

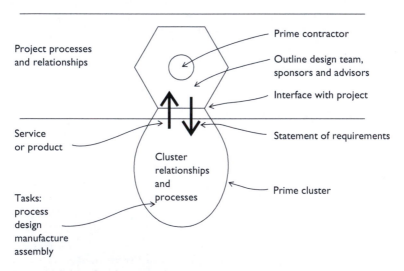

Figure 12.9 Interface between the project and a cluster

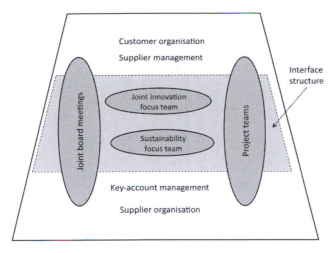

Figure 12.10 An advanced interface structure

Derived from Christopher and Jüttner (2000)[25]

structure and organisation, as summarised in Figure 12.9. This facilitates and supports the sharing of information and learning across the interfaces which are necessary if greater co-ordination and integration is to be achieved.

An advanced form of interface structure between a customer and a supplier is where organisational boundaries become blurred. Such an interface structure allows for both joint strategy development and day-to-day problem solving, and the co-ordination of operational activities. Joint teams can also be formed to co-operate on specific issues affecting the performance of the relationship such as innovation, R&D, anticipating shifts in the market environment, environmental sustainability and enhancing end user satisfaction. This is shown in Figure 12.10.

Shifting from inter-firm competition to network competition

In SCM, participating organisations seek to make the supply chain as a whole more competitive through the value it adds and the costs it reduces overall. Real competition is not then between organisations but rather between supply chains. Reducing costs and increasing value while maintaining or improving profitability are important parts of gaining competitive advantage. The traditional Fordist and Taylorist approach sought to achieve cost reductions or profit improvement at the expense of other participants in the supply chain. In SCM, performance improvement should not be bought at the expense of reducing the profits of suppliers, but through efficiency gains from the reduction of waste. Also, the SCM philosophy recognises that simply transferring costs upstream or downstream does not make the supply

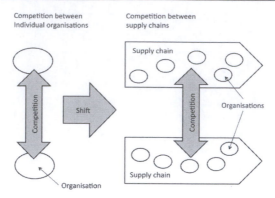

Figure 12.11 The shift to supply chain competition

chain any more competitive as ultimately all costs will make their way to the end customer or user. The shift from classic inter-firm competition to supply chain competition is illustrated in Figure 12.11.

Performance improvement requirements and management

As we have already seen, implementing and sustaining effective SCM requires significant investment in time and resources. The return on this investment will include reductions in waste and the adding of value. This means that continuous improvement and performance measurements are usually key drivers of an SCM strategy. Robust measures of performance and targets for improvement ensure that all parties are comfortable with the performance of the supply chain and the return on their investment of time and resources in creating and sustaining the inter-organisational relationships and processes. They are also necessary to combat the danger of complacency developing in long-term relationships. Some of the UK's leading supermarket chains have been adept at managing their construction supply chains to bring about significant improvements in performance that exceeded some of the Latham and Egan targets. For example, one supermarket chain reduced the cost of their stores by 30 per cent and construction time by 50 per cent. Specifically, organisations need measurement systems to identify:

- The overall effectiveness of the approach to SCM in improving performance and adding value.
- Individual and collective supplier performance and improvement opportunities.
- Trends in relation to key performance indicators such as cost, time and quality.
- Areas for supplier development and the time and the type and scale of resources needed.

As can be seen from the above, supplier development is another key factor in successful SCM and its ability to unlock additional value and competitive advantage residing within suppliers in the supply chain. Supplier (or for that matter customer) development is where a partner in a relationship modifies or influences the behaviour of the other partner with a view to increasing mutual benefit. Cross-functional teams from the organisations work closely together, sharing learning, solving problems and actively seeking improvements in their processes within the partner organisations and at their interfaces.

Having the appropriate governance, ethics and the exercise of power

Governance is the act of governing. It relates to decisions that define expectations, grant power, and verify performance. It consists of either a separate process or part of management or leadership processes. Governance in corporate and supply chain environments consists of the set of processes, customs, policies, laws and institutions affecting the way people direct, administer or control an individual organisation. In the case of supply chains it includes the relationships among the players involved and the goals of the supply chain.

It is important to understand the role of ethics and the exercise of power in the context of developing an SCM strategy. Being an ethical company isn't enough anymore. These days, leading organisations and brands are judged by the company they keep in their business and supply chains. Consumers, clients, investors, business partners, regulators and media organisations now expect a company and its entire supply chain to be ethical.

As we have seen, a customer organisation buys unfinished inputs provided by their suppliers before refining them and sending them through the supply chain to their ultimate end-users. Simply put, the main ethics problem in the supply chain is that clients and end users can blame their first tier suppliers for ethical lapses that were actually committed by other suppliers in the supply chain. Blame then attaches to the purchasing company even though its suppliers are legally (and factually) distinct and independent corporate entities. Despite the pervasiveness and seriousness of the problem, organisations can take a positive step toward protecting themselves against damaging headlines by implementing a credible supplier ethics management (SEM) initiative approach in their supply chains. SEM is the management of suppliers and supply relationships with strategies, programmes and metrics that better align supplier business conduct with purchaser standards, with the goal of reducing the purchaser's overall risk of damage to corporate integrity in the supply chain. SEM typically performs several basic (but important) risk management functions, including the following:

- Agreeing and regularly updating a code of conduct setting out the ethical standards expected in the supply chain.
- Documenting supplier and customer engagement in setting and maintaining these requirements.
- Facilitating two-way communication between customers and suppliers on critical matters relating ethics and compliance.

SEM should not be seen as a new risk management discipline. It simply involves the application of existing risk and supply chain management approaches and tools to identify ethical issues that might arise in the supply chain and agreeing the ways in which they are to be managed. The next three case studies illustrate SEM codes. Case study 12.3 is an example of the code of conduct aimed at shaping the relationships between organisations in the defence industries with the aim of increasing competitiveness.

Case study 12.4 illustrates how a main contractor in the UK has developed a code of practice for the management of their supply chains. It acknowledges the importance of being both a good customer and supplier. It also illustrates the mutual concern between customers and suppliers evident in SCM.

Case study 12.3 World-class supply chains for the 21st century (SC21)

AWM, the West Midlands Regional Development Agency, approved a three-year, £1.5 million funding package for the region's aerospace and defence sectors to engage on the SC21 programme. The programme, delivered by MAS-WM and endorsed by all UK prime contracts and the MOD, is a supply chain improvement programme intended to make the UK aerospace and defence industries more competitive internationally. The initiative offers a range of cost-saving, quality improvement and relationship-building benefits to aerospace and defence industry suppliers. The programme focuses on three work streams:

- Accreditation and quality improvement. All companies in the scheme will recognise the AS/EN9100 quality system standard and the Nadcap special process accreditation programme and audit.
- Development and performance. SC21 signatories are committed to working together to improve performance through a common diagnostic framework. Members are also expected to encourage innovation and investment throughout the supply chain.
- Relationship-building. Signatories are expected to subscribe to a supply chain relationships code of practice that promotes openness, honesty, integrity and trust.

Case study 12.4 Contractor code of practice

A major UK contractor has developed a code of practice for their supply chains, which sets out their purchasing philosophy. It recognises their obligations in being a 'great' customer as well as those of their suppliers. It acknowledges the importance of their suppliers in delivering their projects, given that they spend two-thirds of their revenues in procuring goods and services from suppliers. They recognise that in many ways, they are defined by the performance and behaviours of their supply chain. As a large contractor they manage some of the largest and most complex projects in the world and they see their supply chain as an invaluable resource to their customers. They argue that successful relationship management with their supply chain is fundamental to the delivery of all their projects. The following extracts set out their views on being a great customer and supplier.

Being a great customer

It is our objective to become the customer of choice to the best suppliers. We are achieving this by increasingly concentrating our business with the best performing suppliers and supporting continuous improvement in areas important to us and our stakeholders.

Our philosophy for working with suppliers is:

- Safety and sustainability are of paramount importance
- By sharing information in a constructive and open manner, we aim to motivate suppliers to improve performance
- We will recognise and reward high performance
- We think innovation, learning and teamwork should be positively encouraged
- There should be continuous measurable improvement in the service provided
- The partnership should seek to create an environment of trust, integrity and open communication
- Both parties should always strive to exceed customer expectation
- Being a great supplier

A great supply chain can be a source of competitive advantage. We want to raise the bar and help our suppliers to consistently deliver excellence.

Being a great supplier means:

- Providing a safe and competent workforce

- Incorporating safety and sustainability into solutions and methods of working
- Delivering to agreed specification, time and cost
- Reducing cost and eliminating waste through continuous improvement
- Always display the right behaviours
- Collaborate to deliver innovative solutions

Case study 12.5 illustrates how a group of construction suppliers have developed a code of conduct with regard to their behaviours and procedures. They refer to this as the 'ten commandments'.

Case study 12.5 Supply chain alliance ten commandments

The organisation Baufairbund (BFB) is an association of small- and medium-sized companies in the North Rhine-Westphalia region of Germany. Members co-operate in the provision and co-ordination of building and related services. The 15 members have agreed 'ten commandments' to shape their collective behaviours and procedures.

1 We trust and respect each other.
2 We oblige ourselves to fairness and honesty in our cooperation-community.
3 We attach great importance to quality in consultation as well as in the application of building material.
4 We always inform and advise our customers [in an] integrated, competent and inter-sectoral [manner].
5 We inform transparently about costs and potential savings.
6 We want energy-optimised and eco-friendly construction and redevelopment. We oblige ourselves to use renewable resources.
7 We inform our customers in a professional and detailed way about sustainable and energy-saving products.
8 We oblige ourselves to constant further education in the field of renewable energy sources, energy efficiency and ecological building material.
9 We commit to discuss all kinds of conflicts in an open-minded way to find fair and suitable solutions.
10 We do our best to achieve our goals and to create a positive, fair and pleasant atmosphere.

Power in supply chains

Power needs to be exercised appropriately in supply chains as it needs to be inside individual organisations. Power in the context of supply chains can be seen as the ability of one individual or organisation to influence the behaviour of another individual or organisation, in order to achieve their desired situation and outcome. To have power is relative not absolute, since it is contingent upon the context and relationships within the supply chain.

The more types of power that an individual or a group of organisations possess, the greater their influence in the supply chain. For example, a supplier which is indispensable from the perspective of the customer will be more powerful and will have potentially more influence than a supplier undifferentiated from its competitors. Key customers who buy large volumes of products or services on a regular and continuous basis are also more likely to have greater power over their supply chains than those who are small or only occasional customers. We see considerable power being exercised by big-spending clients of the UK construction industry such as the main supermarket chains. This is a significant factor to be considered when determining the appropriateness of SCM for construction clients in general and the irregular and infrequent clients in particular.

It is important to understand how the exercise of power in supply chains can determine whether the relationships between suppliers and customers are either loosely or tightly coupled. For example, if supply relationships are tightly coupled they may impinge upon the operation of another firm in the network and paralyse its strategic actions and innovativeness. On the other hand, if relationships are too loosely coupled it can affect the management of the interfaces and reduce the overall effectiveness of the supply chain through unco-ordinated actions and activities. Obtaining the appropriate balance of power between organisations is a major factor in the success of a supply chain.

So what are the features of successful SCM? This brief review of SCM suggests that its successful implementation is associated with the following key actions:

- Choosing the best and most appropriate suppliers (and customers) and integrating them into a rationalised supply base.
- Developing long-term, close, stable, 'win–win' and trusting relationships.
- Breaking down the barriers between internal departments, business processes and between companies to improve the management of interfaces.
- Co-ordinating the supply chain to manage fluctuations in demand and provide more predictability and flow of work.
- Sharing information, learning and resources to develop the capacities and competencies of the whole supply chain.

The potential benefits of SCM

A review of literature in a number of sectors of the economy, including construction, provides evidence of the significant benefits to be gained from the synergies developed from the successful implementation of SCM. These include:

- Waste and risk reduction.
- More scope for the application of value management to add value.
- Cost reduction through flow of work (economies of scale) and reduction of transaction costs based on true costs.
- Increased flexibility and responsiveness to change in markets and the wider environment.
- Shorter cycle times including design to use.
- Better understanding of resource requirements and more effective utilisation.
- Quality improvement through a better understanding of requirements and the capabilities in the supply chain.
- Better and longer-term planning ensuring a more even flow and inventory reduction.
- Clearer understanding of the roles of individuals and firms in the supply chain and the contribution they make.
- Greater 'contractual', 'competence' and finally 'goodwill' trust.
- Greater focus on the needs of internal and external customers.
- The development of 'greener' products and processes throughout the whole supply chain.

Examining some of these benefits in more detail, Sako et al. suggest that, for instance, closer collaboration in European vehicles manufacturing has helped speed up the rate of performance of improvements in order to meet global competition.[26] The success of the Japanese car manufacturers in introducing new models faster and with fewer labour hours has been attributed to the early and more proactive involvement of their suppliers in product design.[27] Shorter lead times, reduced total cycle time and smoother and more responsive flow of materials and products are also achieved through an effective management of interfaces of the entire supply chain.[28] Towill argues that time compression through SCM has had a major impact on the accuracy of demand forecasting, the time taken to detect defects, the time to bring new products to markets, and the amount of work in progress.[29] He also identifies a number of other benefits including: elimination (removing a process); compression (removing time within a process); integration (changing interfaces between processes); and concurrency (operating processes in parallel).

Collaboration and shared learning

This greater collaboration between organisations within the supply chain can also result in increased focus on internal and external customers.[30] The collaborative SCM approach, based on frequent and direct communications between suppliers and customers, can help in defining and agreeing the customer requirements and the means by which they can be satisfied. These new types of relationships are increasingly perceived as a means to better utilise resources throughout the whole supply chain.[31] There are further examples of where SCM is delivering significant performance improvements and increased competitive advantage.[32] It can be an important element in innovation in products, processes and organisation,[33] as information can be more readily shared and knowledge identified, captured and disseminated throughout the organisations in the supply chain.[34] This sharing of information and knowledge has led to joint learning resulting in significant improvements in a number of aspects of the supply chain. This shared learning can also lead to better problem identification and joint solving, and allows the more ready application of techniques such as value analysis, engineering and management . It can also improve understanding of changes in markets and the external environment as a result of more effective communication and greater synergy. Greater compatibility and integration of processes and systems achieved through collaboration and partnership within the supply chain are more likely to lead to increased flexibility and more responsiveness to changes in the external environment. All these potential benefits are linked to the development of trust through greater mutual understanding, increased mutual competitive advantage, greater transparency in transactions and more commitment.[35]

Increased trust

As we have already discussed, the development of trust between organisations is critical to the success of SCM and indeed a major and highly valuable output. However, the concept of trust is very complex, not clearly defined and interpreted differently by the literature. Most literature relates trust to predictability, risk, vulnerability, co-operation, good intentions, personal traits and confidence that a partner will demonstrate mutually beneficial behaviour. This vision challenges the prevailing view of economic theory that organisations and individuals are self-interested and will always pursue their interests opportunistically and indeed often with guile. On the other hand, proponents of more trusting business relationships believe that the right partner will be reliable and will fulfil the perceived obligations of the relationship. This greater trust leads to a reduction of the risk that they will perform an action detrimental to the relationship. Trust is also defined in terms of good intentions and confidence. There is a clear connection between predictability of behaviour and trust which are seen as means to manage and reduce uncertainty and risk. However, it must be recognised that risk can be increased in the early stages of building closer relationships as partners increase their vulnerability as a precondition to the formation of trusting behaviour. Trust is therefore related to a degree of expectations which, if not fulfilled, may lead to disappointment. This implies the need to recognise and accept that risk exists in developing more trusting relationships.

When trust is low, other control mechanisms need to be employed to underpin the relationship. Typically, legalistic remedies (for example, accrediting organisations, insurance, bonds and guarantees) are used either to compensate for the lack of trust in exchange, or to create the conditions under which trust might be restored. However, these kinds of mechanisms can be counterproductive and lead to even higher levels of mistrust.

In addition, these remedies to high levels of mistrust are costly and can impede performance. When trust is present within and between organisations the cost of transactions can be lower as fewer controls are needed to measure, monitor and control performance. Forecasts of future events are more realistic, as are budget projections thus increasing confidence in the competence of the partners and building trust. Greater trust and confidence improves co-operation along the value chain, productivity improves, profitability is enhanced and the sharing of learning and innovation is encouraged.

It is also important to recognise that trust must go beyond predictability if it is to produce effective and lasting relationships and partnerships based not only on contractual and competence trust but also in the longer term on good will trust.[36] *Contractual* trust emerges when each partner will follow written or oral contracts. *Competence* trust refers to the ability of the partner to complete tasks to a given standard whilst *goodwill* trust corresponds to the situation where a partner driven by mutual benefits does more than expected and goes beyond predictability and the agreements of a contract. As illustrated in Figure 12.12, trust is cumulative and takes time to build. *Contractual* trust needs to be established prior to *competence* trust which in turn will lead to the development of *goodwill* trust. The latter can continue as the partnership and the relationships develop.

What distinguishes 'goodwill' trust' from 'contractual' trust' is the expectation in the former case that partners are committed to take initiatives (or exercise discretion) to exploit new opportunities over and above what was explicitly promised. The key difference here is that partners are not only looking after their own interests but are also seeking to offer their partner a competitive advantage. If both partners do the same then the combined efforts of both customer and supplier will lead to a mutual competitive advantage, which will help in selling their collective product or service to the end user. However, it is unrealistic to describe the development of trust as a linear process, as the three types of trust are interlinked. In general

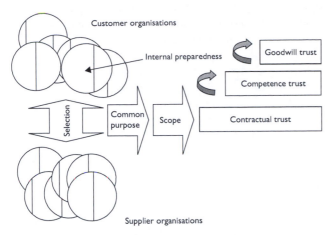

Customer organisations

Internal preparedness

Goodwill trust

Competence trust

Selection

Common purpose

Scope

Competence trust

Contractual trust

Supplier organisations

Figure 12.12 Development of trust
Source: Jones *et al.* (2002)[37]

contractual trust is often the starting point for the development of trust. *Competence* trust cannot develop without contractual trust and goodwill trust cannot exist if the former two are not present. Developing and maintaining trust is a long and ongoing process and takes time and resources.

Although the literature identifies trust at the individual, organisational and collective level, there is consensus that the common denominator or rather the starting point for the development of trust is indeed individual or *interpersonal* trust. Interpersonal traits are considered as forming the foundation of trust.[38] Interpersonal trust is associated with the level of expectation that the word, promise, verbal or written statement of another individual or group can be relied upon.[39] However, the propensity of groups and individuals to trust others is often determined by different developmental experiences, personality types, and cultural backgrounds.[40]

Ongoing interaction allows individuals to know and understand each other and develop higher levels of trust.[41] This is why Kanter and Myers claim that personal relationships and personal connections are crucial in forming effective relationships and greater trust between groups or organisations.[42] Jones and Said argue that these personal and often informal relationships develop trust by exerting pressure for conformity to expectations.[43] These interpersonal and informal ties can increase over time and lead to greater trust in managing partnerships. In such situations, where trust emerges as a consequence of effective and informal relationships rather than a pre requisite to the relationship, then informal contracts are used rather than adversarial, detailed and formal contracts. It is clear that trust between organisations is progressively and incrementally built as organisations and individuals repeatedly interact with each other and the outcome of these interactions is seen as mutually beneficial and satisfying.[44] Repeated personal interactions across firms can encourage higher levels of courtesy and consideration whilst the prospect of ostracism among peers can discourage opportunism. Mutual expectation of repeated trading over the long run is a major incentive for co-operation and effective partnerships.[45]

There is however a risk associated with predicting behaviour and sustaining trust merely through informal relationships. A wider level of trust needs to be achieved within and between the organisations in order to ensure an acceptable degree of predictability and sustainability in the relationship. Emphasis needs, therefore, to be placed on organisational traits, learning and culture which can be conducive to an organisation's identity, image and reputation upon which trust can be based. This can be embedded within the relationship through a clear common purpose, mutual objectives, openness, transparency and the sharing of information and knowledge. Hence internal preparedness based on a culture of trust can be seen as a vital pre requisite for effective and sustainable inter-organisational relationships.

The degree of internal preparedness, the selection of customers and suppliers and the degree of common purpose should help determine the type and nature of trust to be achieved in the relationship. In the case of contractual trust this can be based on a contract which sets out the expected behaviour of each party. Such contractual trust should lead to the elimination of disputes, conflict and adversarial relationships. Improved relationships coupled with more appropriate internal preparedness should lead to more sharing of information, greater joint problem solving, joint learning and innovation, and competence trust. Feedback on the behaviour of the individuals and organisations involved will influence the scope of the common purpose. Negative feedback may result in reverting back to contractual trust or even opportunistic behaviour, while positive outcomes would take the organisations towards the next step on the journey: goodwill trust. Achieving goodwill trust is a long process which evolves slowly and is based on a continuous improvement approach.

Greener construction

The concept of well managed environmental quality fits well with the trend towards leaner production in the supply chain. According to Womack *et al.* a truly lean enterprise manages to produce goods with radically reduced inputs of materials, labour, space, movement and time.[46] The same can be said for a lean supply chain.[47] Given that effective SCM reduces waste and increases efficiency along the whole supply chain, this means that it is seen as part of the strategy for more sustainable production. Increasingly, supply chains are seen as a way for customers to reduce the environmental impact of their products and operations. Through a collaborative approach suppliers may be able to help customers understand the environmental effects of their activities and their causes. At the same time, customers can help suppliers understand issues such as what constitutes competitive advantage in a more environmentally friendly oriented market. Since each has a vested interest in the other's success, this joint working should create more cost effective environmental solutions and better market opportunities for both customers and suppliers. Moreover, this focus on environmental standards and improvement provides a basis for constructive dialogue between customers and suppliers allowing the opportunity to develop a joint commitment to innovation and learning leading to more environmentally friendly, yet competitive, solutions.[48]

Diffusing environmental management approaches backwards and forwards through the supply chain is an effective way of improving the environmental performance of each of its business units and the supply chain as a whole. The broad aims of any environmental procurement policy include:

- Purchasing goods and services to environmental specifications, considering alternative, greener suppliers if appropriate.
- Evaluating the commitment and performance of suppliers.
- Working with suppliers to make improvements possible or more cost effective.

The collaborative relationships and transparent processes associated with SCM allow a more effective use of Environmental Management Systems (EMS), which can provide some affirmation that the required environmental standards are being maintained in the supply chain.

The indicators of an effective EMS within a supply chain might include:

- A well communicated policy built on the results of continuous reviews of environmental performance but particularly at the end of projects.
- Records of performance against compliance issues.
- Well defined roles and responsibilities at different management levels and stages in projects and in different parts of the supply chain.
- Environmental improvements linked to targets.
- Regular internal auditing in relation to legislative and policy standards. Case study 12.6 shows how a UK main contractor has included performance targets for their suppliers in relation to sustainability criteria.

The concerns associated with SCM

As in all management approaches, SCM has some significant weaknesses and inhibitors to its successful adoption and implementation. Successful

Case study 12.6 Targeting the supply chain sustainability performance

A UK main contractor sets demanding targets for sustainable development. Some of these targets are entirely within their own control, such as energy use in their offices or emissions from company vehicles. Others, however, can only be met by working with the right suppliers of goods and services.

They acknowledge that their suppliers represent the greatest impact on their construction sites. The actions by their suppliers largely determine the resource efficiency and the amount of waste generated, the energy used and the resulting emissions, and the impact on biodiversity and local communities. Their aim is to provide leadership in social and environmental performance and they are committed to raising awareness of sustainable development and its importance with their supply chain partners.

Their approach to sustainable procurement includes providing training and advice for their suppliers so that they can help the main contractor to achieve their environmental aims and aspirations. They argue that it is only with their suppliers that they can hope to effectively manage and reduce the impact of their construction activities. Identifying members of their supply chains with whom they can work to meet long-term targets on sustainable development is a key activity for their commercial and procurement teams. They look to develop lasting relationships where partners understand each other's ways of working and appreciating the strengths each separately and jointly bring to projects. They recognise, encourage and reward their suppliers' performance with annual awards in a number of categories including 'best consultant', 'best supplier', 'most improved supplier' and 'sustainability supplier'.

implementation is a long-term, complex and dynamic process requiring a thorough understanding of the concept.[49] It is also seen as closely dependent upon the ability to create, manage and reshape relationships between individuals and organisations within the supply chain.[50] The implementation of SCM requires new intra- and inter-organisational arrangements and culture all of which require considerable commitment and resources over time to develop. Time is required for learning, achieving greater mutual understanding, agreeing a common purpose, selecting the appropriate customers and suppliers, building trust, negotiating objectives, co-ordinating activities and learning processes. Difficulties which can emerge include multiple and often hidden goals; power imbalances; different cultures and procedures; incompatible collaborative capability; the tension between autonomy and accountability; over dependence; complacency leading to a reduction of long-term competitiveness; and continuing lack of openness and opportunistic behaviour.[51] It needs to be recognised that over tight networks can reduce agility and responsiveness to new market conditions and opportunities and lead to the risk of contravening legislation aimed at promoting competition. It can also result in having to continually pass inappropriate levels of value to the customer and customers may miss out on technological and organisational innovations in other supply networks.

There is an emerging recognition that a major difficulty associated with SCM is what actually constitutes effective relationships and how their effectiveness can be assessed. SCM is based on soft or less tangible aspects which current metrics of performance are not sufficiently adequate to measure. The main tools currently include the business excellence model, competitive positioning matrices and the partnership sourcing model. A number of aspects of the business excellence approach relate specifically to customer–supplier relationships as set out in Chapter 6.

There are a number of barriers to the successful implementation of SCM attributable mainly to insufficient internal preparedness by the participating organisations. These include:

- lack of commitment to SCM by the partners
- the partners having insufficient understanding of the concept of SCM
- lack of strategic leadership at all levels and in all functional departments of the partners
- lack of understanding of the concept of supply chain or network competition
- partners having inappropriate organisational structures, but particularly at the interface between the organisations involved
- lack of leverage by any of the members of the supply chain to affect change and modify behaviour throughout significant parts of the supply chain
- unwillingness to embrace win–win thinking or mutual competitive advantage
- partners allocating insufficient of time and resources to building internal and external relationships
- lack of common purpose and transparent and mutually beneficial goals
- resistance to openness and transparency and the sharing of information, procedures and processes
- inappropriate distribution of risk
- inappropriate exercise of power
- partners lacking the necessary commitment to innovation and learning.

The barriers to be overcome in construction

Given that the main objective of SCM is to increase mutual competitive advantage through improved relationships, integrated processes and increased customer focus, it may well be highly relevant to construction with its tradition of adversarial relationships, fragmented processes and lack of internal and external customer focus. However, contingency theory suggests that there may well be specific issues related to implementing SCM in the context and culture of construction that need to be addressed. These include:

- The 80 per cent of the industry's clients with their small or infrequent construction programmes – this means that less than 20 per cent of construction's clients have the commitment, the knowledge and the necessary leverage to adopt network competition.
- The transient and short-term nature of construction projects, processes, teams and relationships.
- Present procurement strategies with their emphasis on contracts and competition based on price.
- Deeply embedded adversarial relationships and opportunistic behaviour.
- Fragmented demand, supply systems and processes.
- Lack of possible partners with the appropriately developed collaborative capability.
- Multiple and hidden goals.
- Major power imbalances between supply chain members.
- Lack of the contractual and competence trust needed to underpin relationships.
- Insufficient resources and time to build relationships, integrate processes and manage logistics within a one-off project environment.
- Differences in professional language, culture and procedures.
- Lack of experience of innovations such as JIT and TQM, which are seen as important prerequisites in adopting SCM in other sectors.

Clearly there are considerable difficulties in applying SCM in construction. Addressing these differences provides the basis of the considerable benefits to be derived from SCM, but these differences also explain the reluctance and slowness of some construction organisations to meet the challenges and complexities of working more closely together.

Examples of successful use of supply chain in construction

Case study 12.7 shows how a main contractor uses the relatively stable relationships with its supply chains to adopt SCM, maintain its competitiveness and deliver more environmentally friendly products and services.

A UK main contractor has developed a management approach that addresses the challenges it faces by integrating its strategies for SCM and sustainability. The overall purpose of the integrated strategy is shaped by the main contractor's commitment to continue to:

- add value for its clients and shareholders
- build upon its history and reputation for contributing to UK society as a whole
- undertake its activities in a sustainable and responsible manner
- take a leadership role in the built environment.

The objectives of the strategy are:

- being clients' first choice in specific sectors of the market and delivering real satisfaction for them through the highest quality performance
- putting people first and providing its staff with the opportunity to develop and reach their career potential
- engaging the supply chain in a manner that reflects its values
- making a positive difference in the community both at local and at national level
- tackling climate change and energy efficiency by aiming to be carbon neutral by 2012
- making smarter use of natural resources and commitment zero waste to landfill by 2012
- aiming to continue year-on-year growth in profit
- aiming to continue to protect its business against variable economic cycles.

The company claims to engage with its supply chain in ways that mirror its commitment to undertaking activities in a sustainable and responsible manner. In order for this to be effective supply chain members are expected to work in collaboration with the main contractor to develop innovations and share best practice. The company commits to working in partnership with its supply chain, consultants and clients to deliver wider social, economic and environmental objectives, in ways that offer real long-term sustainability benefits.

In turn, its supply chain members are expected to demonstrate their commitment to partnership by:

- continuous development and improvement
- having a full and relevant Environmental Policy
- working towards a full EMS (Environmental Management System)
- meeting targets as set out in the supply chain sustainability criteria.

Case study 12.8 demonstrates how a major client of the UK construction industry is able to benefit from working with Principal Supply Chain Partners and their integrated supply chains to deliver better value.

Case study 12.8 ProCure21+ and Principal Supply Chain Partners

The ProCure21+ National Framework is a framework agreement with six supply chains on behalf of NHS clients. The Principal Supply Chain Partners (PSCPs) provide a complete construction and refurbishment solution for an NHS client or joint venture. The PSCPs have been selected via an OJEU Tender process for capital investment construction schemes across England up to 2016. An NHS client or joint venture may select a supply chain for a project they wish to undertake without having to go through an individual OJEU procurement process for the project. It is a suitable procurement route for the following types of work:

- service planning or reconfiguration reviews
- major works schemes (or refurbishments)
- minor works programmes, in which each task value does not exceed £1m
- refurbishments
- infrastructure upgrades (roads, plant, etc.) and non-health buildings (car parks, etc.)
- feasibility studies.

ProCure21+ can be used by any NHS organisation or any non-NHS organisation collaborating with an NHS organisation for the provision of a facility that has a health component. One of the six PSCPs maintains it allows them to combine the best ideas and practices from their own in-house teams with those in their integrated supply chain, which includes some of the best design and cost consultants, and specialist contractors to deliver healthcare facilities from minor works to complex new builds. They maintain that they can achieve excellent results for the NHS across the whole health economy, from primary care and mental health to acute facilities. The benefits of working with their supply chain are demonstrable, they argue, including achieving a consistent standard of quality and cost effectiveness.

Conclusion

Project managers need to be aware that the effective management of construction's supply chains can result in performance improvement provided the nature of demand is sufficient to support stable and long-term relationships between the organisations in supply chains. Clearly this basic requirement exists in the case of regular construction clients such as Sainsbury's, ASDA*Walmart, Defence Estates, housing associations, and some local authorities where effective long-term relationships have been developed. In the case of significant and frequent customers of construction services, elements of SCM, including closer long-term relationships, can lead to greater transparency in transactions, trust and commitment, which are seen as being central to the development of competitive advantage.

However, this is very difficult or impossible for infrequent clients of construction and their one-off projects. Such clients have little opportunity or indeed motivation to stabilise, shape and improve their construction suppliers and supply chains. Centrally co-ordinating and integrating their process with project processes is difficult where the client is unwilling, or unable, to exercise this degree of leadership and co-ordination. In these circumstances leadership and management of supply chains has to be undertaken by other project participants such as main or principal contractors. However, many main contractors have shown a reluctance to develop longer-term relationships with their first tier suppliers.

Although the barriers to implementing SCM in construction are substantial, they also indicate that it offers a highly relevant approach to improving construction's performance. As case studies show, if appropriately implemented, SCM can offer a way forward for improving relationships, integrating processes and increasing customer focus. The benefits to all participants can include improvements in quality and delivery, more repeat work, reductions in the overheads associated with obtaining work, increasing profitability, and the acquisition of new specialist knowledge and skills in significant sectors of the construction market.

SCM can help project managers manage the relationships and processes between the different project participants by providing a more inclusive environment where construction organisations such as specialist and trade subcontractors and suppliers can be given a more participative role in order to ensure significant integration of an appropriate proportion of the project process and its associated supply chains. Both the upstream processes to the client and the end user, and downstream processes involving the specialist and trade subcontractors and their suppliers, can be more effectively and fully involved as shown in Case study 12.9.

Despite the clear benefits to be derived from SCM, extending a comprehensive SCM approach to all types of construction projects has proved to be difficult and challenging. A greater part of the industry needs to be more adept at segmenting the market, identifying and developing critical supply

Case study 12.9 Pavement team framework agreement

Customers and suppliers reap the benefits of long-term relationships.

The pavement team agreement was an early partnering agreement set up over five years between a main contractor and the client, an airport operator and design consultants to deliver aircraft pavement aprons and ancillary works. It was worth a total of £130m and included more than 50 contracts over the five year time period. This includes conceptual and feasibility studies, but the majority of the value was for construction works. The partnership consisted of four primary suppliers. The framework agreement under which the partners operated was not legally enforceable and contracts were placed for each project.

The project teams are collocated so that staff are seconded and the team is led by a general manager from the client utilising client skills in project management. The benefits of keeping the best teams together have met the objectives to reduce cost by 30 per cent over five years and to reduce the programme time. They reduced programme time by one-third between the first and second jobs. The team also had a particular interest in meeting some other headline improvements which include a reduced accident frequency rate compared with the construction industry average and an enhanced productivity rate measured by an improvement in turnover per head of its employees.

Recently the second tier suppliers have accepted management roles in the team. Current challenges are to integrate the third tier supply chain into the partnering agreement and to use skills to further blur management roles in the integrated team, including client and specialist managers.

chain assets, being aware of market conditions, and aligning all project and supply chain processes to meet the needs of internal customers and the client, despite the difficulties of infrequent demand by most clients.

There is a growing awareness of the principles and practice of SCM amongst project managers and other construction practitioners. Indeed most main contractors now employ supply chain managers. It is also clear that the specificities of construction are shaping the form it takes within the industry as practitioners seek to address the complexity, fragmentation, interdependency and uncertainty which characterise the construction market and project processes.

In addition, construction has had to work hard at developing the inter-organisational relationships necessary to enable and support SCM. This is partly attributable to construction's long tradition of adversarial relationships and fragmentation but also to the fact that it moved to the adoption of SCM without having benefited from earlier innovations, such as JIT and TQM, which paved the way for SCM in other sectors and industries. It is only since the 1990s, with the emergence of partnering, that the UK construction industry has started developing the more collaborative relationships and integrated processes that can be seen as laying the foundations for SCM. Even where partnering has been adopted, it is not always fully understood and has not yet led to a widespread change in the industry's culture, which remains essentially adversarial with arms-length relationships and significant use of price-competitive procurement approaches, within an environment shaped by rigid contracts. In addition, partnering is mainly being adopted at the front end of project processes and essentially between clients, consultants and main contractors and has yet to be fully extended to many of the specialist contractors and suppliers downstream of the main contractor.

There are significant difficulties which need to be addressed if SCM is to be effectively implemented in construction. These include the lack of preparedness and readiness of construction organisations to fully adopt SCM; the limited understanding of the concept and the prerequisites associated with its implementation; the unwillingness to rationalise supplier and customer bases; the difficulty in establishing a clear common purpose, exchanging information and sharing learning. This can be interpreted as a further indication of the challenges in embracing the culture associated with the relationships in SCM. Although learning is perceived as increasingly important in supporting innovation in construction, the types of learning being undertaken do not always match the competencies and the cultural changes needed for such a complex, multi-factor and dynamic innovation.

References

1 Holti R., Nicolini D. and Smalley M. (1999) *Prime Contracting Handbook of Supply Chain Management Sections 1 and 2*. London, Tavistock Institute.
2 See London K. and Kenley R. (2001) 'An industrial organisation economic supply chain approach for the construction industry', *Construction Economics and Management*, 19: 799–788; Arbulu R., Ballard G. and Harper N. (2003) *Kanban in Construction*, IGLC11. VA, Blacksberg; Cox A. and Ireland P. (2002) 'Managing construction supply chains: the common sense approach', *Engineering Construction and Architectural Management*, 9(5/6): 409–418.
3 Egan J. (1998) *Rethinking Construction*. London, DETR; Latham M. (1994) *Constructing The Team*. London, HMSO.
4 Cox A., Ireland P. and Townsend M. (2006) *Managing in Construction Supply Chains and Markets*. London, Thomas Telford.
5 Cox A. and Townsend M. (1998) *Strategic Procurement in Construction: Towards Better Practice in the Management of Construction Supply Chains*. London, Thomas Telford.

6 Blismas N., Gibb A.G.F., Pasquire C.L. and Aldridge G. (2003) 'Changing perceptions of value in construction standardisation and pre-assembly', *Proceedings of the 2nd International Conference on Innovation in Architecture, Engineering and Construction.* 25–27 June 2003, Loughborough University, UK, pp.649–660.

7 Cox A., Ireland P. and Townsend M. (2006) *Managing in Construction Supply Chains and Markets.* London, Thomas Telford.

8 Cox A., Ireland P. and Townsend M. (2006) *Managing in Construction Supply Chains and Markets.* London, Thomas Telford.

9 Saad M., Jones M. and James P. (2002) 'A review of the progress towards the adoption of supply chain management (SCM) relationships in construction', *European Journal of Purchasing & Supply Management*, 8: 173–183; Jones M. and Saad M. (1998) *Unlocking Specialist Potential: A More Participative Role for Specialist Contractors.* London, Thomas Telford.

10 Burnes B. (2004) *Managing Change.* 4th edn. Harlow, FT Prentice Hall.

11 Ball M. (1988) *Rebuilding Construction: Economic Change in the British Construction Industry.* London, Routledge.

12 Christopher M. (2010) *Logistics and Supply Chain Management.* 4th edn. London, Financial Times/Prentice Hall.

13 Derived from Slack N. *et al.* (2001) *Operations Management.* 3rd edn. London, Financial Times/Prentice Hall; and Hill T. (2000) *Operations Management – Strategic Context and Managerial Analysis.* London, Macmillan Business.

14 Mowery D.C. (Ed.) (1988) *International Collaborative Ventures in US Manufacturing.* Cambridge, MA, Ballinger.

15 Christopher M. (2010) *Logistics and Supply Chain Management.* 4th edn. London, Financial Times/Prentice Hall.

16 Stadtler H. and Kilger C. (2004) *Supply Chain Management and Advanced Planning: Concept Models, Software and Case Studies.* 3rd edn. Heidleberg, Springer, p.11.

17 Constructing Excellence (2004) Supply Chain Management Fact Sheet. p.1.

18 Tan K.C. (2000) 'A frame work of supply chain management literature', *European Journal of Purchasing and Supply Management*, 7: 39–48.

19 New S. and Ramsay J. (1997) 'A critical appraisal of aspects of the lean approach', *European Journal of Purchasing and Supply Management*, 3(2): 93–102; Christopher M. and Jüttner U. (2000) 'Developing strategic partnerships in the supply chain: a practitioner perspective', *European Journal of Purchasing and Supply Management*, 6: 117–127.

20 Vollman T., Cordon C. and Raabe H. (1997) 'Supply chain management'. In *Mastering Management.* London, FT Pitman, pp.316–322; Koskela, L. (1999) 'Management of production construction: a theoretical view', *Proceedings of the Seventh Annual Conference of the International Group for Lean Construction* IGLC-7, Berkeley, 26–28 July, pp.241–252.

21 Jones M. and Saad M. (1998) *Unlocking Specialist Potential: A More Participative Role for Specialist Contractors.* London, Thomas Telford.

22 Vollman T., Cordon C. and Raabe H. (1997) 'Supply chain management'. In *Mastering Management.* London, FT Pitman, pp.316–322.

23 Hines P. (1994) *Creating World Class Suppliers.* London, Pitman Publishing.

24 Holti R., Nicolini D. and Smalley M. (1999) *Prime Contracting Handbook of Supply Chain Management Sections 1 and 2.* London, Tavistock Institute; Cain T.C. (2003) *Building Down Barriers: A Guide to Construction Best Practice.* Abingdon, Routledge.

25 Christopher M. and Jüttner U. (2000) 'Developing strategic partnerships in the supply chain: a practitioner perspective', *European Journal of Purchasing and Supply Management*, 6: 117–127.

26 Sako M., Lamming R. and Helper R.S. (1994) 'Good news – bad news', *European Journal of Purchasing and Supply Management*, 1(4): 237–248.

27 Lamming R. (1993) *Beyond Partnership: Strategies for Innovation and Lean Supply*. New York, Prentice Hall; Lamming, R. and Hampson, J. (1996) 'The environment as a supply chain management issue', *British Journal of Management*, 7, Special Issue: S45–S62.

28 Hines P. (1994) *Creating World-Class Supplier: Unlocking the Mutual Competitive Advantage*. London, Pitman Publishing; Womack J. P. and Jones D.T. (1996) *Lean Thinking: Banish Waste and Create Wealth in Your Corporation*. New York, Simon and Schuster.

29 Towill D.R. (1996) 'Time compression and supply chain management – a guided tour', *Supply Chain Management*, 1(1): 15–27.

30 Lipparini A. and Sobrero M. (1994) 'The glue and the pieces: entrepreneurship and innovation in small-firms' networks', *Journal of Business Venturing*, 9: 125–140.

31 Dubois A. and Gadde L. (2000) 'Supply strategy and network effects – purchasing behaviour in the construction industry', Supply Chain Management in Construction – Special Issue, *European Journal of Purchasing & Supply Management*, 6: 207–215.

32 Burgess R. (1998) 'Avoiding supply chain management failure: lessons from business process re-engineering', *International Journal of Logistics Management*, 9: 15–23.

33 Holti R. (1997) *Adapting Supply Chain for Construction*. Workshop Report, CPN727, Construction Productivity Network, CIRIA.

34 Edum-Fotwe F.T., Thorpe A. and McCaffer R. (2001) 'Information procurement practices of key actors in construction supply chains', *European Journal of Purchasing and Supply Management*, 7: 155–164.

35 Ali F., Smith G. and Saker J. (1997) 'Developing buyer-supplier relationships in the automobile industry. A study of Jaguar and Nippondenson', *European Journal of Purchasing and Supply Management*, 3(1): 33–42.

36 Sako M. (1992) *Prices, Quality and Trust: Inter-Firm Relations in Britain and Japan*. Cambridge, Cambridge University Press.

37 Saad M., Jones M. and James P. (2002) 'A review of the progress towards the adoption of supply chain management (SCM) relationships in construction', *European Journal of Purchasing & Supply Management*, 8: 173–183.

38 Dasgputa, P. (1988) 'Trust as a commodity'. In D. Gambetta (Ed.) *Trust: Making and Breaking and Co-Operative Relations*. London, Sage; Farris G.F., Senner E.E. and Butterfield D.A. (1973) 'Trust culture and organisational behaviour', *Industrial Relations*, 12(2): 144–157.

39 Rotter J.B. (1967) 'A new scale for the measurement of interpersonal trust', *Journal of Personality*, 35(4).

40 Hofstede G. (1980) *Cultures' Consequences, International Differences in Work Related Values*. Newbury Park, London, Sage.

41 Shapiro D., Sheppard B.H. and Cheraskin L. (1992) 'Business on a handshake', *Negotiation Journal*, 8(4): 365–377.

42 Kanter R.M. and Myers P. (1989) *Inter-Organisational Bonds and Intra-Organisational Behaviour; How Alliances and Partnerships Change The Organisations Forming Them*. Paper presented at the First Annual Meeting of the Society for the Advancement of Socio-Economics, Cambridge, MA.

43 Jones M. and Saad M. (1998) *Unlocking Specialist Potential: A More Participative Role for Specialist Contractors*. London, Thomas Telford.
44 Good D. (1988) 'Individuals, interpersonal relationships and trust'. In D. Gambetta (Ed.) *Trust: Making and Breaking Co-operative Relations*. Oxford, Blackwell Publishing, pp.31–48.
45 Axelrod R. (1984) *The Evolution of Co-operation*. London, Penguin; Kreps D.M. (1990) 'Corporate culture and economic theory'. In J.E.A.A.K.A Shepsle (Ed.) (1990) *Perspectives on Positive Political Economy*. New York, Cambridge University, pp.90–143; Telser L.G. (1987) *A Theory of Efficient Co-Operation and Competition*. Cambridge, Cambridge University Press.
46 Womack J., Jones D. and Roos D. (1990) *The Machine that Changed the World*. New York, Harper Collins.
47 Lamming R. (1996) 'Squaring lean supply with supply chain management', *International Journal of Operations & Production Management*, 16(2): 183–196.
48 Lamming R. and Hampson J. (1996) 'The environment as a supply chain issue', *British Journal of Management*, 7: S45–S62.
49 Akintoye A., Mcintosh G. and Fitzgerald E. (2000) 'A survey of supply chain collaboration and management in the UK construction industry', Supply Chain Management in Construction – Special Issue, *European Journal of Purchasing & Supply Management*, 6: 159–168; Whipple J.M. and Frankel R. (2000) 'Strategic alliance success factors', *The Journal of Supply Chain Management*, 36(3): 21–28; Edum-Fotwe F.T., Thorpe A., and McCaffer R., (2001) 'Information procurement practices of key actors in construction supply chains', *European Journal of Purchasing and Supply Management*, 7: 155–164.
50 New S. and Ramsay J. (1997) 'A critical appraisal of aspects of the lean approach', *European Journal of Purchasing and Supply Management*, 3(2): 93–102; Spekman R.E., Kamauff Jr. J. and Myhr N. (1998) 'An empirical investigation into supply chain management: a perspective on partnerships', *International Journal of Physical Distribution and Logistics Management*, 28(8): 630–650; Harland C.M., Lamming R.C. and Cousins P.D. (1999) 'Developing the concept of supply strategy', *International Journal of Operations and Production Management*, 19(7): 650–673.
51 Huxham C. (Ed.) (1996) *Creating Collaborative Advantage*. London, Sage Publications; Cox A. and Townsend M. (1998) *Strategic Procurement in Construction: Towards Better Practice in the Management of Construction Supply Chains*. London, Thomas Telford.

Chapter 13

Quality and customer care

The project manager's ultimate concern is to provide customer satisfaction. The primary link to the customer is through the project manager who also needs to understand the needs of the end user. For the contractor, repeat business from a knowledgeable, satisfied client is good business, because it allows a relationship to be set up and a better understanding of the product to be gained. The contractors are the key provider of customer care, which includes their quality requirements.

The objectives of this chapter are to:

- understand the nature of quality
- evaluate the Egan targets for quality and customer care and subsequent attainments
- apply Handy's principles of customer care as a model for working with construction customers and their relevance in a project environment to care and quality
- implementation of quality and care principles and their impact on traditional contractual relationships
- nurture the principles and culture of quality management to specifically meet a client's needs where many organisations are working together
- illustrate the use of relational marketing techniques in developing sustainable customer relations.

Introduction

Quality has been defined in at least three ways which are:

- a product or service which meets its specification
- fit for purpose
- meeting the expectations of the customer.

The first two definitions are critical, but the last is an important concept because it connects the project objectives of quality firmly with customer care and has potential for optimising value by stripping out waste from over specification. It also will only work efficiently if there is a strong integration of the client with the project team so that their expectations can be managed and desires separated from key functional purpose, in order to ensure keeping to budget. It also begs the question about who *manages* quality. There is a tightrope to walk between 1) granting the customer their every whim, but what they needed might not be what they got, and 2) giving the customer what you thought they needed and the result was that it was not what they expected. Delighting the customer is about *just* exceeding expectations (see commentary on Maister's law in Chapter 2). It helps to balance imperfection. Waste is closely connected to quality also as shown in Case study 13.1.

BSI defines quality as 'the totality of characteristics of an entity that bear on its ability to satisfy stated or implied needs'.[1] Gardiner differentiates the need to distinguish *grade* which is the level of quality – low grade or high grade – and *reliability* which is an entities ability to sustain need.[2]

Case study 13.1 UK construction waste

In the UK construction industry there was 86.9 million tonnes of basic waste in 2008.[3] Apart from demolition waste, it was due to quality problems such as design faults, inadequate quality of materials and components, lack of management of the process and workmanship problems. It may be worsened by the culture of the organisation. 55 million tonnes was recycled through crushers/screens. Reducing waste by 5 million tonnes (i.e. better quality control) and diverting 5 million tonnes from landfill sites (recycling) are the equivalent of the waste generated by 200,000 people at home each year.

The threshold for construction quality

Egan has identified a focus on the customer as a key driver in the improvement of the construction industry.[4] The report believes that in the best companies

> the customer drives everything. These companies provide exactly what the end customer needs, when the customer needs it and at a price that reflects the product's value to the customer. Activities which do not add value from the customer's viewpoint are classified as waste.

The report claims that recent studies show that up to 30 per cent of construction is rework. This provides a challenge for the fragmented supply chain in the construction industry, particularly in the separation of the design and assembly functions, to improve outputs. There is a need for the project manager to identify the customer in customer care. In many cases it is necessary to look beyond the client to the stakeholders who have a continuing interest in the finished project.

Quality is another driver which is closely associated with customer care and can be defined as fit for purpose and the elimination of defects. In the context of customer care however it is also connected with best value and the cutting out of waste. Egan recognises a quality driven agenda as a more integrated package, which is

> not only zero defects, but right first time, delivery on time and to budget, innovating for the benefit of the client and stripping out waste . . . it also means after-sales care and reduced costs in use. Quality means the total package – exceeding customer expectations and providing real service.[5]

The Latham Review also recommended improving quality by the adoption of quality/value procedures in tendering selection to replace widespread lowest price practice.[6] The review looked at ways of achieving this whilst not contravening the competitive approach required for public organisations.

The Chartered Institute of Purchase and Supply (CIPS) mission statement as communicated to the Latham Review in 1994 suggested that the key objective was 'to improve the quality of the UK construction industry and reduce the average costs by at least 30 per cent within 5 years and to introduce continuous improvement programmes'.[7]

One of the main Egan targets is to reduce defects by 20 per cent year on year and to challenge the concept that it is not possible to have zero defects. The Constructing Excellence benchmark of 8/10 is a few defects and none that substantially affect the clients' business.[8] When you look at the chart for this only 58 per cent of the benchmark population achieved this scale of performance in 2002. The Movement for Innovation (M4i) demonstration projects, which were committed to run with Egan principles, were scoring significantly better with 86 per cent achieving this level of quality.

On a broader base the Atkins Report (1994)[9] for the EU proposes that the level of defects should be reduced together with the cost of 'non quality' and that the level of specifications should gradually be raised on buildings and infrastructure projects. They also recommend that QA systems should be more appropriate for construction companies and that construction firms should move from QA to full quality management.

In this wider definition it is necessary to move away from the model of inspection to a model of quality planning which is focused on clients' needs and quality improvement. This will not work in a climate of selecting designers and contractors on the basis of lowest cost. Clients it seems are prepared to look at best value approaches as long as it is clear that the supply side are making a joint effort to affect changes in practice. To this end the CCF (1998),[10] who represented 80 per cent of construction investment in the UK, produced a 'Pact' document in response to Egan. This pact set out areas of good practice to the construction industry to:

- Present objective and appropriate advice.
- Introduce a right first time culture, finishing on time and to budget.
- Eliminate waste, streamline processes and work towards continuous improvement.
- Work towards component standardisation.
- Use a competent workforce, kept up to date.
- Improve management of supply chains.
- Keep abreast of changing technology and innovation and investing in R&D.

As it is a pact the clients also agreed to set clearly defined objectives, benchmark their own performance, communicate better, promote teamwork and relationships based on trust, share anticipated savings gained through innovation, appraise whole life costs, influence statutory legislation favourably, support training, educate decision makers, not exploit purchasing power, apportion risk fairly and improve management techniques.

Eleven years after the Egan Report, in the Wolstenholme Report (2009),[11] 80 per cent of respondents thought that top management was committed to quality and efficiency and 52 per cent thought a quality driven agenda was very important and had design quality high up their list of issues. It did find however that a lack of integration in the delivery process impeded continuous improvement (p.18). Many projects still run on a significant snagging list at the end of the project. Clients see construction as a commodity purchase with main interest in the upfront costs in their business plans and have no incentive for improving quality. PFI has however improved the quality of buildings with incentives to keep maintenance cost low from the suppliers (p.14). However the report is still unclear about the progress made in reducing quality defects in accordance with the 20 per cent year by year reduction. It is clear though that the UK industry does not yet reach the optimising focus of the Japanese construction industry to get it right first time on the basis that people like to do a good job (p.20). The report also highlights that 73 per cent suppliers still believe a focus on the customer to be very important and that management were committed to this. This leads us on to the principles of customer care which are closely related to quality as defined by meeting the expectations of the customer.

Handy's principles of customer care

According to Brown,[12] customer care is an attitude of mind and the way it is carried out is foundational to every aspect of the relationships which have developed between the supplier and the customer. 'Customer care aims to close the gap' between what the customer expects and what the customer actually gets.

Charles Handy (1991)[13] has three principles, which he believes will transform ability to attract customers:

- 'Customers are forever' which reinforces the need to provide full customer satisfaction so that they return again and again.
- 'Customers are everywhere' encourages us to develop a culture so that the whole workforce has an acute sense of who the client and the stakeholders are so that individuals feel accountable for the standard of service they give. It also applies to the *internal* user of a service output in a value chain within the organisation or project. It moves away from concentrating customer satisfaction in the hands of the project manager.
- 'Customers come first' assumes that we listen to what the customer wants and then design the product to suit. This is on the basis that all businesses need customers. Handy argues it may drive people away if we change the language before we follow the principle.

Principle I Customers are forever

Attracting and keeping customers can be an illusive process if you are not sure who your customer is. A project has a number of stakeholders. The key issue in customer care is to provide a good quality of work on time and to budget. Benefits will be gained from a good track record and the possibility for tendering future work by remaining on tender lists, or preferably negotiating further work.

As a project manager with responsibility for design and planning applications there is a public element to the project as to how it impacts on the community. Construction organisations have to recognise the issues which affect the core business of the client commissioning the construction project and not just concentrate on the efficiency of the project. In the commissioning of the Skye Bridge the joint venture had a responsibility to themselves for ensuring that they had a viable income to cover their expenses and the payments to their shareholders, but also a core responsibility to the community and the environment. They would therefore have to consider the quality of the design and the effect on the lives of the people. Cost is important, but cannot override these public responsibilities. This meant that when the joint venture company was pushed on the planning appeal more money had to be spent on a redesign.

Principle 2 The internal customer

This principle involves everyone in customer care and it introduces the concept of the internal customer. The internal customer is another member of the project team. This involves asking questions such as 'What is the standard required?', 'Am I supplying it on time?' and 'How can I contribute to the process of adding value to the service?' Case study 13.2 indicates this.

The internal customer concept has several implications:

- There is accountability of each person in the project organisation for the quality of their output which means there is nowhere to hide.
- There is a knowledge by all of the overall requirements of the process and the final client.
- Everybody is looking for ways of improving what they do, believing that it makes a difference to the big picture.
- Everybody is an ambassador for his or her organisation and the project.

To achieve this, 'ownership' changes in culture have to be made and comparatively junior personnel have to be given exposure and trusted with clients at their level of operation. A new respect for other members of the project team with accountability has to be set up.

Case study 13.2 Customer care in private housing

One housing developer arranges three progress visits for their customers and this puts site managers directly in the front line with the buyers. Their internal customer is the marketing department and they require reliable promises for completion dates, clean and preferably segregated access to their show homes, with reduced noise and dust for those who are already moving in.

As an ambassador for their own organisation they are showing new owners around their partially completed houses and the way they do this affects how the customer thinks about the organisation. Their communication skills, the site conditions and the structural quality will make a strong impression on the customer. This in turn may change the site manager's attitude to a safe and clean environment and the promises they make about interim progress, because they are now directly accountable to the customer. Customers also notice the mistakes that are made in progress like the blocking up of a door in the wrong place.

Principle 3 Customers come first

Putting customers first means listening to what the customer wants and then trying to enhance the value that is added. They will respond to this in different ways depending upon their experience. Customers come first in construction not by giving away contractor profits, but by:

- Including the client on the team and making sure that the user and the maintenance staff are involved in the early stages of the design and giving time for adaptation.
- Giving time to add value to the original specifications and iterate the design.
- Using feedback from other projects and from current user problems and concerns, which may be relevant for design and occupation of the new building. Lessons learnt should be incorporated into the design.
- Implementing outstanding defects and adjustments in deference to the occupier's priorities. An aftercare and training programme may be appropriate.
- Obtaining feedback from the customers to inform future work.

Putting the customer first is often given more credibility in some of the newer forms of procurement. PFI procurement allows, by definition, a negotiation period when preferred bidders sit down with the client to agree what can be achieved within the available budget framework and how value may be enhanced. This offers the chance to build up relationships and to more formally value and manage the risks. The occupation stages are more integrated with the continuing facilities management role. In design and build there is an opportunity to have a single point of contact and closer relationships between the contractor and the client which helps an understanding of the client's business. In strategic partnering there is the opportunity to use a no blame contract, to develop collaboration with more 'transparent' pricing and to hold out incentives for sharing the benefits of innovation and ongoing improvements from project to project.

The Strategic Forum for Construction has developed a maturity assessment grid, which recognises different stages of client involvement and with increasing levels of customer satisfaction.[14] These stages are listed by the degree of their integration with the client and are differentiated by the sequence of the four main components of development.

- Historical (traditional)   Need > develop > *procure* and implement.
- Transitional   Need > *procure* > develop and implement.
- Aspirational   *Procure* > Need > develop and implement.

These models can be mapped with different procurement systems which essentially bring in a working relationship of a project team with the client earlier and earlier as is indicated by the procurement activity. The last one allows the greatest amount of integration because of the more fundamental integration of client and project team clarifying the need together.

The Strategic Forum claims there are significant financial benefits to integration as well as having more comfortable relationships with the whole supply chain and they have developed an 'integration toolkit'.[15]

The smaller contractor is likely to operate within the competitive tendering field and new ways of integrating with the client will be over laid on model one. However many small contractors are used to provide a turnkey package for the client and have relied on close collaboration and repeat business for many years. It is still important to be able to benchmark performance so that progress may be measured. Case study 13.3 considers working together.

Case study 13.3   The use of a charter to gain mutual commitment

The Client Pact (CCF 1998) calls for a contractor commitment to right first time and zero defects.[16] From the contractor's side some of these aspirations are being worked out by the use of more customer focused policies and the use of quality KPIs for clients to differentiate at tender stage. To encourage progress clients are promising to share savings resulting from innovative methods, to choose contractors on the basis of whole life costs and to not unfairly exploit their purchasing power by squeezing the competition, but to look to form lasting relationships and to educate their decision-makers in 'clientship'. They in return agree minimum standards of good practice through the Construction Client Charter,[17] which are leadership and focus on the client, working in integrated teams, whole life quality and respect for people. The client agrees to make an improvement plan to make sure these are visibly feasible.

Models of quality

Quality management has famously been sponsored in recent years by Juran who developed the quality trilogy of quality planning, quality improvement and quality control, Crosby who coined 'do it right first time' and 'zero defects', and Deming who devised the Plan, Do, Check, Act cycle (Figure 13.1) now PDSA replacing 'check' with 'study'. The so called gurus were responsible for systems which hugely improved the quality and productivity of Japanese industrial products giving them a competitive edge which they still lead in and demonstrating that quality pays.

There is a difference between the culture of quality control and the new paradigm of quality management. Both presuppose some preplanning to reduce incidents which cause defects, but management formally tries to eliminate defects:

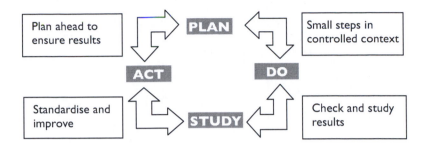

Figure 13.1 Deming's PDSA cycle

- Quality control is monitor output, compare, correct and possibly new processes.
- Quality assurance is previous experience, standards, quality managers, well-defined process, health checks (system) and a stable process to get it right first time.

Chapter 5 considered the importance of planning for construction quality to get a leaner process by eliminating the waste of abortive work and it is now important to take this a stage further and look at the systems for improvement. Quality management of projects depends on the continuous improvement philosophy of the organisation. This section discusses the merits of the various quality regimes commonly used and their relationship to productivity and client satisfaction for the project.

Turner and Huemann suggest that it is less expensive to pay for managers and systems to prevent many quality defects than it is to pay for many failures, scrap, rework and repairs under guarantee.[18] However the cost will be recouped in the medium term as start up costs are high and there is a learning curve as causes of poor quality are unravelled. The Pareto principle that 20 per cent of the causes produce 80 per cent of the faults gives a focus for investigation (Figure 13.2), but also suggests that the last 20 per cent of defects are going to be more expensive to eliminate.

It is assumed that causes are consistent from project to project, so that there is a cumulative driving down of defects occurring as they are understood and prevented. This is likely in similar projects and supply teams, but may vary in other contexts. Management commitment is required to fund upfront quality systems as a central overhead and staff commitment is required to believe that quality can be improved.

The Atkins Report suggests that there are five tools to be used to control quality in construction:[19]

- Product standards and design codes.
- Technical compliance with regulations, standards and specifications.

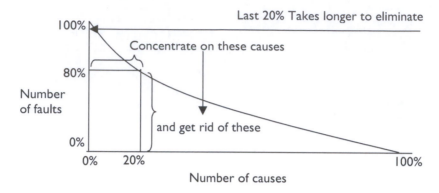

Figure 13.2 The Pareto principle for defects and causes

- Liabilities and guarantees covering risk of defects.
- Registration and qualification of contractors and consultants.
- Quality assurance and quality management systems.

We shall deal with the last category as the others are mainly self-explanatory and are not the main concern of this chapter.

Quality management

In the past quality has been achieved by inspection of the finished output (quality control). This is considered wasteful due to abortive costs and on one off projects mistakes are less easy to retrieve so should be fewer. Clients are demanding a higher standard of process management, design and manufacturing quality that will reduce defects.

There are two approaches which promote the 'right first time' philosophy:

- quality assurance (QA)
- total quality management (TQM).

These approaches both depend upon an early planning approach to quality in order to save on abortive costs. Quality systems need to be set up each time a new project is to be started to ensure that the project quality proposed by the supplier is equal to the level expected by the client for the project.

Before we proceed we must recognise the difference between product, service and process quality and two of these are recognised in the two KPIs in the construction KPI zone on the Constructing Excellence website for:

- client satisfaction for product (the effective working of it)
- client satisfaction for service (the responsiveness of the management team in delivering it).

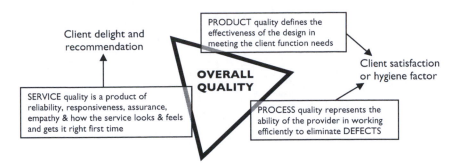

Figure 13.3 Components of quality and their outcomes

The process is represented by the defects KPI which indicates how efficiently the product has been delivered and does not reflect on final product capability as defects are the responsibility of the provider to put right and are usually a management or workmanship problem.

Gardiner quotes Berry *et al.* (1994) for the development of the SERVQUAL measurement system which was developed for the service sectors, but can be used as a measure of satisfaction of the client with the designer and contractor to their project.[20] It is a weighted assessment of the five dimensions in the service quality box in Figure 13.3. It weights reliability (32 per cent of the weighting) as their research indicated that customers rated doing what was promised and '*doing things right first time*' very highly which together with responsiveness '*within the time scale promised*' (22 per cent) made up over half of rating. Assurance (19 per cent) and empathy (16 per cent) were also recognition of the trust built up in the relationship and the understanding of the unique requirements of the client. If done well the client is likely to remember service best in their recommendations to others (evidence a small contractor cleaning the house really well at the end of each day's building work to lessen the nuisance factor to the occupants). It represents the delight factor.

QA is a check on the quality process rather than the product and service provided and can be considered to be the hygiene factor in Herzberg's theory. It is recognised that check listing service is more difficult, as it is to do with the quality of interpersonal relationships and would require an observation based audit.

Quality assurance

QA procedures are well established audit based systems and require a production process to be properly defined so that checks can be made at key points in the process to ensure compliance with procedure. Assurance does

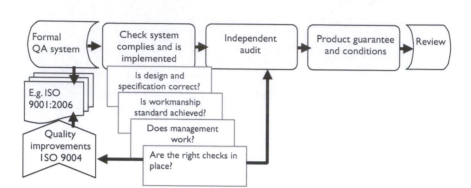

Figure 13.4 Quality assurance system

not make up the procedure, but checks that the procedure is being used correctly. ISO 9001:2006 is the main international standard with third party accreditation.[21] In theory the cost of the quality checks is less than the saving in abortive costs of not getting it right first time. A model of QA is shown in Figure 13.4.

The quality system must be sound otherwise QA will simply 'seal in' the faults. This may be worsened if there is only a periodic review of the system to make improvements so the system in assumes regular management of the assurance process and in addition there is a quality improvement module. For this system to be successful it needs to change the culture so that people understand why they are checking systems. One problem may be that people do not feel ownership of the system, so senior management only gets limited feedback for improvement. Another problem is that a quality system has no senior management leadership – it has been deserted. For the client, third party QA is important evidence of potential, but does not guarantee achievement.

The ISO 9000:2006 series third party accreditation with an improvement module 9004 has been set up for single manufacturing systems and thus assumes a repeatable production process with the quality of outcome depending on one organisation. This is clearly not the case in construction projects. Hellard argues that there is a need to co-ordinate the different QA systems operated by the different suppliers to the project and to overlay a *project quality system* or a quality plan.[22] He claims that the process of self-assessment and client assessment of quality is the most effective line of attack, but requires a senior commitment to drive through changes in attitude. The principles of ISO 9001 and 9004 emphasise leadership, people involvement and mutually beneficial supplier relationships.

Taguchi's model of loss function identifies loss as costs incurred and profit lost and tries to optimise the quality control cost benefit. Loss should be minimised by quality management. He identifies offline (pre-production) and online (production) quality management. Offline management covers:

- systems design (reflecting appropriate technology)
- parameter design – where it is easier to design a product insensitive to manufacturing variances (e.g. brickwork) than it is to control those variances (e.g. rain screen curtain walling which must be 100 per cent impervious)
- tolerance design – the degree of variance permitted in assembly.

Online (production) quality management should aim to minimise losses due to variations and to reduce control where possible. Case study 13.4 looks at a quality control system in practice.

This case study indicates the intricate nature of quality control and the need to supervise it arises out of the tight tolerances required by the client, but value has been built into the process by using the supervisor to also co-ordinate other work. This is an example of feedback control that is used effectively.

Case study 13.4 Quality control of piling

This site consisted of a piling mat of 1,523 twelve metre and fifteen metre long piles to be placed in a series of concentric circles across the foot plate of the building. Continuous flight auger piles were used. The tolerance for the piles was critical as the scientific use of the building required high floor level tolerance and foundation movements were critical to this. High loadings were expected. An inspection and test plan (ITP) was devised and supervised by the managing contractor. First line responsibility for quality and setting out was the responsibility of the specialist contractor. The managing contractor acted autonomously on small discrepancies, but any discrepancies discovered in the piling outside a band of tolerance required structural engineer involvement and possible redesign of the piling pattern to compensate. To ensure the quality of the piles several checks were made:

- a testing programme for the integrity and movement of the piles
- a testing programme for the strength and slump of the concrete
- a record of the piling operations indicating the pile concrete profile and wet concrete pressures
- an interpretation of complex clay and chalk ground conditions to predict different piling methodologies and concrete slumps.

Each of the above gave clues as to the success of the pile and it was the last two measures which might be partly preventative and reduce the cost of remedial action by the more immediate action that was possible. All data and test results identified the pile.

Concrete could be rejected if it was not within certain slump limits and cube tests under strength. The slump needed to be higher if cast into chalk which sucked out the water from the concrete making it stiffen quicker and cause problems for the lowering of the reinforcement cage into the concrete. The seven day results were tied to other data about the pile from the computerised piling log and the 28 day strength predicted. A bad result or prediction would lead to re-boring the pile.

Proof testing was carried out on 5 per cent of piles in situ, by the use of 23 hour load tests to check the friction resistance. Cheaper integrity testing was carried out on all cured piles using non-destructive testing and measuring toe seat deflection. Any substandard piles found would be supplemented with further adjacent piles.

This regime needed specialist contractor to carry out routine testing and a quality checker on site on behalf of the main contractor who was also a setting out engineer. The cost of this was offset by also giving them a planning and monitoring role to predict the rate of piling progress and to co-ordinate safe, uninterrupted access for three piling rigs and their associated equipment and to ensure access for concrete lorries. The checker also carried out supervisory roles for ensuring the progress of other works and progress with the master programme. All work was completed in 11 weeks at an average rate of 30 piles per day and a maximum of 48 piles per day.

It could be argued that in an enlightened role the accountability for non-compliance could have been put upon the piling contractor and there would be no need for inspection. However the huge delay and financial consequences of the foundation works being found faulty at the user stage justified the extra cost of a single well trained checker with adequate management backup.

Quality function deployment

Design quality can be established by the quality function deployment (QFD) method. It starts by listening to the customer and assessing in detail the customer's needs and including them in detail in the design specification as relevant technical requirements. You continue by translating customer demands, 'the roof will be flat and not leak', into quality requirements, 'the roof will use a high performance single skin polymer membrane, which will be installed by trained operatives and tested every 10 years'. Note that the re-testing is an important part of the quality deal in the light of current technology.

Production quality is best served by preventing mistakes and minimising loss. A number of techniques can be used to anticipate problems such as statistical process control (SPC) and QA to ensure that a process is more watertight.

Quality culture

The corporate philosophies, traditions, values and commitment create the culture of an organisation and provide a basis for the mission statement. Theory Z is a term coined by William Ouchi, which describes an ideal culture, based mainly on the successful methods and approach used by the large Japanese companies. This culture is less bureaucratic and hierarchical and encourages responsibility for quality and its improvement at all levels of the organisation by encouraging a consensus in decision-making, reached by agreement with peers and subordinates. He argues that self-direction and mutual trust lead to high levels of performance and job satisfaction. This may also lead to less paperwork and less supervision.

Effective communications and teamwork

All three components of quality are tied up with effective communication between parties and in building up trust and team attitudes. It is important for seamless communications between the client, the designer, the contractor and their supply chain. Only one of these links needs to break down for quality to be at risk. For example:

- Poor quality drawings from the architect to the contractor with missing or wrong dimensions which have not been checked or communicated cause consternation and delays.
- A client's lack of communication to the designer about asbestos in the building being renovated causes regulatory delay, costly removal and redesign to meet budget restrictions.
- The contractor sitting on specialist information requests so that layouts are not designed properly.
- A subcontractor fails to inform about delayed materials so that work is sequenced incorrectly and other subcontractors lose trust in them, closing ranks in terms of co-operation.

Each redesign or delay is inextricably tied up with loss of quality due to pressure on time or strained relationships. The planning and checking system may also be compromised or break down. The system can easily proceed from a win–win to a win–lose for the client, designer or contractor. Problem solving is less effective as the teams become less co-operative. The service quality to the client may also break down if they are not kept informed of delays so they lose *their* assurance.

Teamwork is also a motivator of synergy and depends upon sharing information, building up trust and transparency. Conflicts because of these things mean that vital information is withheld which puts quality of the

Case study 13.5 Quality improvement – sparklemeters

In a rolling programme of workshops for a government agency quality had not been up to the expected client standards. This especially applied to finishes which were very visible and the team had become demotivated, because a lot of remedial work was taking place near the end of the project which delayed handover and was expensive. There was a feeling that they were struggling constantly to time, cost or quality targets.

The programme instituted to overcome this was termed the sparklemeter. The project manager awarded points on parallel contracts for visual appearance of various subcontract elements on the basis of a 'sparkle factor' giving a breakdown score out of 100 for each contract. The overall and element scores for each contract were compared to introduce an element of competition to improve their scores and to up the overall scores. A prize was awarded each month for the best contract teams.

As some of the quality was suffering due to the quality and timeliness of the drawing information, construction teams were also allowed to score the *design* efforts at the progress and design meetings. This allowed a two way competition and included the designers in the delivery team score. Discussion was held at each of these meetings to discuss improvements.

The method proved successful in improving quality service and standards in a fun way and quality assurance was taking place naturally between teams to insure preventative methods to increase scores. The improvement also was motivated by the waste and cost PFIs to be achieved.

- 50 per cent to be improved on construction waste
- 50 per cent to be improved on whole life cost by reducing maintenance.

overall product at risk in its interface with others. If the synergy is lost then the work slows down and demotivation sets in affecting the standard of workmanship and design problem solving. For the operation of TQM (see below) the confidence of workers to get involved in quality depends on a good team spirit. Quality circles require workers from different disciplines to come together and rely on a team spirit and the confidence to try new things which will not happen if goodwill and co-operation have been withdrawn. Case study 13.5 is a practical example.

Reputation and costs

It has been well established in the industrial sector the importance of quality as an essential factor of competitiveness, progress and reputation, and construction organisations are now recognising this. It is also an aid to the development of the firm. Leading clients also recognise the impact of quality faculties upon their business returns. The rationale for a quality system is that it saves money for all parties because:

- search for quality makes it possible to reduce production costs by less abortive work due to defects
- quality is part of the brand image of the firm
- client has more confidence in suppliers and so less interruption
- less disruption due to latent defects.

Applying this cultural approach to quality improvement is the basis of total quality management, which seeks to involve the whole workforce thinking about quality inputs. It advocates not only 'right first time', but continuously improving the process by encouraging feedback and review.

The cost efficiency of quality depends on balancing the cost of appraisal systems and work improvement with the cost of failure.

$$\text{Quality cost} = \text{quality assessment} - \text{quality improvements} - \text{failure costs}$$

TQM and customer focus

The basic principle of TQM is that quality responsibility is delegated to the whole workforce and senior management is committed to resource incentives for worker ideas to be tried out. It is a big change in culture and some managers may feel threatened as workers gain ownership and question instructions. Amongst other things it is spreading the marketing function to the whole workforce and exposing them to client scrutiny to justify their area of quality.

TQM also encourages more self-regulation which leads to less supervision and control and good quality guarantees are available at low cost. This gives worker empowerment and accountability.

It is acknowledged that TQM is an evolutionary process that works in conjunction with the other 'tools' and should be adapted to suit the smaller organisation. It provides a holistic response to the Latham, Atkins and Egan reports which also have a lot to say about integrating structures to remain internationally competitive. It tries to mimic successful manufacturing systems. Case study 13.6 is an example of integrating a product with a service.

Case study 13.6 Bekaert fencing

One customer-focused approach is to provide cast-iron guarantees on the performance specification of the product. For example, Bekaert fencing offer the client a guaranteed level of security for 15 years to the real estate that is surrounded by the fence.[23] Here they are selling a service (security) for a given sum and not a product (fencing). Any worries the client has about the durability and continuing efficacy of a fence is taken away from them and the effort has been invested in building a relationship with Bekaert who install and maintain the fence in a suitable state with compensation for any breaches of security. This is assured through their contractor certification scheme and the insurance called Beksure.

Total quality management (TQM) is difficult to define, but its objective according to Hellard is to set out a framework for action to fully satisfy customer requirements.[24] From a survey of the literature, including Deming, Crosby, Juran and Ishikawa, he also derives common principles of:

*   management leadership to develop a quality culture
*   continuous improvement of processes and product (not just better motivation)
*   wide ranging education programme for management and workforce
*   defect prevention rather than inspection
*   data use and statistics tools for benchmarking
*   developing a team approach between departments and between all authority levels.

The TQM approach, by definition, is customer focused and must identify customer needs and expectations first and set standards consistent with customer requirements. This will provide the basis for establishing quality standards and enabling and empowering middle management and the workforce to achieve it. For it to work in construction, lessons learnt need to be taken to the next project and if possible the same core team used.

In construction projects TQM is not seen as an easy approach and it is not surprising that it is resisted. Some are not convinced that they can make the most money this way as profits are often supplemented by contractual claims. It is also clear that there are several complications that make it harder to adapt principles that were first devised in the context of manufacturing to the one off nature of projects. For example, building up a culture of worker empowerment in the short time a project team is together and blending the system to cut across separate specialist organisations. The entrenched 'them and us' attitudes fostered by the segregation of the client and the supply chain are unhelpful to this. Case study 13.7 gives an example.

Case study 13.7 The developing TQM culture

Turvey, in a study of a major private contractor, looked at their programme to make the company more customer focused and identified various 'key inhibitors' to the adoption of such a policy due to the nature of the industry itself.[25] These included the project based nature of the building industry, which means that the focus tends to be on the building and not on the customer. The sheer number of different organisations involved who may have never worked together before, all trying to make a profit in whatever way possible, made selling the idea difficult. There was also a tendency to revert to old confrontational attitudes when a significant problem occurred. The company however remained optimistic and identified enablers that were remarkably similar to the principles that have been outlined for the development of a TQM policy. These were establishing customer requirements at the outset and operating an open book policy with the client, including a measure of collocation of project staff and developing partnerships with suppliers and training programmes to change the culture.

Hellard mentions several more issues of difficulty in the construction project culture and puts his biggest emphasis on working together in partnership with the client and all contractors.[26] In this respect the client becomes part of the 'senior management' commitment that enables a TQM approach to proceed on their project. This will initially involve a financial commitment, which should pay dividends later by adding value to the project. This sort of collaboration has developed where contractor teams have temporarily been merged into the client organisation.

Hellard and Turvey mention the importance of meeting the needs of your internal customer in achieving TQM. The internal customer is the next person in the workflow chain. For example, this makes the estimator the customer of the quantity surveyor and design team as they put the tender documents together. It also makes the painter the customer of the plasterer and the M&E contractor the customer of the services consultants and main contractor. Customer focus turns the traditional attitude of selling contracting or consulting services on its head and represents a major change in culture. Obligations through the work flow process such as reliable, timely information in an understandable format should serve to cut out waste and help make the internal client's input more effective. The efficiency of the workflow will definitely benefit the project client also.

The sponsoring client represents the external project customer, but in a holistic project management approach the end user of the facility is important. The end user is the client of the project manager responsible for the operation of the various services and building systems. This is close to Handy's second principle of customer care.

Implementing a construction project quality plan

Performance concerns the assurance that quality standards have been met and equipment works. The purpose of a plan is to have a strategy for the implementation of the procedures chosen, which can work and which identifies the right quality tools to control production and monitor them for improvement. The CIOB Code of Practice indicates a hierarchy of quality policy based on the brief and setting standards, quality strategy and quality plan.[27] It is important to allocate responsibilities to specific members of the management team and to appoint leadership in each of the packages and overall during design and construction. They also suggest the periodic review of quality actions for effectiveness. Benchmarking the service and product is possible in order to ensure that a standard has been reached and can be improved.

Other issues in the plan are the administration of guarantees for product and building performance and the timely use and preventative maintenance of the building and equipment which may mean regular servicing, training and replacements. Breakdown maintenance should also be efficient to maintain building functionality, efficiency and safety. A client may choose to pay for this upfront.

Customer satisfaction is centred round their confidence in the system and the effective operation of the system and they will interfere less where there is evidence of effective controls. Zero defects should apply to the performance of the building and equipment and some expectation of minor problems can be tolerated where business productivity is not at stake as suggested in the Constructing Excellence benchmarks. The client experiences more frustration when key components and systems fail to function or business is disrupted than when it is delivered late.

Relationship marketing

Construction services are traditionally bought through hiring consultant time or by arranging a competitive tender, but how are they sold? Getting onto a tender list or in a position to negotiate is a full time job for an estimating department, but it is critically about building up relationships between the buyer and the seller. The price is dependent on competitive market forces or, as suggested previously, may be adjusted strategically to enhance value and whole life costs. In both cases firms are compelled to build up relationships in order to win work. For the client and project manager it is important to recognise the need to work with the supplier. Previous chapters have discussed procurement alternatives for building in best value into the price by, for example, having a two stage tender, where specifications for performance and value are negotiated within the initial price framework. The business relationship is a means to choose the best supplier.

Smyth has differentiated between 'selling to people' and 'selling through relationships'.[28] The latter implies getting to know the client beforehand in order to optimise what you can offer. A client needs good reasons for considering your product and this may be achieved by a number of issues including reputation, value, quality, ability to deliver and attention and differentiation to show understanding of the client's needs. However he claims the industry suffers negatively from low 'switching costs' which mean low cost open tenders which preclude building up relationships where continuous improvements, including quality, can take place at a competitive price. It may also mean substandard buildings. Smyth's four foundations for relationship selling are vision, expectations, authority and trust. This allows a two dimensional approach of getting more for the money and/or enhancing quality for the same amount. Success bodes well for ongoing contractual relationships and in developing strategic partnership.

In construction the vision and expectations are often generated in isolation between the key designer and the client and the contractor comes along with a higher price or a reduced specification. Where there is no mechanism for negotiation, trust is lost by the client. This points to the need for an earlier involvement by the contractor especially where there is less certainty and more choice of options. Smyth emphasises the parallel activity of clinching the sale through the stepwise process of encourage, exhort, enlighten, engage and empower to bring the client on board. These five steps interact with the foundation principles in relationship selling with an ultimate aim by both parties to do repeat business so that the trust relationship might be further

Case study 13.8 Client accounts

It is not unusual for contractors to develop special relationships with clients. A medium sized contractor originally offering a service in the southwest divided their services into three business units which represented retail, leisure and general construction and within this client accounts were created. Organisationally although the company kept a head office in the south, they were now servicing client buildings nationally as the focus of each unit was to build up their relationship with particular clients in the niche market and follow their orders across geographical boundaries. Staff in each unit were trained to get to know the market and the particular client and an account manager rather than a contract manager was responsible for allocating teams and resources who were trained to understand their needs. Clients ask for teams to stay intact and move from contract to contract, which impacted on the expectations for staff to be mobile and changed the staff profile. This impacted on the management of the supply chain and the equal willingness for SCM staff to be mobile to maintain the team. Clients encouraged innovation and set new lower time and cost targets whilst maintaining quality and value.

The contractor client base for the specialist markets also shrunk substantially to match the capacity of the contractor to develop ongoing relationships and their potential to secure their best profit margins in a business they knew intimately. The synergy from better professional teamwork was also released.

developed. Case study 13.8 illustrates how a contractor has developed its business to focus on the client for better value.

Smyth points out the relationship between more emphasis on relationship marketing as described above and the more integrated approach to procurement as indicated in Figure 13.5. Interestingly there is a greater transactional cost, but more compliancy with relationship marketing, which means that savings must be offered in other places and longer term.

One argument against this is the risk to the supplier of a client running out of work after substantial investment in working with the client. Influential larger clients may also squeeze smaller contracting partners so that they operate on reduced margins and at the same time have created dependency by specialising the product and demanding more exclusive supply. Relational marketing should therefore remain in the control of the supplier.

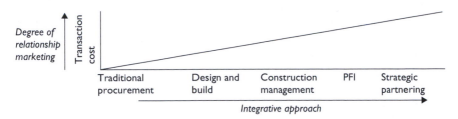

Figure 13.5 Integration of procurement vs transaction
Adapted from Smyth (2000)[29]

The contractor appointment is based on long term partnering or the procurement route chosen and should be made at the earliest stage possible. The CIOB Code of Practice suggest that early involvement of the contractor may usefully contribute to:

- resolution of buildability issues at design stage
- choice of the most efficient materials to be used
- pricing issues to support value management
- giving an opportunity to involve specialist contractors in fundamental design issues
- giving the contractor a better chance to work with the client and understand the client's requirements
- inputs on health and safety issues that influence the design.[30]

These things are relatively limited in traditional procurement and yet if incorporated by more integrative procurement methods can produce valuable inputs that are likely to boost quality and save money.

Conclusion

Customer satisfaction is an important part of the process of project and quality management. Three construction KPIs are relevant to quality and these are strongly tied into client expectations of service product and defects. The service quality is what makes you stand out in the crowd. Quality is also a function of teamwork and communication and this is critical whatever quality management system is chosen. To have a quality control system on its own is wasteful unless it also includes a measure of defect prevention to cut down wasteful rework and to spread the responsibility amongst the workforce.

Starting from a low base in this model is not an option as clients become more confident in their requirements. They are already using industry KPIs in many countries as a way of benchmarking tender qualification to industry average for qualification. They further use KPIs as a way of comparing performance, especially quality as an influential part of a balanced score card in selecting winning tenders. This means they will be looking for excellent service at a reasonable price. The culture of quality improvement will also inflate expectations as time goes by, so sitting still is not an option either even where suppliers start best in class.

Some clients operate their own client charter system,[31] which as well as indicating their expectations of suppliers requires them to keep up to reciprocal standards of 'clientship' which enables contractors to assess clients when choosing who can provide sustainable profitability for them. Contractors in good economic times have taken the line to restrict tendering to 'good clients' and build up relations with them for repeat work to cut down their tendering costs.

Future pathways could include the longer term commitment for a contractor to a service rather than to hand over a facility so that clients may be guaranteed a certain level of performance over a period of time and not just buy a product. This will come at the price of a proper guarantee scheme, but not necessarily by year on year payments as in the case of Bekaert. This already happens with PFI procurement which provides facilities management over a concession period guaranteeing quality components to reduce life cycle and maintenance costs. Private clients can proceed with other similar joint offerings that allow the client to have built in quality and extended economic life for their buildings and infrastructure.

The use of client accounts is more associated with partnering, but reducing the client base with the intention to agree a programme of repeat work which can incorporate the continuous improvement of quality over a period with the same teams involved. Relationship marketing has played a role here in introducing the concept of engaging the client's vision and expectation, building trust and adopting its culture in best meeting their needs in a trusting relationship. It has an element of selection for the contractor/client in the type of and breadth of clients the contractor works with.

TQM is a step further in generating a culture of improvement that pervades throughout the project at all levels of the supply chain to ensure that everyone is on board to work for quality, making the need for a third party quality accreditation system less critical. This is a model which is not yet mature in the industry, but has claimed success for other industries such as retail, car and steel. It also requires a much more integrated project team for its ideal to be released tied up with partnering and SCM. It is related to the culture of lean construction so that waste is reduced, but new ideas are constantly coming from the frontline worker for improved quality in a productive context.

References

1 BSI (1991) *Quality Vocabulary Part 2 Quality Concepts and Related Definitions*. London, BSI.
2 Gardiner P. (2005) *Project Management: A Strategic Planning Approach*. Basingstoke, Palgrave Macmillan.
3 Knapman D. (2008) *Updating Data on Construction, Demolition and Excavation Waste Report*. CON900-001. Banbury, WRAP.
4 Egan J. (1998) *Rethinking Construction*. UK, Department of Environment, Transport and the Regions.
5 Egan J. (1998) *Rethinking Construction*. UK, Department of Environment, Transport and the Regions.
6 Latham M. (1994) *Constructing the Team. Final Report of the Government Industry Review of Procurement and Contractual Arrangements in the UK Construction Industry*. London, HMSO.
7 The Chartered Institute of Suppliers. Final report. *Productivity and Costs* (as reported in appendix vi of the Latham Review 1994, p.128).
8 Constructing Excellence (1992) KPIzone, All construction KPI.
9 Atkins W.S. plc (1994) *Secteur. Strategic Study on the Construction Sector, Final Report*. European Union.
10 CCF (1998) *Constructing Improvement. The Client's Proposals for a Pact with the Industry*. London, CCF.
11 Wolstenholme A. (2009) *Never Waste a Good Crisis*. London, Constructing Excellence.
12 Brown (1989) Quoted in ICSA Study Text (1994). *Professional Stage 1 Management Practice*. BPP Publishing.
13 Handy C. (1991) *Inside Organisations: 21 Ideas for Managers*. London, Penguin.
14 Strategic Forum for Construction (2003). Available online at: http://www.strategicforum.org.uk/sfctoolkit2/help/maturity_model.html (accessed 31 August 2004).
15 Strategic Forum for Construction (2003). Available online at: http://www.strategicforum.org.uk/sfctoolkit2/home/home.html (accessed 21 December 2011).
16 Construction Clients Forum (1998) *Constructing Improvement. January*. London, CCF.
17 Client's Charter Improvement Programme. Available online at: http://www.clientsuccess.org/Improvement.htm (accessed 29 December 2011).
18 Turner R. and Huemann M. (2007) 'Managing quality'. In R. Turner (Ed.) *The Handbook of Project Management*. 4th edn. Aldershot, Gower.
19 Atkins W.S. plc (1994) *Secteur. Strategic Study on the Construction Sector, Final Report*. European Union.
20 Berry L., Parasuraman A. and Zeithami V. (1994) 'Improving service quality in America: lessons learned', *The Academy of Management Executive*, 8(2): 32–52 in P. Gardiner (2005) *Project Management: A Strategic Planning Approach*. Basingstoke, Palgrave Macmillan.
21 International Standards Organisation 9001 (2006) *Quality Management Requirements*. (9001 covers design, manufacture and installation, 9002 just manufacture and installation, and 9003 covers inspection and testing and 9004 gives guidelines for improvements.)
22 Hellard R.B. (1993) *Total Quality in Construction Projects: Achieving Profitability with Customer Satisfaction*. London, Thomas Telford.

23 Constructing Excellence (2004) *Best Practice Case Study*. Available online at: http://www.constructingexcellence.org.uk/resourcecentre/publications/document. jsp?documentID=116220 (accessed 1 December 2004).

24 Hellard R.B. (1993) *Total Quality in Construction Projects: Achieving Profitability with Customer Satisfaction*. London, Thomas Telford.

25 Turvey J. (1999) *Customer Focus in the Construction Industry*. Unpublished thesis report. University of the West of England.

26 Hellard R.B. (1993) *Total Quality in Construction Projects: Achieving Profitability with Customer Satisfaction*. London, Telford.

27 CIOB (2010) *Code of Practice for Project Management: Construction and Development*. Oxford, Blackwells, p.56.

28 Smyth H. (2000) *Marketing and Selling Construction Services*. Oxford, Blackwells.

29 Smyth H. (2000) *Marketing and Selling Construction Services*. Oxford, Blackwells.

30 CIOB (2010) *Code of Practice for Project Management: Construction and Development*. Oxford, Blackwells, p.132.

31 Construction Clients Confederation (2002). Available online at: http://www. clientsuccess.org.uk (accessed 29 December 2011).

Chapter 14

Project close and systems improvement

P. Fewings and M. Jones

The final stages of the project life cycle are to commission, handover and review the project. It is common for the importance of these stages to be overlooked by project managers. The APM Body of Knowledge identifies a handover stage which involves the completion of the project to the satisfaction of the sponsor. It also involves introducing the user to the operation of the facility and compiling documentation that provides information on the safe, efficient and effective use of the building.

Post project evaluation includes a benefits assessment which compares performance with the original client and business objectives. Was the project a success in the spirit of continuous improvement, improvements should have been sought throughout the project. However, the end stage does provide a particularly appropriate point at which to review the outcomes of the project and to reflect on what improvements might be made in subsequent projects.

Post occupancy review considers the impact of the design on the user experience and the success of the wider project objectives for the users. It is the ultimate place to ask whether the project was a success.

The objectives of this chapter are to:

- consider the requirements and management of commissioning
- look at the requirements for successful handover and close out
- consider the completion stages and the process of post project appraisals and benefit evaluation
- discuss system improvement measures that are relevant to construction.

Completion stages

The final stage of the project includes commissioning and handover to the client. Commissioning ensures that the building functions according to specification and that should be tested so that all the different services and building fabric interfaces work together. The sequence of activities is shown in Figure 14.1.

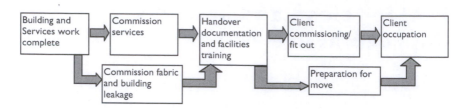

Figure 14.1 The chain of commissioning and completion

Commissioning

Commissioning is a critical operation that takes place at the end of a project but which needs to be planned at the beginning of the project so that adequate time for full, unobstructed access to the whole building, or distinct service zones, is provided in order to make sure that all elements of the building are working satisfactorily. Figure 14.1 indicates the commissioning process. A strategy for commissioning to suit requirements is incorporated in the master plan and the construction plan from the beginning of the project and may be phased if there are many discrete units in the structure. As well as playing an important part in the latter part of the construction phase, commissioning also plays an important role beyond construction into fitting out and occupation, as services and fabric will only be deemed to be operating as intended when the building is loaded and occupied. This has become more important with the increasing emphasis on reducing energy consumption in buildings and CO_2 emissions, and client demands for 'soft landings'.

Final commissioning controls the methodology of the main programme as it is the last activity before handover. Adequate time should be left for adjusting the fabric and/or services to deal with any unexpected deviations from performance. This can include testing for leakages in the building fabric and ensuring BREEAM predictions are met as well as the more traditional adjustment or balancing of heating and air conditioning controls. The fabric of the second fix and finishings needs to operate well so that doors and windows operate properly, units are not leaking, and water, gas and air arrive at the right pressure across seasonal variations. Elements may also be damaged by following trades undertaking subsequent work, such as driving nails through pipes under the floor, punctures in roof membranes or dislodged underground pipes or services.

Despite the greater demands being placed on clients and the construction supply chain to deliver buildings that are right first time and limit energy consumption and CO_2 emissions, many buildings are still being put into service without full commissioning or sufficient fine-tuning to ensure basic performance requirements are being met. As the relationships between most clients and their design and construction teams are short-term and infrequent the latter are rarely called upon to assess how their buildings actually perform in use, whether or not they meet the needs of users and management, and how they might be improved. Some new buildings still have problems with providing the necessary air-tightness, insulation and shading, while many control systems do not work properly, waste energy and have poor management and user interfaces. Environmental assessment systems are mostly concerned with design features and management processes, and do not identify these problems in the functionality of buildings. This situation is further compounded by the increasing use of new and complex technologies designed to help make new buildings use less energy and reduce CO_2 emissions.

This increasing complexity in the functioning of buildings is forcing facilities managers into maintenance and management regimes that they are neither expecting nor trained for, and which the building owners may not be able to afford to run, and which consequently may never work as intended. Such complexities include increasing use of on-site generation, particularly through renewable sources of energy such as biomass boilers and solar power, which can be difficult and expensive to run.

Clients, government and society generally now expect design and building teams to be able to predict more precisely how their buildings will perform in use during the period of their life expectancy. However, in most procurement approaches design and construction teams usually disband at practical completion and do not follow through into occupation of the building to learn how their building performs in practice. This has been changing to some extent with PFI projects and other procurement routes where the design and construction team are tied into the project beyond the traditional point of handover. 'Soft landings' is an approach in which teams can be involved in a more thorough and longer commissioning process that allows follow-through in the project and aftercare. It also allows greater scope for the practitioners involved to pass on their knowledge, undertake post-occupancy evaluation, capture and disseminate information on the effectiveness of the design and construction, and learn lessons – for themselves, their firms, subsequent projects and the industry as a whole.

'Soft landings' provide a process where the professional team can remain engaged in the project beyond practical completion to help guide the building through the first crucial months or indeed a number of seasons of building operation. According to BSRIA,[1] they assist the client and their facilities managers during the first months of operation and beyond, to help fine-tune and de-bug the systems, and ensure the occupiers understand how to control and derive maximum value from the use their buildings. They argue that 'soft landings':

- provide a way of working, and is a new professionalism that responds to the challenge of delivering better buildings

- foster greater mutual understanding between clients, project managers, designers, builders and occupiers about project objectives
- reduce tensions and frustrations that occur during initial occupancy, and to ensure clients and occupiers get the best out of their new asset
- allow greater investment in problem diagnosis and treatment, and in monitoring, review and post-occupancy evaluation
- provide a framework of activities for the entire project team
- bring about greater clarity at inception and briefing about client needs and operational outcomes
- require the early setting of performance targets (such as energy use) and a method of checking whether they have been achieved
- place greater emphasis on building readiness
- require a dedicated soft landings team to be on site during the initial settling-in period
- require the project team to be involved for up to three years to fine-tune the building and monitor its performance.

Case study 14.1 describes the research which led to soft landings.

In civil and infrastructure projects there may be important engineering checks which are dependent on structural strength, water tightness and loading criteria such as maintaining water levels and putting pipe work under

Case study 14.1 Centre for Mathematical Sciences

'Soft landings' was developed by an architectural consultancy and subsequently picked up and developed by the Director of Estates at a leading UK university. The phased development of a new mathematics centre and a 'no blame' attitude adopted by the client permitted a continual assessment of the emerging design in actual physical performance and user expectation. Following completion of the first phase of construction a post-occupancy evaluation was carried out to measure the building performance of the recently occupied buildings. As part of this study an occupant survey and a full building pressure test was also conducted. Many of the results were incorporated into design changes for the subsequent building phases. The results revealed the importance of adequate user feedback from automatic environmental systems. For example, should a user wish to override a window, the controls should be capable of acknowledging the user's input. Without such feedback the controls can be operated repeatedly, leading to frustration. The final appraisal revealed that the occupants and the university viewed the project as a considerable success.

full pressure testing during pumping. Dams and harbours need strength and leakage testing. Roads and bridges require strength, wind and noise tests and tunnels checked for water tightness and performance in fires. Nuclear power stations are safety tested over a period of time with progressive start up and electrical and gas installations are tested for third party certification to IEE and gas safety regulations.

Much of the major structural commissioning is done at interim stages in construction before services and foundations are covered up. Piling is load tested on completion and failing piles are moved or reinforced. Drainage and mechanical pipe work is pressure tested in sections so that subsequent work is not disturbed. Air conditioning, toilet pods and lift units are tested before they are built in to eliminate structural problems. Building regulations work on progressive inspections. Figure 14.2 indicates a system for putting responsibility on each subcontractor for correct performance on completion of their work, prior to final inspection.

The major stages of building commissioning are connected with approvals by the client and statutory authorities often delegated to consultants on non-domestic buildings. The fabric needs several 'OKs' connected with the building regulations such as a fire certificate, energy efficiency compliance, health, structural stability. It also needs compliance for the connection of services and for insurance purposes. Health and safety issues apart from fire will include contamination of the site, asbestos clearance and sensible and safe access for maintenance and repairs during operation. Under the CDM Regulations this comes as a designer responsibility. It is also co-ordinated in a document called the Health and Safety File (HSF). Building envelope leakage tests as required in some building regulations need to be carried out prior to handover. Devices which supply and extract air need to pass health and safety tests such as legionella bacteria. All may generate some significant remedial works that need to be carried out before services can properly be commissioned.

The *client commissioning* is distinguished from engineering commissioning as the stage after practical completion and handover when the facilities are prepared for occupation. The CIOB Code of Practice distinguishes three parts to the task of client – accommodation works, operational commissioning and migration of the workforce.[2] In considering a new office building, this could be the setting out of furniture and second fix ICT installation, the moving in and familiarisation of office staff, their files and personal equipment and the adjustment of heating and cooling and services to meet the live load of the building, which includes computer heat outputs. The client is likely to consider this as a separate project with an in-house project manager to liaise with the main project manager.

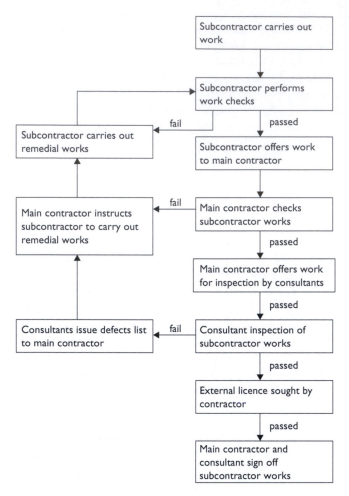

Figure 14.2 Construction inspection and commissioning

Adapted from CIOB (2010)[3]

Handover

The *handover* is the direct interface with the client, user and facilities management team to ensure that full information about operation and use, including health and safety procedures, is passed on. Documentation includes as-built drawings, a range of component specifications, manuals for use and the health and safety file. It also needs to record warranties, contact numbers, spares availability and product codes. It should include training in the use of equipment, familiarisation with maintenance cycles and fire practices for users. Sometimes there is minimum fitting out managed by the project team such as shell and core buildings handed over to developers, but the principle still applies. Liaison with the client for broader fitting out is important to ensure demountable partitions and ceilings will fit and that services connections, fire zone compartmentation and fire escape routes are understood for floor by floor fitting out. Good practice is likely to involve direct contacts between the user and the managing contractor as would be the case in the selling of houses. An example of a combined commissioning and handover is shown for housing in Case study 14.2.

Contractors may also wish to get feedback from their customers, who will have a unique perspective as to how effective the processes are and whether their expectations have been met. A feedback scale needs to identify specific areas of improvement, and perceptions and expectations need to be unravelled. Research indicates that customers can tolerate reasonable waiting periods for putting things right, but are less tolerant of promises that are not kept, such as delayed moving times or noisy or messy sites.

The CIOB advocates a closely managed handover and suggests that many larger projects have phased sectional completion to give clients access. These

Case study 14.2 Stage visits on housing project

One housing developer has three progress visits for the buyer connected to stage completions. The arrangement clearly states cut off points for making certain decisions, such as the choice of kitchen or wall colours, but goes further in letting the buyer see the quality of the work. This is a problem where there is uncertain or late progress on the site, but it expresses confidence by the developer that there is nothing to hide and seeks to build up relationships. Buyers may point out defects. Practical completion is marked by the production of an insurer's final certificate to confirm conformity and in return the customer agrees to a short financial and legal close period following this. Drawn out post occupational defects of any substance that is put right by the developer may result in nuisance compensation under the industry insurance scheme. In turn the developer keeps their insurance contributions down by eliminating significant defects.

A pro forma indicating the checks carried out is handed to the customer making the quality process more transparent. The argument against this practice is that many customers do not understand building practices. However indirect marketing is gained by recommendations made to others. A clean site is often associated with an efficient and healthy site which is also of benefit in building up a reputation, where clients are on site or near the site before or after occupation.

parts may receive a practical completion certificate and then clients insure the building phase for fitting out. Outstanding defects, training and contractor access for post handover works and the procedures for managing payment of outstanding monies and reviewing progress in fixing defects during the defects liability period are important parts of post occupation.

Documentation needs to be comprehensive, but also needs to be easy to access. It often takes up considerable amounts of time to gather the documents, especially if it is left to the end of the project, when items are less accessible. Key documents are:

- As-built drawings for the fabric and structure indicating exact services routes and access points.
- Manuals for the safe and effective use of the equipment and the upkeep of the fabric.
- Test certificates and warranties that give underlying confidence in quality and functioning efficiency.
- Information that gives knowledge for future projects such as the as-services access points, contacts and spares availability. Spares may also be handed over.
- Information contained in the HSF for safe operation and about hazardous materials and their safe use and dismantling.

Project systems need to be closed down so that personnel may be reassigned and responsibilities redistributed or phased out. A continuing interest in settling the final account becomes the main concern of the quantity surveyor. Senior project managers may wish to reassign finishing off activities and commissioning into the hands of their deputies. Final project reviews need to be carried out before key personnel disappear so that lessons may be learnt and passed on.

Occupation

Client occupation is a project in itself and needs a steering committee, inclusive of key user and department personnel, to deal with the move and to liaise with all the relevant departments who begin to take a specialist interest in their new accommodation and facilities. Migration planning involves:

- determining how the building will be occupied (space planning)
- establishing the timing and phasing of the move
- identifying key activities and assigning managers
- determining sequencing of each group to least disrupt business
- keeping the client's staff informed
- identifying potential risks to the plan.[4]

Any organisation will need to appoint a steering group inclusive of key user and department personnel. The PM will provide organisational skills for the move and co-ordinate any contractor activity. Feedback from users needs to be controlled through recognised channels. Typically further construction work needs to be carried out such as provision of services and other structural builder's work.

Project reviews

Traditionally projects have been reviewed according to their ability to deliver on time, budget and quality targets. These are generally termed the project objectives and refer to a closed system unaffected by external factors. Clearly construction projects are open systems and consider the impact of external factors which are beyond the control of the project team and also the impact of the project on the environment and community.

Benefits evaluation

Benefits evaluation is the culmination of the client's benefits management process. In a client's business context the main purpose is to objectively evaluate the successes/failures of the project, to ensure that business objectives have been achieved and benefits are optimised in operation and improvements are carried forward. The client may well be helped by the project team in this review, but the review is initiated by the client. From a project context there may not be direct parallels between this project and the next, or the next project team may work differently. It is therefore important to distinguish more generic lessons and to make the report accessible and helpful to a new team. Buttrick's key areas for evaluation shown in Figure 14.3 refer to issues that may be raised in both types of review.[5]

In addition outstanding accountabilities are handed over, so that the project can be carried on in an appropriate manner, as well as providing accurate and objective information for future client projects.

The NI Department for Finance and Personnel (2011)[6] gives these objectives:

- Evaluate the effectiveness of the project in realising the proposed benefits as outlined in the business case.
- Compare planned costs and benefits with actual costs and benefits to allow an assessment of the project's overall value for money to be made.
- Capture and document any lessons learned – identify particular aspects of the project which have affected benefits either positively or negatively and make recommendations for future projects.

Business objectives	Were they met? Did they change?
Benefits	• A restatement of the tangible benefits, how they will be measured in the future and by whom.
Outstanding issues	• These are listed, together within the necessary action, accountability and deliverables.
Project efficiency	• Actual costs and resources used are compared with the plan. • Indicators such as cost/m² or cost/bed may be compared with a benchmark.
Lessons learned	• Identify areas for improvement as well as noting improvements to plan.
Close out	• Note contributions, pass on responsibilities and celebrate success.

Figure 14.3 Buttrick's key areas for evaluation

- Reveal opportunities for increasing the project's yield of benefits and make recommendations on actions required for these to be achieved.

This checklist shows potential as part of a benefits management system so that the business case is fully met. Case study 14.3 shows how detail planning needs to be synchronised with different user groups in order to not derail business benefits.

Project evaluation

Post project appraisal is more related to the review of project objectives, but also for continuous evaluation to the control of the project. CIOB endorses post completion review as a basis for the assessment of fitness of purpose and satisfaction of requirements, but they also suggest that certain lessons can be carried forward by measuring performance in assessing the strengths and weaknesses of the project and to encourage appropriate action in the future. This is one of the main arguments for long term client project team partnerships. In auditing the project they also suggest summarising amendments and their reasons, commenting on appropriateness of contract conditions, organisation structure and project milestone achievements and identifying any unusual developments and difficulties. These will be used in bringing forward improvements.

Case study 14.3 Heathrow Terminal 5[7]

Heathrow Terminal 5 was a very successful project from the project team point of view, finishing on time and within its £4.3 billion budget and with architectural merit. However when it was handed over to the client the automated baggage handling systems were ineffective and were unable to cope with the level of use. The system required a different approach for the check-in and baggage handling crew who needed to use an automated system to find out where they were allocated. Check-in and handling staff were unable to find car parking and had problems with getting into the building on the first day. This was unfortunate as queues got steadily worse at Heathrow T5 with a high level of lost luggage which caused a serious outcry from customers, who had their departure flight plans disrupted and were unable to access their luggage to enjoy their holidays. Arrivals were delayed in collecting their bags and clearing the terminal.

The business case was sound, but the objective to get the airport up and running on the first day with insufficient practice, untested security checking and insufficient information to staff about the new building was disastrous and significant lessons were learnt for future complex user changes. The PM's relationship with the BAA project staff was excellent, but communication had broken down with other BAA managers and the need to train users.

The main improvements which can be passed on from project to project will relate to process and management systems in time, cost, quality and health and safety control. Generic learning points may be used by the project manager to apply to any future project and specific ones to projects of the same nature, client and team. Both are called systems improvement.

The idea of continuous project evaluation has been developed by Timms.[8] In this model there is a regular meeting to review the main benchmarks for the project in a spirit of continuous improvement. In Constructing Excellence, benchmarking measurements are made on a continuing basis to provide control and to keep on target. These benchmarks need to be chosen for their relevance to the particular project so that what is compared is considered to be a critical indicator of success. They also need to measure parameters which are measurable, can be fed back quickly and clearly point to corrective action. Timms advocates that the whole team should be involved in the evaluation meetings and that this should lead to integrated construction and design action on a regular basis to benefit that project.

Post occupancy reviews

These involve the users of the building or facility. They typically rate satisfaction in terms of the functional working of the building and their experience of the comfort, convenience and impact on their productivity. At a simple level a new tenant may be asked to rate their satisfaction and to point out any issues for praise or improvement. In more complex buildings there will be a range of stakeholders from the facilities team to each of the departments. It is possible that there will be contradictory feedback and these issues may need further investigation. There is a lot that can be done in the spirit of continuous improvement.

Systems improvement

Systems improvement bases itself on a project maturity model which measures the level of consistency, repeatability and optimisation which is operating in the project management organisation. Systems improvement itself is an incremental improvement programme possible because of continuous review of processes to improve performance, facility and reliability.

Capability maturity models

The capability maturity model (CMM) is more competently defined as 'a structured collection of elements that describe characteristics of effective processes'.[9] It can be applied to a consultant or a client organisation for the purpose of improving regular systems. The CMM provides a scale of improvement, but also needs an objective assessment of the level at which an organisation, consultancy or team is operating its management systems if systems are to be compared. Figure 14.4 indicates that used by the SEI Capability Maturity Model which was originally developed for government control of computer projects. It describes five levels from initial to optimised. It is adapted here for construction projects.

Construction projects need complex management of time, cost and quality and incorporate major changes of scope and sometimes design. This means that at level 1 where little experience is implicated projects can be adequately managed by those who know the construction industry, but have developed ad hoc tools for managing small projects. More risk develops if this level of maturity is applied to more complex jobs.

At the next level up construction projects require some type experience and this means some expertise and more delegation of tasks and co-ordination with design. There may be a partnership or framework agreement and repeat work drives up project type experience and protocols or even process maps are developed to record efficient methods. Training in techniques such as quality management will be necessary.

Maturity level	Characteristics of effective process	Examples in construction
Level 5 Optimised	Continuous improvement Continuous collection of data Analysis of defects for prevention	As 4, with an intention to tweak the system for maximum productivity and quality in context of building type interpreting appropriate improvements through wide experience and innovation
Level 4 Managed	Quantitatively measured e.g. KPI Metrics for quality and productivity Collection of process experiences	As 3, but KPIs are used that can compare performance between projects managed by the organisation. Some knowledge management through centralised project reviews
Level 3 Defined	Defined processes for control Process groups defined	The principles have been widened and been defined for many clients as robust for different situations and adjustable within parameters
Level 2 Repeatable	Depends on individuals Minimum of process guidance Highly risky in case of new challenges	This is as 1, but through repeat experience systems have been agreed between client and PM which work well and will be used again with same team e.g. partnering/frameworks
Level 1 Initial	Ad hoc, not formalised No adequate guidance No consistency in product guidance	This equates to a one off project model where individuals interpret construction project management principles

Figure 14.4 SEI Capability Maturity Model adapted for construction from SEI 2012 Capability Maturity Model Integration

At level 3 some definition of procedures which will remain robust on a range of projects is necessary. Systems such as the OGC gateway review or the PRINCE2 methodology may be used in public projects or a procedure may be tailor-made. At level 4 there is a move to measure performance more consistently and a recognition that systems by themselves need to tailored for clients so that improvements in productivity can be made and more innovative approaches tried. This may occur where the client requires fast track construction to complete early, manufacture offsite or deal with exceptional quality levels. Health and safety initiatives to reduce accidents substantially or to deal with a very flexible change system may be necessary.

At level 5 any project can be tackled and will be at a demonstrator level and innovative approaches to technology such as needed in achieving BREEAM excellent or zero carbon will require the highest use of teamwork between the design and construction teams. Teamwork will operate synergistically with early contractor involvement and integrated teamwork with the client to enhance value and to manage risk proactively so that budget and timescales are bought down.

The scale as the level rises is to reduce risk and to increase productivity and quality even though there might be complexity, new technology and tight schedules.

Continuous improvement

Continuous improvement, or '*kaizen*', consists of measuring key quality and other process indicators in all areas, and taking actions, normally small and incremental, to improve them. It focuses on *processes* and should be pursued in all areas. Maylor distinguishes improvement between '*learning by doing*' as those elements which can be learned from previous experience and '*learning before doing*', using results from external inputs when we break new ground or experiment or develop other people's ideas.[10] The latter may have an element of training, research or development.

Continuous improvement is a fundamental element of total quality management (TQM). It is seen as a very challenging approach because it is based on developing a learning culture with the aim of improving continuously through an endless search for excellence. It is a bottom-up approach in that workers are the recognised experts because they have the detailed knowledge of how the work is done, thus they are the best equipped to improve the process. This view contrasts with the Tayloristic approach in which the experts were the engineers, managers and supervisors and an operator was a skilled 'pair of hands'.

Top management must initiate and support and co-ordinate continuous improvement, but it is a democratic process. The approach is built around teams who are responsible for the individual operations they perform and for improving the process. The role of the supervisors and managers changes to that of team coach, making sure that the teams have the resources they need to fulfil their missions to improve performance. In construction, project teams are in the 'learning before doing' category in the sense that the projects may be working to unique designs, unique locations and with unique clients, but are 'learning by doing' category, in the sense that a lot of construction applies traditional solutions and familiar technology to provide a tried and tested solution. Case study 14.4 is an example of information exchange for learning before doing.

There are several tools that have been developed in order to achieve better productivity. These are continuous improvement, lean construction and benchmarking. Each of these presents challenging changes in a culture that requires a sustained effort. In the context of productivity long lasting and not short term improvements are needed. Improvements will be lasting if they continue to respect and motivate the workforce whose co-operation is needed at all points.

Case study 14.4 Constructing Excellence – improvement by
information exchange

Constructing Excellence is a UK programme to promote construction
excellence by creating continuous improvement through the exchange
of best practice.[11] It does this by organising events, usually locally, to
exchange information on best practice through

- workshops and seminars
- demonstration projects
- feeding back project information for benchmarking
- a knowledge portal.

Demonstration projects organise site visits, embrace particular aspects
of sustainability, an integrated team approach and promoting respect
for people. The demonstration projects also look at business improvement in the organisation as a whole and feedback data for benchmarking. Workshops provide a structure by which to drive change more
widely by raising the awareness of the benefits of cultural changes as
seen in leading edge companies. This leads to the three objectives of
improving performance, improving industry image and engagement to
take action. The Constructing Excellence website is a knowledge portal
to support best practice, but is also a sharing resource. It is updated by
feedback from best practice and maintains KPIs on average industry
performance in its KPIzone.

Case study 14.5 indicates the important role that the client has in
facilitating change in the construction industry. Many of the improvements
above need at least tacit understanding by the client for implementation.
Construction companies then also need to work with the client to ensure
maximum utility.

Business systems re-engineering

Other ways show up as a more informed assessment is made. According to
Towill there is dramatic evidence of real savings in time spans by using what
he calls business systems engineering.[12] He recommends that a 'quick hit'
should be made by focusing on the basis of the 'Pareto' principle (20 per
cent of the problems cause 80 per cent of the waste). The solutions must be
implemented and monitored properly otherwise regression to the old level
is likely.

Case study 14.5 Fusion[13]

Fusion stands for six values which are considered critical in developing a new culture of openness. These are Fairness, Unity, Seamless, Initiative, Openness and No blame. It is a system that has been used by a large pharmaceutical client to carry out refurbishment projects of the size range £12–20m. They developed the system to turn around the traditional procurement system from

Client need > develop > procure > implement

to

Client need > procure > develop > implement.

This allows all partners to come on board early, take part in the development process and contribute to systems improvement by giving them equal status. The system starts with a 'stock take' which identifies strengths and weaknesses in each organisation for the development of each of the seven values and commits that organisation to be a contributor to the FUSION culture. For the members of the construction team this means a commitment to non-confrontation methods and an open books approach to accounts. The projects where this has been applied have ended up more innovative and more flexible and have met the time, cost and quality targets very easily. The essence of the culture has been driven by the client who has only engaged partners who are prepared to use this approach. This has meant that single project partnering has had more major savings than usual.

Partnering is a key aspect in systems improvement as it presents the opportunity for innovation and extended relationships over a series of projects and allows a culture of improvement to develop within a fixed team. It is not however the only method by which improvements may be made as can be seen by Case study 14.6, which looks at the profile of a main contractor that has developed its own culture and has given its committed business support working with a lot of different clients.

Integration of business processes makes the process of review sturdier in passing on lessons learnt. Knowledge management for lessons learnt needs to be easily accessible. Changes to re-engineer the process have to be accessible and good communications established.

Case study 14.6 Integrating business improvements

The corporate group employs 3,000 staff in 40 offices around the UK. It has grown over the past 10 years through expansion and acquisition. This profile represents a continuous improvement process that has been integrated to build upon each of the 11 initiatives that have been incorporated.

Process re-engineering. This was a mapping exercise to capture the flow of activities and to improve the process where there were problems. It began to look at increasing value for money for the client by introducing initiatives such as supply chain management and customer service. Task forces were set up with a director to champion the cause and to put forward proposals that might be adopted by the whole group. This resulted in a need to develop a partnering culture and they used workshops in order to develop ideas and to streamline their supply chain, which they were able to sell to their clients.

Customer service involved bringing in an outside consultant to tune up their customer approach. This resulted in coaching for 600 members of staff so that customer care could be extended in depth across its projects and culture.

The key issues in making progress from a traditional approach contractor to an integrated supply, customer focused one has been a sustained management commitment which has involved the staff at all levels, through the workshops and the task force. These workshops also help build the culture widely. The development of initiatives has been done by task forces represented by all parts of the group and company performance has been benchmarked in these areas. As a result the group has substantially increased its turnover and profit margins in five years. It has also won a number of industry awards and increased repeat business to 85 per cent.

The company has been able to make a change of culture because it has involved its staff and made it a top down commitment and a bottom up change to meet the challenges of new client expectations, whilst becoming more competitive.[14]

Conclusion

The commissioning stages of a project should be planned first so that sufficient time is allocated throughout the project for checking functionality and putting defects right. The traditional culture has looked to others to provide a checking service which comes in the form of a list of snags or defects, for which there is an intensive programme of remedial works. Snagging and re-testing in front of a third party is wasteful as work is being carried out twice. Pre-inspections should be carried out in the spirit of 'right first time' for finished work areas. It is also clear that a formal test should pass first time as a pre-test should be done to identify any malfunctioning which is then already corrected. Continuous control of the quality standards is required to ensure that defects are not compounded by successive trades. Occupation and handover is often overlooked as a process, but it needs to be well planned so that the facilities management and the users understand the building and benefit from the full benefits. In helping a smooth handover a system called soft landings is used to ease a handover to the users and facilities team so that good practice can be developed with the systems operated during use. It may include ongoing user training, tweaking of controls and more efficient 'debugging' of systems.

The process of *benefits management* allows the client to review the achievement of their business plan objectives in different aspects such as budget, functional requirements, space layouts and user reactions. This may take the form of an audit trail that takes each goal and looks at the outcomes from the new project. Findings may guide the valuation and allocation of risk if facilities managers and users are consulted in ongoing developments. A client needs to assess the overall value for money, which may need an assessment of economic profit as well as financial profit derived from the project. For a new factory building this takes into account productivity gains made from more efficient layouts, better staff motivation and more integration. The project manager is unlikely to be involved in this process unless they have responsibility for a programme of client projects.

Systems improvement is connected to the culture of continuous improvement or '*kaizen*' and is proactive rather than reactive to iron out past problems and to reduce system waste still further the next time around. It does not assume blame, but looks to learn from mistakes. It will also depend on using a system of measurement so that the improvement is based on known performance. The measurement system is hard work and needs to be selective to suit the key parameters of the project in hand; otherwise it increases the paper chase without the benefit of identifying causes of poor performance. Constructing Excellence is a good example of what contractors are doing to improve themselves, but it is also important for clients to take the lead in better processes as they will directly gain. Systems improvement also needs to identify issues and feed forward changes during the same project if possible. This might be achieved by a having a series of evaluation meetings at key stages of the project so that lessons learnt may be used to adjust systems immediately or passed on down the supply chain. The no blame culture means that mistakes are admitted, conflict is resolved as it arises and problems are treated as challenges. More traditional post project evaluation has often been neglected because the project team breaks up before the end of the project and it is difficult to communicate lessons between projects unless the project team follows on. Lessons learnt from a *project review* include process and design issues for the project team such as schedule control systems, sustainable technologies and life cycle running costs, relationships and team building; health and safety help guide management of the new building and feed forward into other projects.

Integration is being promoted as a way of setting up project teams that are often termed 'virtual organisations'. The rationalisation for such teams is to cut out the waste caused by poor communication and interaction between organisations and systems. At the heart of such a system is a determination to build a common culture and expectation together with an integrated communication system that allows an instantaneous distribution of information through a dedicated website or similar.

References

1 BSRIA, a UK organisation, provides independent, objective and practical help in design, construction and occupancy of the built environment.
2 CIOB (2010) *Code of Practice for Project Management: Construction and Development*. Oxford, Blackwells.
3 CIOB (2010) *Code of Practice for Project Management: Construction and Development*. Oxford, Blackwells.
4 CIOB (2010) *Code of Practice for Project Management: Construction and Development*. Oxford, Blackwells, p.94.
5 Buttrick R. (1997) *The Project Workout*. London, FT Pitman Publishing.
6 NI Department for Finance and Personnel (2011) *Benefits Evaluation*. Available online at: http://www.dfpni.gov.uk/benefits-management-mainsection3-6 (accessed 30 December 2011)
7 BBC News (2008) *What Went Wrong at Heathrow's T5?* BBC news bulletin 31 March. Available online at: http://news.bbc.co.uk/1/hi/uk/7322453.stm (accessed 30 December 2011).
8 Timms S. (2003) *Project Evaluation*. Unpublished dissertation. Bristol, University of the West of England.
9 Humphrey W. (1993) *Capability Maturity Model for Software Technical Report* CMU/SEI-93-TR-93-177. SEI. (Now developed into the Capability Maturity Model Integrated for Development.)
10 Maylor H. (2003) *Project Management*. 3rd edn. Harlow, Pearson.
11 Constructing Excellence (2004) *Best Practice Knowledge*. Available online at: http://www.constructingexcellence.org.uk/bpknowledge/default.jsp?level=0 (accessed 27 December 2011).

12 Towill D. (1998) 'Business System Engineering: a way forward for improving construction productivity', *Construction Manager*, April: 18–20. CIOB.
13 Fusion (2004) *Team Stocktake*. Available online at: http://www.fusion-approach.com (accessed 20 November 2004).
14 Adapted from the Construction Best Practice Profile. Constructing Excellence (2001) *Mansell Best Practice Profile*. Available online at: http://www.constructingexcellence.org.uk (accessed 15 October 2004).

Chapter 15

PPP models

PPP which stands for public private partnership can take many forms, but essentially it is a way of involving private sector investment into public projects to reduce the public borrowing requirement. It also allows the public sector to transfer more risk to the private sector for creating and looking after or even funding public facilities so that they can concentrate on their core business such as healthcare, transportation policy or managing justice. Some of the principles for integrated project management which have been emerging from previous chapters such as early involvement of contractors, the committed use of risk and value management, integrated design and construction, life cycle costing and collaborative relationships are an integral part of PPP procurement. It is a flexible model and comes in several forms according to client need and might be expanded further.

The purpose of this chapter is to introduce the principles and practice of the PFI which uses total private funding as used by public bodies in the UK and to assess its successes and failures as a PPP model and to look at other possible models of PPP and development of PFI because of its holistic life cycle approach. The discussion in this chapter will be limited to the provision and operation of building and physical infrastructure such as roads and bridges, but PPP can be used in most public sectors. The objectives of this chapter are to:

- consider the special needs of public sector procurement
- introduce the PPP process and rationale
- discuss the principles for PFI and to assess its impact on the culture of contracting and define its specialist vocabulary
- evaluate some of the real and perceived problems that have beset the practice of PFI and look at the solutions which are being offered and developed as best practice
- identify the differences in practice and application between the public and the private sectors
- consider the current issues and the future of PPP in construction projects.

The need for public sector expansion

In the early 1990s the need for major infrastructure renewal was becoming an increasing burden for many governments and there was recognition that public purchasing was not producing the value for money that was expected in the private sector. Public construction projects were not performing well under traditional procurement and older property was in need of refurbishment, but left to dilapidate for lack of proper maintenance. This led to the introduction of partnerships for funding with the private sector to relieve pressure on public borrowing and to maximise private skills in the management of efficient building. Opportunities for providing and running public facilities over a concession period made long term investment and income for the private sector.

Public private partnership (PPP) is an alliance which leverages some private funding and the strengths of private entrepreneurship and management, for the maximum provision of public services in a climate of scarce public resources. The EC refers to them as 'forms of co-operation between public authorities and the world of business which aim to ensure the funding, construction, renovation, management or maintenance of an infrastructure or the provision of a service'.[1] PPP has been used selectively in the past by many governments globally as a joint venture in public projects where private innovation and partial or full finance can reduce costs and enhance services. The Private Finance Initiative is a PPP special case where *all* the finance needed for the capital funding and its basic operation is supplied by the private sector in return for a service charge and capital repayment over a concession period when the private sector looks after the asset. It is a clearly defined commitment.

PFI in the UK

The rationale for the introduction of PFI in 1992 by the Conservative government in the UK was to reduce the public borrowing requirement by the use of private funding and to reduce the risks of time and budget overruns. It was expected that value for money might be enhanced using private sector management skills and innovation and by transferring the specialist design, construction, facilities management of which the private sector has a better control than the public sector. In 1997 the 'New' Labour government agreed with the principles of PFI and adopted the policy and ironed out many of the structural problems that had delayed the process of signing up contracts with private sector consortiums. By July 2011 there were more than 900 contracts signed with a further 60 in the pipeline,[2] and £20 billion of private capital had been levered into public infrastructure projects such as schools, roads, bridges and prisons, which look set to continue. The building or other asset is returned at no cost to the public body (the Authority) at the end of the agreed operating period. The advantage for the Authority is that substantial public sector borrowing requirement (debt) is removed from the balance sheet and converted into an ongoing expenditure underpinned by central government. A recent requirement in international accounting standard (IAS) protocols adopted in the EU requires some of the PFI asset to be shown on the public balance sheet which makes PFI less attractive.

Much central guidance has been developed by a series of UK government agencies,[3] but it is the Treasury that has the veto on all larger PFI projects through its assessment panel. There is a large volume of material available on PFI publicly, which makes it a good case to analyse. The criterion for using PFI is whether more VFM than other forms of procurement options, more expensive private finance costs can be offset by:

- opportunities for private gain such as chargeable use for community facilities in schools or the release of land for development so that significant discounts may be given
- transferring significant risks can be managed much better by the private sector
- facilities can be made more attractive.

Since 2011 there is a recognition that PFI may need to be used more scarcely as repayment for services rendered is cumulative on the public purse in the long term and there is a danger of future over-commitment in times of public stringency when services have to be cut back. If this is to be avoided PFI flexibility needs to be built into the design of the facility, for PFI is to be favoured over other forms of direct procurement where there is less borrowing cost. The NAO (2006)[4] has produced guidance for evaluation providing a matrix for key issues of sponsor concern and the Treasury (2009)[5] has considered the impact of a difficult private funding environment. PFI contracts have limited incentives to cut down cost when the level of planned/agreed services is reduced. PFI should only be used when best value can be proved against public financing which is cheaper. This is offset by private risk management efficiencies and by realizing extra value to the private sector such as those on page 277. But in difficult credit conditions private finance costs can soar and public borrowing costs can go up less and the immediate value of opportunities such as releasing development land can drop dramatically so a new balance of VFM is created. This must be evaluated before the decision to go ahead with PFI is agreed. However PFI still is attractive where there are funding gaps for essential services in recession, which will only get more acute if left. An example of this is an ageing school asset base in the UK.

PFI take up globally

Other countries such as Canada and South Africa have followed the PFI model in the UK to procure certain public services such as running building and running prisons. Grimsey and Lewis list 28 countries with such arrangements including advanced countries such as Canada, Australia, Italy, France, Germany, Hong Kong, Spain and the USA and also emerging countries such as Argentina, Poland, South Africa, India and the Philippines and Romania.[6] PPP has been particularly successful in Argentina where large infrastructure projects have been let and VFM is achieved by competitive bidding and in the Philippines where BOT (Build, Operate, Transfer) projects have been well below the current tariffs for water and sewerage projects. Other countries such as Australia and the UK use PSCs to control VFM and India uses it for power projects bases VFM on bidding which allows tariffs below existing rates. Japan does similar things using project based companies like the UK. The USA has project based concessions also but the market is fragmented and no federal rules exist.

BOOT (Build, Own, Operate, Transfer) is a form of PPP which is similar to PFI providing a turnkey facility for the provision of infrastructure such as toll bridges and roads and is often connected with the operation of a concession to collect tolls or fees for an agreed period in order to recoup cost in return for private capital. As in PFI, profit for this service needs to be factored in and the infrastructure is eventually transferred back to the public authority. The choice of PFI or BOOT will be discussed later and additional forms of PPP where more public capital is available.

BOOT and other derivations with more or less contractor ownership are also used in international competitive bidding and are subject to the rules of the World Bank or other major providers of international capital.[7] The World Bank, in relation to developing countries where they offer partial finance, is particularly interested in ensuring fair evaluation criteria to control conditions such as:

- cost and magnitude of the private finance offered
- performance and specifications offered
- cost effectiveness of benefits relevant to the user to ensure value for money.

The OECD has also produced guidance on the use of PFI in international development and shows that over 60 countries have used this method.[8]

How does PFI work?

In the provision of built assets, PFI provides for 100 per cent capital funding and building running costs by using a private joint venture special purpose vehicle (SPV) for a concessionaire period of 20–60 years, which may be renewed. During this concession the Authority pays monthly or annual unitary payments when the building or facility becomes available for use. They do not have to show the total building value as an asset on their balance sheet.

The unitary charges payable cover capital repayment including fixed or variable interest on cost of capital, service charges for user services and maintenance costs. The service charge is usually inflation indexed to cover rising costs as necessary. The concession period is adjusted to make the unitary charge affordable to the Authority. An example is given in Case study 15.1.

Case study 15.1 **A hospital PFI**

> For instance in the case of a hospital the maintenance of the buildings and provision of heating, lighting, building services, cleaning, catering, laundry, incineration/waste disposal and a help desk is usual, but much more may be supplied including the core clinical service itself as in the prison service. The Authority will receive the hospital back at the end of the initial capital repayment period, or renegotiate the service provision. If the asset is received back, it may be written off or valued by an actuary to go back on the balance sheet as an asset. The hospital may also have some land that is 'landlocked' within existing hospital grounds which can be released and sold for a new development to ease the unitary charge.

Penalties are levelled by the Authority for any loss of facility use, or downgrade of the service quality. These are operated within tolerances. As charges are only received on occupation a funding partner is normally involved with a contractor and facilities managers in the special purpose vehicle (SPV) which is set up to deliver the services. If the SPV renegotiates loan conditions the Authority is normally able to share in any gains due to reduced interest rates. The main principles are:

- Puts risk where it can be best managed and assesses it properly.
- VFM by harnessing private management efficiency and may also release land and assets for wider use. It has to prove better VFM to the taxpayer with a business plan over the contract period.
- Affordability, which means it should be financially viable over the contract period.

PFI allows the Authority to release the facilities management function and concentrate on its core business, for example public health services in a hospital.

The structure of agreements

The structure of an SPV in larger PFIs is a joint venture company consisting of a funder, contractor and facilities company. The enterprise is financed by debt and equity capital with a 10–20 per cent equity contribution from the members. Alternatively a contracting company may choose to fund smaller PFIs 100 per cent off its own balance sheet. The SPV contracts directly with a public authority who commissions the project on behalf of their stake-holders. The Authority will need to appoint a project leader – quite often the chief accountant, or a specialist member of their project staff, who will engage specialist advice as required. This shown as found by the Audit Commission in Figure 15.1.

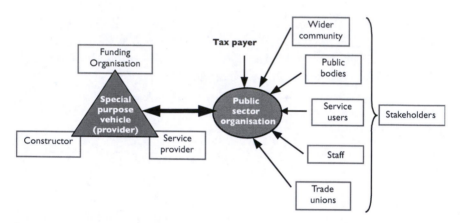

Figure 15.1 Structure of the special purpose vehicle
Adapted from Audit Commission (2000)[9]

Case study 15.2 shows the organisational structure of a typical SPV and indicates the main cross relationships with the NHS Trust.

Some research by the RICS indicates that this model was used by all the case studies they investigated.[10] The most common PFI projects (48 per cent) in this research were in the NHS and schools sector. In the NHS the Trust chief executive is commonly the project sponsor. In schools PFI, it is likely to be the Local Education Authority.

The procurement, commercial and construction managers are directly involved in the production process and need to liaise closely as a construction team which overlaps the detail design, procurement and construction stages. Package managers are appointed to cover specialist contractor control.

Case study 15.2 Redevelopment of major NHS teaching hospital, part one

The new hospital budget was agreed between the NHS Trust and the SPV as £100m construction costs for a three year construction programme. The SPV consisted of an FM provider, a design build contractor and a financing partner. The special purpose vehicle (SPV) in Figure 15.2 consists of a joint venture between a contractor who is responsible for the construction and the facilities management, a medical services enterprise and sponsoring banks.

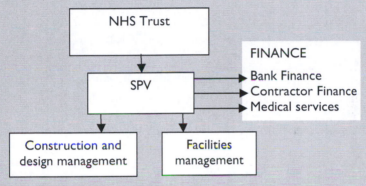

Figure 15.2 Structure of a special purpose vehicle (SPV)

The structure of the construction team varied from six to eight people in July 1997 involving six key disciplines as the project developed towards its construction start date in 1999. These departments were populated to take on full responsibilities for the tendering and construction stages and an external design company worked in partnership on a design and build basis. Figure 15.3 shows the detail.

The structures indicate the organisation of the Trust project team and the SPV project team. The Trust team has interaction points at both the user and the facilities level in order to determine the design requirements. There are regular design meetings and weekly user group meetings that are attended by the design consultants and the design manager. The ultimate aim of these meetings is to determine scope and not detail design of the departments. Progress meetings are called by the PM who will also approve the affordable scope.

The project manager is responsible for the delivery of the new build design and construction and maintaining the budget, programme and

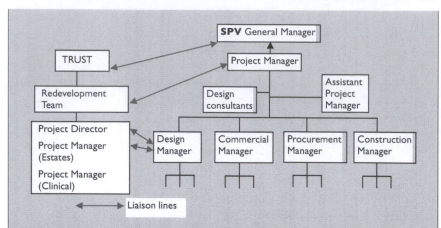

Figure 15.3 NHS Trust project structure – project new build hierarchy

quality targets. He liaises at the level of Trust project director in order to establish project strategy and overall design and construction co-ordination. He is at the hub of the communication system and will be involved in the negotiations with the preferred bidder to reach financial close.

The SPV general manager liaises directly with the Trust Board at a strategic policy level. He is responsible for the liaison with the financiers, signing off financial close, the performance of the construction arm and on an ongoing basis throughout the operation phase of the hospital.

The services provider operates the facilities management function and may also be consulted as to the life cycle costs and maintenance issues at this stage. The FM function is provided by another business unit within the same contractor.

Contract and tender

Authorities in the EU advertise in the Official Journal European Union (OJEU) in accordance with European public procurement rules and need to show that fair competition has taken place in choosing a preferred bidder with whom they negotiate the best deal. The provider, which is the *special purpose vehicle* (SPV) joint venture, has vested interest in their own risk and value assessments for the capital and whole life costs of the building. Case study 15.3 shows typical time frames.

Case study 15.3 Lincolnshire schools PFI

The Lincolnshire scheme consisting of three primary and four special schools was a successful scheme completing in just two years and nine months from the outline business case produced by the local education authority (LEA) for the Building Schools for the Future funding of the Labour government in the UK. The programme below indicates a lead in time from Treasury approval of 13 months.

Outline business case	January 2000
PFI credits agreed	August 2000
Bids submitted	October 2000
Preferred bidder announced	June 2001
Contract signed	September 2001
First school opened	September 2002

The LEA appointed an SPV (see Figure 15.1) consisting of a main contractor to do the design and construction and the facilities management and a bank to provide private funding to build the school. Risk is mostly transferred to the SPV for the planning permission, the soundness of the design and the provision of schools to suit purpose which would be continuously fit for use over the concession period of the management of the school facility. The client has to keep the risk for the scope and the educational objectives enshrined in the outline business case. In this case some degree of detail beyond an output specification was given by the client who gave them more security in getting the centre of excellence they wanted.

The lessons learnt by the client were to

- decide the risk allocation very precisely before the contract was signed to avoid later conflict
- hold a contingency fund, held centrally, for necessary changes which occurred after the signing of the contract and the need for transparent open book accounting in relation to make such changes
- have tight clauses regarding the completion dates to get schools ready for critical start dates
- work closely with the schools and to manage expectations in the climate of major local school reorganisation and to raise understanding amongst stakeholders

> • appreciate the budget and time pressures which force your hand into certain solutions – the Project Board needs a good level of delegated power to make these decisions in commercial timescales
> • foster good relationships with the provider so that trust can be maintained in these situations.
>
> The SPV and the client have to work well together and the client has a better chance of keeping value for money on the project if they are able to negotiate maximum detail of the design without change. Change management must be efficient for inevitable developments.

The Standardisation of PFI Contracts version 4 (SoPC4) recommends two end arrangements, either:

• asset is automatically transferred back at no cost; or
• transferring the asset back, whereby the authority makes final payment to the SPV based on the market value – this may induce better care of asset.

The NAO has produced guidance for tender evaluation providing a matrix for key issues for bringing down the cost of tenders.[11] Two of these are to reduce negotiation times and associated inflationary costs and adjust contract conditions to claw back later improvements in profits by the provider.

PFI competition

Competition for significant public contracts in the EU is governed by the Public Contracts Regulations 2006,[12] which require open and timely advertisement of the tender for forthcoming contracts (€5.278 million at time of publication). The contract may be in open competitive tender using prequalification rules or it may be for negotiation, as in case of the PFI contract due to the complexity of PFI projects. These come under negotiated concession contracts in the Regulations. Some restricted work such as work to do with confidential defence may be subject to exclusion from Europe wide tendering. Utilities supply contracts come under different regulations.

PFI contracts are almost certainly going to come over the threshold value by their nature. The procedure most used for PFI is the negotiated procedure, which tries to ensure an open competitive first phase followed by a negotiated phase to squeeze value from the original proposals (Figure 15.4). The Commission position on PFI has recently hardened indicating the need to

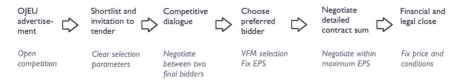

Figure 15.4 Hierarchy of PFI selection

ensure that competitive processes with the negotiated procedure are maintained.

The competitiveness of projects is affected if there are many interested parties, because of the cost of tendering. A selection criteria to reach a short list has to be transparent to all bidders and reasons for exclusion given if requested. Bidding costs in the *detailed* stages are very significant and the numbers of bidders being invited to tender (ITC) on complex projects should not be more than six shortlisted to develop their proposals. Tenderers may need to be reassured of some remuneration for second stage bidding costs where they are shortlisted along with others. The third stage is to work with two preferred bidders to move into detailed design options and negotiation.

Competitive dialogue

This process is called competitive dialogue which helps to make the target cost called the estimated prime sum (EPS). The client will need to work closely, but separately, with each bidder – see Case study 15.4. When an EPS is in place it is acceptable to move to a preferred bidder in the negotiation. A preferred bidder is only chosen when there is no real chance of significant amendments being made to the contract scope and conditions. Prior to competitive dialogue, PFI negotiation at the detailed stage was open to criticism that the final bid agreed at financial close could agree on a significantly different specification, which denied other bidders alternative approaches to improve their competitiveness. The dialogue to agree the design with two bidders gives the opportunity for innovations to be made and competitively to improve VFM, but acknowledges that compensation should be made to the losing bidder at this detailed stage, prior to preferred bidder status.

Risk transfer

Risk transfer is a critical part of the equation in assessing value for money. Under PFI, many risks that are normally carried by the public sector are transferred to the provider. For instance it is normal to transfer design risks

Case study 15.4 Competitive dialogue on hospital PFI

A large hospital PFI in the UK was eventually closed financially at a capital cost of £450m with a £1 billion value of services and capital repayment over the life cycle of the concession. This hospital was subject to a three year lead in period from outline business case to start on site and approximately one year of this was competitive dialogue followed by approximately six months of financial negotiation, post preferred bidder. The two bidders presented two fundamentally different designs to suit the client's capacity and brief and choice was made on a matrix of cost, quality and design fit to client performance requirements and life cycle business costs. During this period both bidders were able to reduce their bid significantly as they entered a detailed design stage and a regular client weekly meeting took place with both bidders separately to develop the design and to increase value for money. The bidding costs of the winning bidder alone were substantial. The costs of the other bidder were compensated.

There are many other factors apart from cost of the design and efficiency of construction and maintenance. These revolve around client functionality, but also around the closeness of fit of other client values such as a sustainable and attractive hospital building. These issues affect the business case and the sustainable nature of that business in the light of future policy and expectation. Inevitably flexibility of design is important to a client in a long term concession such as a hospital. Private funding commonly reaps benefits in the rescheduling of cheaper funding once the major risks have been passed. PFI contracts now have a requirement to share subsequent benefits which result from reduced provider risk. These may manifest themselves in favourable refinancing deals.

and schedule overrun risk and even planning delay risk. In order to determine value for money, it is normal to calculate the cost of a conventional procurement method called a public sector comparator (PSC) using standard government funding costs and to add to it an estimate of risks that under PFI procurement are transferred to the private sector. A PSC is determined by the unit cost of similar project type that has been procured under a conventional non-PFI route and adjusted to allow for the floor area and differences in other factors such as inflation, location and level of quality. To minimise the subjectivity of the value of the risk the Treasury has laid

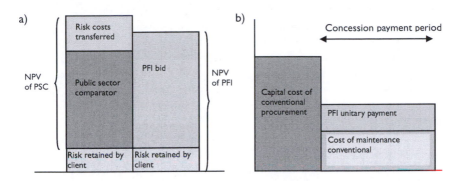

Figure 15.5 (a) Public sector comparator, (b) Comparison of payment cash flow
Based on Grimsey and Lewis (2005)[13]

down certain rules in its calculation. The client risk of delayed use of the facility is also improved by reserving any payment till the completion of the facility and its readiness for use. Figure 15.5 shows a favourable situation for a PFI bid against the PSC, where risks are transferred and included in the bid. A contractor is also incentivised to reduce the level of maintenance (Figure 15.5b) as they are responsible for paying for it for a fixed unitary charge.

The principle of putting risks where they can *best* be managed is illustrated by Figure 15.6 and shows an optimum point where unreasonable risk transfer causes declining VFM because an unaccustomed risk liability is expensive to manage. For example the occupancy rate of prisons, if transferred as a risk to the SPV, is not easy to manage as the prison governor and the judiciary

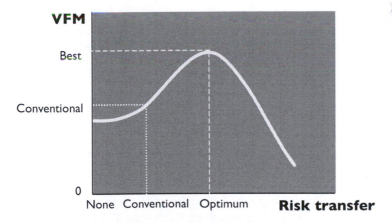

Figure 15.6 Value for money and risk transfer[14]

make the decision for prison allocation and safe occupancy levels. Without this knowledge a private risk assessment might err towards a low occupancy rate being priced in the contract, when in fact the prison may normally be full, creating economies of scale for services, which are not passed on to the Authority. Figure 15.6 shows this effect.

The main risks that are transferred are construction delay, cost overrun, design problems which are not to do with changing the scope, some planning risks and even some occupancy risks which are controllable. All risks which can be controlled by the provider should be transferred, but also care is needed to check the value for money factor. Case study 15.5 indicates how a PFI for a large teaching hospital was set up and illustrates the relationship between the structural and risk transfer. It provides some further information about how budgets were agreed on the same project.

Case study 15.5 Redevelopment of major NHS teaching hospital, part two

The budget was agreed between the NHS Trust in Case study 15.2 and the SPV over a three year inception programme from 1996–1999 when financial close took place. The Trust provided the project team with a performance specification and clinical constraints from which a proposal outline design was developed by a major architectural practice working for the contractor on a design and build basis. This design was developed with the design and build contractor (D&B), the structural engineer and the architect to change the footprint of the outline design so that a more efficient floor to wall ratio and circulation could be achieved. During negotiation detailed 1:100 design drawings were developed to ensure functionality and space planning and agreed to a given budget.

This budget was constrained by an acceptable annual unitary service charge, which covered the capital costs and yearly running costs and this could not be exceeded. This allowed for adjustments between capital costs and whole life costs. Planning was obtained by the contractor for their developed 1:200 scheme design after sign off by the Trust. After planning approval a detailed design was developed with 1:50 plans, which were signed off by the client if compliant and these became the reference point for the structural, services and complete clinical functionality of the design. This was important so that detailed services design and performance specifications could be agreed with the Trust. In the case of components the contractor insisted on choice wherever possible, with a commitment to meet client performance requirements. Design cost was controlled by the D&B by elemental cost planning checks with the architect.

In order to achieve sign off of the 1:50 design drawings the architect attended the weekly clinician's project meetings to agree detail with the D&B design manager. This allowed for innovations to be developed which would be acceptable to users and was supplemented by wider less frequent user groups where specific areas of interest were being discussed. Each 1:50 floor plan was split into four zones to aid the more specialised inputs of individual departments and to finalise service requirements within the overall framework of the original 1:100 layouts. In order to co-ordinate the hospital appointed an internal redevelopment project team. This team liaised with the contractor project manager and with the general manager of the SPV in order to ensure the continuity of relationships when the hospital was occupied and to co-ordinate building operational procedures with the clinical, administration and support services supplied by the Trust.

The risk allocation in the hospital allowed for the contractor to take on design, construction and planning approval and maintenance risk and to run some of the more general services such as help desk and car parking and incineration. The occupancy and the supply of critical medical equipment and the administration of bookings and confidential medical records was retained by the hospital as core to their own skills.

It is common to offset the cost of the car park against a contractor concession period within agreed charges they collect.

Value for money considerations

PFI projects are approved to proceed on grounds of value for money potentials: viability, desirability, achievability.[15] VFM is about improving on the public sector comparator and making significant improvements to it. The formula below represents this:

$$\text{VFM} = \frac{(\text{PSC} - \text{PFI})}{\text{PSC}} * 100\%$$

PFI = Total net present value (NPV) of PFI bidder's cash flow will equal the unitary charges over the life of contract

PSC = Total NPV of public sector comparator represents the cost of delivering the same output over the life of contract without the risks transferred to the SPV

Case study 15.6 Value for money calculation

Brent Emergency Care and Diagnostic Centres (BECaD) in North West London was a PFI contract to provide emergency care centres throughout the Trust. The comparisons for the PSC used and the PFI for a 60 year payback period are shown in Table 15.1.

EAC is the equivalent annual cost; the PFI contract is for 30 years and there are four years of construction; design life for buildings is 60 years. Before risks transfer PFI was poor VFM by 1.06 per cent and after risk transfer it was good VFM by 1.23 per cent. The risk costs are assumed to be less when transferred to the private sector and the non-financial benefits less. The actual costs before risk adjustment are more because of extra finance costs, but the efficiency of the private sector maintenance will offset these.[16]

Table 15.1 Brent hospital NPV for PSC and PFI

(Discounted at 6%)	PSC (£m)		PFI (£m)		Saving (£m)		Saving (%)	
	64 years	34 years	64 years	34 years	64 years	34 years	64 years	34 years
NPC before risk adjustment	1222.60	1083.90	1233.36	1095.34	(10.8)	(11.4)	−0.88	−1.0
NPC of the risk	42.13	38.96	16.88	13.70	25.3	25.3	59.93	64.84
Risk adjusted NPC	1264.73	1122.86	1250.24	1109.04	14.5	13.8	1.15	1.23
Non-financial benefits	1.59	1.41	1.41	1.25	0.2	0.2	11.32	11.35
Risk adjusted EAC	73.35	73.73	72.51	72.82	0.8	0.9	1.15	1.23

The figure used here should be tested for sensitivity as the actual case for VFM is quite narrow given the large figures involved. This could be tested on the discount rate and the budget shift to see where at what stage the benefit would be lost. Costs as shown have to be protected from optimism bias where a business case is proven to have optimistic assumptions.

There is considerable debate about the VFM balance of PFI and Figure 15.7 indicates the balance of extra costs such as the inflated cost of capital and the need to offer returns to equity holders.

A discounted cash flow calculation (Chapter 3) for the concession period generally estimates the impact cost of risks transferred and also the probability of those risks occurring. However good the arithmetic, there is bound to be some subjectivity in both of these estimates, especially as there is a

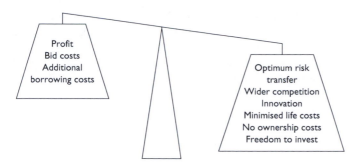

Figure 15.7 The VFM balance
Treasury Taskforce (1996)[17]

very limited track record to run on for PFI long term costs. The temptation to 'make the figures work' by incorporating optimism is compounded by the fact that PFI credits granted from the Treasury may be the only way for some public authorities to gain borrowing credit, as other public borrowing credits are more scarce especially when government income is limited. For the Authority PFI repayments are a binding commitment for the concession period and the Treasury needs to be sure they are affordable. Treasury assessment procedures are therefore rigorous and provide a check on irresponsible or inexperienced commitments. The Treasury is particularly cautious about the impacts of poor credit availability, stability of funding sources, coverage of failure risk and favourable refunding opportunities being shared with the Authority when some major risks are passed. Since 2003 they also apply an *'optimism bias'* which comes in the form of percentage reduction to the discount value to make future costs more potent. However, as discussed earlier, there are credible intangible benefits such as

- better maintained, available and longer life for buildings
- focus on the whole life cost savings
- use of an open performance specification to encourage innovative and efficient solutions from providers
- clear division in the allocation of risks so that responsibility can be enforced
- some degree of flexibility to allow changes in specification, without undue penalty
- sufficient incentives for the contractor to ensure delivery of quality and maintain full capacity managing the scale of the procurement so that tender costs are not disproportionate to value.[18]

Projects which are small and have a dynamic or short term requirement which are well understood by the client are better managed by them. Research by Henjewele has indicated that extra scope creeps in (with associated cost) up to the preparation of the full business plan (FBC) when the budget has to be rationalised to keep VFM and viability at the financial close.[19] Shedding specification/capacity is inefficient and is not directly proportional to the average square metre cost of capacity. He argues for a sustainable VFM which is attained by controlling escalation of the specification during the earlier stages and controlling operational and maintenance costs.

Other concerns have been expressed with the value for money by the NAO and PAC. For example, the Dartford and Gravesham Hospital PFI had a poorly researched PSC and it was recommended that the PSC should be properly priced and risks from traditional contracts should be calculated accurately from several previous contracts whilst maintaining a competitive climate and by ensuring sufficient bidders.[20] Likewise in another contract it was felt that the long negotiation period of 21 months with a preferred bidder had created non-competitive escalation in costs of 9 per cent.[21] Although this was considered not to be out of scale for the time period the period itself could have been reduced if fundamental matters such as the suitability of possible project sites had been more thoroughly researched prior to going to tender.

Affordability

Affordability is the long term ability of the authority to pay the unitary charge which covers the combined capital and service charge (Case study 15.7). It is calculated by comparing the long term discounted cash flow (DCF) for all services and benefits and comparing it with the money available to pay that charge. A repayment period can theoretically be lengthened to make it viable, but this actually increases the total repayment.

The business case is normally vetted by the Treasury to insure a level playing field has been used and normally a 3 per cent discount rate is used to check viability. In PFI the outline business case will have valued the benefits as well as the financial income. Benefits must be valued consistently and presented to the public funding source as a convincing case of need. Non-financial benefits are factored in such as transport and efficiency in the core process, such as fewer staff, but must be proven. A project will not go ahead unless the figures stack up and income has a reasonable chance of reaching expectation. Operating costs are value tested from time to time to see if the given rates in the unitary charge are exceeding the market rate. If they are some contracts allow renegotiation on a shared savings basis. Case study 15.6 illustrates affordability on a small project.

Escalation costs and changes

The length of the negotiation period has been a cause of escalating the contract price as it increases the resources required and the tender suffers

Case study 15.7 New college building, affordability and funding

Due to the holding of old buildings, expensive to run and situated in the suburbs awkward for student access, the college needed a more central modern campus. They looked around for a development partner who could offer a turnkey building development that could be maintained for an annual fee. The capital value of the new development offered was £13m. As a public authority the college carried out a PSC with a conventional approach to ascertain value for money and advertised in the EU to ensure broad competition. An OJEU notice was issued in June 1998 and the preferred bidder was chosen in October. Financial close was achieved in April 1999 on a 25 year lease with the developer responsible for FM of the building according to the college performance specification prepared mainly prior to the OJEU advertisement. The developer agreed to find a site and to carry out the planning design and finance risks for the project. At the end of the 25 years the building would be transferred to the college for a nominal sum of £1. The agreement also allowed for an escape clause for the college to buy back early on building completion for an agreed price. The developer added value by finding a site suitable to develop a hotel as well as the college in the city centre and they completed construction according to the agreed performance specification. This defined the scale and scope of the accommodation, but gave flexibility to progress the design and to build in value savings. A small project management team was formed by the college headed up by the principal.

The college saved itself the job of buying land and controlling a major project and was able to spread the cost of the project over quarterly unitary payments for the 25 year period to cover capital and running costs. Value for money was passed on by the developer sharing opportunity savings on the land costs. The pre-project period was slightly extended, but a quick construction stage was mutually acceptable to the developer to gain hotel revenue. Credits were obtained from the Further Education Funding Council (FEFC) on the production of a properly assessed full business case assessing the risks and value in a favourable comparison with a PSC.

The college learnt that it was important to allocate risks more closely in the negotiation period and it was able to take advantage of the current low interest rates to seal a long term contract more competitively.

from the inflation of building and facilities management prices and the negotiation phase becomes less competitive and competitive dialogue does not help this. Many NHS projects took more than two years of negotiation in the early days which effectively 'hung' the negotiations and frustrated tenderers. It is important to manage this phase effectively to encourage win–win agreements and to ensure client changes at this stage do not change the nature of the contract. One large PFI has been criticised by the NAO[22] and the PAC[23] for the extensive time that elapsed in the negotiation stage with a preferred bidder. This had the effect of escalating the bidder price by 21 per cent before agreement was reached and could have been avoided by the better consideration of the problems of location and technical transition costs which were still being sorted at this stage and impinging on the negotiations and giving the contractor a price advantage. The NAO (2008)[24] report on managing change in PFI projects indicated that up to £180m of changes were made in 2006 by public authorities and that many of these could have been more efficient. They recommend that changes should be bundled up and that SPV overhead costs should be reduced. One example of this is the Treasury itself wanted to make changes in 2010 to its offices which were a PFI contract in 2000. This older contract is proving difficult to renegotiate.

When is it right to use PFI?

This is not an easy question, but some general guidelines may be given. Authorities that are providing facilities with little room for innovation or shared use or who have very specialist use which requires limited public access are unlikely to prove value for money, for example a laboratory building on a university campus. Public facilities which depend heavily on large and uncertain visitor numbers attract premium borrowing rates. These need very careful scrutiny and sensitivity analysis to look at worst case scenarios and backups. The Royal Armouries, Leeds, is a case in point where the SPV had to be bailed out by the substantial use of public money.[25] Facilities that levered substantial voluntary help would also be disadvantaged by the tied up professional provision.

On the other hand, authorities may enhance their business case figures by the release of unused assets such as buildings or land that have commercial value. They may also encourage joint private use of income generating facilities such as public use of school swimming pools at the evening or weekends and these are called *opportunity* benefits to be shared with the authority by discounted provider offers in reducing the capital or service charge. A large part of the success of PFI is responsible negotiation with a preferred bidder to reduce costs to the tax payer and to enhance afford-ability by innovative design and use, and economic running costs. Again, independent guidelines are needed to protect long term assets and offer

Table 15.2 Comparative costs and time prediction for PFI compared with non-PFI

	Pre-2003 PFI %	2003–2008 PFI %	2003–2008 Non-PFI %
To time	63	69	66
To agreed cost	54	65	50

sustainable services to the public that are at least as good as they have experienced previously. It is also important for providers to gain a reasonable profit margin to cover the additional risks they incur. Keeping this balance right throughout the project is critical for public trust in the system.

The NAO (2009)[26] report compared with the earlier NAO (2003)[27] report seems to suggest that there is good efficiency of execution. In Table 15.2, the 2003 report had 37 projects and the 2009 study had 114 projects. This suggests an improving performance on PFI projects over time and on cost efficiency especially. Only 10 per cent were more than 5 per cent over the cost when letting the contract. This is encouraging for the efficiency of the PFI management process.

There is however an increasing concern (2011) by the taxpayer of the affordability that 'instant facilities' provided by finance that are more expensive than a public authority over a long payback period might be a cumulative disaster. The NAO estimates of the 500 projects let by PFI in the UK up to 2008, worth £44bn capital costs, that future payments up to 2030–2032 will amount to £91bn, or more than twice as much over a typical 25 year concession period. The key issue is whether the other benefits of efficiency on life cycle cost and risk transfer discounts against these extra borrowing costs are enough to cover the required investment profits of the private sector. This has led to a review by the UK 2010 government to tighten up affordability appraisal even more. However it has become obvious that other benefits are important such as the cost of fast deteriorating existing stock in the slower renewal scenario of direct public finance and the scarcity of such finance. Inefficient operating costs will also persist on this older stock. A positive balance has made PPP type models popular in many countries where the problem of scarce public funds is particularly significant.

The Confederation of British Industry (CBI) urges continued commitment to PFI, in order to maintain infrastructure in financially stringent times, though it is cautious about learning lessons from past projects so that PFI is used selectively. They see advantages in the control of time and budget slippages, value for money savings and the control of life cycle costs and regular maintenance because the private sector is incentivised in cutting down maintenance costs and presenting fully operating buildings before they receive capital and service payments on an ongoing basis. Like the government they urge some caution as well:

The Private Finance Initiative (PFI) remains a financial tool which, along with other forms of financing, is essential to solving the infrastructure funding gap. The *CBI* accepts that PFI and other private investment in public services aren't the right solution for every problem. However, better use of private finance will require significant improvements in the ability of government agencies to get the right product, at the right price and on terms which capitalise on the skills and experience of private sector partners to deliver better services, more efficiently and sustainably.[28]

The Treasury (2010)[29] have considered the impact of a difficult private funding environment to be up to 12 per cent more on unitary charge payments and set up the Infrastructure Financing Unit to try and ease credit jams and revive market confidence for loans to PFI projects.

The culture and leadership of PPP

The culture of PPP is more open and flexible than conventional procurement and uses a performance specification expressed in outputs and not prescribed. It allows the provider to be much more innovative in its approach to the design and provision of the facility and it is important not to be straitjacketed by design rules or guidelines. Entrepreneurial approaches to design mean that alternative ways of providing space may be discussed in the spirit of enhancing value and the fundamental definition and inclusion of client values.

The tendering culture is competitive in its initial stages as required by the EU procurement competition rules, but it allows a final negotiated stage in order to bring discussions to the table with a preferred bidder who has passed the initial competitive stages. The culture is collaborative and seeks to build trust in order to optimise the project objectives within the budget which is available and with competitive dialogue may help to sharpen the mind. The client brings their outline business case to the table and the contractor seeks to bring agreement to their design proposals by transparent pricing of alternatives in order to build up a detailed set of specifications, which are agreed as meeting the output specifications. With this culture it should be possible to lever in extra value and to allow some flexibility for the client as user groups get involved.

Important to the success of the PFI contract is the formation of a project team and a leader. For the client this leader may well be the chief finance officer of the Authority in question, but the key issue is that they should have the teambuilding and negotiating skills to provide an effective lead and response for the procurement process. Skills in negotiating, leadership and knowledge of the whole PFI contractual process need to be built up. It is also clear that the leader must have good abilities in building relationships, communications and dealing with conflict. They must also be able to set a timetable and oversee the production of a project brief and have authority to make the resources available in a timely manner to resist delayed starts and unnecessary scope changes.

At completion a proper evaluation needs to be organised and feedback arranged. Lessons need to be learned and a continuing benefits evaluation against expected benefit be completed. The role and authority of the project leader needs to be formally established to deal with a cross section of department heads, advisors, users and other stakeholders so they need to be empowered to make decisions.

The skills required in PPP

Research by the RICS amongst surveyors has indicated that project managers are often not adequately skilled at driving PPP projects forward and that in the best PPP projects a partnership of skills between the public and private sectors is required.[30] The NAO Report (2001)[31] identified the need to thoroughly understand the project, be familiar with contract terms and how they work in PFI and have good relationship and communication skills. The RICS identified the mismatch between the technical skills of the consultants and the political and specialist knowledge of the client, but identified the need for more political, negotiating skills and client understanding. Overall it became clear that a new culture was emerging which for PFI to be successful needed to integrate the consultants and the client more. The NAO (2010)[32] management model for complex capital investment programmes using PFI suggests six functions to manage, including influencing commercial policy, programmes and quality management, operational benefits, continuous improvement and market operators.

Relationships in PPP contracts

In a survey of authorities, the nurturing of good relationships is considered one of the most important issues for successful outcomes on PPP contracts.[33] The relationship between the Authority and the contractors at the construction and operating stages is long term and the groundwork is laid at the negotiating stages with the preferred bidder. The key question is whether authorities manage their PPP relationships to secure a successful partnership. A happy contract is one where win–win solutions have been identified to the mutual benefit of both sides. This means that an authority gets good value for money and a contractor gets reasonable returns. It is also clear that poor communications can sour what was a good relationship and that building open communications is an ongoing process during construction and operating so that value for money is built on innovation and flexibility. The report found that 72 per cent of authorities and 80 per cent of contractors on PFI believed that relationships were good and many believed that relationships were improving since the contract letting rather

Case study 15.8 Colfox school PFI relationships[34]

In this early, first school PFI project, the involvement of a comprehensive client project team helped to speed up decision making, which depended on agreement between the headmaster, Dorset County Council (DCC) and the school governors. The project team consisted of the deputy county treasurer, as project manager, and included the school head and one school governor, officials of the local education authority (LEA) and outside financial advisors. A smaller specialist core team carried out the negotiation process and reported periodically to the project team. The 4Ps was available for advice.

The use of all the school governors in the project decision making process brought them on board and their enthusiasm and confidence was a help in convincing the funders that a viable operating phase was possible. The report also emphasised the success of open days in which potential bidders were introduced to the vision for a new school and the management imperatives for its operations and attracted 12 bidders to submit schemes. Eight of these were invited to detailed discussions with the team and four were invited to tender. It was believed that this improved the quality and value for money of the bids. A project room was also set up at the county hall as an accessible source of material to aid the bidding process.

Typical non-contractual relationships exist with the parents and the community and these were nurtured by occasional press and parents' meetings on progress.

Key negotiation points with the contractor developed around the long term agreement between the LEA and the school governing body for financial contributions, the compliance with the then Department of Employment and Education (DfEE) guidance for accommodation and contractual provisions

The case study gives an insight into the extensive thought which went into the management and communication systems to bring all parties on board and to enhance relationships between the various parties to the contract and to bring in other stakeholders.

Relationships with the SPV are likely to be between different levels such as the contractor project manager and the project board, the design team and the school governors and the manager of the SPV and the LEA. It is quite likely that each will change as it moves into the occupation stage.

than getting worse. This has been maintained and in the IPSOS Mori (2008) survey 83 per cent of managers believed that good relations were also in place during the operational stage.[35] Managers considered key issues to be:

- A good contractual framework with the correct allocation of risks.
- Value for money mechanisms which encouraged rather than discouraged innovation.
- Benchmarking and monitoring the quality of service without being intrusive.
- Building in arrangements to deal with change.

Seventy per cent of authorities chose to exercise their rights to impose performance penalties.

The recommendations for the report were directed at sealing in the right skills for contract management by measures that encourage training and rewards that retain staff and pass on the experience gained in past PFI contracts by key staff. These experiences might well be shared between authorities and need to be developed during the project by the authorities having a greater commercial awareness.

Relationships need to be nurtured with the stakeholders of a project as their views may differ because of different perspectives. It is important to manage the priorities on the basis of the degree of influence (power) and the need to differentiate between consulting and informing. Communication systems can be used positively to break down barriers based on the belief that it will be worse than it is. The aim is to bring stakeholders on your side by confronting perceived problems. Case study 15.8 is an example.

Risk identification and allocation

The RICS categorises the typical risks in PFI as project cost and completion time, commissioning, future maintenance, cost predictions, residual risk such as the value assigned to an asset at the end of a contract, functional and technical obsolescence, changed regulatory and legal standards and cost of project finance.

A survey carried out by Akintola *et al.* (1998)[36] indicated that contractors, clients and lenders viewed the priority of the risks quite differently. Table 15.3 shows the top three risks out of a list of 26 risks for each party.

The top three risks are covered in the top 10 of all three parties so it is interesting to see from the third column top risks that are not prioritised by the others, but may significantly determine the behaviour of that party individually. Some of the least prioritised risks may be because they are well known and easily managed by those who normally manage them and not because they have little impact if things go wrong. The way that the SPV prices the risk depends upon the reaction of the individual SPV parties.

Table 15.3 PFI risk priorities

	Top 3 risks	Bottom 3 risks	High risks not considered high by either of the others
Contractor	Design, construction cost and risk of cost overrun	Euro legislation, credit and land purchase	Contractual
Client	Commissioning, performance and delay	Credit, bankers and debt risk	Operating and health and safety
Lender	Payment, volume and risk of cost overrun	Project life, land purchase and development	Credit risk

Adapted from Akintola *et al.* (1998)[37]

Henjewele lists key common risks headings in PFI hospital projects as: design risks; construction and development risks; operating cost; variability of revenue; termination of contract; technology and obsolescence risks; control risk; residual value and others.[38]

The *allocation* of these risks should be with a view to reducing the cost of managing that risk and every project will be different. From a public body point of view the risks associated with procuring, financing, designing and constructing and maintaining the facility are considered to be a non-core business and they would like to see the private sector bear these risks. In many cases the client sees the risk associated with the service being offered within the facility, such as clinical health services, education or policing, as a public responsibility and therefore retains the volume risk and functional obsolescence. The public body is also in a better position to predict as yet unplanned changes in regulatory and legal standards as a service provider. However changes to the building regulations are better understood by the provider used to dealing with local authority planning and building regulation departments. Shared risk is important where the impact caused by a failure is out of proportion to the impact cost of the risk itself. They are characterised by the need to collaborate closely so that risks are mitigated by sharing information fully. An example would be the risk of a delay in getting planning permission.

Sector differences in PFI

Different public sectors have developed alternative provisions to reflect their particular needs. Fewings compares some key criteria which indicate the differences between the higher education, health, prison, local government (schools) and infrastructure sectors.[39] Interestingly there is very little standardisation between the different sectors. Table 15.4 gives a summary of the findings of the research in the areas of tender length, authority for go-ahead, risk allocation, affordability and value for money. A further comparison indicates the difference of procurement methods:

- Prisons – DCMF Design, Construct, Manage and Finance. Sign off by HM Prison Service (HMPS).
- Health – DBFO Design, Build, Finance, Operate. Sign off by National Health Service Executive (NHSE).
- Schools – DBFO. Sign off by Department for Education and Employment (DfEE).
- Roads and bridges – BOOT Build, Own, Operate and Transfer. Sign off by Department of Transport (DoT).
- Higher Education – there are a variety of models possible but generally DBFO. Guidance by Higher Education Funding Council (HEFCE).

The efficiency of the tendering process from OJEU to financial close varies from 12 months to 36 months and there is an indication of expected improvements to reduce this to nine months in some cases. This would cut down the escalation of capital costs through inflation. In most cases approval in these case studies is shown to be by the relevant government department. The affordability does not always use a PSC. The risk analysis is normally against a pre-arranged matrix, but no indication is given of the way in which the risk is valued. Value for money is assessed generally by doing an NPV calculation for an option appraisal. The value of saving varies from 1.1 per cent to 10 per cent. This VFM is different for the operation period.

The risk that the asset value at contract end is not as anticipated has not been tested by the completion of a PFI service period, but is closely associated with the maintenance of the asset during the concession period. Some of the risks taken on by the provider may eventually return to the Authority and these risks should be properly priced.

Issues affecting the operation of PFI projects

Where core services are offered by the provider they affect allocation of risk. The DCMF form of PFI suits some custodial services in prisons. In this case a public interest is maintained through the oversight of a prison governor. Other examples exist in infrastructure such as BOOT contracts or museum provision, where an agreed charge can be levied on the public, which goes directly to the operator to offset all capital and operating costs over a concession period. These risks are not always handled well. For example the Ashworth Young Offenders' Institution originally failed because of an inability to recruit skilled staff locally at the rates of pay budgeted. On the other hand the Skye Bridge was delayed because it was recognised that re-negotiation of the bridge charges and concession period were needed to make the PFI viable.

Table 15.4 Comparison of key sector case studies

Sector	OJEC to financial close	Approval & non-financial appraisal	Risk allocation and cost	Affordability	Value for money	Improvements
Higher Education (hypothetical case study)	12 months minimum	Independent university decisions Institutional with HEFCE sign off if requested[40] Subject to OJEU Weighted criteria	Estimated cost impact and probability Covers design, construction, operations, finance costs and demand	Test against known financial income before and after procurement choice	NPV expressed over contract period compared with other options[41]	NAO report suggests improvement of option appraisal and valuation of risk
Education a) Colfox School b) Bristol Schools for the future 2005–2009	16 months	DfEE sign off and full council approval Consultation with the community and Governors 10 year Local Enterprise Partnership. 4 new schools open 2007–2009 value £119m. 25 year operating contract A managed service for ICT. Focus on sustainability[42]	Risk allocation proposal and value from bidders compared against LEA risk matrix at short list stage	Initial feasibility study test against PSC capital and revenue grant	2% savings. Use PSC to compare initial bids and estimated risk transfer. Benefit was usually worth 50% of risk transfer[43] Later report concerned with diminishing VFM in the operating stage and closure of schools	Bring in financier earlier, more detailed PSC & risk valuation before shortlist bids Reduce external advisor costs by bundling schools in single PFI
Prisons a) Bridgend	17 months 17 months	HM Prison Service (HMPS) approves	Option appraisal including HMPS daily	Compared each option with police	10% saving. Compare with artificial PSC	Nine months OJEU to start

b) Fazakerley c) Ashfield		design and build specification Political imperative for expanded provision from 1995 with PFI custodial private service. 25 year concession	contract cost based on number of places available. Service risk allocated to private supplier including occupancy Deductions for poor custodial services and maintenance up to 5%	cell option	and estimated risk transfer. 4 PFI prisons perform in the highest performing group for decency and purposeful activities and flexibility, but worse on security	Early identification of non-transferable risks *Equitable custodial services KPIs with public prisons Flexibility for policy change to rehabilitation Financial viability*
Health a) Large SW Hospital 2010 b) Dawlish c) Bodmin	36 month with competitive dialogue 36 months 12 months	Department of Health approval including variations at the preferred bidder business case 76 projects worth £890m/year. FT, NHS Trusts and PC	Standard checklist in outline business case. Public/private/shared	The full confirming business case is used to confirm affordability	Compare with PSC in a discounted cash flow model and sensitivity analysis Most are well managed and deliver expected VFM NAO (2010)[44]	NAO Report 2010. Reduce tender period Provide scope for sharing contractor efficiencies in refinancing gains and maintenance operation over multiple hospitals. Manage operations – possible reduction in scope of PFI services

Fewings (1999) [40]

The claims for VFM in operation have still to be tested in the full term of the contract as there is also escalation built in through raising service charges. The research by Henjewele in healthcare and transport sector operational projects showed that more than two-thirds of the projects are paying extra unitary charges due to increase in operation costs. In healthcare projects over 40 per cent increased facilities and had to incur additional capital cost. The 2007 report by the NAO shows that major changes account for less than 10 per cent of all changes made but they contribute over 90 per cent of total cost overruns. There is a feeling that financial commitment to a unitary charge cuts down the flexibility for changes in policy and that the risk of obsolescence should properly be valued in the business case or suitable flexibility in the design for change. For example, if the Millennium Dome had been built under a PFI procurement then a 25 year ongoing charge for a building redundant after one year would have to be totally unacceptable. However some efficiencies may also be gained by bulk purchase of several PFI concessions by the same contractor e.g. Premier Prison Services operate over the 5 prisons. Other efficiencies may come through project refinancing.

Members of the SPV in their turn may wish to sell on their interest and authorities need assurances about the quality of their new partners. Authorities are generally given powers to terminate contracts which are substantially in breach of the terms. There are a growing number of reports that have dealt with various aspects of PFI identifying problems with escalating operation costs, the sharing of savings gained by rescheduling finance and the degree of design flexibility that can be built in to make the facility sustainable for long term future use. New contract conditions help to share future savings made by the operator with the client, but more needs to be done to share the risk of policy change on facility provision. For example the current UK schools designs incorporate quite open plan facilities which may get out of date.

At the centre of the arguments the concerns lead back to properly assessing value for money over the operating term. Whilst some believe that paying out for profit to private providers and the less favourable borrowing terms cancel out benefits for almost any project over the long term. Many believe that the proper assessment and the credible use of tools in that assessment will help decide which projects are suitable for this form of procurement. A veto by the Treasury to guard against uninformed use of PFI by public bodies as a 'cure all' is critical. PFI projects may not have a better life cycle cost, but they do need to show additional credible intangible benefits such as improved maintenance, flexibility, longer economic life and remain affordable over the concession term. Value for money needs to be properly tested for each project against the PSC and other political or extra project benefits have to be properly assessed if the project is to proceed.

Further NAO (2003) recommendations suggest that continuous improvements may be made to PFI in general in making examples of good practice available, to stimulate further innovation in future projects.[46] Other problems in performance were identified by PAC (2003)[47] as the need to reduce major changes and additions in the early part of the operation of the facility. Where necessary they recommend benchmarking them and rigorously testing the PFI service contractor costs by having conditions which allow transparency of pricing or competition from other contractors.

Track record

The other fundamental problem is the lack of track record for PFI and the huge commitment made for public funds on untested risk analysis in a wide range of sectors. Table 15.4 only gives a limited view which needs to be updated by regular reports. This however has led to one of the merits of PFI as extensive public assessment is an audited requirement and private assessment is good business, so in partnership the risk is significantly better planned and managed than standard procurement. Conventional procurement has often led to poor maintenance regimes and the premature demise or expensive renovation of public buildings when insufficient maintenance or poor quality buildings have been accepted to reduce capital borrowing. This puts risk factors in on the opposite side of the balance sheet as a saving and a promoter of integrated best practice. In more recent fluctuating conditions the NAO (2010)[48] has warned of the increasing cost of new private finance compared with public sector facilities in difficult credit environments. They suggested that the 2008 conditions increased the annual charge of PFI projects by 6–7 per cent.

For the project manager PFI and DBFO is a pecuniary responsibility and thorough assessment must be seen to be done. In addition later undermining of the business case should not occur through poor negotiations in the preferred tenderer period, where the pressure to proceed or to heavily compensate tender costs in the case of pulling out may unbalance the negotiation leverage for a good deal. On the other hand a fair deal for the provider is critical in order to engender the partnership that is so important to value for money in long term relationships.

Funding

Funders generally build in step-in-rights to protect their interests in the case of non-performance and may, 'in extremis', replace contractors who are under performing. These rights are part of the standard contracts in most sectors. Funders also take an interest in the finalisation of terms in the negotiation stage. Clients and providers will push for competitive long term interest rates as this substantially increases the competitive nature of any PPP.

Refinancing the SPV debt has potential for savings and was first controversially carried out on Fazakerley Prison by the contractor,[49] which one report quotes as trebling the contractor's profit from 13 per cent at the letting of the contract to 39 per cent after refinancing.[50] Savings may be gained from reducing the interest repayments, by lengthening the term of payment, or by combining the debt to get better terms. This led to new OGC guidance to build in contractual provisions for sharing any refinancing windfalls which may be gained on the original deal with the Authority. Refinancing may also lead to transferred liabilities for the Authority in the case of termination of the contract or selling on of interest and it is now usual for contractors to require consent for any refinancing that may affect third parties.

For instance the Calderdale Hospital gained a windfall saving of £4 million for the taxpayer when it was refinanced after the more risky construction stage and a deal between the contractor and their financiers was struck to reduce the interest rate levels. 30 per cent was voluntarily shared with the NHS Trust in accordance with the OGC code for historic contracts and recommendation for a 50 per cent sharing in standard contract conditions.[51]

Design quality

Design quality has often been criticised in PPP by its opponents and its users. Building design quality has been defined as the extent to which the asset has a high standard of space, light and sensory comfort, whilst retaining all functional requirements and relating to the surroundings.[52] The government however is anxious not to straitjacket the process and to allow some innovation into the system to achieve results. Later experience in Building Schools for the Future has introduced innovative teaching environments and more durable products.

The guidance note on PFI indicates the following key principles:

- Functionality must be agreed by the sponsor to satisfy the existing use, but also allow for expected future change, growth and adaptability.
- Provision for service enhancement refers to an airy, clean and well lit environment in which customers and staff feel valued and respected.
- Particular architectural requirements should be given a strong steer at the bidding stage so that bidders may make comparable proposals. For example, the Berlin Embassy PFI was put out to an architectural competition prior to the bidding proposal. The winning design was then made available to the bidders so that they could take account of the key features in their own detailed design.
- A consideration of the social and environmental impact of the building.
- Design responsibility is handed over to the SPV on the basis that an output specification provides greater freedom for innovation. There is also a need to give the contractor an incentive to provide high quality design without affecting their ability to win the contract.

Supplier risks

The supplier risks are well outlined in CIC (1998),[53] and CIC indicate the importance of asking the question 'Is this project for me?' from a business strategy and financial point of view. This is because of the significant involvement which is required and for a major partner of the SPV there is also a large financial commitment. It is also an extended commitment and if the terms are not carefully negotiated and the partnership is not right, a haemorrhaging of resources is also possible with poor financial returns over a long period.

The wider supplier risks that are cited as one of the main rationales for PFI are indicated as bidder risks, design, construction, commissioning and operating risks, the consequences of delay or defective quality, operating costs and indexation, demand risk, residual value risk, maintaining constant quality risk and finance risk. The latter risks are unaccustomed risks in conventional contracting and arise out of a longer involvement. Once the contracting phase is over however the most risky components are seen to be past by the lenders and it is possible to favourably refinance on the basis of the successful achievement of the risky phase.

The compensation for taking on more risk is that the remuneration should be greater and returns of 7–15 per cent should be expected. Risks under priced or overlooked can offset the compensation and realistic pricing of new risks comes from experience. It may also be possible to sell on an equity share in the PFI, if it is found that the conditions are not suitable to the trading profile of a future business strategy. As yet, the market in buying and selling PFI equity is not firmly established and significant 'write offs' may need to be made.

The other more significant cost compared with traditional projects is the tender bidding cost. These are relatively low in the first round, but increase significantly in the shortlist and even more during the two track competitive dialogue and negotiation phase to financial close, because of the degree of financial planning and design detail which is required to convince the client that they have value for money. The use of an output specification by the client means that quite a bit of design is done to determine the functional form before a price can be worked up. More work is also carried out on forming the SPV, bringing a funder on board with the right financing deal and determining the basis of the business relationship. The formula for a competitive bid is also more complex with a balance of non-financial factors such as design, team partnerships and innovation arising out of value management.

Opportunities may also arise out of asset offers by the Authority (e.g. excess land), or the widening use of the premises for private as well as public use. Each of these takes time and costs money for consultancy advice.

The financing of a PFI project is generally a banking facility with a small amount of supplier funding off the contractor balance sheet required.

Alternatively project equity financing can be taken out against the cashflow of the project. In practice on a project of any size some of each is taken out with an 80–90 per cent debt funding being the normal range. The influence of the lender on that amount of debt stretches to contractual requirements for step-in-rights to retrieve debt in the case of project difficulties and the presentation of a viable business case and risks contingency plan. In the early days the validity (vires) of certain quasi-government organisations (e.g. NHS trusts and local authorities) to place and underwrite such long payments was questioned. This has been solved latterly by the passing of the Local Government Contracts Act and a similar legislation for the NHS Trust.

Conclusions

The development of PPP has meant that a large proportion of public work (15 per cent) in the UK and significant amounts in other countries is now carried out using PPP procurement. The long term relationships which are necessary encourage partnership and the long term commitment encourages business planning and risk management at the beginning of the project where the problems can be ironed out with less conflict. Value for money exercises have led to greater discussion between the user and the provider. Integration between the various parties of the SPV and the client is much stronger than when funders, contractors, clients and consultants are all operating under conventional procurement. The project manager has the potential to play a leading role to drive the project particularly where they have a good understanding of the client business.

There are still concerns about the justification of PFI projects that are induced by the lack of alternatives to other finance. It is possible in these cases that public institutions are motivated to make up cases that are affordable because they attract the funding, but are not proven value for money. Collapses in these types of schemes have been illustrated where over optimistic incomes have been forecast. Some go as far as to argue that PFI as a concept is less efficient and argue that we are just delaying the extra costs. Others are also concerned that handing over design responsibility on landmark developments and pressure on the planning authorities may lead to compromised building and urban design quality. SPVs often operate in name only with contractors keen to be paid off at the end of the construction phase, nullifying the commitment to long term success predicted above.

Evidence collected so far by the National Audit Office and others suggests that capital cost savings and occupational efficiencies are being achieved. These same reports stress that conclusions about the full life cycle costs cannot be proven until projects' concessions are complete as other factors such as inflation, obsolescence and handover costs are uncertain. It is also clear that certain projects are more suitable for PFI and these include schemes that are more easily accessible to the general public so that multiple use can be made of capacity to generate more income. The release of land and other assets that can only be redeveloped because of the PFI scheme is also another way of reducing capital and service charges.

Since the major world credit crisis there is a heightened concern over the comparative cost of private and public finance for new projects widening disproportionately so that more PFI projects are pushed out of the VFM zone. This could be countered in the long term on the basis of refinancing deals when the balance reverts. The public sector also has an eye on sharing any subsequent efficiencies that emerge from grouping project maintenance to make bulk savings.

PFI type arrangements such as DBFO and BOOT are occasionally used by the private or quasi-public sector, but have yet to see a real uptake outside the public realm.

References

1 EC (2008) *Commission Interpretative Communication on the Application of Community Law on Public Procurement and Concessions to Institutionalised Public-Private Partnerships (IPPP)* C(2007)6661. Available online at: http://ec. europa.eu/internal_market/publicprocurement/partnerships/public-private/ index_en.htm (accessed 14 November 2011).

2 Public Accounts Committee (2011) *Lessons Learned from PFI*, 44th Report of PAC. HM Parliament, 18 July. Available online at: http://www.publications. parliament.uk/pa/cm201012/cmselect/cmpubacc/1201/120102.htm (accessed 12 December 2011).

3 The Private Finance Panel (1992–1997), The Treasury Taskforce (1997–2000); Partnerships UK (2000–2010) was formed as a 50 per cent equity with the government to provide consultancy services to central government and government agencies, 4Ps gives support to local authority PFI projects and other large agencies like the NHS have their own support units. In addition PFI projects have been scrutinised regularly by the National Audit Office (NAO), the Audit Commission (AC) and Public Accounts Committee (PAC). Verdicts and guidance are published in the public realm.

4 NAO (2006) *A Framework for Evaluating the Implementation of PFI Projects.* Available online at: http://www.nao.org.uk/publications/0506/pfi_framework. aspx (accessed 10 October 2011).

5 HM Treasury (2009) Application Note PPP Projects in Current Market Conditions. Available online at: http://www.hm-treasury.gov.uk/d/market_dislocation_ 280809.pdf (accessed 10 October 2011). The Treasury department estimates that payments for the current commitments will increase from £8.6 billion for year 2011/12 to £9.5 billion for 2020/21 before they can start dropping.

6 Grimsey G. and Lewis M. (2005) 'Are Public Private Partnerships value for money? Evaluating alternative approaches and comparing academic and practitioner views', *Accounting Forum*, 29(4), December: 345–378.

7 World Bank (2004) *Guidelines: Procurement under IBRD Loans and IDA Credits.* Available online at: http://siteresources.worldbank.org/INTPROCURE MENT/Resources/Procurement-May-2004.pdf (accessed 10 October 2011), p.43.

8 OECD (2006) A Policy Framework for Investment. Available online at: http://www.oecd.org/document/61/0,3746,en_2649_34893_33696253_1_1_1_1,00.html (accessed 10 October 2011).
9 Audit Commission (2000) *Taking the Initiative. A Framework for Purchasing under PFI.*
10 RICS Project Management Forum (2003) *PFI and the Skills of the Project Manager.* Project Management Research Papers on CD. July.
11 NAO (2007) *Improving the PFI Tendering Process.* HC149, March.
12 These derive from the Public Contracts Regulations 2006.
13 Based on Grimsey D. and Lewis M.K. (2005) 'Are Public Private Partnerships value for money? Evaluating alternative approaches and comparing academic and practitioner views', *Accounting Forum*, 29(4): 345–378.
14 Treasury Taskforce (1996) *Treasury Taskforce Partnerships for Prosperity.* London, HM Treasury.
15 HM Treasury (2006) *Value for Money Assessment Guidance.* London, HM Treasury, p.8.
16 Henjewele C. (2010) *Value for Money Uncertainties in the Planning, Delivery and Operating of PFI Projects.* Unpublished thesis for the University of the West of England.
17 Treasury Taskforce (1996) *Treasury Taskforce Partnerships for Prosperity.* London, HM Treasury.
18 HM Treasury (2006) *Value for Money Assessment Guidance.* London, HM Treasury, p.8.
19 Henjewele C. (2010) *Value for Money Uncertainties in the Planning, Delivery and Operating of PFI Projects.* Unpublished thesis for the University of the West of England.
20 NAO (1999) *The PFI Report for the New Dartford and Gravesham Hospital.* HC423. 19 May.
21 NAO (2003) *GCHQ New Accommodation Programme.* HC955. 16 July.
22 NAO (2003) *GCHQ New Accommodation Programme.* HC955. 16 July.
23 PAC (2004) Government Communications Headquarters: New Accommodation Programme HC65. 15 June.
24 NAO (2008) *Making Changes in Operational PFI Projects.* Available online at: http://www.nao.org.uk/publications/0708/making_changes_operational_pfi.aspx (accessed 9 September 2011).
25 NAO Guidance Notes (2001) *The Renegotiation of the PFI Type Deal for the Royal Armouries Museum in Leeds.* HC103. 18 January.
26 NAO (2009) Performance of PFI Construction. Available online at: http://www.nao.org.uk/publications/0809/pfi_construction.aspx (accessed 12 March 2012).
27 NAO (2003) *PFI Construction Performance.* HC371. 3 February.
28 CIB (2011) *Building Strong Foundations Financing Britain's Infrastructure.* Available online at: http://www.hm-treasury.gov.uk/ppp_pfi_stats.htm (accessed 21 November 2011).
29 HM Treasury (2010) *Financing PFI Projects in the Credit Crisis and the Treasury's Response.* Report by the Comptroller and Auditor General. HC287. 27 July.
30 RICS Project Management Forum (2003) *PFI and the Skills of a Manager.* London, RICS.
31 NAO (2001) *Successful Partnerships in PFI Projects.* London, NAO.
32 NAO (2010) *Managing Complex Capital Investment Programmes Utilising Private Finance.* Available online at: http://www.nao.org.uk/publications/1011/complex_pfi_projects.aspx (accessed 12 March 2012).

33 NAO (2001) *Managing the Relationship to Secure a Successful Partnership in PFI Projects.* HC375. 26 November. London, HMSO. This survey was based on 121 PFI projects with views from contractors and sponsoring clients.

34 OGC (1998) *PFI Material Colfox School Case Study.* Available online at: http://www.ogc.gocv.uk (accessed 18 August 2004).

35 IPSOS Mori (2008) *Investigating the Performance of Operational PFI Contracts.* On behalf of Partnerships UK and HM Treasury. 151 responses and a wide range of project sizes and types. Available online at: http://www.partnershipsuk. org.uk/view-news.aspx?id=96 (accessed 5 December 2011).

36 Akintola A., Taylor C. and Fitzgerald E. (1998) 'Risk analysis and management of Private Finance Projects', *Engineering, Construction and Architectural Management,* 5(1): 9–21.

37 Akintola A., Taylor C. and Fitzgerald E. (1998) 'Risk analysis and management of Private Finance Projects', *Engineering, Construction and Architectural Management,* 5(1): 9–21, Table 2.2: Ranking of PFI risks.

38 Henjewele C. (2010) *Value for Money Uncertainties in the Planning, Delivery and Operating of PFI Projects.* Unpublished thesis for the University of the West of England.

39 Fewings P. (1999) *The Implementation of Public/Private Partnerships: The UK Experience.* Proceedings of the CIB W55/W65/W92 Joint Symposium – Customer Satisfaction: A Focus for Research and Practice in Construction. Cape Town.

40 Russell T. (2004) *Practical Guide to PFI for Higher Education Institutions.* HEFCE.

41 HEFCE (2003) *Investment Decision Making: A Guide to Good Practice.* HEFCE.

42 Making the Partnership Work (2011) Available online at: http://www.bristollep. co.uk/3_pages/stories/partnership.htm (accessed 24 November 2011).

43 Davies J. and Ghani A. (2006) *PFI School Procurement: Analysis of a DTI survey of English Local Authorities.* DTI Industrial Economics and Statistics Directorate.

44 NAO (2010) The Performance and Management of Hospital PFI Contracts. Available online at: http://www.nao.org.uk/publications/1011/pfi_hospital_ contracts.aspx (accessed 12 March 2012).

45 Fewings P. (1999) *The Implementation of Public/Private Partnerships: The UK Experience.* Proceedings of the CIB W55/W65/W92 Joint Symposium – Customer Satisfaction: A Focus for Research and Practice in Construction. Cape Town.

46 NAO (2003) *PFI Construction Performance.* HC371. 5 February.

47 PAC (2003) *PFI Construction Performance.* HC567. 15 October.

48 NAO (2010) *Financing PFI Projects in the Credit Crisis and the Treasury's Response.* Available online at: http://www.nao.org.uk/publications/1011/pfi_in_ the_credit_crisis.aspx (accessed 24 November 2011).

49 NAO (2000) *The Refinancing of the Fazakerley PFI Prison Contract.* HOC Session HC584. 29 June.

50 CIOB (2003) *PFI Value for Money Two.* International News. 25 August. Available online at: httop://www.ciobinternational.org.com (accessed 28 July 2004).

51 OGC (2002) *Voluntary Code of Conduct.* Speech by Peter Gershon, Chief Executive of Office of Government Commerce, at the UK Government Conference on Public–Private Partnerships in London on 16 October 2002.

52 HM Treasury (2006) *Strengthening Long-term Partnerships.* Available online at: http://www.hm-treasury.gov.uk/d/bud06_pfi_618.pdf (accessed 21 March 2012), p.105.

53 Ive G. and Edkins A. (1998) *CIC Constructors' Guide to PFI.* London, Thomas Telford.

Chapter 16

Towards a more integrated approach

I have found no greater satisfaction than achieving success through honest dealing and strict adherence to the view that, for you to gain, those you deal with should gain as well.

(Alan Greenspan)

The final chapter of this book is an attempt to pull together the varied strands of project management which have been portrayed and integrate them as a whole. Project management is often portrayed as a set of tools and it is important to see that different actions need to be related in the whole process of delivering a successful project to a client. There is a need to develop an integrated team, because there are so many fragmentary processes in the construction industry.

There are now seven key themes that have emerged throughout the book and are described in terms of promoting the project manager's job:

- **P** lanning and controlling for lean and safe performance.
- **R** ealising an ethical and sustainable project for life.
- **O** ptimising information flow, risk and value in the supply chain.
- **M** eeting a client's requirements sustainably for all.
- **O** rganising flexible structures for continuous improvement and communication.
- **T** otal quality and safety management.
- **E** ngineering an integrated project team and leading effectively.

Each of the chapters has promoted at least one of these themes and they are important to the overall delivery of the project. Project management is the integration of these themes.

The umbrella principle of integrate is the bringing together of disparate parties into a holistic approach to the delivery of a project. The first half, 'inter', has the sense of mutual, reciprocal, between and among. The second half has the opposite intonation, conjuring up pictures of friction; being too close together and rubbing up people the wrong way, sounding a note of caution for the principle relating to words such as interfere, interject and interrupt. Integration in the project context is not a toolkit, but a whole philosophy promoting togetherness for more productive ways to bring more value to the project.

The opposite to integration, differentiation,[1] is not a bad thing and has come about as a cyclical phase in construction when projects have been subject to specialisation and project activities have been divided up to cope with the increasing complexity of projects, because of the range and sophistication of technologies expected and the wider division of labour. Walker describes the need for pooled interdependence where lots of specialist contributors from independent organisations have interfaces where their work overlaps, but depends upon the work of others.[2] Interdependence has also occurred sequentially through each stage of the life cycle during:

- Project definition (client and project team members and external stakeholders).
- Design (design team members).
- Construction (contractor, design team and downstream supply chain).

He gives an ascending hierarchy of pooled, sequential and reciprocal interdependency. The hierarchy is defined by the degree of interaction and overlap of the interdependent tasks. Reciprocal interdependency is needed in the construction process where decision making is iterative between many parties and across the life cycle stages, including external stakeholders. As an industry construction is fragmented between many different professions and suppliers with complex relationships and an independent and uncoordinated approach is not appropriate. In particular the divide between construction and design is almost unique in the delivery of a product. This is further compounded by the difficulty in identifying a single customer for the finished product – developers, owners, users, taxpayers and shareholders are all likely to input their brief as stakeholders. The PM has a unique role to recognise interdependency and to ensure that it has good communication channels and to manage any conflicts.

A further concept Walker introduces is sentient groups, a term given relevant context by identifying the strong commitment of individuals in the industry to commit themselves to their own professional or disciplinary groups.[3] This sentience within groups triggers strong loyalties and causes conflict at the task interface between two sentient groups, such as architect and contractor. The group norms and educational traditions create differences in problem solving approaches and priorities. These need to be resolved by finding common ground around the conflict. Known contractual conditions help commonality but often fall short. Here an integrated approach has much importance because it involves the application of leadership.

Integrated leadership does not negate specialisation, but looks to develop creative means of getting people to work together with maximum synergy to intertwine processes that help to cut out waste. It could more accurately be described as managing a change process. The preceding chapters have indicated current and best practice, but have also discussed the emerging integrated processes of sustainable construction which requires innovation as well as signing up to longer term aims beyond a single project's outputs. At this stage it is worth looking at the case study background of an integrated project (Case study 16.1).

Integration

Procurement

Prime contracting procurement provides an output specification to the prime contractor (PC) who mobilises the design and construction team to deliver the estate. The principles of prime contracting are to establish an integrated project team which levers in the direct involvement of the supply chain in strategic decision making and to ensure a high standard of management by the PC to create conditions that respect the workforce. Some of the mechanisms suggested by Defence Estates (DE) for achieving long term value for money are incentivised payments, improved supply chain management, continuous improvement and partnering.[4] Other principles that contractors take on board are a three stage selection process including negotiation, a single point of contact, clear allocation of risk, through life costing, open book accounting, prompt payment of suppliers and the early appointment of a disputes review board to provide timely interventions to manage disputes. The contract has standard core provisions and will be tailored to suit specific project circumstances. It is connected to the Ministry of Defence programme called Smart Procurement of which Building Down Barriers (BDB) has been a successful pilot run for its application to construction. Andover North Site Redevelopment was one of the BDB pilot projects.

In practice integration and differentiation take place on a project simultaneously. Simplistically different skills and roles are differentiated in the OBS and responsibilities are assigned to associated task completion through the WBS. At the same time those tasks are integrated by the sequencing of the tasks in a programme or a network so that a logical relationship is agreed and programme slots are allocated. If there is some common working on the task so that one task affects another in terms of quality or health and safety then more detailed integration is required. As the process gets more complicated it may be necessary to integrate construction methodology and design and this is the rationale for procurement systems that move away from the traditional approach and bring in the contractor at an early stage. The capital cost and the operating costs are

Case study 16.1 Andover North Site Redevelopment

Andover North Site Redevelopment is one of the first tranche of *capital* prime contracts rolled out by Defence Estates (DE), post 2000, to improve performance in DE procurement as an alternative to PFI.

The site is the headquarters for defence logistics on a brown field site and consists of the development of a technical building, offices for 780 people, a separate training facility, a crèche, sports facilities, a mess with accommodation, a gatehouse and 34 acres of landscaping. It is worth £40m and was started in January 2000 and completed in October 2002. An additional operating budget was agreed for a six-year maintenance (compliance) period. We shall be using this case study throughout the chapter and refer to it as the case study. See Brown (2002) for more detail.[5]

Figure 16.1 The ANSR external view and inside of office facility. Used with permission of A. Brown[6]

integrated to assess the balance for maximum value. Whole life costs allow alternative design options to be properly evaluated and decisions made properly. It will also affect the selection process for contractors as the lowest tender for capital cost may not be the best value. One possibility is the letting of a contract on a two stage tender so that a competitive first stage leads to a negotiation stage where value is squeezed out on the basis of sharing the savings within a target budget. Emerging out of this is the integration of design and facilities management. A maintenance contract with a design build contractor ensures this. BIM is an emerging technology that encourages interdependence early on merging design with costs, construction time constraints and facilities and asset management.

ANSR is procured on a prime contract with a seven year maintenance period and this involves a core team including the key designers and three cluster leaders co-ordinating the supply chain. The core team tendered as a joint consortium and were chosen on a balanced scorecard basis with a three stage selection process. Factors taken into account when selecting a prime contractor are project management ability, technical competence, financial standing, supply chain management, willingness to share risk and under-standing of the MOD culture.[7] The winning prime contract team signed an integrated project agreement giving a project bank account and shared PII to bind the team together with common risk and shared profit. The Manchester Central Library is an example of the advanced use of BIM in Case study 3.3.

Leadership and teamwork

Further integration is a factor of managing people and getting the project team performing by getting some synergy over and beyond the sum of the parts. This requires effective leadership, motivation and teambuilding exercises that make the team aware of its strengths and weaknesses and the generic roles that are required. Integration may also mean resolving conflict and having mechanisms in place that encourage 'win–win' situations without a culture of blame, using principled negotiation techniques that support this philosophy. Trust has been raised in the context of supply chain management and collaborative methods such as partnering recognise the need to develop deeper, preferably long term relationships that allow straight talking and cut conflict by not attaching blame, but concentrate on trying to sort out the problem. There are several areas for building trust such as the use of joint ventures, transparent project accounting and the abolition of retention. Leadership flourishes in the integrated approach because it encourages a wider range of leadership, taking away the bottleneck of channelling all outputs through one person. It requires a clear sharing out of responsibility amongst cluster leaders.

In the case study the prime contractor and core team agreed to go with a fixed margin for all core members making them jointly liable for performance with direct leadership responsibilities for their supply chain and executive responsibility for project decision making (Figure 16.2). They established themselves as a virtual joint venture making them financially accountable to each other. A core team of cluster leaders have the democratic right to challenge the prime contractor who holds the largest share, but has less than a controlling 50 per cent share of the combined vote (Figure 16.3). Empowerment is a buzz word, but allows a widening of ownership and commitment.

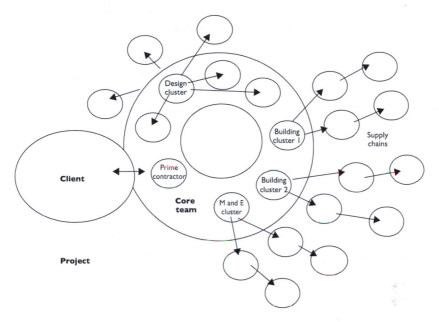

Figure 16.2 Diagram showing leadership through clusters and supply chain

Team building is another possible benefit of integrated teams as it allows for teams to build up an understanding. Collocation has been discussed as a way of improving communications in both quality and speed of response. A team may also choose to develop generic roles on the basis of natural tendencies such as the Belbin or Myers Briggs assessments to profile the team better, but also to understand the way others work, either in meetings or in problem solving situations, when inputs are needed. Recognition of ways to increase synergy within the team can be carried out by using the team to create its own motivation.

Integrated decisions

The stage gate system is a way of integrating the client approvals with the tasks to be carried out by the project manager through the life cycle of the project. The OGC Gateway Review™ shows six gateways for decisions from inception through to completion for application to major project management. These provide a framework which can be related to the RIBA Plan of Work. The framework identifies the client approval to achieve gateway target dates. This may be harmful to the project morale if these gateways are associated with bottlenecks or stop points in the project and this requires committed leadership on both sides to identify the potential problems and to ensure approval requirements are known and planned and achieved. For example, planning permission is an unknown final committee decision, but the planning officer and others may be integrated into the team and should be used to reduce uncertainty to ensure better local knowledge and hence submissions. The gateways may be further broken down into milestones that are relevant to the particular project and provide targets for particular people down the hierarchy.

Integrated decision making is about getting inputs from the whole team and assigning the final decision to the team members who are accountable for the outcomes. This provides mutual support and a no blame culture encourages more experimentation without taking away accountability. One model indicates the important of generating ideas, evaluating them from a team and task point of view and then identifying clearly the way forward based upon the evidence.

Risk and value

The early involvement of contractors in the design team to ensure that there is an input on constructability at the early stages of the project allows a wider and more integrated consideration of risk and value from all parties that are going to be expected to carry risk. It also allows contractors to gain a better understanding of the values and objectives of the client in managing those risks appropriately. Contractors under traditional forms of procurement are excluded until construction, which is too late for them to make any significant contribution to design. Buildability is an important part of design construction integration. There is also an incentive at this stage to introduce alternative products and methodology which may induce value for money. Contractors may identify health and safety problems with a design methodology, and stop erosion of value by stopping late changes to the design and abortive costs.

In the ANSR, Figure 16.3 shows how the power was shared among the cluster leaders and the risk apportioned and passed down the supply chain. The percentages shown indicate the derivative weightings of risk sharing and profit sharing. The client rewarded savings made to the prime cost target in the same weighting to encourage team generated value improvements. Cost over the prime cost target was paid for in the same proportions. They saved £2m on the budget.

Innovation

Innovation emerges out of an integrated approach, because it requires a collaborative no blame approach and the allowance for experimentation. Time is also required because innovation is iterative like design and needs time for research, reflection and testing. Innovation usually starts off as a

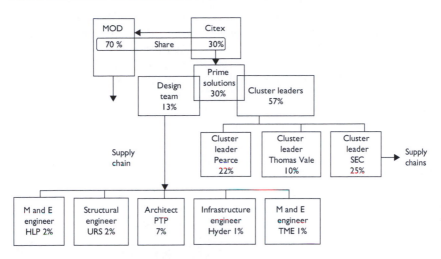

Figure 16.3 Pain-gain sharing amongst core team

creative brainstorming process around a particular idea so the focus of innovation becomes wider. The next step is to narrow the ideas down by a process of screening. This might happen in a value management workshop and the ideas can be narrowed down by risk management. At this stage an integrated team can decide to give space for rapid prototyping as Maylor calls it.[8] The use of an innovation fund which compels management to develop all reasonable ideas received from the workforce through a suggestions box could improve the rate of ideas. These ideas could be patented to the workers and any savings over and above the costs of development shared. This encourages the ownership of ideas and the harnessing of knowledge and creativity in the project.

In the case study it was agreed to join the training building to the office block, making a key saving in capital costs by reducing external cladding and in FM by sharing registration and staffing levels. In the case study, incentives for innovation emerged from sharing risk and gains as Figure 16.3, so that if savings were made, then all benefited from a proportional share of the gain agreed at the beginning to represent the degree of risk which each organisation was prepared to take. In a project for a series of vehicle maintenance workshops (Case study 13.5) workers were invited to workshops to generate new ideas for savings based on improvements to previous projects and a reward system was used.

A different type innovation is connected with adjusting the project process. An example of this would be instituting weekly payments instead of monthly payments for suppliers to enhance their cashflows, in return for a reduction in price. The idea has the potential to get the subcontractor to give the project priority treatment (fewer delays), as it improves supplier motivation. It could also improve productivity. It has a cost because the cashflow of the main contractor will be worse, but the contractor is already motivated. Safeguards need to be in place by monitoring the sustainability of the productivity improvement assumed. Overall it has a good chance because it is a 'win–win' with a likely profit increase for the contract, a cashflow worry off the mind of the supply chain and better relations. It has a chance of failure if extra quantity surveyors are used, eroding the profit gain through overheads and exasperating relationships by paring supplier profits in the cash back.

In ANSR prefabrication workshops were set up on the site for early preparation of mechanical and electrical elements in the 'rapid prototyping' model. This was complemented by the use of an M&E wholesaler on the site who carried out all the purchasing JIT on behalf of the contractor, in return for being the sole supplier. Weekly payments were used for smaller suppliers in return for standard prompt payment discount.

Sustainable construction

Sustainable construction strives for the more durable operation of a building so that there is longer life and less use of non-renewable resources with the restriction of emissions that promote climate change or development that harms ecological balance. It also looks at promoting social change to promote well-being, all at an economic cost. In theory it will pay back greater capital cost to the owner through use of fewer resources, in the production and use of the building, and the better motivation of the workforce and benefit to the community. It promotes a changed culture and is not an add-on and requires the leadership of the client in their policy and strategy for the building. For designers there is the challenge of innovative design that produces better value for money for the new paradigm, for the project manager it is the challenge of making sure predicted performance is delivered keeping within affordable budgets, and for the contractor it is the risk of making new technologies work within a 'rapid prototype' environment. A collaboration and integration of client entrepreneurship, design innovation, experience, cost control and testing is a key aspect of making sustainability perform. The case for a considerate contractor has already been espoused for increased partnership with the public during the construction period for employment and for neighbourly and community action. Schemes are already in position for this and have often led to other benefits for the contractor and better acceptance of building schemes in general.

In ANSR there was an objective to reach BREEAM excellent standard and an emphasis by the client on improving the ecological balance of a brown field site (old RAF base) by creative landscape design incorporating all existing trees on site and preserving chalk downland especially as it included living quarters on site.[9] For the project manager he instituted a 'balanced scorecard approach to the choice of components encompassing through life costs, programme risk, health and safety, environment and aesthetic'. The contractors reduced the soil and demolition materials to landfill by the use of cut and fill and reused topsoil and maximised recycling. The designers introduced a flexible open span structure with 'truly demount-able' partitions, using heat pumps for cooling and heating, low energy lighting and building fabric cooling. They estimated savings of £375,000 per annum on the running cost of buildings. Health and safety was a proactive approach and they staged two rescue scenarios with the emergency services whilst on site. They also carried out training for occupiers on the economic use of water and energy and the BMS monitors use to allow economic adjust-ments.[10]

Information systems and BIM

An integrated information system is not something that exists at the beginning of the project. This requires the building of a system which delivers information in a timely and user friendly way. Information can be delivered to people in different formats to suit each user, or it may be delivered in a common format that all users understand and can receive when and if they need it both to focus their attention and where necessary to understand the 'whole' picture. The latter is an integrated approach, and on larger projects it may well mean the development of a common extranet with proper access to information portals. Consistency is more easily achieved with a single format. Too much information can be just as confusing as information that is non-existent or incomplete. Late information is disruptive and if response to queries is delayed this is also disruptive. What are the priorities and who decides? Information that is delivered instantaneously can help to reduce late or partial access to information, but can also hide priorities. An integrated system needs to be developed to the extent that it is economic for the benefits gained and will also encourage face to face communication amongst the team. BIM is an emerging technology that encourages up front problem solving by integrating design and construction time and cost into a single source. This means more preplanning and certainty, which saves disruptive late information scenarios. The extra work involved needs a culture change from JIT information, but helps whole picture understanding.

Meetings are an interactive forum and so should be used to gather information and make joint decisions; they also build up trust face to face. Other alternatives such as teleconferences – phone or video – may save time. Collocation is a continuous opportunity for feedback and quick decisions, but may be expensive to organise, so virtual collocation (electronic alternatives), or shorter regular periods of collocation may be an alternative for less formal interaction and teambuilding.

In ANSR virtual collocation was used by the development of an extranet, which provided full information to parties, including the client, and was properly managed to prioritise distribution and alert the most relevant parties. No BIM was available.

Conflict resolution

Integration is required between the leadership of different organisations so that the responsibility for key decisions such as payments, programme, cost and quality control, information synchronisation are integrated as a whole for the overall project with objectives to manage conflict. The quality of one organisation impinges upon the quality, profits and programming of subsequent work. Strategic leadership is proactive and is needed to plan and anticipate conflict and make strategic decisions for its mitigation and to exploit opportunities for combined solutions to problems. This has been discussed as feed forward control.

In ANSR prime contracting makes provision for a conflict panel that has a standing brief to mediate as a third party, but could also have a monitoring role in anticipating non-productive conflict and providing an additional eye for the project manager. The case study also indicates the existence of a peer review team consisting of the prime contractor and directors not directly involved in the project, who examine critically the systems and progress as a way of auditing and giving early warning where there are conflict factors which threaten the commercial integrity of the project and therefore by implication the success of the project as a whole. The client was invited onto this review team, but declined in order for it to carry out, without inhibition, a necessary process which was seen to protect their own profits.

Integrating personal and project objectives

The respect for people movement is an example of a more integrated approach between the construction employer and potential employees and organisations in the industry. More and more main contractors and consultants facilitate production rather than do it. The CBPP (2003) report reviewed this initiative,[11] but it is only one of several methods for identifying the wider picture of integrating people into the business and having access to the best talent available for the project. Various initiatives including female recruitment, diversity, equality, better site conditions, CSCS cards for health and safety, and career development and lifelong learning have been introduced. The initiatives need to be integrated together to harness new ideas.

The EFQM model is a more general model for excellence used by 800 organisations across Europe mostly as a self-assessment device. It puts a greater emphasis on people issues and awards prizes for good performances overall, integrating the context of people management and people satisfaction with the business reality and proving that good business is looking after your workforce.

In ANSR it was a contract requirement to do a 360° peer review, which was carried out by two core team directors for each of the core team managers, including one from the employees' own organisation. These were to be found to be more hard headed than the third party coaching approach and gave opportunities for the directors to tour the site and access documents so that personal success in delivering the project was directly under the spotlight. This retrieved a direct connection between the performance of the individual and their ability to deliver on the job as well as bringing in a view from an outsider.

Integrated planning and control

Programme scheduling and control is carried out by planners who have limited communications with the cost managers responsible for budget control and cashflow and the design team. Control measures for time and cost need to be integrated across the supply chain, otherwise they may work against each other. Earned value is often cited as providing a link with the time schedule prediction by comparing the budget spent with the budget that should have been spent at time now and predicting the slippage or advance of the programme. This link is tenuous and assumes that the cost of an activity and the time it takes and its criticality are directly related. This is patently not true and although it may give some approximate guide especially if the cost of materials is stripped out of the cost, more is required to integrate the two. Object oriented systems are beginning to be rolled out through BIM, where basic cost estimates are being linked to design components (objects) which are related to the way that buildings are constructed so that time allocation may be given when the resources are allocated or normal. For example, it would be possible to use this as a powerful 'what if' tool for value management as well as in change management in assessing the sensitivity of different designs and methodologies with an immediate view of its effect on cost and time. After letting the contract market prices could be fed into the software to give early warning and it would aid the use of instantaneous control systems for change control, such as the NEC system, for compensation events is an attempt to put this into practice. With or without the presence of such a tool, integration of corrective action and the reason for problems should be integrated with the team to meet project objectives on a regular basis. A project process map may be helpful to longer term projects and strategic partnering or measured term contracts.

The integrated supply chain in the case study was effective in value management and organised workshops for value management. These use the buildability and technical knowledge of prequalified first and second tier supply chain to provide time and cost information to make better programmes and cost planning. BIM is beginning to be used on projects such as the Manchester Central Library which is an old building that needs a co-ordinated approach to the design with a common digital measurement of the Victorian structure and a database of cost and time implications to make the design buildable and optimise cost and time efficiency.

Design quality

Integration of the project design with the urban planning process requires more attention to the wider impact of the development on the built environment. These issues have led to the setting up of organisations such as the Commission for Architecture in the Built Environment (CABE) and the use of design quality indicators (Chapter 9). Aspirations are to better assess this impact such as is currently carried out on large projects in Europe in doing an environmental impact analysis (EIA). However the emphasis should be to lever in greater value for the community by the provision of attractive and inspirational structures that not only provide employment for the community, but may supplement the enhancement of other social solutions such as crime reduction (security by design), a sense of place by the use of attractive spaces and landscaping and building up collective community pride as a spin off in its functional provision. These may be imposed legally by imposing planning agreements on permissions, but perhaps they are better delivered by seeing them as challenge for added value that maintains the client budget constraints, but also enhances their reputation locally. ANSR won an award for its attractive landscaping and use of existing trees.[12]

Integrated culture

The culture is an integrating tool for bringing people together with the same objective and needs to be led by the client or the project manager. The core project team of consultants and the key contractors need an open no blame culture to build confidence to admit mistakes and take forward a solution to put it right as a team. The culture of openness can be extended to the client and the supply chain so that trust is built into the contract and the opening up of partnering with value savings over and above the agreed price to reduce the cost and to share the gain. Trust is a critical part of the relationships right down into the supply chain and rests in the resolve to treat each other fairly, have more transparent accounting and keep to prompt payment when work is completed.

It depends on a predictable environment and some contracts try to build this in, though good faith is not enforceable in all legal systems[13] because it can be used too widely, but application of the law is sensitive to self-interest and deceit that completely ignore it at expense of weaker parties, in all European courts.[14]

The closer culture at ANSR meant a partnership charter (Figure 16.4) was established to unify objectives. This meant that the project could operate a

Figure 16.4 The pact that was signed by the ANSR team

single project bank account for joint profit sharing and single professional indemnity insurance for joint accountability. This applied peer pressure on team members to perform well and to give each other a hand to perform well if profit was to be maximised and not eroded. The major challenge for developing this type of culture is to pass this down the supply network (chains suggest no interdependency) so that the weakest link is also performing well. SCM is not well developed in construction (Chapter 12), because of short term relationships and the traditional attitudes that have meant that suppliers down the chain have gained special relationships without the chance for 'back to back' value inducing contracts. However more partnering in the industry has been a vehicle from which SCM can get more stability. There is also a requirement for public organisations to regularly widen their partnerships and to market test or to share out work between an approved list.

In ANSR the second and some of the third level of supply chain were brought into the training regime by the cluster leaders. They also had to be prequalified as willing to embrace a partnering culture. The risk regime also required a good rationale to pass risk down the supply chain.

Health and safety, quality and the environment

Integration has already been flagged between the areas of health and safety and the environment as a SHE or sustainability approach. In addition it is possible to integrate quality systems with the SHE systems to cut down documentation, but primarily to recognise the close relationship there is between a continuous improvement quality approach and health and safety and good business practice. Many project managers point out that health and safety is good business because a safe site, where employers are 'looking after their workforce', often produces more productive work and so the investment in safe systems reaps a reward. The relationship between better welfare, a clean site and health and safety is important and pays back the welfare measures. In ANSR and in many other projects awards were available for individual ideas for improving health and safety. If this was connected with quality and the development of a behavioural approach then it might be feasible to make quality or 'first time right' awards and to develop a culture of worker collaboration and continuous improvement towards zero accidents. This could be connected to the *reduction* of supervision levels and the simplifying of quality assurance paperwork to offset the initial costs.

Recent legislation such as the Construction (Design and Management) Regulations has encouraged a more integrated approach to health and safety, by bring together designers, contractors and clients in assessing health and safety risk. It is recognised that the blame for many accidents is interconnected and a poor design consideration can lead to unnecessary lifting and injury problems, due to poor management and organisation on site, indirectly caused by breakneck competition and client choice of lowest price tenders from work hungry poorly organised contractors with poor resourcing. It is also recognised that the safe design and use of products is interconnected and effective handover with client training is required to ensure safe use and maintenance of the building.

In ANSR the prime contractor was proactive in staging simulations of likely accidents in order to be able to improve efficient warning and evacuation procedures.

Customer satisfaction

Customer satisfaction is the starting theme for Chapter 14 as an integrating factor in the delivery of projects and gives a basis for providing sustainable profitability for the contractor and the consultant. Relationship marketing has also been discussed as a way of developing an understanding of the client's business and involving all levels of the workforce in building relationships with the client as a starting point for better satisfaction and customer care. An integrated approach would include the setting up of an internal and external customer system so that there is a culture of continuous improvement. This takes the strain off project management. It is important to recognise the differentiation principle to give a unique service to the external customer in accordance with their particular aims and objectives. Customer accounts have a single key contact over a series of projects and a dedicated core team may be required by the client. This however will need to be balanced with the need to offer variety for the supply chain and to maintain the challenge and variety that many people join the industry to attain. Maister's principle of customer satisfaction is to meet specification by slightly more than agreed to delight the client and get the customer loyalty and referral to other potential customers. In ANSR it is very clear that the customer is very satisfied and used the template of the contract for future delivery.

The integrating principle is fundamental to success and examples of best practice represented by the case studies are inspiring, though there is plenty of room for further improvements. The potential for a wider application is obvious and even though many other best practice cases have not been mentioned here plenty of them exist.

References

1 Lawrence P.C. and Lorsch J.W. (1967) 'Organisation and environment: managing differentiation and integration'. Graduate School of Business Administration. Boston, Harvard University. In A. Walker (2007) *Project Management in Construction Projects*. 5th edn. Oxford, Blackwells.

2 Walker A. (2007) *Project Management in Construction Projects*. 5th edn. Oxford, Blackwells, pp.92–93.

3 Miller E.J. and Rice A.K. (1967) *Systems of Organisations*. Tavistock Institute. Reprinted London, Routledge 2001.

4 Defence Estates (2001) Prime Contracting Initiative 'Spearheading Innovative Procurement Initiatives', Briefing Pack. PCI IPT Communications.

5 Brown A. (2002) Andover North Site Profile: Bucknall Austin Solutions Team. Available online at Strategic Forum for Construction website: http://www.strategicforum.org.uk/sfctoolkit2/help/andover.doc (accessed 20 March 2012). The case study material is derived from the write up on ANSR by A. Brown. (Used with kind permission of Ryder Levitt Bucknall.)

6 Brown A. (2002) Andover North Site Profile: Bucknall Austin Solutions Team. Available online at Strategic Forum for Construction website: http://www.strategicforum.org.uk/sfctoolkit2/help/andover.doc (accessed 20 March 2012). The case study material is derived from the write up on ANSR by A. Brown. (Used with kind permission of Ryder Levitt Bucknall.)

7 The Prime Contracting Initiative Integrated Project Team (2001) Prime Contracting: Spearheading Innovative Procurement Solutions, Briefing Pack, November.

8 Maylor H. (2003) *Project Management*. 3rd edn. Harlow, Pearson Educational, p.79.

9 Defence Estates (2004) Better Defence Buildings North Site Development Andover Landscape Design Case Study. Sutton Coldfield, Defence Estates.

10 Bucknall Austin Prime Solutions (2004) Constructing Excellence Andover North Site Initial Presentation to Constructing Excellence, March. Available online at: demos.constructingexcellence.org.uk/userimages/Attachment1587.pdf (accessed 2 January 2012).

11 Rethinking Construction Respect for People Working Group (2004) *Framework for Action*. May. 3rd edn. London, Constructing Excellence.

12 Defence Estates (2004) Project Case Study, North Site Development Andover. Sutton Coldfield, Defence Estates.

13 Musy A.M. (2000) *The Good Faith Principle in Contract Law and The Pre-contractual Duty to Disclose: Comparative Analysis of New Differences in Legal Cultures.* Scientific Paper. Università del Piemonte Orientale Facoltà di Economia, Novara, p.7.
14 Fewings P. (2008) *Ethics for the Built Environment.* Abingdon, Taylor and Francis, Chapter 11.

Glossary of important terms

BOOT projects. See public private partnerships.

Building information modelling (BIM) is an object oriented database with multiple dimensions which models 3D CAD objects with time, cost, quantities, building geometry, geographic information and building component properties. It can be used as an integrated information system to generate multiple factor information and graphics solutions useful to the whole building design, construction time and cost control and facilities/asset management life cycle costing and durability.

Business plan is a client based proposal for formalising the development of the business case for a new part of a business or facility. It consists of some basic objectives and feasibility assessments and may be developed as the project planning gets underway. On receiving director outline approval it moves the project from conception to inception, which is the arbitrary beginning of the project life cycle as defined in this book.

Client is the commissioning sponsor of the facility and is the main business decision maker in the project team. The client is the key customer, but not the only one. In contract terms the client is often called the employer, although they may have an agent which separates them from any decision making. The client will have the responsibility for paying the bills.

Commissioning is the process of checking and testing the efficacy and compliance of a building and its equipment and adjusting it for maximum performance.

Concept designer is used in many countries to provide an overall concept design without construction detail, but to meet functional and detailed planning permission requirements.

Construction manager is manager of the construction and commercial interests of the contractor on site. In construction management procurement the construction manager provides construction advice and procurement management direct to the client on a fee basis.

Consultants are payable by fee as advisors to the client and members of the project team. An executive project manager is a consultant, as are registered consultants such as architects, cost managers, civil engineers or building services engineers (mechanical and electrical). The client may use a range of non-design consultants such as legal, funding, estate managers etc.

Contractor/subcontractor/specialist contractor. Contractor or main contractor generally refers to the co-ordinating role of construction activities in which there are sub or specialist roles. In the modern context of supply chain there is still a hierarchy, but co-ordination may be operated in clusters, accountable direct to the project manager, client or construction manager.

Cost planning is the scheduling of costs into categories in order for the purpose of estimation and control.

Design team is responsible for all aspects of the design. It consists of specialists in specific areas of design such as a building services engineer (mechanical, electrical and other services), structural engineer (building structures), civil engineer (ground engineering and civil structures), acoustic engineer, landscape architect etc. They may work individually for the client or indirectly for the lead designer (below), design and build contractor or PFI provider.

Earned value analysis is a classic contractor control for comparing actual costs with progress adjusted earned value at any point in time and provides a simple mechanism to predict end cost.

Environment assessment method is a standard for testing sustainability level of a complete building or facility. Well known methods are BRE Environmental Assessment Method (BREEAM) and Leadership in Energy and Environmental Design (LEED).

Environmental impact assessment is a formal system of checking out the impact of a project on the neighbourhood ecology and inhabitants for emissions, geotechnical, noise and visual pollution.

Facilities management is the provision of non-core services to the client's facility. It is the planning, organisation and managing of a facility on a day-by-day basis, to maintain the physical assets. A facility is a building or an infrastructure.

Feasibility generally refers to the testing of alternative solutions in order to optimise value and ensure affordability without destroying functionality. Affordability is different and is the test of issues which might terminate a project. It must include a financial appraisal.

Handover sometimes called **close-out** is the process of practical completion when the building or facility is given up for use by others and involves training, documentation, health and safety file and maintenance manuals and readiness for use.

Integration or integrated team is the concept of seamless working between the client and their consultants and contractor supply chain. It assumes a moving away from an interface management in support of a virtual organisation, where individual organisational objectives are subjugated (integrated) to project objectives. It should lead to an early appointment of the whole team to take part in the development process.

Lead designer co-ordinates the design team effort. In traditional procurement (see below) they have access to the client, receive the brief at inception stage, develop the design, arrange to tender for construction and exercise a quality checking role during construction. They are the key communication channel which gives them a project management role as well as a design role in the absence of an executive project manager. The lead designer is usually the architect in building projects or the engineer in civil projects.

Lean construction is constructing buildings by driving down the waste using the principles of more efficient design, productive layouts, less wasteful and better use of materials to reduce waste and rework.

Life cycle costing is similar to **whole life costing** but is normally confined to the costs in use and the capital cost rather than the broader business issues of access and market and ongoing productivity. It is important for assessing sustainability.

Master plan in our context is similar to the **project execution plan**. However it is often a pre-planning permission, design based document used to provide stakeholder analysis, spatial planning and blocking layouts, typical elevations, transportation routes, landscaping, signposting, access and planning use zoning for complex redevelopments of sensitive or large development sites. It may also be a single development layout plan.

Planning co-ordinator is a technical term for co-ordinating a suitably integrated health and safety approach in the pre-tender health and safety documents and ensuring key as-built construction information is collated and handed over to the facilities team. They play a statutory role that is complementary to the lead designer. Now called a **CDM co-ordinator**.

Portfolio management is the management balance of a portfolio of projects within an organisation to provide the assets and facilities required for the strategic health of that business.

Principal contractor is the equivalent co-ordinator in the construction phase to produce a health and safety plan and integrate safe methods of working and plan and control a safety management system.

Private Finance Initiative (PFI) is a special form of PPP (see below) which leverages all the capital funding from private finance so that banks or other funding instruments are also brought into the partnership. The facility is transferred back to the public body at the end of a concession of 20–30 years' maintenance and ancillary functions.

Procurement specifically applies to the method of tendering and contracting between the client and their suppliers. The different methods vary from traditional separate procurement of design and construction to turnkey arrangements with a single point of contact. The choice determines the balance of risk allocated between the client and members of the project team. The project manager needs to fit procurement type to client needs.

Programme management is the management of several related parallel or sequential projects. A programme is often identified to provide co-ordinated management and may be many related small or large projects. It helps strategically for clients to manage resources, standardise outputs and allow supply chain management to look at issues which are common between projects to develop value and to capitalise on past knowledge gained. Alternatively the programme will relate construction project/s to IT projects, moving people projects and managing space projects that they need, setting up new systems to make everything work. To manage these separately would court disaster.

Programming, sometimes called **scheduling,** is the breaking down of project time frames into activities, milestones and logical sequence to determine durations, resources and responsibilities. The terms work breakdown structure (WBS) and organisation breakdown structure (OBS) are classically connected with this.

Project is defined in BS6079-1:2000 as 'A unique set of co-ordinated activities with definite starting and finishing points, undertaken by an individual or organisation to meet **specific objectives** within defined schedule, cost and performance parameters'.[1]

Project board. The joint body responsible for decision making. See also project sponsor.

Project co-ordination. The non-executive functions which make up the project either after appointment of the consultants or after the appointment of the contractor.[2]

Project definition is the development of the project brief and scope up to the planning application stage, so that the risks have been identified and the value for money has been optimised to suit business needs. (Chapter 3 develops this theme.)

Project execution plan is a master strategy that develops the details of the strategic stage of the life cycle and provides a baseline schedule in areas such as stakeholder analysis, cost plan, quality plan, environmental assessment and programme, together with strategic control documents for organisation, health and safety, control systems, ethics and policies. It also recognises the need for continuous development of the plan to meet the dynamic conditions of the project. Procedures are sometimes separately put together in a **project handbook.**

Project management is 'the overall planning, co-ordination and control of a project from inception to completion, aimed at meeting a client's requirements in order to produce a functionally and financially viable project that will be completed on time, within authorised cost and to the required quality standards'.[3]

Project manager (Annex A, BS6079). The professional individual with the responsibility for managing a project to specific objectives. An executive project manager manages the project on behalf of the client from inception to completion. Experienced clients occasionally provide this role direct. It can be combined with other roles such as lead designer in traditional and less complex projects.

Project sponsor (Annex A, BS6079) is the individual or body for whom the project is undertaken and who is the primary risk taker. The individual represents the sponsoring body and is accountable for the spending of funds.

Project team. CIOB refers to those who have a professional role such as design, funding and cost planning and the key first tier contractors led by the project manager. It does not include the project sponsor.

Public private partnerships (PPP) These partnerships between a public authority and a private contractor or developer involve the leveraging of private funds with all or some of the finance needed to provide a built facility. This facility usually houses a public service, but this service may also be run by the private sector on behalf of the public body, for example a toll bridge allows capital financing, maintenance and collection of the tolls over a concession period. The latter is often termed a BOOT (build, own, operate, transfer) project.

Quality management is the process of developing a quality plan and managing the quality of the product and service of components and elements of the building.

Quantity surveyor is a specialist in the financial and contractual areas of project management, utilised by the client side for preparation of bills of quantities, auditing payment and providing project financial advice, utilised by the contractor for cost planning and control, procurement preparing payment claims and paying subcontractors.

Quality systems refer to the *techniques* such as quality function development/quality assurance, or *principles* such as quality assurance, continuous improvement or total quality management.

Risk management is a proactive approach to the identification, assessment, allocation and mitigation of risk factors to rationalise and optimise contingency arrangements or take considered opportunities.

Stakeholders of the project are those who have the potential to influence the course of a project. The most direct influences are likely to come from the direct parties to the contract and the project team, with influence from the statutory permissions that are required on behalf of the community. The shareholders have a much less direct influence in projects.

Strategic planning is the parallel process with feasibility to ensure that optimum systems for delivery such as programme, funding, procurement and project control processes give expected cost benefit.

Sustainable construction is a term to describe environmentally and socially attractive solutions to the delivery and maintenance of facilities and the limitation of climate change, but as its name implies should consider the impact of the building with scarce resources in the community and globally.

Traditional procurement means the separation of design and construction. Design is awarded on a fee basis and construction is awarded on a 'fixed' price basis to a main contractor, using design and specification documents prepared by others. Specialist contracts are sub awarded by the contractor to their supply chain. There is limited whole life project management, but key client contact is through the lead designer. Standard contracts are well known such as FIDIC, JCT and NEC.

User is the ultimate occupier/operator of the facility and may have nothing to do with its commissioning.

Value engineering is the tactical application of value management in the later design stages when element and components are being optimised in their design and procurement.

Value management (VM) or analysis is the strategic analysis of functional requirements and the optimisation of provision within the physical asset provision, its maintenance and operation throughout its life cycle. It should be specifically separate from cost saving which has the sense of cutting quality or provision and is a reactive activity.

Whole life costing is the estimation of capital and subsequent costs for maintenance, replacement and use as well as regulatory charges applicable over a business cycle, set out on a time adjusted basis such as net present value (NPV). If it is a short cycle then demolition and refurbishment are important. Sometimes it may be able to estimate business efficiency cost benefit and intangible benefits.

References

1   BSI 6079-2 (2010) *Guide to Project Management*. UK British Standards Institute.
2   Chartered Institute of Building (2002) *Code of Practice for Construction and Property* (Table 1.1). 3rd edn. Oxford, Blackwells.
3   Chartered Institute of Building (2002) *Code of Practice for Construction and Property*. 3rd edn. Oxford, Blackwells.

Index

Numbers in bold denote a more substantial reference page or a case study page if followed by 'cs'